Biorefinery Co-Products

Wiley Series
in
Renewable Resources

Series Editor

Christian V. Stevens – Faculty of Bioscience Engineering, Ghent University, Ghent, Belgium

Titles in the Series

Wood Modification – Chemical, Thermal and Other Processes
Callum A. S. Hill

Renewables-Based Technology: Sustainability Assessment
Jo Dewulf & Herman Van Langenhove

Introduction to Chemicals from Biomass
James H. Clark & Fabien E.I. Deswarte

Biofuels
Wim Soetaert & Erick Vandamme

Handbook of Natural Colorants
Thomas Bechtold & Rita Mussak

Surfactants from Renewable Resources
Mikael Kjellin & Ingegärd Johansson

**Industrial Application of Natural Fibres – Structure, Properties
and Technical Applications**
Jörg Müssig

**Thermochemical Processing of Biomass – Conversion into Fuels,
Chemicals and Power**
Robert C. Brown

Forthcoming Titles

**Pretreatment of Plant Biomass for Biological and Chemical Conversion to Fuels
and Chemicals**
Charles E. Wyman

Introduction to Wood and Natural Fiber Composites
Douglas Stokke, Qinglin Wu & Guangping Han

Bio-Based Plastics: Materials and Applications
Stephan Kabasci

Cellulosic Energy Cropping Systems
David Bransby

Biobased Materials in Protective and Decorative Coatings
Dean Webster

Biorefinery Co-Products

Phytochemicals, Primary Metabolites and Value-Added Biomass Processing

Edited by

CHANTAL BERGERON

Tom's of Maine

DANIELLE JULIE CARRIER

Department of Biological and Agricultural Engineering

SHRI RAMASWAMY

Department of Bioproducts and Biosystems Engineering

A John Wiley & Sons, Ltd., Publication

Library of Congress Cataloging-in-Publication Data

Biorefinery co-products / edited by Chantal Bergeron, Danielle Julie Carrier, Shri Ramaswamy.
     p. cm.
   Includes bibliographical references and index.
   ISBN 978-0-470-97357-8 (cloth) – ISBN 978-0-470-97559-6 (pdf) – ISBN 978-1-119-96788-0 (ebk.)
  1. Plant biomass. 2. Biomass energy. 3. Renewable energy sources. 4. Phytochemicals. I. Bergeron., Chantal, 1967- II. Carrier, Danielle Julie, 1959- III. Ramaswamy, Shri, 1957-
   TP248.27.P55B568 2012
   333.95′39–dc23

                                                                                           2011044478

A catalogue record for this book is available from the British Library.

Print ISBN: 9780470973578

Set in 10/12pt Times New Roman by Thomson Digital, Noida, India
Printed and bound in Malaysia by Vivar Printing Sdn Bhd

1   2012

Contents

Series Preface

Renewable resources, their use and modification are involved in a multitude of important processes with major influences on our everyday lives. Applications can be found in the energy sector, chemistry, pharmacy, the textile industry, paints and coatings, to name but a few.

The area interconnects several scientific disciplines (agriculture, biochemistry, chemistry, technology, environmental sciences, forestry, etc.), which makes it very difficult to have an expert view on the complicated interaction. Therefore, the idea to create a series of scientific books, focusing on specific topics concerning renewable resources, has been very opportune and can help to clarify some of the underlying connections in this area.

In a very fast-changing world, trends are not only characteristic of fashion and political standpoints; science, too, is not free from hype and buzzwords. The use of renewable resources is again more important nowadays; however, it is not part of a hype or a fashion. As the lively discussions among scientists continue about how many years we will still be able to use fossil fuels, with opinions ranging from 50 years to 500 years, they do agree that the reserve is limited and that it is essential not only to search for new energy carriers, but also for new material sources.

In this respect, renewable resources are a crucial area in the search for alternatives for fossil-based raw materials and energy. In the field of energy supply, biomass and renewable resources will be part of the solution, alongside other alternatives such as solar energy, wind energy, hydraulic power, hydrogen technology and nuclear energy.

In the field of materials science, the impact of renewable resources will probably be even bigger. Integral utilisation of crops and the use of waste streams in certain industries will grow in importance, leading to a more sustainable way of producing materials.

Although our society was much more (almost exclusively) based on renewable resources centuries ago, this disappeared in the Western world in the nineteenth century. Now it is time to focus again on this field of research. However, it should not mean a "retour à la nature", but it should be a multidisciplinary effort on a highly technological level to perform research towards new opportunities, to develop new crops and products from renewable resources. This will be essential to guarantee a level of comfort for the growing number of people living on our planet. It is "the" challenge for the coming generations of scientists to develop more sustainable ways to create prosperity and to fight poverty and hunger in the world. A global approach is certainly favoured.

This challenge can only be dealt with if scientists are attracted to this area and are recognized for their efforts in this interdisciplinary field. It is therefore also essential that consumers recognize the fate of renewable resources in a number of products.

Furthermore, scientists do need to communicate and discuss the relevance of their work. The use and modification of renewable resources may not follow the path of the genetic engineering concept with regard to consumer acceptance in Europe. Related to this aspect, this series will certainly help to increase the visibility of the importance of renewable resources.

Being convinced of the value of the renewables approach for the industrial world, as well as for developing countries, I was myself delighted to collaborate on this series of books, focusing on different aspects of renewable resources. I hope that readers become aware of the complexity, the interaction and interconnections, and the challenges of this field and that they will help to communicate on the importance of renewable resources.

I certainly want to thank the people of Wiley from the Chichester office, especially David Hughes, Jenny Cossham and Lyn Roberts, in seeing the need for such a series of books on renewable resources, for initiating and supporting it and for helping to carry the project to the end. Last, but not least I want to thank my family, especially my wife Hilde and children Paulien and Pieter-Jan for their patience and for giving me the time to work on the series when other activities seemed to be more inviting.

Christian V. Stevens
Faculty of Bioscience Engineering
Ghent University, Belgium
Series Editor *Renewable Resources*
June 2005

Preface

We are entering a very interesting period in our history. It is more than likely that renewable energy will complement, or even in some instances replace, fossil-oil-based energy systems. Within the renewable energy portfolio, the conversion of biomass to electricity and liquid fuels is certainly touted as a credible possibility. It is recognized that lignocellulosic biomass can be converted to fuels, chemicals and other bioproducts, using thermochemical or biochemical strategies, or a combination of both. Biochemical conversion technologies for biomass are centered on the hydrolysis of cellulose and hemicellulose in biomass, followed by fermentation of the resulting sugars to biofuels and/or bio-based chemicals. Thermochemical biomass conversion strategies involve high temperatures, such as the gasification of cellulose, hemicellulose and lignin, to produce synthesis gas (syngas), followed by the conversion of CO, CO_2 and H_2 into liquid fuels by catalyst-based processes. Biomass also can be converted through fast pyrolysis to a dark-brown liquid, which can then be combusted for energy.

Depending on the conversion technology, 10–25 million tons of dry biomass feedstock will be required to produce 1 billion gallons of liquid fuel. It is estimated that approximately one billion dry tons of biomass will be required annually to ensure that the US can produce up to 30% of its liquid fuel demand from renewable resources. It is also estimated that assuming a conversion yield of 80 gallons of ethanol per ton of biomass, approximately 2000 tons of dry biomass per day are required to produce 50 million gallons of liquid biofuels. It should be recognized that even though corn- or grain-based biofuel has been in commercial production for some time, cellulosic ethanol is yet to become a prevalent commercial reality.

It is important to note that the biomass discussed in this book is generally not food-destined biomass, but rather non-food agricultural residues, waste sources or energy crops. Biomass conversion facilities, often referred to as biorefineries, will most likely be located in rural areas close to biomass sources, respecting sustainability needs and diversity issues. Locating biorefineries in rural settings will most likely result in the stimulation of rural economies. Given the seasonal nature of this nascent industry, it is more than likely that biorefineries will use various sources of biomass throughout the year. Supply chains that will deliver 2000 tons per day to facilities may be composed of mixed species. Depending on the season, herbaceous, annual and woody biomass will most likely be biorefinery feedstocks.

It is more than likely that some of this biomass will contain interesting phytochemicals, proteins or other value-added products which could be extracted prior to or post processing.

Biorefining, bio-based products and alternatives to petroleum products are becoming increasingly important. In addition to biofuels and bioenergy, value-added biomass processing for bio-based co-products is an important area that should be simultaneously considered in an integrated biorefinery in order to achieve the future biovision. Phytochemicals could find use in human and animal healthcare products, cosmetic applications and as essential ingredients in green cleaning products. According to market research surveys, there is a growing preference among consumers for phytochemicals in the foods they consume, as well as other personal care and household products they utilize. Growth in the use of phytochemicals is predicted in the flavour industry, which includes beverages, confectionery, savoury, dairy and pharmaceuticals.

This book was prepared with the intention of introducing the reader to the concept of using biomass to produce energy, but when it is possible, to also extract value-added bio-based chemicals. The first three chapters are intended to give the reader an overview of biomass conversion technologies, an introduction to phytochemicals and an introduction to separation processes. Afterwards, the reader will be presented with various feedstocks that, in addition to being excellent biomass sources, also contain useful value-added chemicals. Existing grain supply chains, algae and waste residues are examined in light of their phytochemical contents. Given the importance of the Brazilian ethanol industry and the prevalence of sugar around the world, sugarcane is also inspected as a phytochemical source. The possibility of extracting proteins from switchgrass is also presented. Woody feedstocks, grown in Canada and Scandinavia, are surveyed for their phytochemical content. Finally, current streams from paper mills and pyrolysis processes are assessed as interesting sources of value-added chemicals.

We hope that this book will continue the discussions on using biomass for fuels and energy production, and for extraction of value-added components, increasing the sustainability and economic feasibility of the biorefinery and helping to make this area a commercial reality.

Chantal Bergeron, Danielle Julie Carrier and Shri Ramaswamy

List of Contributors

Mamdouh Abou-Zaid   Canadian Forest Service, Great Lakes Forestry Service, Natural Resources Canada, Sault Ste. Marie, Ontario, Canada

Venkatesh Balan   Biomass Conversion Research Laboratory, Department of Chemical Engineering and Materials Science, Michigan State University, Lansing, Michigan, USA

Bryan D. Bals   Biomass Conversion Research Laboratory, Department of Chemical Engineering and Materials Science, Michigan State University, Lansing, Michigan, USA

Chantal Bergeron   Tom's of Maine, Kennebunk, Maine, USA

Liam Brennan   Charles Parsons Energy Research Programme, Bioresources Research Centre, School of Agriculture, Food Science and Veterinary Medicine, University College Dublin, Belfield, Dublin, Ireland

Danielle Julie Carrier   Department of Biological and Agricultural Engineering, University of Arkansas, Fayetteville, Arkansas, USA

Edgar C. Clausen   Ralph E. Martin Department of Chemical Engineering, University of Arkansas, Fayetteville, Arkansas, USA

Michelle Co   Department of Physical and Analytical Chemistry, Uppsala University, Uppsala, Sweden

Bruce E. Dale   Biomass Conversion Research Laboratory, Department of Chemical Engineering and Materials Science, Michigan State University, Lansing, Michigan, USA

Nurhan Turgut Dunford   Department of Biosystems and Agricultural Engineering and Robert M. Kerr Food & Agricultural Products Center, Oklahoma State University, Stillwater, Oklahoma, USA

Abigail S. Engelberth   Laboratory of Renewable Resources Engineering, Department of Agricultural and Biological Engineering, Potter Engineering Center, Purdue University, West Lafayette, Indiana, USA

Navam Hettiarachchy   Department of Food Science, University of Arkansas, Fayetteville, Arkansas, USA

Hua-Jiang Huang   Department of Bioproducts and Biosystems Engineering, Kaufert Lab, University of Minnesota, Saint Paul, Minnesota, USA

Arvind Kannan   Department of Food Science, University of Arkansas, Fayetteville, Arkansas, USA

K. Thomas Klasson   USDA-ARS Southern Regional Research Center, New Orleans, Louisiana, USA

Jean Legault   Laboratoire d'Analyse et de Séparation des Essences Végétales (LASEVE), Département des Sciences Fondamentales, Université du Québec à Chicoutimi, Chicoutimi, Québec, Canada

John A. Manthey   USDA-ARS, U.S. Horticultural Research Laboratory, Fort Pierce, Florida, USA

M. Angela A. Meireles   LASEFI/DEA/FEA/UNICAMP, University of Campinas (UNICAMP), Campinas, SP, Brazil

Anika Mostaert   School of Biology and Environmental Science, University College Dublin, Belfield, Dublin, Ireland

Cormac Murphy   School of Biomolecular and Biomedical Science, University College Dublin, Belfield, Dublin, Ireland

Philip Owende   Charles Parsons Energy Research Programme, Bioresources Research Centre, School of Agriculture, Food Science and Veterinary Medicine, University College Dublin, Belfield, Dublin, Ireland and School of Informatics and Engineering, Institute of Technology Blanchardstown, Dublin, Ireland

André Pichette   Laboratoire d'Analyse et de Séparation des Essences Végétales (LASEVE), Département des Sciences Fondamentales, Université du Québec à Chicoutimi, Chicoutimi, Québec, Canada

Juliana M. Prado   LASEFI/DEA/FEA/UNICAMP, University of Campinas (UNICAMP), Campinas, SP, Brazil

Shri Ramaswamy   Department of Bioproducts and Biosystems Engineering, Kaufert Lab, University of Minnesota, Saint Paul, Minnesota, USA

Kent Rausch   Department of Agricultural and Biological Engineering, University of Illinois, Urbana, Illinois, USA

Srinivas Rayaprolu   Department of Food Science, University of Arkansas, Fayetteville, Arkansas, USA

Ian M. Scott   Agriculture and Agri-Food Canada, London, Ontario, Canada

Mahmoud A. Sharara   Department of Biological and Agricultural Engineering, University of Arkansas, Fayetteville, Arkansas, USA

François Simard   Laboratoire d'Analyse et de Séparation des Essences Végétales (LASEVE), Département des Sciences Fondamentales, Université du Québec à Chicoutimi, Chicoutimi, Québec, Canada

Charlotta Turner   Department of Chemistry, Centre for Analysis and Synthesis, Lund University, Lund, Sweden

G. Peter van Walsum   Forest Bioproducts Research Institute, Department of Chemical and Biological Engineering, University of Maine, Orono, Maine, USA

1

An Overview of Biorefinery Technology

Mahmoud A. Sharara[1], Edgar C. Clausen[2] and Danielle Julie Carrier[1]

[1] *Department of Biological and Agricultural Engineering, University of Arkansas, Fayetteville, Arkansas, USA*
[2] *Ralph E. Martin Department of Chemical Engineering, University of Arkansas, Fayetteville, Arkansas, USA*

1.1 Introduction

Fossil fuel resources are being depleted and, whether we have 50, 100 or 200 years' worth of petroleum reserves, irrefutably, at some point in time, there will be no more oil to extract in an economical fashion. The world is gradually adapting to this new paradigm, and the price of petroleum-based energy systems is steadily increasing. The developed and developing nations alike are seeing a transition to renewable-based energy forms, with solar panels installed on buildings and wind turbines part of the landscape. Liquid fuels that are produced from renewable feedstocks, in the form of ethanol or biodiesel, are now commercially available. Fuels containing 10–85% w/w ethanol are available to power internal combustion engines throughout the US. Most of the commercially available ethanol, produced from corn in the US and from sugar cane in Brazil, is often referred to as first-generation biofuel. In 2011, nearly 14 billion gallons of ethanol were sold in the US. A thorough discussion of corn, other grains, and sugarcane-to-ethanol processes are described in the chapters prepared by Rausch, Dunford and Prado, and Meireles, respectively. Thus, this chapter will concentrate on second-generation biofuels.

Second-generation biofuels are characterized as fuels that are produced from non-food biomass systems, such as forestry and agricultural residue, and dedicated herbaceous and

Biorefinery Co-Products: Phytochemicals, Primary Metabolites and Value-Added Biomass Processing, First Edition.
Edited by Chantal Bergeron, Danielle Julie Carrier and Shri Ramaswamy.
© 2012 John Wiley & Sons, Ltd. Published 2012 by John Wiley & Sons, Ltd.

wood energy crops. Some of the second-generation fuels are in the mid to late stages of becoming commercially available. As examples, pilot- and demonstration-scale production facilities are now operational at POET, LLC in South Dakota and Iowa, and DuPont in Tennessee. The conversion technologies for these second-generation fuels use non-food crops, or lignocellulosic feedstocks, and are centred around:

1. the hydrolysis of plant cell wall polysaccharides into their single sugar components, followed by fermentation of the resulting sugars to fuels like ethanol or butanol;
2. the gasification of the plant material to produce syngas (synthesis gas), followed by the conversion of CO, CO_2 and H_2 to ethanol or other alcohols by fermentation or catalyst-based processes; or
3. the conversion of organic compounds in biomass through fast-pyrolysis to a dark-brown liquid, called pyrolysis oil, which can be upgraded for transportation fuels.

Depending on the conversion technology, 10–25 million tons of dry cellulosic biomass will be required to produce 1 billion gallons of liquid fuel. Partially replacing our consumption of gasoline with renewable fuels will require huge quantities of feedstock that will be obtained from both cultivated and collected biomass resources. Regardless of which conversion platform is selected, colossal masses of feedstock will pass through the door of a conversion facility. As an example, a 50 million gallon second-generation cellulosic facility will require about 2000 dry tons per day of biomass for processing. To complicate matters, it is very likely that the composition and quality of the feedstock stream will vary throughout the year. An annual biorefinery feedstock cycle could consist of agricultural residues in the fall, woody residues or crops in the winter, cover crops, like rye, in the spring, and energy crops, like sorghum or switchgrass, in the summer. Some of these bioenergy-destined feedstocks will contain valuable compounds that can be extracted prior to or after the conversion process. These compounds or products are often referred to in the literature as value-added biorefinery co-products.

In order to appreciate the array of possible biorefinery-related co-products presented throughout this book, this chapter will provide an overview of various feedstocks, as well as the steps involved in biochemical and thermochemical conversion processes. This chapter will also provide insight as to where co-product generation can be integrated in the biofuel manufacturing process.

1.2   Feedstock

If the US is to produce, as a target, more than 21 billion gallons per year of second-generation biofuels, more than 250 million dry tons per year of biomass will be required. Biomass will be available annually in the form of forest residues, mill residues, dedicated woody and herbaceous energy crops, urban wood waste, and agricultural residues (Bain *et al.*, 2003; Perlack *et al.*, 2005). It is important to note that most dedicated energy crops need time to be established. As an example, perennial warm season crops will take two to five years to establish (Propheter *et al.*, 2010). Herbaceous energy crops include, amongst others, switch-grass (*Panicum virgatum*), sorghum (*Sorghum bicolour*), and miscanthus (*Miscanthus* spp). Woody energy crops include, amongst others, maple (*Acer saccharinum*), sweetgum (*Liquidambar styraciflua*), sycamore (*Platanus occidentalis*), and hybrid poplar (*Populus* spp.).

Switchgrass is a warm season perennial, metabolizes CO_2 via the C4 cycle and remains an important component of tallgrass prairie that has a wide distribution from southern Canada to northern Mexico. There are two main switchgrass ecotypes. The lowland ecotype grows better in the wetter southern habitats, whereas the upland ecotype develops in drier mid and northern latitudes. Kanlow and Alamo are prevalent lowland ecotype cultiavars, while Cave-in-Rock is a common upland switchgrass ecotype cultivar. Switchgrass yields of $8.7 \pm 4.2\,Mg\,ha^{-1}$ and $12.9 \pm 5.9\,Mg\,ha^{-1}$, respectively, for upland and lowland ecotypes were reported (Wullschleger *et al.*, 2010). Propheter *et al.* (2010) reported the yield of the Kansas-grown Kanlow of $9.2\,Mg\,ha^{-1}$. Wullschleger *et al.* (2010) reported that, of the agronomic parameters, ecotype, temperature, water supply, and nitrogen fertilization affected the switchgrass yields.

Miscanthus is a perennial rhizomatous grass that also metabolizes CO_2 via the C4 cycle. Heaton, Dohleman, and Long (2008) established miscanthus seedlings in 2002, and monitored the yields for three years, 2004, 2005, and 2006, at three Illinois locations, north, central, and southern. Their reported overall three-year state average was $38.2\,Mg\,ha^{-1}$. The maximum miscanthus yields in the trials located in Sweden, Denmark, England, Germany, and Portugal were 10.7, 7.9, 12.1, 17.0 and $26.9\,Mg\,ha^{-1}$, respectively, where it was observed that the trials in northern latitudes tended to mature slower than those located in more southern locations (Clifton-Brown *et al.*, 2001). In the Illinois trials, yields were independent of rainfall, nitrogen fertilization, and growing degree days (Heaton, Dohleman, and Long, 2008), whereas irrigation was a key in obtaining the maximum yields in Portugal (Clifton-Brown *et al.*, 2001). Further European trials showed that the peak yields in Austria, Belgium, France, Germany, Ireland, The Netherlands, Portugal, Spain, and England were 23, 25, 21, 25, 18, 25, 15, 18 and $23\,Mg\,ha^{-1}$, respectively (Clifton-Brown, Stampfl, and Jones, 2004). Similar to switchgrass, miscanthus translocates its nutrients to roots and rhizomes at the end of the growing season. The clone selected for the trials in Illinois survived at $-30\,^{\circ}C$ in the winters, while the clone used in the 2001 European study did not survive Danish and Swedish winters, indicating that clonal selection is critical to overcome the harsh winter. It is important to note that although miscanthus is an attractive energy crop because of its high biomass yields, there are some concerns with respect to its invasive potential (Raghu *et al.*, 2006).

Sorghum (*Sorghum bicolour* L. Moench) is a drought-tolerant crop that has wide distribution in the world; in the US, sorghum is cultivated in the Midwest where rainfall is often scarce (Ananda, Vadlania, and Prasad, 2011). There are different *S. bicolour* cultivars which are commonly referred to as grain, sweet or forage sorghum; what sets these cultivars apart is the flowering period, which is highly dependent on the photoperiod. As an example, the photoperiod-sensitive forage sorghum, capable of producing high lignocellulosic yields without the production of grain, will not flower until the day length becomes shorter than 12 h and 20 minutes, and this, in the Midwest, occurs when seasonal frosts takes place, impeding the reproductive flowering stage (Propheter *et al.*, 2010). Sweet sorghum is desirable because readily fermentable carbohydrates can be extracted from its stalks and processed into biofuels following the sugarcane-processing methods. Sugar yields from sweet sorghum of 4.1 and $4.8\,Mg\,ha^{-1}$ were obtained in Nebraska and Kansas, respectively (Wortmann *et al.*, 2010; Propheter *et al.*, 2010). Grains obtained from grain sorghum can be processed into biofuels following the same starch-based methods used for corn or wheat. Wortmann *et al.* (2010) reported a maximum grain yield of $6.62\,Mg\,ha^{-1}$. Propheter *et al.* (2010) reported on sorghum grown side by side in two Kansas locations for two years using photoperiod-sensitive, sweet, dual-purpose forage and brown midrib. The yields were 26.8, 32.6, 20.7 and 14.8 Mg ha^{-1}, respectively, indicating that, in this study, sweet sorghum showed the highest potential.

With respect to woody energy crops, hybrid poplars, *Populus deltoides*, are hardwoods that often grow in the forest areas that are logged for softwood, and can be cultivated with a short term horizon, 3 to 6 years, or for a longer term period for 15 to 20 years. Plantation densities can average 18 000 stems ha^{-1} and yield, within a six-year time frame, 16.9 Mg ha^{-1}year^{-1} (Fortier *et al.*, 2010). In Sweden, the productivity of poplar stands could be in the range of 70 Mg ha^{-1} after a 10- to 15-year time frame (Johansson and Karačić, 2011). After seven years of growth, yields of 17.5 and 41.6 Mg ha^{-1} were reported for sweetgum (*Liquidambar styraciflua*) and sycamore (*Platanus occidentalis*), respectively (Davis and Trettin, 2006).

1.3   Thermochemical Conversion of Biomass

In the thermochemical conversion processes elevated temperature and pressure are utilized to break the chemical bonds in the biomass matrix to either release its energy content directly, as in combustion, or to be chemically converted to fuel precursors, as in pyrolysis, gasification and liquefaction. These outputs are usually considered fuel precursors (or crude) as they require further cleaning, reforming, and fractionation to generate stable drop-in fuels. Generally speaking, fast pyrolysis or hydrothermal liquefaction are the terms used to describe processes that directly produce a liquid bio-crude, while gasification is used to describe processes that produce a synthesis gas consisting primarily of CO, CO_2, and H_2. The following sections will briefly review the principles, limitations, and state-of-the-art developments in these technologies.

1.3.1   Fast Pyrolysis and Hydrothermal Liquefaction

Pyrolysis can be understood as the thermal depolymerization of biomass in the absence of oxygen followed by rapid quenching to produce char, gas, and, more importantly, condensable oils. In fast pyrolysis, the temperature, around 500 °C, the heating rate, over 100 °C min^{-1}, and the residence time, 2–3 seconds, are instrumental in minimizing cracking of condensable vapours into permanent gases; hence the name "fast" or "flash" pyrolysis. The bio-oil, also known as bio-crude, is essentially a mixture of over 100 chemical species (Evans and Milne, 1987) with a wide range of molecular weights that are formed by primary and secondary reaction mechanisms. The oils are highly oxygenated, which is characteristic of the biomass feedstock, and also high in water content. Bridgwater, Meier, and Radlein (1999) present the typical elemental composition and properties of fast pyrolysis oils, which are also shown in Tables 1.1 and 1.2. As is noted, the liquid is relatively viscous, with a medium heating value.

Numerous reactor configurations have been investigated for biomass pyrolysis, with an emphasis on high heat transfer rates and short residence times for both solids and vapours. Examples of such reactors include bubbling fluidized bed and circulating fluidized bed, both of which have high heat transfer rates. Other reactors under investigation are ablative, vortex, rotating cone, and rotating blade reactors (Meier and Faix, 1999).

An important area of research associated with thermochemical conversion is catalysis. The use of suitable catalysts significantly increases the yields and also optimizes the composition

Table 1.1 *Typical elemental composition of fast pyrolysis bio-oils (Bridgwater, Meier, and Radlein, 1999). Reprinted with permission from Bridgwater et al., 1999 © Elsevier (1999).*

Element	Wt %
Carbon	44–47
Hydrogen	6–7
Oxygen	46–48
Nitrogen	0–0.2

of the output products. Catalysts can be incorporated during or after the production process, or in both stages. In pyrolysis, the use of catalysts can help overcome the problematic qualities of bio-oil, mainly thermal and temporal instability, through the hydrodeoxygenation process. This step essentially converts oxygenates in pyrolysis oil, aided by catalysts, into more stable species. Several studies have investigated the effects of catalyst addition to a pyrolysis reaction; mainly activated alumina, silicate, beta, Y-zeolite and ZSM-5 (Williams and Horne, 1995; Carlson *et al.*, 2009). ZSM-5, an acid–base zeolite, is a prime example of this class of catalysts which was reported to increase the yield of aromatic species in the bio-oil by up to 30% wt (Carlson *et al.*, 2009; Zhang *et al.*, 2009). Aromatic species such as naphthalene and benzene are more stable and industrially significant, with a higher calorific density than un-catalyzed bio-oil oxygenates, phenols and carboxylic acids. Another approach is the downstream treatment of bio-oils in an aqueous, high-pressure, catalyzed environment (Elliott, 2007). Catalysts used in such techniques include sulfide NiMo, sulfide CoMo, and, more recently, Ru/C catalyst (Wildschut *et al.*, 2009; Mercader *et al.*, 2011), which was reported to yield more bio-oils and higher deoxygenating levels.

According to reports, pilot-scale and full-scale pyrolysis facilities are gradually going into uninterrupted operation worldwide. Ensyn Technologies has constructed six circulating fluidized bed plants, with the largest having a nominal capacity of 50 tons/day (DOE, 2005).

Table 1.2 *Typical properties of fast pyrolysis bio-oils (Bridgwater, Meier, and Radlein, 1999). Reprinted with permission from Bridgwater et al., 1999 © Elsevier (1999).*

Property	Analysis
Moisture content	25%
pH	2.5
Specific gravity	1.20
Elemental Analysis (moisture free basis)	
Carbon	56.4%
Hydrogen	6.2%
Nitrogen	0.2%
Sulfur	<0.01%
Ash	0.1%
Oxygen (by difference)	37.1%
Higher heating value (moisture free basis)	22.5 MJ kg^{-1}
Higher heating value (as produced)	17.0 MJ kg^{-1}
Viscosity (at 40 °C)	30–200 cp
Pour point	−23 °C

In 2008, Ensyn and Honeywell's UOP LLC established a joint venture (Envergent Technologies, LLC) to commercialize pyrolysis conversion facilities (trademarked as rapid thermal processing; RTP™). DynaMotive Corporation (Vancouver, Canada) has established two fluidized-bed pyrolysis plants in Canada; first in West Lorne in 2002 at a capacity of 100 tons day^{-1} and then in Guelph in 2007 with 200 tons day^{-1}capacity. BTG (The Netherlands) developed the rotary cone reactor (RCR) system for biomass-to-liquid (BTL) conversion with an operational capacity of 50 kg h^{-1}, that was later scaled up to 250 kg h^{-1} capacity in 2001 (Venderbosch and Prins, 2010). In Finland, a 2 MW integrated pyrolysis plant was built in collaboration with Metso, UPM, Fortum, and VTT. This plant went into operation in 2009–2010, with sawdust and forest residue as the feedstock, generating around 90 tons of bio-oil by the summer of 2010 (Lehto *et al.*, 2010).

Fast pyrolysis occurs at moderate thermal conversion temperatures, 500 °C or so, and produces a liquid product, pyrolysis oil or bio-oil, which may be used as a chemical intermediate, or directly used as a liquid fuel. The US DOE (2005) reported the reactor technologies, including bubbling fluid beds, circulating, and transported beds, cyclonic reactors, and ablative reactors that can achieve a 75% conversion of the biomass to liquid fuels.

Direct hydrothermal liquefaction also produces an oily liquid by bringing biomass into contact with liquid water at moderate thermal conversion temperatures such as 300 °C. The water is maintained in the liquid phase by elevating the pressure. In contrast to fast pyrolysis, residence times of up to 30 minutes are required. The oily liquid is once again available as a liquid fuel, but large quantities of water are also present. The technology is being developed for use on algae, as well as waste biomass. According to US DOE (2005) technology developers include Changing World Technologies (West Hampstead, NY), EnerTech Environmental Inc (Atlanta, GA), and Biofuel B.V. (Heemskerk, Netherlands).

1.3.2 Gasification

Historically, gasification was developed more than 100 years ago to generate gaseous fuel, town gas, from coal and peat. To compensate for petroleum shortages during World War II, gasification was implemented in Europe by converting woodchips to gas to operate vehicles and to generate electricity. Similarly, in the aftermath of the energy crisis in the seventies, interest was renewed in biomass gasification, particularly wood and forestry residue, as an alternative energy strategy. Since then, studies have investigated a wide range of biomass feedstock, including agricultural and crop residue, farm and livestock wastes, industrial and processing by products, municipal and landfill wastes, and, more recently, marine and aquatic biomass.

In principle, biomass gasification is simply the incomplete combustion of biomass, due to low oxygen, to produce CO, CO_2, and H_2 instead of CO_2 and water. In reality, the process is much more complex as it involves overlapping reactions and stages such as drying, pyrolysis, char gasification, and oxidation. This operation converts biomass chemical energy to energy-carrier gases, such as CO, H_2, CH_4, and other hydrocarbons that are known as syngas, while minimizing formation of tar, pollutant oxides, and polycyclic aromatic hydrocarbons.

This conversion typically occurs at elevated temperatures, varying from 500 to 1400 °C, and pressures ranging from atmospheric pressure to as much as 30 atm. Much like combustion, an oxidant is used: air, pure oxygen, steam or a mixture of these gases. Air-blown gasifiers are most common due to their simple design, and relatively low operating costs. However, syngas yields are diluted by the atmospheric nitrogen, with a typical heating value of 100–150 Btu ft^{-3} (3–5 MJ Nm^{-3}). Oxygen- and steam-blown gasifiers produce syngas that is richer in CO and H$_2$, with a heating value of 250–500 Btu ft^{-3} (8.4–16.7 MJ Nm^{-3}). Van der Drift, van Doorn, and Vermeulen (2001) conducted a gasification study, using a fluidized-bed gasifier, on ten types of biomass residues, and reported their syngas composition, heating value and conversion efficiency, as shown in Table 1.3.

The physical and chemical properties of the feedstock are some of the major factors affecting the quality of syngas and the gasification process efficiency in general. The fact that there are large variations in moisture content, ash, and organic species between different biomass types adds to the complexity of this process configuration. In general, high moisture and ash content are considered problematic, as they consume a fraction of the heat supplied without contributing positively to syngas formation. Hughes and Larson (1998) illustrated, through modelling, that the gasification process efficiency in integrated gasification combined cycle (IGCC) increases with a decrease in moisture content of the input biomass down to 30%, where further moisture decreases had negligible influence. However, Brammer and Bridgwater (2002) showed that the high overall efficiency (at a moisture content of 35% in their case) is due to the sensible heat carried by the water vapour. They demonstrated that the cost of an electrical power unit will be higher, compared to a biomass feed with 10% moisture, since the output gas heat will not contribute to power generation.

Similarly, the presence of ash minerals in the biomass can be problematic to gasification, causing agglomeration, fouling, and corrosion in the gasifier bed. Furthermore, catalyst poisoning occurs as a result of mineral oxides deposition on the active site. Priyadarsan *et al.* (2004) studied the gasification of poultry and feedlot manure in an air-blown gasifier, and reported agglomeration in the bed due to the high alkaline oxides, such as sodium and potassium, in the ash. Leaching of high mineral biomass streams to improve the thermal characteristics under gasification conditions (Garcia-Ibanez, Cabanillas, and Sanchez, 2004) is another approach to minimize ash generation.

An important process parameter is the operational mode of the gasifier. Generally, gasifiers operate in a fixed bed, fluidized bed, or entrained flow mode. However, there are many variations to these basic designs. For instance, fixed-bed gasifiers can be operated with updraft or downdraft configurations. Similarly, fluidized-bed reactors can be either bubbling beds or circulating beds. Each of these designs has its pros and cons: a summary of these considerations is presented by Ciferno and Marano (2002). Also, Reed and Das (1987) presented a comprehensive review of biomass gasification modes with the emphasis on downdraft gasifiers. Generally, fixed-bed reactors are known for their ease of operation, with downdraft preferred to updraft gasifiers as they produce syngas with a lower tar content. Although more complex in operation than fixed-bed gasifiers, fluidized-bed gasifiers are commonly preferred in large-scale conversion plants. Fluidized-bed gasifiers essentially keep the biomass and inert bed material (usually alumina) suspended by air flow (or oxidizing agents), thus ensuring a fluid-like state, which greatly improves the heat transfer rate to the biomass particles.

Catalysis Role in Gasification

As stated earlier, the use of a catalyst greatly improves thermochemical conversion by facilitating a preferred reaction mechanism. In gasification, the main role of catalysts is

Table 1.3 Syngas composition of different biomass residues[a] (Van der Drift, van Doorn, and Vermeulen, 2001). Reprinted with permission from Van der Drift et al., 2001 © Elsevier (2001).

Biomass	CO	H_2	CO_2	CH_4	C_2H_4 (Vol.% dry)	C_2H_6	Benzene	Toluene	Xylene	HHV - calculated (MJ m^{-3})	Cold-gas efficiency (%)
Willow	9.400	7.200	17.100	3.300	1.100	0.100	0.210	0.065	0.013	4.92	66
Demolition wood	11.350	7.050	15.750	3.250	0.850	0.030	0.210	0.031	0.005	4.77	58
Park wood	11.650	6.770	15.510	3.170	0.990	0.040	0.250	0.043	0.009	4.93	66
Chip board materials	9.640	6.420	15.630	2.770	0.870	0.030	0.220	0.040	0.007	4.41	59
verge grass	9.890	7.280	15.610	2.580	1.150	0.050	0.120	0.035	0.004	4.64	58
Demolition wood + paper residue sludge	9.240	6.080	16.110	2.810	1.020	0.040	0.190	0.054	0.070	4.51	56
Demolition wood + sewage sludge	10.530	8.020	15.020	3.190	1.120	0.040	0.230	0.058	0.008	5.13	58
Woody excess fraction of ODW	5.340	1.800	16.090	1.240	0.400	0.020	0.120	0.019	0.008	2.05	37
Park wood (bio-dried)	8.310	5.380	16.040	1.720	0.600	0.020	0.140	0.040	0.007	3.28	47
Railroad ties	10.570	5.850	13.940	2.880	0.950	0.030	0.280	0.110	0.110	4.97	60
Cacao shells	8.000	9.020	16.020	2.340	1.130	0.050	0.160	0.060	0.030	4.61	62

[a] These feedstocks were tested in a 500 kW circulating fluidized-bed gasification facility [Energy Research centre (ECN) in the Netherlands, in co-operation with a Dutch company (NV Afvalzorg) and the Dutch agency for energy and environment (Novem)]. The process temperatures were approximately 850 °C for all tests to facilitate comparison. The process pressure was atmospheric.

the reduction of condensable organic vapours (that later form tar contamination) by favouring their cracking into permanent gases, such as H_2 and CO. Sutton, Kelleher, and Ross (2001) reviewed gasification catalysts and classified them into three groups: dolomite, alkali (metal), and nickel catalysts. Catalysts could either be added to the gasification bed, or to separate catalytic reforming beds after conversion. The use of independent catalytic reactors is advantageous as it minimizes catalyst fouling and facilitates catalyst regeneration in large-scale unit operations. Both dolomite and nickel catalysts are suited for downstream separate catalytic reformers, while alkali metals can be impregnated into the biomass feed directly. Devi, Ptasinski, and Janssen (2003) listed gasification char as an effective bed catalyst that has been found to reduce tar formation. In addition to the catalysts used, a variety of mechanical separation methods are employed to remove entrained particles from the product syngas. The end use of the syngas generally dictates the necessary quality of the syngas, and in turn, the downstream cleaning stages. In general, cyclones and bag filters are used to separate entrained fly-ash from the gas stream. Furthermore, condensers, scrubbers and strippers are used to remove pollutants and contaminants from the gas stream.

Synthesis Gas Usage

The configuration of the gasification facility depends largely on the intended use, or uses, of the syngas stream. These uses include direct combustion in boilers, turbines or internal combustion engines (ICE) to generate electricity, heat or both. More sophisticated, and capital intensive, implementations include gas-to-liquid (GTL) processes to produce various liquid fuels. Also, syngas can be used to produce hydrogen by separation, or a range of other chemical industry products (non-energy carriers), such as ammonia, by chemical synthesis. Examples of large-scale biomass gasification implementations include the following.

A Battelle/FERCO (Vermont) gasifier demonstration plant was built in 1994, using a fluidized bed reactor unit, where the char was combusted in a second bed to provide reaction heat to the gasification bed by heating the sand (Paisley *et al.*, 1989). The output syngas stream was intended for power generation via gas turbines. Unfortunately, the operation of this facility was discontinued due to operational and logistical issues. The RENUGAS® process was developed by GTI (Gas Technology Institute) and essentially consisted of a pressurized, fluidized-bed gasification unit with either air or oxygen as the oxidizer and a hot-gas cleaning unit (Lau and Carty, 1994). A demonstration unit, with a maximum capacity of 91 metric tons day^{-1}, was constructed in Hawaii, using sugarcane bagasse to power a combustion turbine for electricity generation. Unfortunately, the operation of this facility was also discontinued due to lack of funding (Rollins *et al.*, 2002). The U-GAS® process, patented by GTI, was implemented in Tampere, Finland by Enviropower, Inc., now Carbona, Inc. This is a high-temperature ($\sim$1000 °C), high-pressure fluidized-bed conversion process, of which the syngas is used to power combined-cycle turbines for district heating. The facility can run on a variety of feedstocks, including coal, paper-mill waste, straw, and willow. A series of gasification systems were manufactured by Primenergy LLC (US) to convert biomass by products, such as rice hulls and corn fibres, into energy and steam. Riceland Foods Inc., located in Stuttgart, AR, operates three Primenergy-built gasifiers, gasifying 600 tons day^{-1} of rice hulls to produce 68 039 kg h^{-1} of steam and 12.8 MW of electricity (UCR, 2009).

Syngas can also be used as feedstock to produce liquid fuels through the Fischer–Tropsch (F–T) process. This process requires only hydrogen and carbon monoxide as input components, which requires additional purification steps to remove other syngas components such as CO_2, CH_4, and N_2. A H_2 and CO mixture, preferably at a ratio of H_2: CO of 2, are exposed to specific temperatures, in the range 200–350 °C, pressure, in the range 15 to 40 bar, and catalytic agents, such as Co, Fe, Ni. This generates a distribution of saturated hydrocarbons, referred to as alkanes, including: methane (CH_4), ethene (C_2H_4), ethane (C_2H_5), LPG (C3–C4, propane and butane), gasoline (C5–C12), diesel fuel (C13–C22), and light waxes (C23–C33) (Demirbas, 2007). This process has been developed and utilized on an industrial scale by SASOL Ltd. (South Africa) since the 1950s to produce petroleum liquids from coal and natural gas. To our knowledge, no large-scale conversion facilities have been established for biomass-to-liquid via the F–T process. Modelling studies showed that converting biomass into F–T fuels was not economical; however, with the incorporation of green energy premiums, production costs may become competitive (Tijmensen *et al.*, 2002).

1.4   Biochemical Conversion

As opposed to thermochemical processing, biochemical processing does not require temperatures in the realm of 500 °C. On the other hand, this processing platform is based on the saccharification of sugars that make up the plant cell wall into high-quality sugar streams that will then be converted to bio-based fuels or chemicals. To produce fuels using biochemical processing technologies, feedstock must be reduced in size, pretreated, hydrolyzed with enzymes, and fermented (Lynd *et al.*, 2008). A schematic of the plant cell wall and its ensuing components is presented in Figure 1.1. Specific pretreatment methods attack different components of the cell wall, with ammonia and lime treatments resulting in the disruption of lignin, while water and dilute acid cause hemicellulose solubilization (Wyman *et al.*, 2009). The efficacies of pretreatments are rated according to their production of reactive fibre, utility of the hemicellulose fraction and limitation of the extent to which the pretreated material

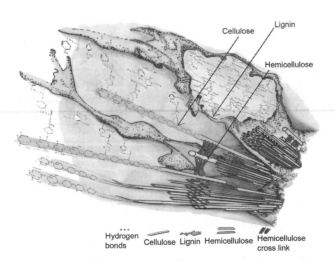

Figure 1.1   *A close up of the plant cell wall where lignin, hemicellulose, and cellulose are intertwined.*

inhibits enzymatic hydrolysis and growth of the fermentation microorganism (Laser *et al.*, 2002). Pretreatment can lead to the generation of inhibitory products such as lignin derivatives and xylose degradation products, such as furfural and formic acid (Du *et al.*, 2010). Leading pretreatments include: ammonia fibre explosion, ammonia recycling, controlled pH, dilute acid and lime.

1.4.1 Pretreatment

Ammonia fibre explosion (AFEX) pretreatment expands the plant cell wall, facilitating the work of the hydrolyzing enzymes (Balan *et al.*, 2009; Teymouri *et al.*, 2005). This disruption is very important because the enzymes during subsequent hydrolysis will cleave the monomeric sugars from the cellulose micro fibrils and the hemicellulose backbones, respectively. AFEX pretreatments are usually performed with an ammonia:biomass ratio of 1:1, 2:1 or 3:1 in high-pressure vessels that can sustain pressures from 1.4 to 4.8 MPa, and temperatures from 60 to 200 °C for 5 to 45 min. AFEX pretreatments usually do not lead to the generation of inhibitory products. On the other hand, because of the use of pressurized vessels, this pretreatment is quite costly.

Lime pretreatments are usually conducted with a 0.5:1 $Ca(OH)_2$:biomass loading, at temperatures ranging from 25 to 150 °C, for reaction times that can vary from hours to weeks in oxidative and non-oxidative environments. As expected, the processing time is inversely related to temperature. When conducting the pretreatment at a temperature of 25 °C, the processing time is in terms of weeks, requiring vast biomass holding areas. On the other hand, processing at temperatures in the range of 150 °C results in reaction times in terms of hours. Because the loosening of the plant cell wall of the feedstock can be carried out at 25 °C, this process can be set up as an on-farm operation. Lime pretreatment solubilizes lignin and results in the production of very few degradation products (Kim and Holtzapple, 2005).

Controlled pH pretreatment requires a loading of 1:6.2 biomass:water, with reaction temperatures between 170 and 200 °C, and reaction times of 5 to 20 min. Under these conditions, xylan is solubilized as oligomers that will later be hydrolyzed into their monomeric components by enzymatic hydrolysis. Very few degradation products are produced during this pretreatment method (Kim, Mosier, and Ladisch, 2009; Mosier *et al.*, 2005).

Dilute acid is a pretreatment that can be conducted at a range of sulfuric acid concentrations, 0.22 to 0.98%, at temperatures varying between 140 to 200 °C for times between 1 min and 1–2 hours. This pretreatment results in the solubilization of hemicellulose, exposing the cellulose microfibrils and the lignin in the plant cell wall. The cellulose is ready to be hydrolyzed by the enzymatic action. The xylose fraction will be released in the form of xylose monomers or as xylose degradation products, which need to be removed or at least minimized for further processing. Although dilute acid pretreatments present many drawbacks, this technology is likely to be adopted at the deployment scale because of its low cost and ease of use (Lloyd and Wyman, 2005; Wyman *et al.*, 2009).

Ammonia recycle pretreatment is usually conducted at a loading of 3.3 mL of ammonia 15% (w/w) per g of biomass, residence times between 10 to 12 min, processing temperatures between 170 and 220 °C and a processing pressure of 2.3 MPa. This process will result in the solubilization of lignin, leaving the hemicellulose and cellulose fractions ready to be hydrolyzed by the enzymes. Because the lignin content of grasses, 15–20%, is lower than

Table 1.4 *Sugar recovery of corn stover and poplar recovered with leading pretreatment technologies. Conversion for the corn stover using 15 FPU enzyme cocktail.*

	Corn stover	Poplar
Dilute acid	92%[a] glucose 94%[a] xylose 160 °C, for 20 min (Lloyd and Wyman, 2005)	87%[a] glucose 72%[a] xylose 190 °C, for 1.1 min (Wyman et al., 2009)
Controlled pH	91%[a] glucose 82%[a] xylose 190 °C, for 15 min (Mosier et al., 2005)	56%[a] glucose 96%[a] xylose 200 °C, for 10 min (Kim, Mosier, and Ladisch, 2009)
AFEX	96%[a] glucose 92%[a] xylose 1,7 MPa 90 °C, for 5 min (Teymouri et al., 2005)	52%[a] glucose 53%[a] xylose 4,8 MPa 180 °C 30 min (Balan et al., 2009)
Lime	93%[a] glucose 76%[a] xylose 55 °C, for 4 wk (Kim and Holtzapple, 2005)	78%[a] glucose 96%[a] xylose 2,3 MPa 140 °C 120 min (Sierra, Granda, and Holtzapple, 2009)
Ammonia recycling	90%[a] glucose 88%[a] xylose 170 °C, for 10 min (Kim and Lee, 2005)	49%[a] glucose 69%[a] xylose (Gupta and Lee, 2009)

[a]Indicates percentage recovery of available xylose and glucose, respectively.

that of hardwoods, 20–35%, this pretreatment is advantageously used with herbaceous materials (Gupta and Lee, 2009; Kim and Lee, 2005).

Table 1.4 presents side-by-side comparison of monosaccharide release from corn stover and poplar that were subjected to these pretreatments. It can be observed that the processing conditions, as well as the monosaccharide release are different for the five leading pretreatments using the same feedstock. It appears to be easier to release the carbohydrates from a feedstock like corn stover as compared to that of a woody feedstock such as poplar.

1.4.2 Enzymatic Hydrolysis

Pretreatment opens the tightly woven biomass plant cell wall by separating cellulose, hemicellulose, and lignin, thereby increasing the internal surface area. With more available surface area, enzymes can have access to the carbohydrate polymers, hydrolyzing hemicellulose into the five-carbon xylose, and cellulose into the six-carbon glucose. The hydrolysis of the treated cellulose is performed with an enzymatic cocktail composed of β-glucosidase, endo-cellulsase and exo-cellulase. These enzymes hydrolyze cellobiose as well as the middle and extremities of the cellulose polymer, respectively. Loadings of 15 FPU g^{-1} glucan and

30 FPUg^{-1} glucan of cellulase and β-glucosidase, respectively, are reported as being utilized (Gupta and Lee, 2009). Xylan cocktails including endo-β-(1,4)-xylanases, β-xylosidases, acetylxylanesterases, and feruloyl esterase are used to cleave the hemicellulose, including their attached monosaccharides (Bauer *et al.*, 2005).The enzyme α-glucuronidase is used to remove glucuronic acid residues from the hemicellulose. Specific activities related to hemicellulose hydrolysis need to be determined because the hemicellulose composition varies from plant to plant. It is critical to digest the plant cell wall monosacharides with the least amount of cellulase and xylanase enzymes; these biocatalysts account for the most costly component of the biochemical process and, if not used sparingly, could forfeit the economic viability of the biorefinery.

Although pretreatment is critical, this processing step unfortunately can lead to the production of inhibitory compounds that restrain the enzymatic saccharification of the carbohydrate polymers. The enzymatic saccharification step is obstructed mainly by three groups of compounds: lignin derivatives that cause non-productive binding of the saccharification cocktail (Berlin *et al.*, 2006); xylose-derived compounds inhibiting the enzyme cocktail (Cantarella *et al.*, 2004); and oligomers and phenolic-derived compounds deactivating the enzymes over time (Kumar and Wyman, 2008; Berlin *et al.*, 2006 and Ximenes *et al.*, 2011). To lessen the effect of these inhibitory compounds and minimize the amount of enzymes used, the inhibitors must be removed by separation methods, for example, washing the pretreated biomass with successive volumes of water (Hodge *et al.*, 2008).

1.4.3   Fermentation

After biomass is pretreated and the ensuing plant cell wall components are saccharified, the released sugars can be fermented into biofuels or other desired bio-based products. It is critical that both five-carbon and six-carbon sugars are fermented to alcohol because the process could never achieve economic viability if only the more easily fermented six-carbon sugars are metabolized. Desirable performance metrics of ethanol titre and rate are 40 g L^{-1} and 1 g L^{-1} h^{-1}, respectively (Lau *et al.*, 2010). To achieve these performance metrics, glucose and xylose must be processed by the microorganism.

Unfortunately, no native organism can metabolize simultaneously both carbon streams. Native *Saccharomyces cerevisiae*, which is a leading microorganism for ethanol fermentations, metabolizes glucose into ethanol, but cannot process xylose. On the other hand, *Pichia stipites* is a native microorganism that metabolizes xylose into ethanol (Jeffries, 2006). There are two scientific approaches to make sure that both glucose and xylose are processed into ethanol: (1) genetically introduce the missing xylose pathway into the target organism, such as *S. cerevisiae*; (2) co-culture xylose and glucose metabolizing organisms, such as *S. cerevisiae* with *P. stipites*.

There are a number of microorganisms into which the xylose-metabolizing pathway was introduced, amongst others: *Escherichia coli*, *Zymomonas mobilis*, and *S. cerevisiae*. Lau *et al.* (2010) reported on the side-by-side comparison of these three strains using AFEX pretreated corn stover; ethanol titers of 30, 40 and 30 g L^{-1} were reported for *E. coli*, *S. cerevisiae*, and *Z. mobilis,* respectively, during 80, 160, and 40 hours of fermentation. From these results, *S. cerevisiae* appeared to be the least troubled by the fermentation inhibitors that are produced during pretreatment.

If desirable ethanol yields are to be obtained via the co-culture route, one must be cognizant of two important considerations: (1) the strains must not be toxic to each other, such as a *S. cerevisiae* and *P. stipites* combination; and (2) the strains must be cultured at similar temperatures and pH, also such as a *S. cerevisiae* and *P. stipites* combination. The *Z. mobilis* and *P. stipites* system is an example of an unsuccessful co-culture combination; although both require a temperature of 37 °C, the former requires an optimum pH 7 and the latter needs an optimum pH 5, illustrating well the impasse. Guidelines as to how to set-up a *S. cerevisiae* and *P. stipites* co-culture system include: (1) initial total sugar concentration between 20 and 100 g l^{-1} with xylose varying between 20 and 50%; fermentation temperature of 30 °C; and 2 mmol l^{-1} h^{-1} of oxygen. These cultivation conditions resulted in a volumetric ethanol productivity of 0.94 gl^{-1} h^{-1} and an ethanol yield of 0.50 g g^{-1} over a 40-hour fermentation cycle (Chen, 2011).

Although the discussion has been centred mostly on ethanol production, it is important to bear in mind that the biorefinery could be centred on other fermentation products. When the biomass plant cell wall is saccharified and the sugars are released, the nature of the fermentation organism determines the produced product. *Clostridium tyrobutyricum* can be used to produce butyric acid from carbohydrates, and then *C. acetobutylicum* can be used to ferment the butyric acid to butanol. With this two-step approach, yields of 0.40–0.47 g acetone:butanol:ethanol products per g sugar utilized (gg^{-1}) are common (Ramey, 1998).

1.4.4   Pre-Pretreatment

As mentioned earlier in this chapter, pretreatment is a critical operation in the biochemical processing operations; without some sort of pretreatment, it is almost useless to perform enzymatic hydrolysis because the cell wall will not be loosened, and the cellulase and xylanase enzymes will not have sites to bind to. In addition to standard pretreatments, the biomass feedstock can undergo a "pre-pretreatment," where the biomass is soaked in water or dilute acid before being subjected to the actual pretreatment operation. The pre-pretreatment step can be, in a sense, compared to the soaking of a stained garment before washing. The wash water from the pre-pretreatment step can sometimes contain phytochemicals; this can be an opportunity to extract high-value speciality chemicals, while improving pretreatment and adding value to the overall biorefinery operation.

Ground early fall-harvested switchgrass var. Cave-in-Rock was extracted in 90 °C water for 20 min prior to dilute acid pretreatment; the flavonoids, quercitrin (quercetin-3-*O*-rhamnoside), and rutin (quercetin-3-O-rutinoside), were detected and their concentrations were 193 and 186 mg kg^{-1} of dry biomass, respectively (Uppugundla *et al.*, 2009). The extracts were subsequently purified by centrifugal partition chromatography, using a ethyl acetate:ethanol:water (2:1:2, v/v/v) solvent system, into quercitrin and rutin fractions; these purified flavonoids decreased by 78 and 86%, respectively, the oxidation of low-density lipoprotein as measured by the thiobarbituric reactive substances (TBARS) assay (Uppugundla *et al.*, 2009).

Ground spring-harvested Switchgrass var. Alamo was extracted in 85 °C water for 2 h; neither rutin nor quercitrin was detected, but quercetin was noticed at concentrations of 140 and 120 mg kg^{-1} of dry biomass, respectively, for leaves and stems (Martin *et al.*, 2011). The pre-pretreated biomass was subsequently pretreated in 1% sulfuric acid in a 2 L Parr bioreactor for 25 min at 130 °C. The coupling of the pre-pretreatment step to that of the pretreatment resulted in xylose increases of 5 and 23% for stems and leaves, respectively (Martin *et al.*, 2011). These results point to the fact that not only can high value chemicals be recovered in the pre-pretreatment water, but that this operation increases carbohydrate recovery.

Sweetgum (*Liquidambar styraciflua* L.) has a widespread distribution in the southeastern United States. Much of its distribution is as understory amongst planted pines that are grown for softwood lumber and pulp for paper products. However, hardwood competition in the pine forest understory is a major impediment to pine forest growth; therefore, southern pine forests are intensively managed. Instead of being a nuisance, this hardwood understory growth could become an important source of biomass for the upcoming biochemical-based biorefineries, especially in view that sweetgum is a fast-growing hardwood. In addition to being a possible biorefinery feedstock, sweetgum contains shikimic acid, which is a precursor for the drug Tamiflu® used in treating avian flu. Water-extracted sweetgum bark and heartwood yielded 1.7 and 0.2 mg g^{-1}of shikimic acid, respectively. The addition of a 65 °C shikimic acid extraction step coupled to pretreatment with 0.98% H_2SO_4 at 130 °C for 50 min resulted in 21% and 17% increases in xylose percentage recovery from bark and de-barked wood, respectively. These results indicate that, in addition to recovering a high-value product, the 65 °C wash step also increases xylose recovery (Martin *et al.*, 2010).

1.5 Conclusion

The goal of this chapter was to present a general overview of what a thermochemical and a biochemical biorefinery would entail in terms of feedstock and of processing. The reader can appreciate the fact that, although both technology platforms can be conducted with biomass, their processing parameters widely differ. The thermochemical conversion platform is conducted at temperatures of at least 300 °C and results in the production of bio-oils or syngas that can be upgraded into further products. On the other hand, biomass processing following the biochemical conversion platform is, in a sense, more mild because of the requirements of cell wall loosening coupled to enzymatic release of the structural sugars. Because the sugars are further fermented into various products, this technology platform offers the possibility of producing various bio-based products. Hopefully, both platforms will be widely utilized in the decades to come, processing colossal amounts of biomass. The goal of this book is to illustrate that speciality chemicals and other bio-based products could be extracted prior to or after the conversion process, increasing the overall profitability and sustainability of the biorefinery. Therefore, subsequent chapters will describe various classes of secondary metabolites that are present in biomass, present strategies for the extraction and purification of these various phytochemicals, and finally describe examples of biorefinery-destined biomass that contains speciality chemicals.

Acknowledgements

The authors would like to thank the University of Arkansas, Division of Agriculture, and the Departments of Biological and Agricultural Engineering and Ralph E. Martin Chemical Engineering for financial assistance. The authors would also like to acknowledge National Science Foundation (NSF) award #0828875; NSF Experimental Program to Stimulate Competitive Research # 0701890, specifically NSF-EPSCoR P3; Department of Energy award # 08GO88035 for support.

References

Ananda, N., Vadlania, P., and Prasad, P. (2011) Evaluation of drought and heat stress grain sorghum (*Sorghum bicolor*) for ethanol production. *Industrial Crops and Products*, **33**, 779–782.

Bain, R., Amos, W., Downing, M., and Perlack, R. (2003) Biopower technical assessment state of industry and the technology. NREL/TP-510-33132. January 2003.

Balan, V., Costa Sousa, L., Chundawat, S. *et al.* (2009) Enzymatic digestibility and pretreatment degradation products of AFEX-treated hardwoods (*Populusnigra*). *Biotechnology Progress*, **25**, 333–339.

Bauer, S., Prasanna, V., Mort, A., and Somerville, C. (2005) Cloning, expression, and characterization of an oligoxyloglucan reducing end-specific xyloglucanobiohydrolase from *Aspergillus nidulans*. *Carbohydrate Research*, **340**, 2590–2597.

Berlin, A., Balakshin, M., Gilkes, N. *et al.* (2006) Inhibition of cellulase, xylanase and β-glucosidae activities by softwood lignin preparations. *Journal of Biotechnology*, **125**, 198–209.

Brammer, J. and Bridgwater, A. (2002) The influence of feedstock drying on the performance and economics of a biomass gasifier-engine CHP system. *Biomass and Bioenergy*, **22**, 271–281.

Bridgwater, A., Meier, D., and Radlein, D. (1999) An overview of fast pyrolysis of biomass. *Organic Geochemistry*, **30**, 1479–1493.

Cantarella, M., Cantarella, L., Gallifuoco, A. *et al.* (2004) Effect of Inhibitors Released during Steam-Explosion Treatment of Poplar Wood on Subsequent Enzymatic Hydrolysis and SSF. *Biotechnology Progress*, **20**, 200–206.

Carlson, T., Tompsett, G., Conner, W., and Huber, G. (2009) Aromatic production from catalytic fast pyrolysis of biomass-derived Feedstocks. *Topics in Catalysis*, **52**, 241–252.

Chen, Y. (2011) Development and application of co-culture for ethanol production by co-fermentation of glucose and xylose: a systematic review. *Journal of Industrial and Microbial Biotechnology*, **38**, 581–597.

Ciferno, J. and Marano, J. (2002) Benchmarking biomass gasification technologies for fuels, chemicals and hydrogen production. Prepared for US Department of Energy National Energy Technology Laboratory (DOE-NETL), June 2002. Available online at: http://www.netl.doe.gov/technologies/coalpower/gasification/pubs/pdf/BMassGasFinal.pdf [last accessed: December 2nd, 2011].

Clifton-Brown, J., Lewandowski, I., Andersson, B. *et al.* (2001) Performance of 15 miscanthus genotypes at five sites in Europe. *Agronomy Journal*, **93**, 1013–1019.

Clifton-Brown, J., Stampfl, P., and Jones, M. (2004) Miscanthus biomass production for energy in Europe and its potential contribution to decreasing fossil fuel carbon emissions. *Global Change Biology*, **10**, 509–518.

Davis, A. and Trettin, C. (2006) Sycamore and sweetgum plantation productivity on former agricultural land in South Carolina. *Biomass and Bioenergy*, **30**, 769–777.

Demirbas, A. (2007) Converting biomass derived synthetic gas to fuels via Fisher-Tropsch synthesis. *Energy Sources Part A: Recovery, Utilization & Environmental Effects*, **29**, 1507–1512.

Devi, L., Ptasinski, J., and Janssen, F. (2003) A review of the primary measures for tar elimination in biomass gasification processes. *Biomass and Bioenergy*, **24**, 125–140.

Du, B., Sharma, L., Becker, C. *et al.* (2010) Effect of varying feedstock-pretreatment chemistry combinations on the formation and accumulation of potentially inhibitory degradation products in biomass hydrolysates. *Biotechnology and Bioengineering*, **107**, 430–440.

Elliott, D. (2007) Historical developments in hydroprocessing bio-oils. *Energy and Fuels*, **21**, 1792–1815.

Evans, R. and Milne, T. (1987) Molecular characterization of the pyrolysis of biomass. *Energy and Fuels*, **1**, 123–137.

Fortier, J., Gagnon, D., Truax, B., and Lambert, F. (2010) Biomass and volume yield after 6 years in multiclonal hybrid poplar riparian buffer strips. *Biomass and Bioenergy*, **34**, 1028–1040.

Garcia-Ibanez, P., Cabanillas, A., and Sanchez, J. (2004) Gasification of leached orujillo (olive oil waste) in a pilot plant circulating fluidized bed reactor. Preliminary results. *Biomass and Bioenergy*, **27**, 183–194.

Gupta, R. and Lee, Y.Y. (2009) Pretreatment of hybrid poplar by aqueous ammonia. *Biotechnology Progress*, **25**, 357–364.

Heaton, E., Dohleman, F., and Long, S. (2008) Meeting US biofuel goals with less land: the potential of miscanthus. *Global Change Biology*, **14**, 1–15.

Hodge, D., Karim, N., Schell, D., and MacMillan, J. (2008) Soluble and insoluble solids contributions to high-solids enzymatic hydrolysis of lignocellulose. *Bioresource Technology*, **99**, 8940–8948.

Hughes, W. and Larson, E. (1998) Effect of fuel moisture content on biomass-IGCC performance. *Journal of Engineering for Gas Turbines and Power*, **120**, 455–459.

Jeffries, T. (2006) Engineering yeasts for xylose metabolism. *Current Opinion in Biotechnology*, **17**, 320–326.

Johansson, T. and Karačić, A. (2011) Increment and biomass in hybrid poplar and some practical implications. *Biomass and Bioenergy*, **35**, 1925–1934.

Kim, T. and Holtzapple, M. (2005) Pretreatment and fractionation of corn stover by lime pretreatment. *Bioresource Technology*, **96**, 1994–2006.

Kim, T. and Lee, Y.Y. (2005) Pretreatment and fractionation of corn stover by ammonia recycle percolation process. *Bioresource Technology*, **96**, 2007–2013.

Kim, Y., Mosier, N., and Ladisch, M. (2009) Enzymatic digestion of liquid hot water pretreated hybrid poplar. *Biotechnology Progress*, **25**, 340–348.

Kumar, R. and Wyman, C. (2008) Effect of enzyme supplementation at moderate cellulose loadings on initial glucose and xylose release from corn stover solids pretreated by leading technologies. *Biotechnology and Bioengineering*, **102**, 457–467.

Laser, M., Schulman, D., Allen, S. *et al.* (2002) A comparison of liquid hot water and steam pretreatments of sugar cane bagasse for bioconversion to ethanol. *Bioresource Technology*, **81**, 33–44.

Lau, F. and Carty, R. (1994) Current status of the IGT RENUGAS process. Paper presented at the 19th World Gas Conference, Milan, Italy, (June 20–23) 1994.

Lau, M., Gunawan, C., Balan, V., and Dale, B. (2010) Comparing the fermentation performance of *E. coli* K011, *S. cerevisiae* 424 (LNH-ST) and *Zymomonas mobilis* AX 101 for cellulosic ethanol production. *Biotechnology for Biofuels*, 3:11.

Lehto, J., Jokela, P., Alin, J. *et al.* (2010) Bio-oil production integrated to a fluidised bed boiler – experiences from pilot operation. *Industrial Fuels & Power.*, Published online, available online at: http://www.ifandp. com/article/006550.html [last accessed: November 30th, 2011].

Lloyd, T. and Wyman, C. (2005) Combined sugar yields for combined dilute Sulfuric acid pretreatment of corn stover followed by enzymatic hydrolysis of the remaining solids. *BioresourceTechnology*, **96**, 1967–1977.

Lynd, L., Laser, M., Bransby, D. *et al.* (2008) How biotech can transform biofuels. *Nature Biotechnology*, **26**, 169–172.

Martin, E., Cousins, S., Talley, S., West, C., Clausen, E., and Carrier, D.J. (2011). The effect of pre-soaking coupled to pretreatment on the extraction of hemicellulosic sugars and flavonoids from switchgrass (*Panicum virgatum* var. Alamo) leaves and stems. *Transactions of ASABE*, **54**, 1953:1958.

Martin, E., Duke, J., Pelkki, M. *et al.* (2010) Sweetgum *(Liquidambar styraciflua* L.): Extraction of shikimic acid coupled to dilute acid pretreatment. *Applied Biochemistry and Biotechnology*, **162**, 1660–1668.

Meier, D. and Faix, O. (1999) State of the art of applied fast pyrolysis of lignocellulosic materials - a review. *Bioresource Technology*, **68**, 71–77.

Mercader, F., Groeneveld, M., Kersten, S. *et al.* (2011) Hydrodeoxygenation of pyrolysis oil fractions: process understanding and quality assessment through co-processing in refinery units. *Energy Environmental Science*, **4**, 985–997.

Mosier, N., Hendrickson, R., Ho, N. *et al.* (2005) Optimization of pH controlled liquid hot water pretreatment of corn stover. *Bioresource Technology*, **96**, 1986–1993.

Paisley, M.A., Creamer, K.S., Tewksbury, T.L., Taylor, D.R., and Schiefelbein, G.F. Gasification of Refuse Derived Fuel in the Battelle High Throughput Gasification System. Prepared for Pacific Northwest Laboratory (PNL), U.S. Department of Energy. Contract DE-ACX06-76RLO 1830 under Agreement 007069-A-H6, Battelle Columbus Division, July 1989. Available online at: http://www.osti.gov/bridge/ servlets/purl/5653025-QRQFYH/5653025.pdf [last accessed: November 30, 2011].

Perlack, R., Wright, L., Turhollow, A. *et al.* (2005) A bioenergy and bioproducts industry: the technical feasibility of a billion ton annual supply. US Department of Energy, Office of Scientific and Technical Information, PO BOX 63, Oak Ridge, TN 37831-0062. Available electronically at www.osti.gov/bridge.

Propheter, J., Staggenborg, S., Wu, X., and Wang, D. (2010) Performance of Annual and Perennial Biofuel Crops: Yield during the First Two Years. *Agronomy Journal*, **102**, 806–814.

Priyadarsan, S., Annamalai, K., Sweeten, J. *et al.* (2004) Fixed-bed gasification of feedlot manure and poultry litter biomass. *Transactions of the ASABE.*, **47**, 1689–1696.

Raghu, S., Anderson, R., Daehler, C. *et al.* (2006) Adding biofuels to the invasive species fire? *Science*, **313**, 1742.

Ramey, D. (1998) Continuous, two stage, dual path anaerobic fermentation of butanol and other organic solvents using two different strains of bacteria. US Patent 5,753,474 May 19 1998.

Reed, T. and Das, A. (1987) *Handbook of Biomass Downdraft Gasifier Engine Systems*, (SERI-1988) BEF Press, Golden, CO.

Rollins, M., Reardon, L., Nichols, D. *et al.* (2002) Economic Evaluation of CO_2 Sequestration Technologies Task 4, Biomass Gasification-Based Processing: Final Technical Report. DE-FC26-00NT40937, Prepared for the US Department of Energy.

Sierra, R., Granda, C., and Holtzapple, M.T. (2009) Short-term lime pretreatment of poplar wood. *Biotechnology Progress*, **25**, 323–332.

Sutton, D., Kelleher, B., and Ross, J. (2001) Review of literature on catalysts for biomass gasification. *Fuel Processing Technology*, **73**, 155–173.

Teymouri, F., Laureano-Perez, L., Alizadeh, H., and Dale, B. (2005) Optimization of the ammonia fiber explosion (AFEX) treatment parameters for enzymatic hydrolysis of corn stover. *Bioresource Technology*, **96**, 2014–2018.

Tijmensen, M., Faaij, A., Hamelinck, C., and Van Hardeveld, M. (2002) Exploration of the possibilities for production of Fischer Tropsch liquids and power via biomass gasification. *Biomass and Bioenergy*, **23**, 129–152.

Uppugundla, N., Engelberth, A., Vandhana Ravindranath, S. *et al.* (2009) Switchgrass water extracts: Extraction, separation and biological activity of rutin and quercitrin. *Journal of Agricultural Food Chemistry*, **57**, 7763–7770.

US DOE (2005) (http://www1.eere.energy.gov/biomass/printable_versions/pyrolysis.html).

(UCR) University of California, Riverside (2009) Evaluation of emissions from thermal conversion technologies processing municipal solid waste and biomass, final report. Prepared for: BioEnergy Producers Association, 3325 Wilshire Blvd, Ste. 708, Los Angeles, CA 90010. Available online at: [http://www.bioenergyproducers.org/documents/ucr_emissions_report.pdf].

Van der Drift, A., van Doorn, J., and Vermeulen, J. (2001) Ten residual biomass fuels for circulating fluidized-bed gasification. *Biomass and Bioenergy*, **20**, 45–56.

Venderbosch, R. and Prins, W. (2010) Fast pyrolysis technology development. *Biofuels, Bioproducts and Biorefinery*, **4**, 178–208.

Wildschut, J., Mahfud, F., Venderbosch, R., and Heeres, H. (2009) Hydrotreatment of fast pyrolysis oil using heterogeneous noble-metal catalysts. *Industrial and Engineering Chemistry Research*, **48**, 10324–10334.

Williams, P. and Horne, P. (1995) The influence of catalyst type on the composition of upgraded biomass pyrolysis oils. *Journal of Analytical and Applied Pyrolysis*, **31**, 39–61.

Wortmann, C., Liska, A., Ferguson, R. *et al.* (2010) Dryland performance of sweet sorghum and grain crops for biofuels in Nebraska. *Agronomy Journal*, **102**, 319–326.

Wullschleger, S., Davis, E., Borsuk, M. *et al.* (2010) Biomass production in switchgrass across the United States: Database description and determinants of yield. *Agronomy Journal*, **102**, 1158–1168.

Wyman, C., Dale, B., Elander, R. *et al.* (2009) Comparative sugar recovery and fermentation data following pretreatment of Poplar wood by leading technologies. *Biotechnology Progress*, **25**, 333–339.

Ximenes, E., Kim, Y., Mosier, N. *et al.* (2011) Deactivation of cellulases by phenols. *Enzyme and Microbial Technology*, **48**, 54–60.

Zhang, H., Xiao, R., Huang, H., and Xiao, G. (2009) Comparison of non-catalytic and catalytic fast pyrolysis of corncob in a fluidized bed reactor. *Bioresource Technology*, **100**, 1428–1434.

2

Overview of the Chemistry of Primary and Secondary Plant Metabolites

Chantal Bergeron

Tom's of Maine, Kennebunk, Maine, USA

2.1 Introduction

There are two different types of plant metabolites: primary and secondary. Primary refers to the metabolites required for growth, development, and reproduction whereas the secondary metabolites are not directly involved in those processes but have important ecological functions like natural defense (e.g., against insects or fungi) or attracting pollinators. In general, the pharmaceutical, cosmetic, and nutraceutical industries are more interested in the compounds from secondary metabolism, while the bioenergy industry will be more focused on the primary metabolites like simple sugars, polysaccharides, starch, cellulose, and hemicellulose for energy production. However, secondary metabolites are a good way to add value to the biomass. Secondary metabolites also confer certain properties to the plant. Some of these properties can be harnessed by humans for their own benefit.

In general, the primary metabolites will be present in great quantities, whereas the secondary metabolites will be present in trace amounts; but if there is elicitation (stimulation) either by environmental factors or the presence of a threatening organism like an herbivore or an insect, the quantities produced by the plant may increase. The localization of the different components will vary with the function of the metabolites and the perenniality of the plant. For

Biorefinery Co-Products: Phytochemicals, Primary Metabolites and Value-Added Biomass Processing, First Edition.
Edited by Chantal Bergeron, Danielle Julie Carrier and Shri Ramaswamy.
© 2012 John Wiley & Sons, Ltd. Published 2012 by John Wiley & Sons, Ltd.

example, storage substances like starch will accumulate mainly in roots and tubers, or seeds. Secondary metabolites tend to be more abundant in organs like the bark, which is the first barrier of defense against the environment.

This chapter will give an overview of primary and secondary metabolites. The general structure of each class is included, with the numbering system of the carbon atoms, to help the reader who is not familiar with a particular class of compounds to better understand the literature. When a class of compounds is not ubiquitous, we will give a brief overview of the phylogeny (a list of families where those compounds are found). It may greatly help to envision what kind of compounds can be expected in the species of interest. We will also give an idea of the ecological significance of the different classes of compounds.

Some compounds of the primary metabolites will be further discussed as important components in biofuel and biodiesel production. The objective of this chapter is to give the language necessary to discuss, understand, and dive into the fabulous chemical world of terrestrial and marine plants.

2.2 Primary Metabolites

Sugars, lipids, amino acids, peptides, and proteins belong to primary metabolism. These substances are ubiquitous in the plant kingdom.

2.2.1 Saccharides (Sugars)

Monosaccharides

Monosaccharides contain one sugar (saccharide) unit. Monosaccharides are divided according to the number of carbons: trioses ($C = 3$), tetroses ($C = 4$), pentoses ($C = 5$), hexoses ($C = 6$), and so on. Most commonly found in plants are the pentoses and hexoses. Important pentoses are L-arabinose and D-xylose. The hexoses D-glucose, D-mannose, and D-galactose are also widespread. They are constituents of hemicelluloses, polysaccharides, pectins and polymers secreted by plants (gum and mucilage).

Monosaccharide

The monosaccharides are further separated into pyranoses and furanoses. A pyranosepossesses a six-membered ring structure while the furanose has only five atoms in its ring. Some sugars, like glucose, galactose, or mannose, exist predominantly as a pyranose, whereas fructose and ribose mainly appear as furanoses (although fructose also occurs in the pyranose form). Some monosaccharides are present as both furanose and pyranose (e.g., arabinose). When a monosaccharide forms a ring, carbon one can have its hydroxyl group in one of two conformations: axial (pointing straight down) or equatorial (pointing slightly up). When it is axial, the glucose is alpha. When it is equatorial, it is beta. This doesn't have as much impact on free, monomeric sugars because they tend to switch back and forth rapidly. When a sugar is polymerized into polysaccharides, though, it matters a lot. If alpha-glucose is polymerized you get starch, which is soluble and digestible by humans. If you polymerize beta-glucose, however, you get tough, indigestable cellulose. Saccharide can be in a free form or polymerized as polysaccharides like starch and cellulose, or linked to proteins to form glycoproteins. Sugars are often linked to various molecules from the secondary metabolism (the "non-sugar" part is referred to as the aglycone) to form heterosides.

Disaccharides

Disaccharides contain two sugar units. These sugars are produced when two monosaccharides are linked by an "oxygen bridge" called an *O*-glycosidic bond. This interglycosidic linkage is an important feature in characterizing disaccharides. Maltose is formed from two α-D-glucose molecules. It is a disaccharide linked by an α (1 → 4) glycosidic bond. The number 1 carbon of one molecule is bonded to the number 4 carbon of the other molecule. Cellobiose is made of two glucose units with a β (1 → 4) link. Trehalose is also formed by two glucose units but linked 1 → 1. Saccharose is composed of glucose and fructose, and lactose of galactose and glucose.

Oligosaccharides

Oligosaccharides are comprised of between 3 and 10 sugar units. Examples are raffinose, stachyose, or lychnose.

Polysaccharides

Polysaccharides are made of over 10 units. Some well-known polysaccharides are starch, cellulose, pectin, and hemicellulose. Red and brown algae contain gel-forming polysaccharides like agar and carrageenan. Sulfated polysaccharides like carrageenan are often found in algae (De Souza *et al.*, 2007).

Cellulose   Cellulose is probably the most universal polymer. Cellulose is a linear polymer composed of D-glucose which has a β (1 → 4) linkage. The degree of polymerization (DP), which indicates the number of sugar units in a polysaccharide, varies between 300 to 15 000, according to the botanical source, tissue age, and the process used during isolation. In the secondary wall of higher plants, the DP varies generally between 6000–10 000. However, it can reach up to15 000 in cotton seed capsules (Bruneton, 1993, Bruneton, 2009).

Hemicellulose   Hemicellulose is a constituent that, unlike cellulose, is soluble in alkaline solutions. It is normally constituted of less than 200 saccharides, which confers this solubility. It is easily degraded by cell enzymes and may form reserve substances (e.g., albumen in date) (Guignard, Cosson, and Henri, 1985).

Distribution   They are present in higher plants where their level can reach 25–30% of wood. Their structure differs if they are part of the primary or secondary wall or if they are present in monocotyledons or dicotyledons. Xyloglucans, galactans, arabinogalactans, arabinans are constituents of the primary wall whereas xylans, mannans and glucomannans are components of secondary walls (Guignard, Cosson, and Henri, 1985).

Starch   Starch is the main substance of stockage in plants. It is present in all plant parts, but is mainly found in cereal grain (wheat, oat, corn, barley, rice), legumes (pea, chickpea, bean, lentil), some fruits (banana, plantain), and underground parts like potato, yam, or manioc (Bruneton, 2009).

Starch is a white substance, insoluble in cold water, but soluble in hot water, where it forms a gel. It is a mixture of two constituents, amylose and amylopectin. The amylose is soluble in water at 80 °C and amylopectin, which is the most abundant, is insoluble in hot water. The proportion of both varies according to species and tissue.

Amylose is a helicoidal long chain formed of 60 to 6000 units of D-glucose linked α 1 → 4. Amylopectin contains between 6000 to 600 000 units of glucose and has some ramifications. The interglycosidic linkage of the ramification is alpha 1 → 6 (Guignard, Cosson, and Henri, 1985).

Ecological Significance   Saccharides play a role:

- As structural elements and cell wall polysaccharides, for example, cellulose
- In energy storage, for example, starch
- As constituents of various metabolites, for example, nucleic acids, coenzymes, and heterosides
- In protecting tissues against dehydration

2.2.2 Lignin

Lignin is a constituent of plant fibers. It is a tridimensional heteropolymer formed from phenylpropane units (see Section 2.3.2) with high molecular weight. Very hydrophobic, it penetrates the cell wall to give rigidity, impermeability, and resistance (Bruneton, 2009).

2.2.3 Amino Acids, Peptides, and Proteins

As their name suggests, the amino acid contains an amine $-NH_2$, a carboxylic acid $-COOH$, and a side chain (R) that varies between the different amino acids.

General Structure

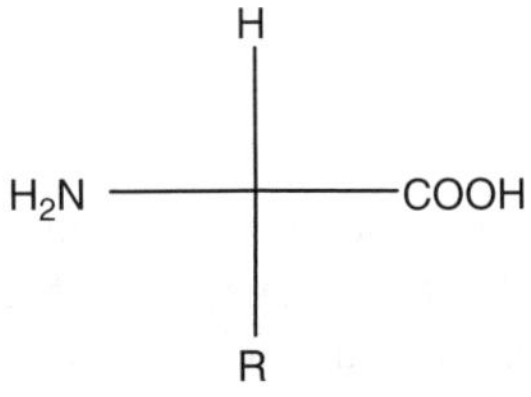

$$H_2N \quad\rule{1.5cm}{0.4pt}\!\!\overset{\displaystyle H}{\underset{\displaystyle R}{\rule{0.4pt}{1.5cm}}}\!\!\rule{1.5cm}{0.4pt}\quad COOH$$

The amino acids possess four different substituents with the exception of glycine, where R=H.

The amino group of an amino acid can be linked to the carboxyl group of the next amino acid to form a peptidic bond. Thus, amino acids form the base units of peptides (consisting of 2–35 amino acids) or proteins (> 35 amino acids). A molecule consisting of two amino acids is called dipeptide. If additional units are added, tripeptides (three), tetrapeptides (four), pentapeptides (five), and so on are obtained.

Ecological Significance

Proteins can be part of a structural element of a cell and are very important in cell signalling. Enzymes and many receptors are also proteins.

2.2.4   Fatty Acids, Lipids

Fatty Acids

Fatty acids are divided in two classes, the saturated and unsaturated. The later possess one or more double or triple bonds between two adjacent carbons.

Lipids include fats, waxes, sterols, monoglycerides, diglycerides, and others. Lipids can be constituents of cells, like the phospho- and glycolipids, or elements of the cell surface, like waxes or cutin (Bruneton, 2009). The main biological functions of lipids include energy storage, as structural components of cell membranes, and as important signalling molecules. They are generally hydrophobic or sometimes amphophilic.

We distinguish normally between the simple lipids (triglycerides, cerides) and the complex lipids (phospholipids and glycolipids). The simple lipids are esters of fatty acids and alcohol. If the alcohol is glycerol, we have a triglyceride. If it is a different alcohol, we obtain a ceride. Phospholipids and glycolipids are complex lipids. They play a role as constituents of membranes (Bruneton, 2009).

Phospholipids

Phospholipids are major components of all cell membranes, where they form a bilayer with the hydrophobic tail reaching inside the cell and the hydrophilic moiety towards the outside. Most phospholipids contain a diglyceride, a phosphate group, and a simple organic molecule such as choline. Lecithin is the common name to describe a mixture of phospholipids. The first phospholipid identified was lecithin or phosphatidylcholine and was isolated from egg yolk, but is generally sourced from soya. Other sources include, but are not limited to, rapeseed and sunflower.

In the pharmaceutical and the cosmetic industries, lecithin is used to obtain liposomes, but can also form a stable emulsion. Its main use is by the food industry in margarine, chocolate, or in instant mixtures, as it can be rapidly reconstituted without forming lumps (Bruneton, 2009).

2.2.5   Organic Acids

Some of the organic acids play a role in the Krebs cycle. These are citric, aconic, malic, isocitric, succinic, fumaric, oxaloacetic, ketoglutaric, and pyruvic acid. This cycle plays a major role in synthesis within plant cells, while it is mainly an energy source in animals (Guignard, Cosson, and Henri, 1985).

Organic acids are found in roots, stems, buds, and fruits. During maturation of fleshy fruits, the organic acids that are synthesized by the leaf migrate to the pericarp of the fruit, where they accumulate. In succulent plants, malic acid increases the osmotic pressure of the cell sap which increases the resistance to dryness (Guignard, Cosson, and Henri, 1985).

2.3   Secondary Metabolites

Unlike primary metabolites, secondary metabolites are organic compounds that are not directly involved in growth, development, or reproduction of organisms. They are often limited to certain classes, orders, families, genera, or species. For a more detailed description of secondary metabolites there are some excellent reference books (Evans, 2009; Hänsel and Sticher, 2007; Heinrich *et al.*, 2004; Andersen and Markham, 2006; Baxter, Harborne, and Moss, 1999; Bruneton, 2009).

There are four different pathways used by plants to produce secondary metabolites and each is responsible for the synthesis of different classes of compounds.

- Shikimic acid pathway: simple phenolic glycosides (arbutin, saliciline), phenylpropanes, lignans, coumarins, and naphthoquinones.
- Acetate-malonate pathway: anthraquinones, flavonoids, tannins, phloroglucinol derivatives.
- Mevalonate pathway: terpenes.
- Amino acid pathway: alkaloids

2.3.1   Simple Phenols and Phenolic Acids

Simple phenols are defined as compounds with at least one hydroxyl group attached to an aromatic ring.

Phenolic acids are organic compounds with at least one carboxylic acid and a hydroxyl group on an aromatic ring. They are hydroxylated derivatives of benzoic acid.

General Structure

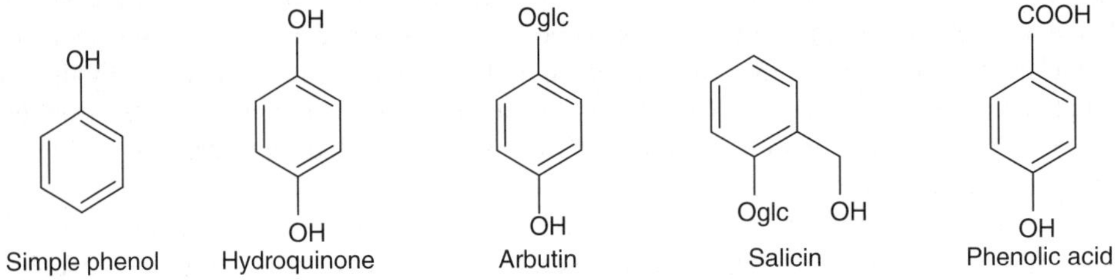

| Simple phenol | Hydroquinone | Arbutin | Salicin | Phenolic acid |

Distribution

Simple phenols are rare in the nature with the exception of hydroquinone, which is widely present in the Ericaceae family (blueberry family) and often found in the glycosidic form, arbutin. Salicin is widespread in the Salicaceae family and is the precursor of salicylic acid.

2.3.2   Polyphenols

Phenylpropanoids

Phenylpropanoids are described as C_6-C_3, C_6 being the phenyl and C_3, the propane.

General Structure

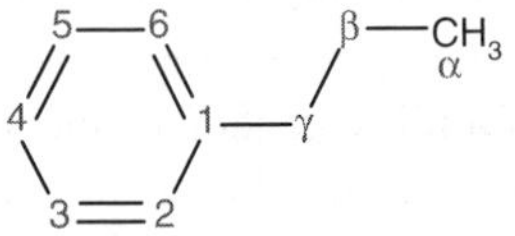

Some examples of phenylpropanes are:

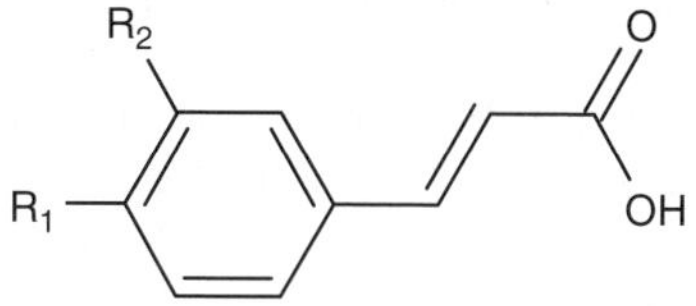

Cinnamic acid: $R_1=R_2=H$
Caffeic acid: $R_1=R_2=OH$
Ferulic acid: $R_1=OH$, $R_2=MeO$
Isoferulic acid: $R_1=MeO$, $R_2=OH$

Caffeic acid derivatives:

- Chlorogenic acid (found in numerous families like the Rubiaceae (e.g., the coffee plant) and Asteraceae (e.g., Echinaceae))
- Cynarine (found in artichoke and echinacea (Asteraceae))
- Rosmarinic acid (present in many members of the Lamiaceae, e.g., rosemary and sage)

Lignans   Lignans are dimers (two units) of phenylpropanes linked by their β carbon, also called C-8 in the lignan. There are more than 500 known lignans, for example, podophyllotoxin and pinoresinol.

There are six different types of lignans, namely:

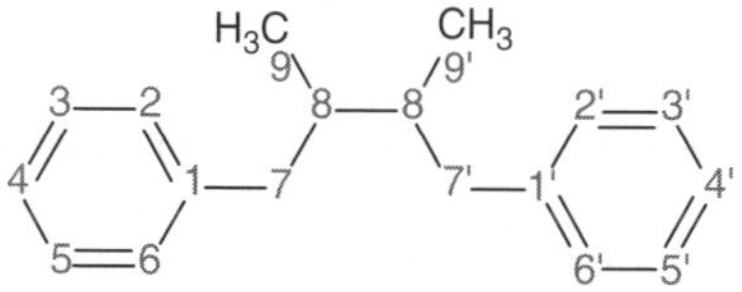

1) Dibenzylbutanes 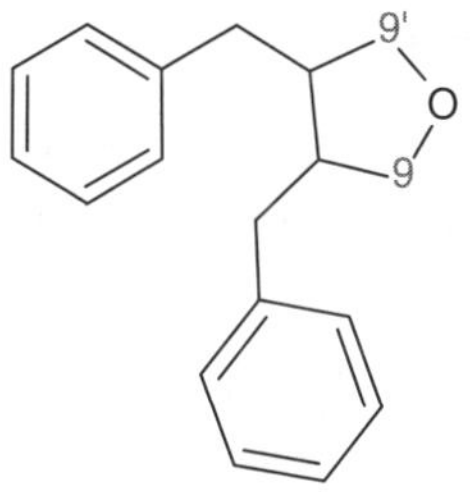   2) Monofuran lignans 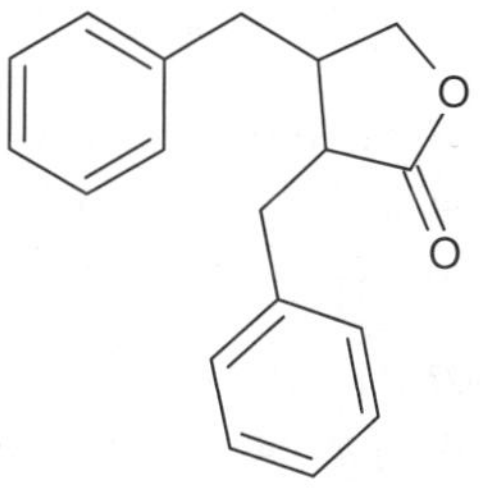   3) Butyrolactones

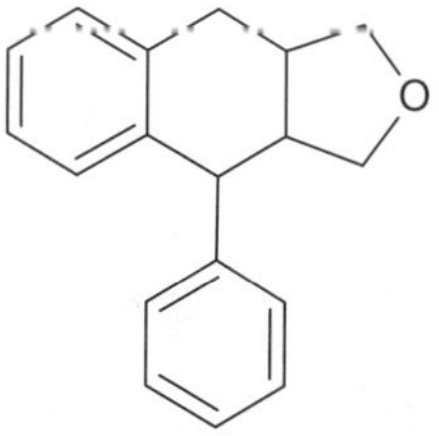

4) Arylnaphtalenes 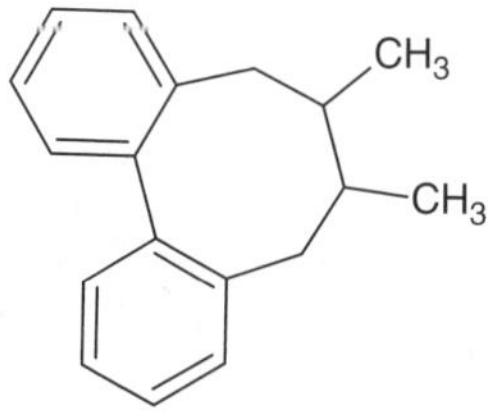   5) Dibenzocyclooctanes 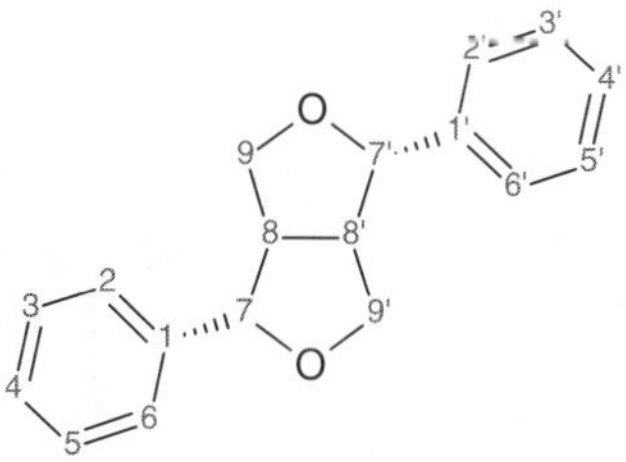   6) Furanofuran lignans

The range of their structures and biological activities is broad. Various lignans are known to have antitumour, antimitotic, and antiviral activities and to specifically inhibit certain enzymes. Toxicity to fungi, insects, and vertebrates is observed for some lignans and a variety of other biological activities have been documented (MacRae and Towers, 1984).

Distribution   Lignans are widely distributed in angiosperms (flowering plants) and gymnosperms (coniferous). Families containing lignans include: Linaceae (*Linum usitatissimum*), Berberidaceae (*Podophyllum peltatum*), Asteraceae (*Silybum marianum*), Schizandraceae (*Schizandra chinensis*), Zygophyllaceae (*Guaiacum officinale*), Magnoliaceae, and Polygalaceae, to name a few.

Neolignans are dimers of phenylpropanes with a link that involves one or no β carbon (e.g. 8-3'-, 8-1'-, 3,3'-neolignans).

Coumarins

General Structure

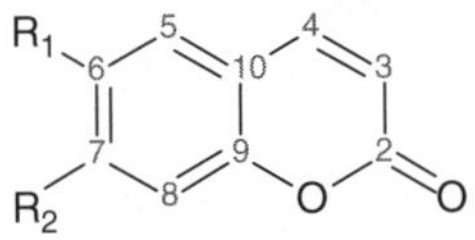

Coumarin	Furanocoumarin
Umbelliferon $R_1=R_2=H$	Psoralen $R_1=R_2=H$
Esculetin $R_1=R_2=OH$	Bergapten $R_1=$ MeO, $R_2=H$
Scopoletin $R_1=MeO$ $R_2=OH$	

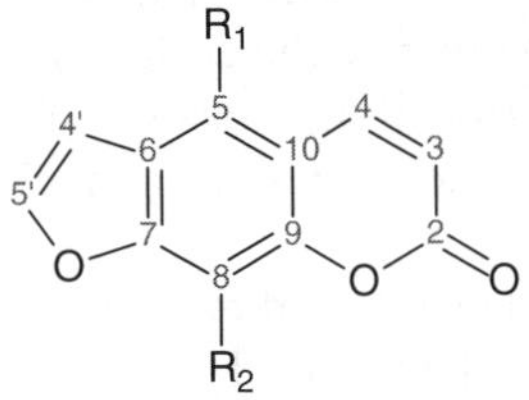

Some furanocoumarins are responsible for photosensibilization which is an erythema of skin after exposure to the sun. Examples are psoralen and bergapten, which are constituents of the essential oil of bergamot oil (Rutaceae).

Distribution   Coumarins are present in many families like the Fabaceae, Rubiaceae, Oleaceae and Apiaceae. Furanocoumarins are well known in the Apiaceae and Rutaceae.

Quinones

General Structure

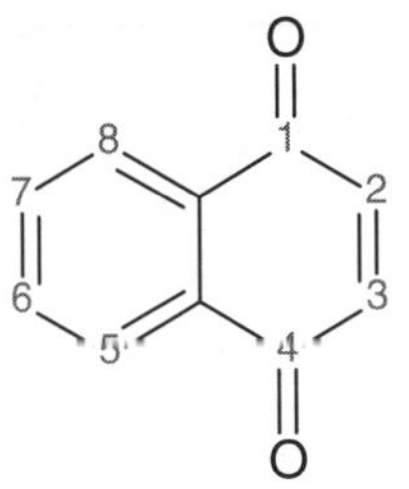

| | |
| Naphthoquinone | Anthraquinone |

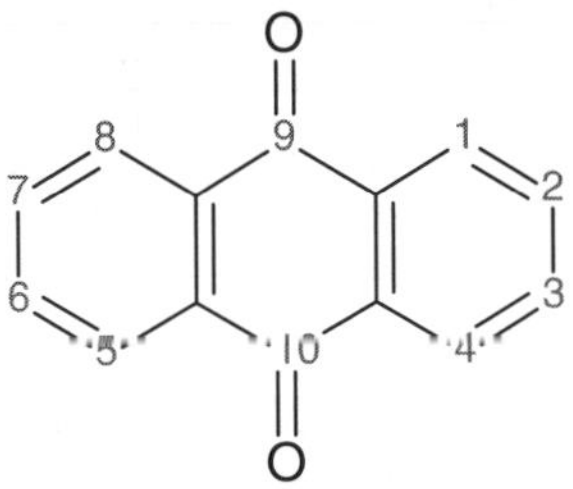

Naphthoquinones   Naphthoquinones are produced by the shikimic acid pathway. The best known member is vitamin K1. It is a colourless liquid, stable in the presence of oxygen but degrades in presence of light. Present in spinach, asparagus, alfalfa, and so on.

Distribution   Naphthoquinones are found in the Fabaceae, Juglandaceae, Droseraceae, Bignonaceae, Ebenaceae, and so on.

Anthraquinones   Anthraquinones are coloured compounds known for their laxative properties.

Ecological Significance   They are involved in plant–animal interactions. Some are anti-inflammatory, antibacterial, antifungal, trypanocide, insecticide, or antiviral (Bruneton, 2009).

Distribution   Anthraquinones occur in fungi, lichen, and in Angiospermae: Rubiaceae, Fabaceae, Polygonaceae, Rhamnaceae, and so on (Bruneton, 2009).

Stilbenoids

Stilbenoids include stilbenes and bibenzyls distinguished by the presence or absence of a double bond between the benzyl groups.

General Structure

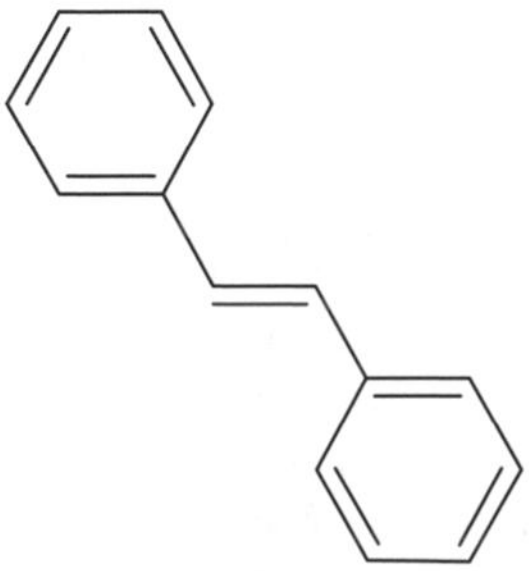

Stilbene 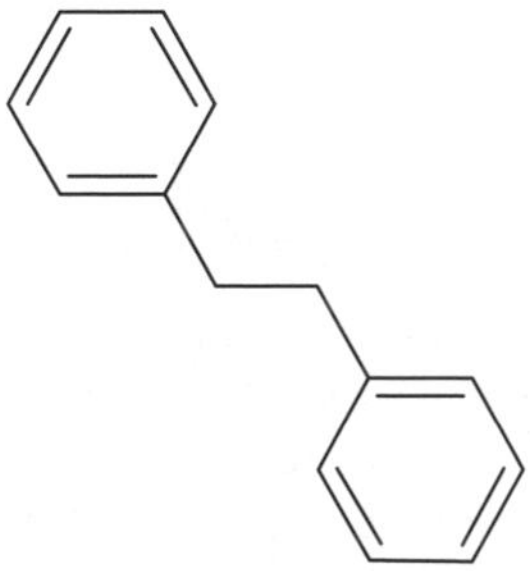   Bibenzyl

Ecological Significance   They possess some antifungal properties and some can act as phytoalexins (Guignard, Cosson, and Henri, 1985).

Distribution   Stilbenoids are present in numerous families. Resveratrol from grape is a stilbene.

Flavonoids

Some flavonoids have a bright colour like the yellow flavonoids (chalcones, flavonols) or the red or purple anthocyanins, and some are uncoloured (e.g., isoflavones).

General Structure

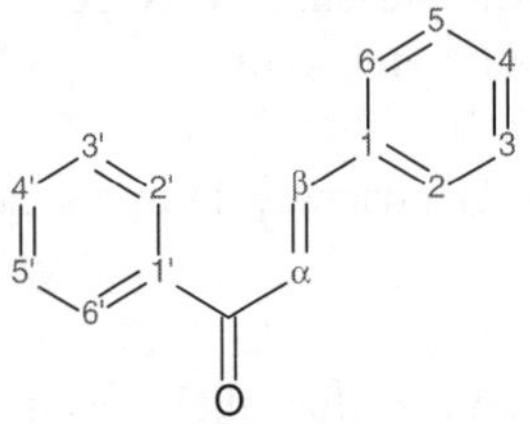

| Chalcones | Flavanones | Flavones |

| Flavonols | Dihydroflavonols (= flavanonols) | Isoflavones |

| Anthocyanins | Flavan-3-ols |

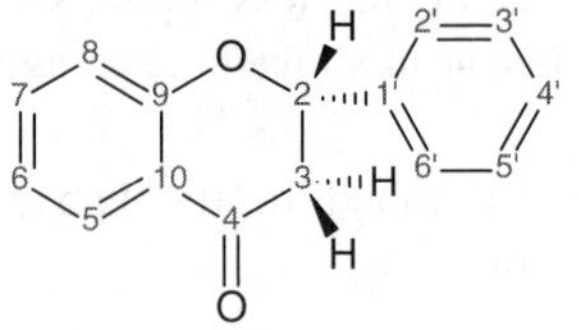

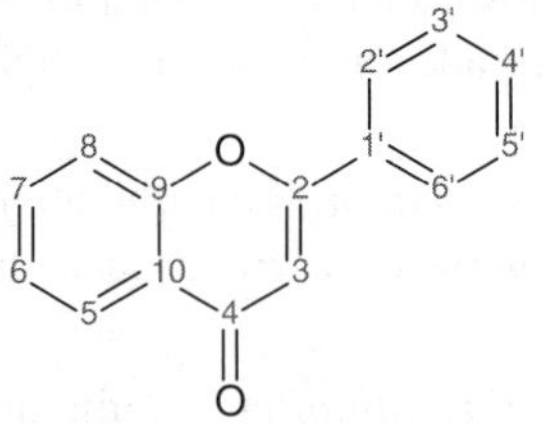

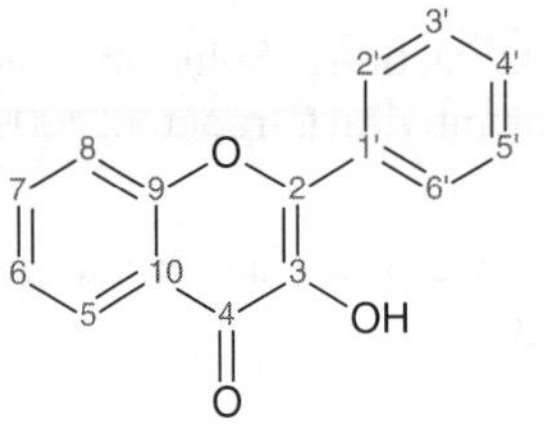

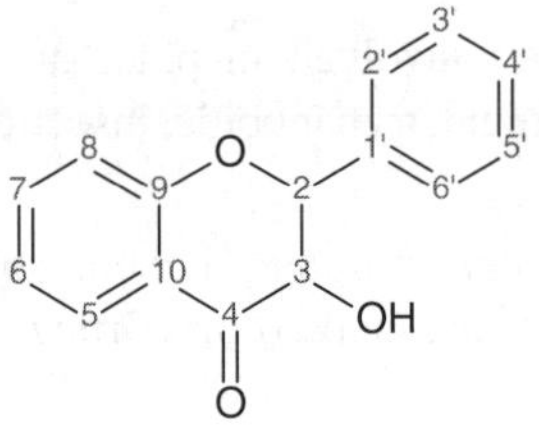

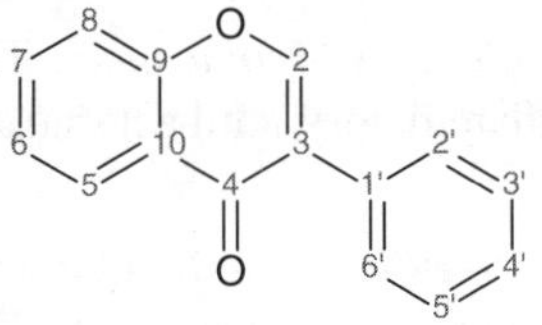

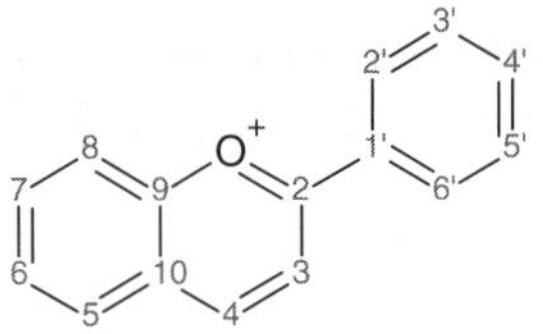

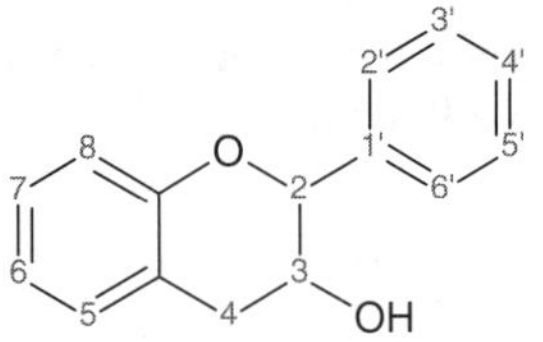

Distribution   Flavonoids are ubiquitous in the plant kingdom (Gould and Lister, 2006).

Ecological Significance   They have key functions in plant growth and development. They play a critical role in survival, such as the attraction of animal vectors for pollination and seed dispersal, the stimulation of *Rhizobium* bacteria for nitrogen fixation, promotion of pollen tube growth, and the resorption of minerals from senescing leaves. They also enhance tolerance to abiotic stressors, are agents of defense against herbivores and pathogens, and they form the basis of allelopathic interactions. Furthermore, they are involved in protection against ultraviolet rays, act as antioxidants, and increase heavy metal tolerance, to name a few (Gould and Lister, 2006).

Anthocyanins

General Structure

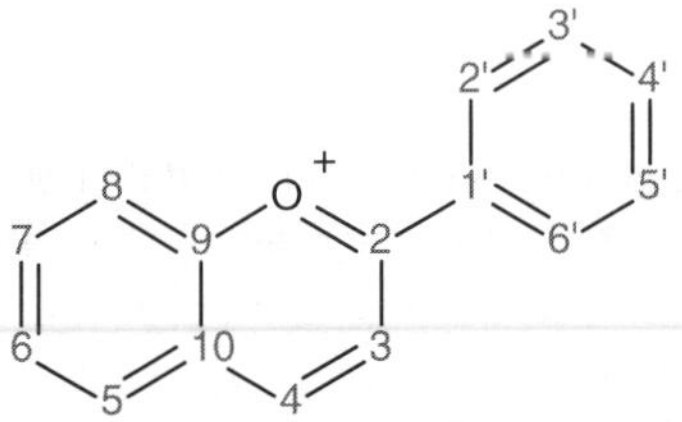

Distribution Anthocyanins are responsible for the colour of most flowers and fruits of angiosperms. They are sometimes present in other tissues like roots, tubers, stems, and bulbils, and can be found in gymnosperms, ferns, and some bryophytes (Andersen and Jorheim, 2006). Other pigments found in flowers and roots are betalains that are present in some mushrooms, such as amanita, and in the families of Centrospermae, including Aizoaceae, Amarantaceae, Basellaceae, Cactaceae, Chenopodiaceae (beet), Didieraceae, Nyctaginaceae, Phytolaccaceae, and Portulaceae (Mabry, 1977).

Anthocyanidin refers to the aglycone and anthocyanin defines the heteroside (aglycone and saccharide(s)).

Tannins

Tannins are divided into three categories, the hydrolysable tannins (gallotannins, ellagitannins), the condensed tannins (non-hydrolysable) and the phlorotannins.

Hydrolysable Tannins Hydrolysable tannins are esters of gallic acid or ellagic acid and glucose or a related polyol.

General Structure of Basic Unit

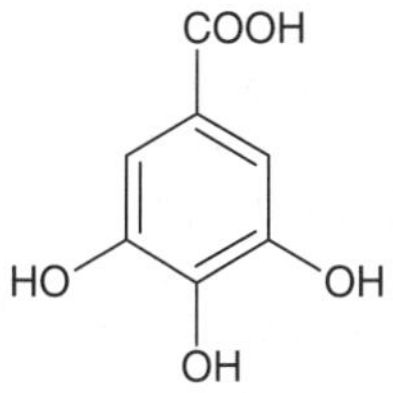

Gallic acid

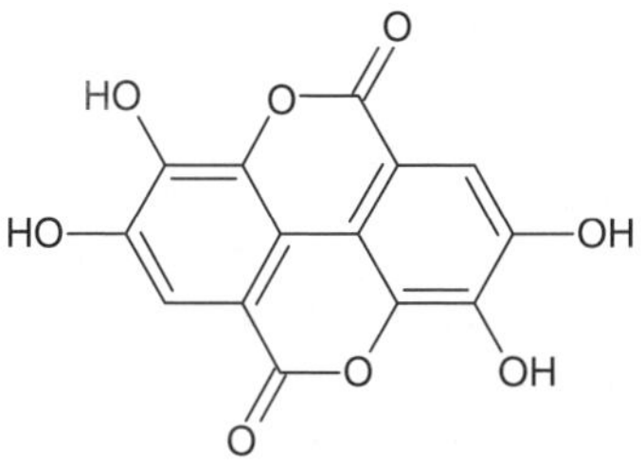

Ellagic acid

Condensed Tannins Condensed tannins, also called proanthocyanidins, possess as a structural element a flavan-3-ol, for example catechin and epicathechin. Their molecular weight varies between 500 and 3000.

General Structure of Basic Unit

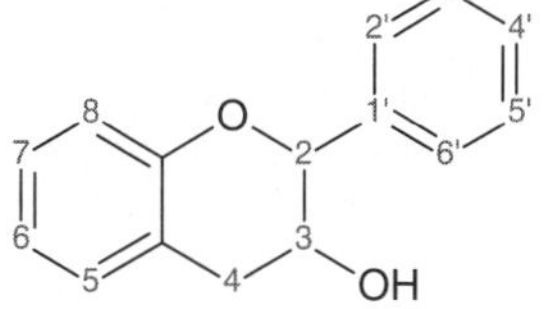

Flavan-3-ol

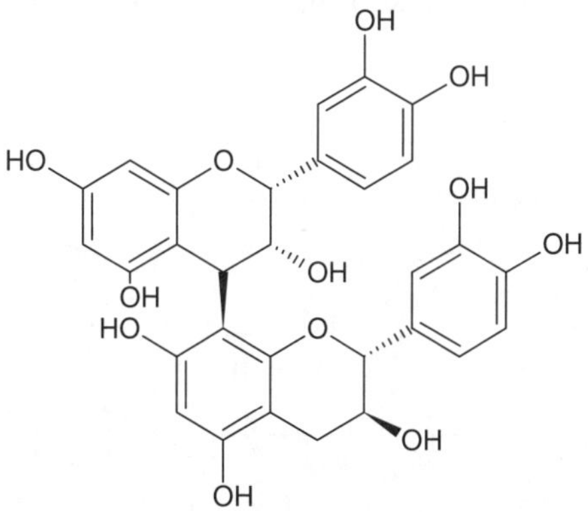

e.g. Procyanidin B-1

Phlorotannins   Phlorotannins are polymers of phloroglucinols with a wide range of molecular sizes (400 to 400 000 kDa) (Shibata *et al.*, 2004).

General Structure – Basic Unit

Phloroglucinol   e.g. Eckol

Ecological Significance   The properties of tannins are linked to their capacity to form complexes with macromolecules, especially proteins (including enzymes). This complex formation may or may not be reversible. It is reversible at physiological pH and in non-oxidative conditions, but irreversible in oxidative conditions where an o-quinone will be formed (Bruneton, 2009). Tannins can coagulate dermal proteins (leather tanning) and also precipitate saliva proteins, mainly those rich in proline. They can also bind to collagen. The hydrolysable tannins have a higher affinity for proteins than the condensed tannins. In nature, this makes the plant tissue unpalatable to herbivores (Guignard, Cosson, and Henri, 1985).

They can also precipitate extracellular enzymes, which are proteins, secreted by infecting microorganisms, resulting in more difficult penetration by bacteria and fungi (Guignard, Cosson, and Henri, 1985).

Brown algae contain a special class of tannins called phlorotannins, which are strong chelators of heavy metals, and have antioxidant and anti-inflammatory properties (Heo *et al.*, 2009; Toth and Pavia, 2000). They can reach more than 10% of dry weight in brown seaweed (Toth and Pavia, 2000).

Distribution   Rich in tannins are members of the Salicales, Fagales, Polygonales, Rosales, Geraniales, Sapindales, Myrtales, Ericales, and Ebenales. They are absent in the Papaveraceae and Brassicaceae. Few tannins occur in the Ranunculales, Rhoeadales, Opuntiales, and Primulales. Phlorotannins are present in brown algae.

Xanthones

General Structure

Ecological Significance Many xanthones are antifungal and antibacterial. Xanthones are monoamine oxidase inhibitors principally acting on MAO A. Certain xanthones possess anti-inflammatory properties (Bruneton, 2009).

Distribution They have been extensively investigated in the Gentianaceae and Clusiaceae, but are also present in other families like the Fabaceae, Loganiaceae, Lythraceae, Moraceae, Polygalaceae, and Rhamnaceae (Hostettmann and Hostettmann, 1989; Bennett and Lee, 1989; Rodríguez *et al.*, 1996; Baxter, Harborne, and Moss, 1999).

2.3.3 Terpenes

Isoprene is a five-carbon structure and is the building block of terpenes. According to the number of isoprene units we distinguish monoterpenes (C_{10}), sesquiterpenes (C_{15}), diterpenes (C_{20}), triterpenes (C_{30}), and tetraterpenes (C_{40}). The mono- and sesquiterpenes together with phenylpropanoids are the major compounds in essential oils. Phytosterols and saponins are triterpenoids and carotenoids are tetraterpenoids.

General Structure – Basic Unit

Isoprene

Monoterpenes and Sesquiterpenes

Below are the major basic structures of monoterpenes found in nature.

myrcane   p-menthane   seco-iridane   iridane   carane pinane   thuyane   bornane   fenchane   isocamphane

artemisane   santolinane   chrysanthemane   lavandulane

General Structure   Iridoids are monoterpenes with a pyrane (oxygenated heterocycle). In secoiridoids, the cyclopentane ring is open. Glycosylation stabilizes the molecule and makes it water soluble. They are found in every part of the plant: the seed, bark, root, leaves, and so on (Guignard, Cosson, and Henri, 1985).

Iridoids   Secoiridoids

Distribution   Monoterpenes are widespread in the plant kingdom, especially in the Rutales, Cornales, Lamiales, and Asterales, and are absent of the Ranunculales, Violales, and Primulales. Iridoids are only found in dicotyledons. Sesquiterpenes are more abundant in the Magnoliales, Rutales, Cornales, and Asterales (Guignard, Cosson, and Henri, 1985).

Ecological Significance   Monoterpenes are involved in plant–animal interactions. Some are anti-inflammatory, antibacterial, antifungal, trypanocide, insecticide, or antiviral (Bruneton, 2009).

Diterpenes

Phytol, vitamin A, and gibberellines are some examples of diterpenes. Phytol and gibberelline, play an important physiological role and are therefore present in most plants. Phytol is a constituent of chlorophyll and is required for chlorophyll to be active (Guignard, Cosson, and Henri, 1985)

One of the most important therapeutic diterpenes is the taxol which is a nitrogenated tricyclic diterpene.

Distribution   Diterpenes are less widespread than mono- and sequiterpenes. They are present in the Fabales and Geraniales, but also in the Lamiales and Asterales (Guignard, Cosson, and Henri, 1985).

Triterpenes

Triterpenes can be found in a free form (squalene, phytosterol, or pentacyclic triterpenes) or in an heteroside form (saponin).

The pentacyclic triterpenes can be classified into lupane, oleane, or ursane groups (Laszczyk, 2009). Betulinic acid and betulin occuring in birch are pentacyclic triterpenoids. Betulin is a lupane with hydroxyl groups in positions C-3 and C-28.

Phytosterols are steroid alcohols with hydroxyl in position C-3 naturally occurring in plants. The most common phytosterols are stigmasterol, and sitosterol. They are constituents of reproductive organs and meristems and are part of the composition of wax (Guignard, Cosson, and Henri, 1985). Generally, phytosterols are sterols with an additional chain of one to two carbons in C-24 and in certain cases there is an additional methyl in C-4 or C-14.

General Structure

Tetracyclic triterpene   Steroid   Pentacyclic triterpene

Different types of pentacyclic terpenes

Oleane   Lupane   Ursane

Ecological Significance   The pentacyclic triterpenes are the main saponin aglycones found in dicotyledons (e.g., oleanolic acid), while steroidal triterpenes characterize the monocotyledons. The pentacyclic triterpenes are antifungal in plants (Guignard, Cosson, and Henri, 1985).

Tetraterpenes (Carotenoids, Xanthophyll)

Carotenoids are widespread naturally occurring pigments, the name carotenoid coming from the carrot. Carotenoids are divided into two groups based on oxygen content (or lack thereof); xanthophylls are oxygenated derivatives of carotenes (Liu, 2007). Carotenoids consist of a 40-carbon skeleton.

Ecological Significance   Carotenoids play a major role in capitation of light for photosynthesis. They also protect against UV light (Guignard, Cosson, and Henri, 1985). Carotene is a precursor of vitamin A.

Below are the structures of two examples of carotenoids, one carotene (lycopene) and one xanthophyll (zeaxanthin).

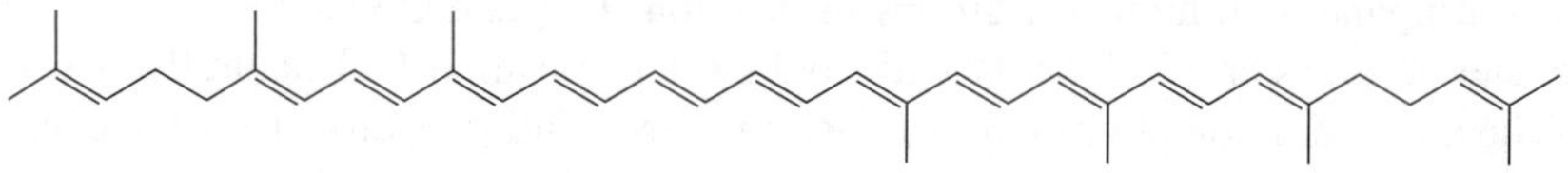

Lycopene

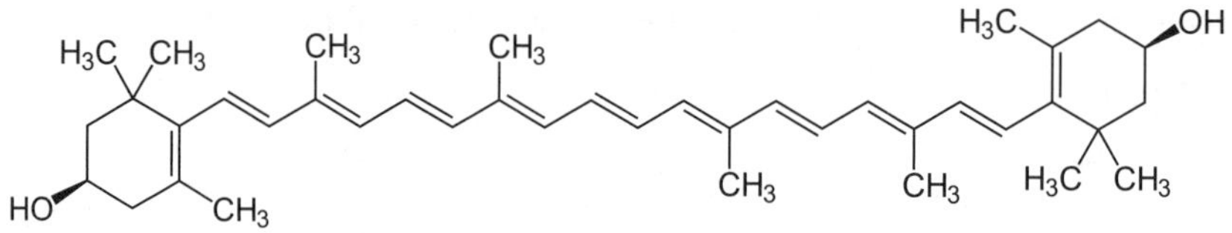

Zeaxanthin

2.3.4   Alkaloids

General Structure

All alkaloids contain nitrogen, and most of the time the nitrogen is included in a heterocycle. In general, they are classified according to the nature of their cycle.

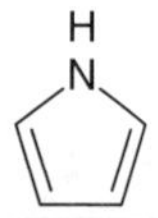

pyrrole 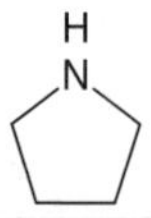   pyrrolidine 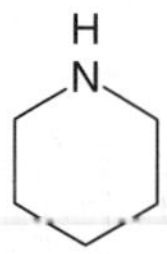   piperidine 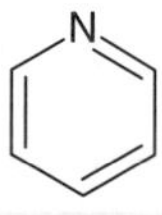   pyridine 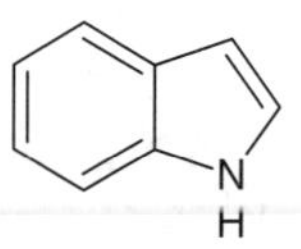   indole 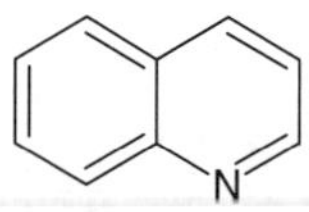   quinoline

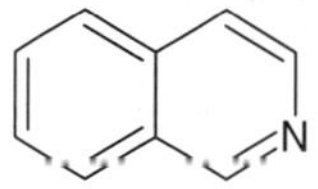

isoquinoline 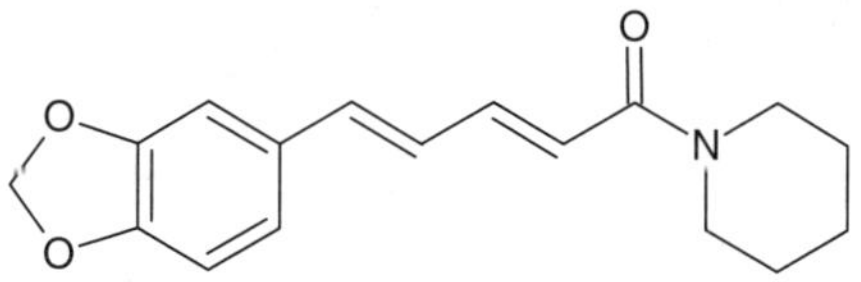   piperine

Distribution

The monocotyledons, with the exception of the Liliaceae, Amaryllidaceae, and Colchicaceae, are poor in alkaloids (Guignard, Cosson, and Henri, 1985). They appear in greater quantities in dicotyledons like the Annonaceae, Apocynaceae, Lauraceae, Loganiaceae, Magnoliaceae, Papaveraceae, and so on (Bruneton, 2009).

2.4 Stability of Isolated Compounds

Just a note on the stability of isolated compounds. Some compounds are unstable under certain conditions. The most common sources of degradation are oxidation, pH susceptibility, water, enzymes, temperature, and presence of metals. Gafner and Bergeron (2005) reviewed each factor that can influence stability and showed some examples of compounds susceptible to degradation. In general, a low pH is more desirable than an alkaline pH. The final form is also important, as some compounds will degrade more rapidly in solution, especially in water, than in a powder form. Enzymes like polyphenol oxidases (PPOs) will contribute to degradation of polyphenols, especially when the water activity (a_w) is high. Water activity is defined as a measure of water that is available for microbial growth and enzymatic reactions. Water activity will also influence the microbiological stability, as bacteria usually require at least $a_w = 0.91$, and fungi at least $a_w = 0.7$ to grow. Metal ions, especially bi- and trivalent cations will act as pro-oxidants and can catalyse oxidation.

2.5 Conclusion

The objective of this chapter was to give a brief overview of the different compounds present in plants and to help the reader unfamiliar with the chemistry of plants, also called "phytochemistry", to understand the literature. There are excellent reference books but each treats a particular aspect and so will focus on, for example, synthetic pathways, primary metabolites, secondary metabolites, roles in plants, or medicinal properties. These books will complete the information given here and answer the many questions that may have risen from this chapter.

References

Andersem, O.M. and Jorheim, M. (2006) The anthocyanins. in Flavonoids. Chemistry, Biochemistry and Applications, *CRC Press*, p. 471–553

Bennett, G. and Lee, H. (1989) Xanthones from guttiferae. *Phytochemistry*, **28**, 967–998.

Bruneton, J. (1993) *Pharmacognosie, Phytochimie, Plantes Médicinales*, 3rd edn, Lavoisier, Paris (FR), p. 915; Also available in English (Bruneton, J. Pharmacognosy, Phytochemistry and Medicinal Plants 3rd ed. Paris: Intercept-Lavoisier; 2000.

Bruneton, J. (2009) *Pharmacognosie, Phytochimie, Plantes Médicinales*, 4th edn, Lavoisier, Paris (FR), p. 1269 (French).

De Souza, M.C.R., Marques, C.T., Dore, C.M.G. *et al.* (2007) Antioxidant activities of sulfated polysaccharides from brown and red seaweeds. *Journal of Applied Phycology*, **19**, 153–160.

Evans, W.C. (2009) *Trease and Evans Pharmacognosy*, 16th edn, Elsevier Saunders, p. 603.

Andersen, O.M. and Markham, K.R. (eds) (2006) *Flavonoids. Chemistry, Biochemistry and Applications*, CRC Press, p. 1237.

Gafner, S. and Bergeron, C. (2005) The challenges of chemical stability testing of herbal extracts in finished products using state-of-the-art analytical methodologies. *Current Pharmaceutical Analysis*, **1**, 203–215.

Gould, K.S. and Lister, C. (2006) Flavonoids functions in plants, in *Flavonoids. Chemistry, Biochemistry and Applications* (eds O.M. Andersen and K.R. Markham), CRC Press, p. 1237.

Guignard, J.L., Cosson, L., and Henri, M. (1985) *Abrégé de Phytochimie*, Masson, Paris (French).

Hänsel, R. and Sticher, O. (2007) *Pharmakognosie - Phytopharmazie, 8*, Auflage Springer Verlag, Berlin Heidelberg New York (German).

Heinrich, M., Barnes, J., Gibbons, S., and Williamson, E. (2004) *Fundamentals of Pharmacognosy and Phytotherapy*, Churchill Livingstone, Edinburgh.

Heo, S.J., Ko, S.C., Cha, S.H. *et al.* (2009) Effect of phlorotannins isolated from Ecklonia cava on melanogenesis and their protective effect against photo-oxidative stress induced by UV-B radiation. *Toxicology in Vitro*, **23**, 1123–1130.

Hostettmann, K. and Hostettmann, M. (1989) Xanthones, in *Methods in Plant Biochemistry, vol. 1, Plant Phenolics, P-493-508*, Academic Press, London.

Laszczyk, M. (2009) Pentacyclic triterpenes of the lupane, oleanane and ursane group as tools in cancer therapy. *Planta Medica*, **75**, 1549–1560.

Liu, R.H. (2007) Whole grain phytochemicals and health. *Journal of Cereal Science*, **46**, 207–219.

Mabry, T.J. (1977) The order of centrospermae. The order centrospermae. *Annals of the Missouri Botanical Garden*, **64**, 210–220.

MacRae, W.D. and Towers, G.H.N. (1984) Biological activities of lignans. *Phytochemistry*, **23**, 1207–1220.

Baxter, H., Harborne, J.B., and Moss, G.P. (eds) (1999) *Phytochemical Dictionary: A Handbook of Bioactive Compounds from Plants*, 2nd edn, Taylor and Francis.

Rodríguez, S., Marston, A., Wolfender, J.-L., and Hostettmann, K. (1996) Iridoids and secoiridoids in the Gentianaceae. *Current Organic Chemistry*, **2**, 627–648.

Shibata, T., Kawaguchi, S., Hama, Y. *et al.* (2004) Local and chemical distribution of phlorotannins in brown algae. *Journal of Applied Phycology*, **16**, 291–296.

Toth, G. and Pavia, H. (2000) Lack of phlorotannin induction in the brown seaweed Ascophyllum nodosum in response to increased copper concentrations. *Marine Ecology Progress Series*, **192**, 119–126.

3

Separation and Purification of Phytochemicals as Co-Products in Biorefineries

Hua-Jiang Huang and Shri Ramaswamy

Department of Bioproducts and Biosystems Engineering, Kaufert Lab, University of Minnesota, Saint Paul, Minnesota, USA

3.1 Introduction

Recently, there has been increasing interest in the conversion of biomass to bioproducts, including biofuels, chemicals, and heat and power, due to the growing demand and increasing costs of fossil fuel and the need for national energy independence, as well as the associated environmental concerns about greenhouse gas emissions. At present, biomass feedstock cost is still expensive, accounting for a large fraction of the total operating cost. In order to reduce the production cost of biofuels such as bioethanol, it is necessary to extract and separate bioactive compounds or phytochemicals as value-added co-products prior to or during biomass conversion. Extraction and use of phytochemicals have been studied for many years, with more focus on pharmaceuticals and nutraceuticals from fruits, vegetables, or other food crops/plants, while less research has been done on separation of phytochemicals from bioresources that can be used as feedstocks in biorefineries, such as woody and perennial plant materials, and microalgae. As the use of different species of biomass for various bioproduct applications, including biofuels and biochemicals, becomes more prevalent, it is important to consider simultaneous production of value-added co-products within the framework of a biorefinery. As has been successfully shown in petrochemical refineries, in addition to manufacturing large-volume commodity

Biorefinery Co-Products: Phytochemicals, Primary Metabolites and Value-Added Biomass Processing, First Edition.
Edited by Chantal Bergeron, Danielle Julie Carrier and Shri Ramaswamy.

transportation fuels, it is important to have high-value co-products such as plastics, solvents, and other chemicals. Plant biomass is especially suitable for co-products in a biorefinery as it is well known that plants contain a great variety of complex compounds, each with specific purpose. As is shown in other chapters throughout this book, in addition to the major components of cellulose, hemicellulose, and lignin for fuel and energy, better understanding of all of the other components and efficient separation and manufacture of bio-co-products are of paramount importance for sucessful biorefnieries and a sustainable bioeconomy in the future.

Switchgrass is a fast-growing biomass species having great potential for the production of bioethanol, renewable energy, and chemicals. The extraction of phytochemicals such as antioxidants and flavonoids from switchgrass prior to the conversion process would produce value-added co-products (Uppugundla *et al.*, 2009). For example, alfalfa has numerous environmental advantages and great potential economic benefits. In the USA, alfalfa is the third most widely grown perennial crop, with over 23 million acres in 2003. The alfalfa stem fraction contains a high cellulose content suitable for bioethanol production, and the leaf fraction makes up about 45% of the total harvested crop. In addition to having 26 to 30% protein, in legumes such as alfalfa, the leaf fraction is a rich source of flavonoid antioxidants and phytoestrogens as potential nutraceutical products, including luteolin, coumestrol, and apigenin, as well as phytosterols (National Alfalfa & Forage Alliance, 2007). Another example is microalgae, which is one of the most promising biomass sources, with high biodiversity and huge productivity, as shown by Mata, Martins, and Caetano (2010), with a potential oil yield of 59 700–136 900 L oil ha^{-1} year^{-1} for microalgae compared with 5366 L oil ha^{-1}year^{-1} for palm oil (which is the second highest). Some algae species, in addition to high lipid content, contain value-added phytochemicals, such as phenolics, terpenes, sterols, enzymes, polysaccharides, alkaloids, toxins, and pigments.

Value-added co-products can also be produced from some current biorefineries. For instance, surfactants, sucrose acetate isobutyrate, polyols, and organic acids, such as citric, gluconic, and lactic acid, can be produced from sugarcane biorefineries (Aalford and Morel, 2006). Woody biomass is another rich resource for phytochemicals. In addition to three main components: cellulose, hemicellulose, and lignin, wood contains a large number of hydrophobic and hydrophilic secondary metabolites (generally called extractives) such as flavonoids, terpenes, phenols, alkaloids, sterols, tannins, suberins, resin acids, and carotenoids. These secondary compounds may constitute 1 to 33% of the dry wood. The active components in the wood extractives can be classified into three major groups (FERA, 2011): phenolics, terpenes, and nitrogen-containing compounds. Phenolic compounds include simple phenol and polyphenols such as flavonoids, anthocyanin pigments, and tannins, rich in the barks of many trees, such as silver birch, cherry, Douglas fir, and so on. Terpenes or terpenoids, which can be essential oils, flavours, fragrances, and plant pigments, are lipophilic phytochemicals. The main nitrogen-containing compounds are alkaloids, mainly present in plants (more rich in bark, seeds, roots, and leaves than stem or trunk) as salts of carboxylic acids, for example, citric, lactic, oxalic, acetic, malic, tartaric, fumaric, and benzoic acids. The major components detected in lipophilic extractives of cork and cork byproducts are triterpenes, including betulinic acid (11.7 g kg^{-1}), cerine, and friedeline. Also, the main triperpenes – friedeline (95.3 g kg^{-1}), betuline (13.1 g kg^{-1}), and betulinic acid (12.1 g kg^{-1}), R-hydroxy fatty acids (115.1 g kg^{-1}) and R, o-dicarboxylic acids (21.2 g kg^{-1}) are detected in black condensate after alkaline hydrolysis. The results demonstrate that these two industrial byproducts can be considered as promising sources of bioactive chemicals or chemical intermediates for the synthesis of polymeric materials (Sousa *et al.*, 2006). A large number of phytochemicals from different wood species have been collected into an online database (Turley *et al.*, 2006).

Phytochemicals from plants can be used in pharmaceuticals, cosmetics, nutritional, and consumer products. Carotenoids, polyphenols, and flavonols are known phytochemicals that can be used for the treatment of a number of disorders (Vinson *et al.*, 1995; Middleton, Kandaswami, and Theoharides, 2000). Flavonoids, the water-soluble pigments, have antioxidant, antiviral, anti-inflammatory, and antihistaminic properties. Alkaloids are well known for potent pharmacological activities, such as analgesic, antimalarial, antispasmotic, and the treatment of hypertension, mental disorders, and tumors (Badami *et al.*, 2003). Tocopherols, phenols, flavonoids, and terpenic acids, such as oleanolic acid from olive tree leaf extracts, are antioxidants. In addition, oleanolic acid has extensive pharmacological activity, and phenols mainly comprising oleuropein and hydroxytyrosol (the major product of hydrolysis) are of great biological interest (Rada, Guinda, and Cayuela, 2007; Savangikar and Savangikar, 2010). The various functions of different phytochemicals are also available in online databases (Turley *et al.*, 2006).

Phytochemicals from plants are usually present in very dilute quantities. The thermolabile or heat-sensitive properties of phytochemicals and the increased difficulty of solid–liquid extraction over liquid–liquid extraction bring a great challenge for efficient separation of phytochemicals from such a dilute biomass matrix. Basically, the whole separation and purification process consists of several large steps: feedstock handling such as grinding, biomass pretreatment, extraction of phytochemicals from ground biomass, and concentration and/or purification of phytochemicals from the extract. The latter two steps of separation and purification are, in general terms, costly processes. Depending on the biomass species and the particular part, there exists a wide variety of secondary chemical compounds. There are approximately 8000 known phenolic compounds from which 4000 flavonoids have been identified; 20 000 terpene structures, and more than 10 000 structures of alkaloids have been reported (Savangikar and Savangikar, 2010). Thus, selection of suitable separation and purification methods using appropriate solvents and processing conditions for targeted active components becomes critically important. In order to help understand various approaches to phytochemical extraction, separation, and purification, and to aid in choosing appropriate methods for the extraction of value-added co-products, this chapter provides a comprehensive review of the major separation methods, drawing on experiences from plant-based value-added products and their extension to co-products in biorefineries. In the following sections, conventional separation approaches, including steam distillation, conventional solid–liquid extraction (leaching), ultrasound-assisted extraction, microwave-assisted extraction, and pressurized liquid extraction or near-critical (subcritical) fluid extraction, are described first. Next, supercritical fluid extraction (SFE) is introduced, followed by the separation and purification of phytochemicals from extracts of dilute solutions that are produced in a biorefinery context. In addition, separation techniques, including liquid–liquid extraction, membrane separation, and molecular distillation are covered.

3.2 Conventional Separation Approaches

3.2.1 Steam Distillation

Steam distillation is a conventional, simple method for separation of volatile organic compounds such as essential oils and perfumes from plant material. In this method, steam is introduced by heating water, and is passed through the oil-containing plant material.

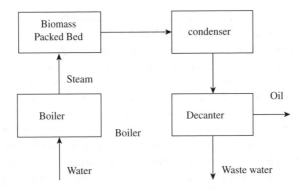

Figure 3.1   *Simplified diagram of steam distillation (Masango, 2005). Reprinted with permission from Masango.,*
2005 © Elsevier (2005).

With the addition of steam, the oil–water mixture boils at a lower temperature (below 100 °C at
1 atm) allowing heat-sensitive compounds to be separated with less decomposition. Steam
distillation has been commercially utilized in separation of essential oils and perfumes that are
sensitive to high heat from plant material.

The steam distillation technique where biomass feedstock is immersed in or mixed with
water is called hydro-distillation. This process can result in the removal of polar compounds
due to their dissolution in water. In another type of steam distillation process (Figure 3.1), a
packed bed of plant materials is placed above the steam source, allowing only steam to pass
through the bed without carrying the liquid water from the boiler. The steam flow is an
important parameter for this process, as water from the steam condensate can dissolve the
water-soluble compounds. The optimum (minimum) steam flow can be experimentally
determined in order to increase essential oil yield and reduce the loss of polar compounds
in the effluent (Masango, 2005).

Coupling steam distillation with maceration in solvent can increase the extraction yields of
plant essential oils and bioactive compounds (Manzan *et al.*, 2003), but thermolabile
molecules can be damaged using this technology (Pasquet *et al.*, 2011). The major advantages
of steam distillation are its simple equipment, easy operation, and relatively low degradation of
essential oils due to the addition of steam. Its major disadvantages are its relatively high
operating temperature, long separation time, and the significant loss of water soluble
components. Steam distillation is suitable for extracting light components whose vapour
pressures are relatively high ($\geq$1.33 kPa or 10 mmHg at 100 °C). For components whose
vapour pressures at 100 °C are between 0.67 kPa and 1.33 kPa, superheat steam is used for the
distillation. Steam distillation is not suitable for recovery of the heavy components whose
vapour pressures at 100 °C are less than 0.67 kPa.

3.2.2 Conventional Solid–Liquid Extraction

In general, there are three major categories of conventional solid–liquid extraction (leaching):
(1) fixed-bed leaching where biomass is fixed while the solvent is moving, for example,
Soxhlet extraction; (2) moving-bed leaching in which solid biomass is moving, for example,
screw extractor, and (3) agitated solid leaching where biomass particles are suspended in
solvent solution, for example, a counter-current multistage system. Some commercial

processes for solid–liquid extraction of plant ingredients include maceration (soaking in solvents), percolation, hot continuous extraction (Soxhlet), counter-current extraction, and pressurized liquid extraction. These techniques have been widely described for extracting naturally occurring lipids and organic pigments from biomass (Wang and Weller, 2006). The extraction agents used for solid–liquid extraction can be organic solvents, edible oils, or water.

Extraction with Organic Solvents

Organic solvents are traditionally used in the extraction of bioactive compounds. The commonly used organic solvents are hexane, ethanol, methanol, acetone, methyl acetate, and dichloromethane. Devi and Arumughan (2007) used solvents of different polarities, such as hexane, ethyl acetate, ethanol and methanol, to extract bioactive compounds, including oryzanols, tocols, and ferulic acid from defatted rice bran. They found that methanol was the most efficient solvent at the optimized conditions of a dry bran/solvent ratio of 1:15 (wt/vol) and an extraction time of 10 h. The dry methanol extract obtained with filtration followed by vacuum evaporation was re-extracted with less polar solvents, such as acetone, ethyl acetate, and ether. The dry acetone extract was then re-extracted with hexane for further purification, separating into a lipophilic phase, which was enriched in lipophilic compounds, and a polar phase, which was enriched in polar compounds. Rada *et al.* (2007) used ethanol to extract phenolic compounds and terpenic acids from olive tree leaf biomass, at ambient temperature for 48 hours with a ratio of solvent volume to weight of dry biomass (ml g^{-1}) of 5. The extraction process was: two successive extractions with ethanol; solid–liquid separation with a sieve to remove the exhausted biomass; filtration of the ethanolic solution with 2% activated charcoal followed by vacuum concentration at 40 °C. In another example, Vaughn *et al.* (2007) conducted an experimental study on the extraction of two polyphenols (flavonoids), hyperoside and quercitrin, from *Albizia julibrissin* wood using water at 85 °C; however, they did not purify the compounds. Uppugundla *et al.* (2009) used 85 °C water to extract two flavonoids: quercitrin and rutin from switchgrass at a ground dry biomass to solvent ratio of 1 g to 30 ml. Using water as solvent, maximum yields of rutin and quercitrin were 186 and 193 mg kg^{-1}, respectively, while maximum rutin and quercitrin yields, using 60% methanol, were 620 and 732 mg kg^{-1}. Although the yields using water are lower, it is worth noting that water is cleaner and more compatible with the subsequent biochemical conversion of the extracted biomass, while methanol requires costly acquisition and disposal costs, in addition to inhibiting the subsequent enzymatic hydrolysis and fermentation steps. More recently, Pawar and Surana (2010) used the solvent mixtures ethanol–water and acetone–water, to extract gallic acid, which can be used as an antioxidant and an anti-inflammatory, from *Caesalpinia decapetala* wood. They reported a maximum gallic acid yield of 17.85% at temperatures of 65–70 °C, extraction time of 48 hours, and ethanol to water ratio of 70 to 30.

Dilute acid or alkali can also be used as extractant. In this extraction process, wet or dried plant material is firstly comminuted and then extracted in hot or cold conditions with dilute acid or alkali solutions (Savangikar and Savangikar, 2010).

Extraction with Water or Oil

Water and edible oil are two clean and green solvents that can be used for conventional solid–liquid extraction. Water can be used to extract hydrophilic phytochemicals, and oil to

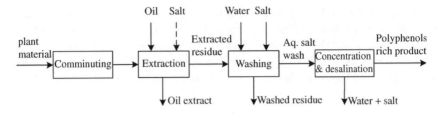

Figure 3.2   *Simplified blocks of oil extraction.*

extract hydrophobolic phytochemicals from the primary products from biomass. An example of this processing scheme is reported by Savangikar and Savangikar (2010) where a dry plant material is extracted with oil, as illustrated in Figure 3.2.

In the oil extraction process, wet or dried plant material is firstly comminuted and then extracted with edible oil and separated into oil extract and residue. For instance, powdered fenugreek (*Trigonella foenum-graocum L.*), the dry leafy vegetable, can be fractionated by two sequential extractions with sesame oil. At the oil to dry plant material ratios of 1.44 (kg kg^{-1}) (1st extraction) and 3.03 (kg kg^{-1}) (2nd extraction), a total of 109.6 mg carotenes, 148.8 mg xanthophylls, and 15.4 mg chlorophylls was extracted into the oil. The residue is then washed with water and salt. The aqueous salt wash contained 3.662 g of polyphenols per kg of dry biomass. The wash can be concentrated and desalted to produce a polyphenol-rich product. The oil extraction can be improved by adding common table salt as a modifier for increasing the solubility of biomass-contained phytochemicals in the oil. In addition, the oil extract can be further fractionated or purified by solvent extraction, crystallization, and chromatography (Savangikar and Savangikar, 2010).

In brief, conventional solid–liquid extraction with organic solvents is usually effective, but it implies the use of large amounts of solvents. As organic solvents are usually toxic, flammable, and enviornmental unfriendly, they are not suitable for extration of nutraceuticals or phytochemicals that are intended for human food or medicinal applications. Though oil is a generally recognized as safe solvent, it is not a solvent of choice because of the difficulty in further separation of the phytochemicals from oil. Water is a green, friendly solvent for biomass leaching for recovery of hydrophilic active compounds. The major disadvantages of hot water extraction are the low extraction yields under mild conditions and the potential degradation of phytochemicals at high temperatures due to hydrolysis.

3.2.3 Ultrasound-Assisted Extraction

Ultrasound-assisted extraction (UAE) is a process where ultrasound waves are introduced for improvement of solid–liquid extraction. In the UAE process, ultrasound waves with a frequency of over 20 kHz travel through a solid particle–liquid mixture, resulting in mechanical vibrations, which favour cell disruptions and improve mass transfer, that is, transfer of solvent into biomass cells and transfer of dissolved solutes from inside the particles to the outside bulk solvent. UAE has been considered an efficient approach for extraction of nutraceuticals, antioxidants, and triterpenoids (Wang and Weller, 2006). The use of UAE technology in the food industry was reviewed by Vilkhu *et al.* (2008), showing that UAE can enhance extraction of herbal, oil, protein, and bioactive agents, such as polyphenols, anthocyanins, tartaric acid, aroma compounds, and polysaccharides, from plant materials, leading to increased extraction yields, extraction rates, and processing output, and decreased extraction time.

Factors influencing UAE performance include ultrasound wave frequency, temperature, sonication time, and extraction time, as well as biomass moisture content and particle size (Wang and Weller, 2006). The UAE process entails wave distribution inside the extractor because of the different distances between the local particles and the radiating surface of the ultrasound source. To eradicate wave distribution, additional stirring equipment is often installed in the extractor.

Lavoie and Stevanovic (2007) studied selective UAE of lipophilic phytochemicals from *Betula. papyrifera* (paper birch) and *B. alleghaniensis* (yellow birch) at low temperatures, using dichloromethane as the solvent. Three families of compounds were detected in the extract: aliphatic, sterols, and triterpenoids. The lipophilic extracts of both birch species contained two major constituents – betulonic acid and squalene (triterpenes). These phyto-chemicals display bioactivity, which will be discussed in further chapters. The UAE process-ing variables that affected the yields of the recovered compounds included ultrasound intensity, extraction time, pulsation, temperature, and solvent to biomass ratio.

Compared to conventional solid extraction, with similar biomass and under the same operating conditions, such as temperature, extraction time, and solvent to solid ratio, UAE has higher extraction yields and faster kinetics; additionally UAE can be operated at lower temperatures, which lowers the degradation of heat-sensitive compounds. Lavoie and Steva-novic (2007) compared UAE and Soxhlet extraction for recovering phytochemicals from birch wood, and reported that UAE with dichloromethane as solvent resulted in a 33% higher yield of phytochemicals in yellow birch and a 20% higher yield in paper birch. Diouf, Stevanovic, and Boutin (2009) compared UAE with conventional maceration using ethanol as the solvent in extracting triterpenes and polyphenols from different tissues of yellow birch. Results demon-strated that UAE produced less toxic extracts, with comparable yields and in a shorter time. In addition, UAE could selectively extract bioactive molecules from foliage and twigs.

3.2.4 Microwave-Assisted Extraction

Microwave-assisted extraction (MAE) is a process in which microwaves, or electromagnetic radiation, are introduced for improvement of solid–liquid extraction. In the MAE process, microwaves with frequencies from 0.3 to 300 GHz penetrate the biomass matrix and the microwave energy is absorbed by polar molecules such as water in the matrix; hence the whole biomass can be heated efficiently and homogeneously from inside. Also, cell disruption is enhanced by internal heating (Wang and Weller, 2006). In principle, the higher the dielectric constant of the solvent, the more strongly it absorbs microwave energy. Polar solvents such as water, ethanol, methanol, and acetone have high dielectric constants, so they can better absorb microwave energy and can be good candidates as MAE solvents. Non-polar solvents such as hexane, on the other hand, have low dielectric constants, thus they are not suitable for MAE. However, a mixture of non-polar and polar solvents, for example, a hexane–methanol mixture can be used for MAE. Therefore, when choosing an efficient solvent for MAE, the microwave-absorbing capacity of the solvent must be considered. The operating variables of MAE include temperature, the power of the microwave equipment, extraction temperature, and extraction time. In addition, like the other separation processes, biomass particle size must be sufficiently small to ensure that large surface area for mass transfer between the biomass matrix and the bulk solution are obtained, resulting in high extraction yields and faster extraction rates.

MAE has been considered as an efficient approach for extraction of natural metabolites from plant materials, for example, antioxidants (Xiao *et al.*, 2009), vegetable oils (Cravotto *et al.*, 2008), and lipids from microalgae (Lee *et al.*, 2010). Pasquet *et al.* (2011) used the MAE process to extract pigments from two marine microalgae: *Cylindrotheca closterium* and *Dunaliella tertiolecta*, and compared MAE with cold and hot soaking and ultrasound-assisted extraction. The major pigments in *C. closterium* are chlorophyll *a* and fucoxanthin, and the main pigments in *D. tertiolecta* are chlorophyll *a*, chlorophyll *b* and β, β-carotene. *C. tertiolecta* presents a frustule, which is a hard and porous cell wall or external layer of diatoms, composed of silica, and is coated with a layer of organic substance. Due to the absence of frustule in *D. tertiolecta*, all processes had rapid extraction rates, and equivalent extraction yields for extracting the pigments because of immediate solvent penetration in the microalgae cells. In contrast, presence of the frustule in *C. closterium* formed a mechanical and transfer barrier to pigment extraction. MAE was identified as the best extraction process for pigments as it has a rapid extraction rate, high extraction yield, and homogeneous heating characteristics. The extraction yield of chorophyll *a* by MAE is up to 8.65 g kg^{-1} dry biomass in 5 min at Rt (room temperature), which is higher than the yields obtained by VMAE (vacuum assisted MAE) (5.35 g kg^{-1} in 3–10 min), cold soaking (7.48 g kg^{-1} in 60 min at Rt), and UAE (4.95 g kg^{-1} in 3–10 min at Rt). The yield of MAE is lower than that of hot soaking (9.31 g kg^{-1} in 30 min at Rt), but the latter requires much longer time and results in significant degradation of chlorophyll *a*, soaking over 30 min. The yield of fucoxanthin by MAE is 4.24 g kg^{-1} dry biomass in 3–5 min, compared to 3.68 g kg^{-1} by VMAE in 5 min, 4.68 g kg^{-1} by cold soaking in 60 min at Rt, 5.23 g kg^{-1} by hot soaking in 60 min, and 4.49 g kg^{-1} by UAE in 3–10 min at 8.5 °C. Overall, MAE is the best amongst the above-mentioned methods in terms of extraction yield, extraction time, and product integrity.

MAE and VMAE have been applied to marine microalgae to extract lipids (Lee *et al.*, 2010), and resulted in the highest recovery for all tested species compared to autoclaving, bead-beating, sonication, and maceration in 10% NaCl solution. The combination of sonication and microwaves was studied to extract lipids from vegetables and microalgae sources. Ultrasonication alone, microwave irradiation alone or a combination of both techniques resulted in excellent extraction efficiencies in term of yields and time. When compared to conventional methods, the yields obtained with these combinations increased from 50 to 500% and extraction times were reduced a 10-fold (Cravotto *et al.*, 2008). This suggests that microwaves could be of interest to extract pigments from microalgae.

In summary, MAE results in higher extraction yields, decreased solvent consumption, and shorter extraction times compared to conventional solid extraction (e.g., Soxhlet extraction) and ultrasonic extraction. Major disadvantages of MAE are the requirement of both polar and non-polar solvents for extracting non-polar phytochemicals. Compared to supercritical fluid extraction, MAE requires additional solid–liquid separation steps to remove the solid residue from the liquid phase (Wang and Weller, 2006).

3.2.5 Pressurized Subcritical Liquid Extraction

Pressurized liquid extraction (PLE) is a process (Figure 3.3) where solid–liquid extraction is carried out under a high pressure of 10–15 MPa and a temperature between 50 and 200 °C (Wang and Weller, 2006), depending on the solvents used and the constituents extracted. PLE is also called pressurized solvent extraction, accelerated solvent extraction, subcritical

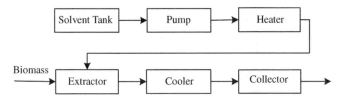

Figure 3.3   *Simplified block diagram of the PLE process.*

solvent extraction, or subcritical fluid extraction. The pressure is elevated, but below the critical point, in order to keep the solvent in a liquid state at a relatively high temperature. The elevated temperature can increase the solubility and the diffusivity of the solutes in the solvent, resulting in enhanced extraction performance. Amongst the solvents for PLE, water is the most commonly investigated extractant. The corresponding process is usually referred to as subcritical water or pressurized hot water extraction. Duan *et al.* (2009) investigated the PLE of five silymarin compounds, taxifolin, silichristin, silidianin, silibinin, and isosilibinin, from milk thistle fruits at extraction termperatures above 120 °C. In their research, no degradation was detected when using pure ethanol, whereas significant degradation was reported when using ethanol–water combinations at temperatures greater than 140 °C. Similar degradation observations were also reported elsewhere (Kawamura *et al.*, 1999; Ibañez *et al.*, 2003). The degradation of compounds suggests that, although increasing temperature can reduce the extraction time and can increase the phytochemical extraction yield within a certain range of temperature, further increase in temperature may cause phytochemical degradation, leading to decreases in the extraction yield. Hence, temperature is an important PLE processing parameter. When the operating pressure is near the fluid critical point, the process is called near-critical fluid extraction. Near-critical water (250–300 °C) has properties similar to those of common polar organic solvents (Patrick *et al.*, 2001), selectively extracting polar (at lower temperatures), moderately polar, and non-polar (at higher temperatures) organic compounds from biomass. The simulation, design, and scale-up of pressurized liquid extractors has been reviewed by Pronyk and Mazza (2009).

PLE can result in high extraction yields due to its elevated temperature and ensuing increase in solute solubility and diffusivity. The major disadvantage of PLE is that its high temperature favours the degradation of solutes, especially in aqueous solution. Hence, PLE operating at high temperatures is not suitable for extracting heat-sensitive active compounds from biomass.

3.3 Supercritical Fluid Extraction

Supercritical fluid extraction (SFE) is a process (Figure 3.4) where a supercritical fluid at its critical conditions is applied. SFE can be used to extract the desired solutes from a solid matrix or a liquid mixture. Supercritical fluids (SCFs) have the physical properties between those of gas and solid states. Compared to the liquid state, SCFs have much lower density, viscosity, and interfacial tension, and much higher diffusivity and thermal conductivity, which favours mass and heat transfer in the critical region (Pereda, Bottini, and Brignole, 2008). Supercritical carbon dioxide (ScCO$_2$) is the most commonly used supercritical fluid in food, pharmaceutical, and chemical industries. Being non-polar, or hydrophobic, ScCO$_2$ is very suitable for extracting hydrophobic constituents from biomass. The density and polarity of ScCO$_2$ can be tuned by changing the operating pressure and temperature. For example, the density and

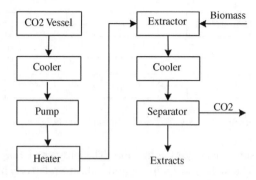

Figure 3.4   *Simplified block of supercritical fluid extraction process.*

polarity of $ScCO_2$ can be increased by increasing pressure; in this way $ScCO_2$ can also extract polar compounds. In addition, $ScCO_2$ is sometimes added alongside polar co-solvents such as ethanol or methanol in order to efficiently extract polar solutes.

SFE with CO_2 has been widely researched for many years, especially in food, neutraceutical, pharmaceutical and chemical industries; for instance, extraction of caffeine from green tea (Kim *et al.*, 2008). Wang *et al.* (2007) studied the sequential SFE process with carbon dioxide, water, and ethanol as solvents for separating high-value chemicals from the dried distillers grains with solubles (DDGS), the byproduct of dry-grind ethanol biorefineries. Some researchers investigated extration of lipids, pigments, phenolics, and carotenoids from microalgae (Mendes *et al.*, 2003; Mendes, 2008; Temelli *et al.*, 2008; Macías-Sánchez *et al.*, 2009a; Macías-Sánchez *et al.*, 2009b; Klejdus *et al.*, 2009). For instance, Halim *et al.* (2011) performed an experimental study on the $ScCO_2$ extraction of lipid from marine microalgae (*Chlorococcum* sp.) for biodiesel production. Results showed that the extracted lipid had a suitable fatty acid composition for biodiesel, and that the lipid yield increased with increasing pressure and decreasing temperature.

The extraction yield and selectivity of solid–liquid extraction can be adjusted by changing the solvent physico-chemical properties, that is, density, viscosity, diffusivity, and dielectric constant through varying the pressure and temperature of the extraction process (Pronyk and Mazza, 2009).

Compared with conventional solvent extraction, $ScCO_2$ has many advantages as a green extractant: higher extraction yield and higher selectivity of desired solutes; milder temperature conditions and hence less degradation of heat-sensitive active compounds; shorter time for equivalent or better extraction performance; tunable physico-chemical properties of $ScCO_2$; no residue or solvent contamination in the products. Also, CO_2 is non-toxic, non-flammable and hence environmentally friendly, inexpensive, and recyclable – using existing CO_2 without adding greenhouse gas effects. A disadvantage of SFE-based processes is their inherant high capital and operational costs.

3.4   Separation and Purification of Phytochemicals from Plant Extracts and Dilute Solution In Biorefineries

3.4.1   Liquid–Liquid Extraction

In the process of liquid–liquid (L–L) extraction, a separate or mixed solvent is added as extractant to selectively extract one or more solutes from the solution. L–L extraction is a

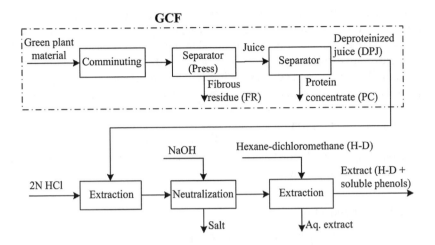

Figure 3.5   *Simplified blocks of green crop fractionation process.*

conventional and efficient method for separting solutes from dilute mixtures. In cellulosic biorefineries, hydrolysate from a biomass pretreatment step, for example, dilute acid hydrolysis of biomass, usually contains value-added phytochemicals such as acetic acid, furural, and antioxidants (phenolics or polyphenolics (Cruz *et al.*, 1999)), in addition to hemicellulose sugars. These phytochemicals can be recovered from the hydrolysate by L–L extraction. For example, antioxidants in the hydrolysates from acid-hydrolyzed *Eucalyptus globules* wood chips with 2.5–5% H_2SO_4 at a liquid to solid ratio of 8:1 g g^{-1} and a temperature of 100–130 °C, were extracted with ethyl acetate. The organic phase was vacuum-evaporated to remove and recycle the ethyl acetate to the extractor, while the aqueous phase containing xylose was transferred to the subsequent fermentation process to produce ethanol or xylitol (Gonzalez *et al.*, 2004).

The primary fractions from the conventional green crop fractionation (GCF) process (Figure 3.5) include water soluble fibre free deproteinized juice (DPJ), water insoluable protein concentrate (PC), and fibrous residue (FR). PC can be utilized for human consumption or as an ingredient for animal feed. FR and DPJ can be used as cattle feed. As an improvement, polyphenols can be extracted from DPJ and PC by a hexane–dichloromethane mixture (H–D, at 1:1 proportion) (Figure 3.5, extraction of PC is not shown in the figure). For instance, the primary fractions PC and DPJ of the GCF process with fenugreek as feedstock were extracted respectively with 2 N HCl; the resulting extracts were neutralized with NaOH and then extracted by the H–D mixed solvent. The H–D soluble phenols in the solvent phase were found to be 0.662 g and 0.214 g respectively per kg of the fraction, while the H–D insoluble phenols in the aquous phase were found to be 2.861 g and 4.545 g respectively per kg of the fraction. In addition, PC also can be fractionated by L–L extraction with oil for recovery of carotenes, xanthophylls, and chlorophylls. When using ethyl acetate (EA) as extractant, 8.487 g phenols per kg extract were recovered from DPJ (Savangikar and Savangikar, 2010).

A salt aqueous solution also can be used to extract polyphenols from biomass. For instance, salt water extraction of dry powdered fenugreek resulted in extract containing 3.662 g polyphenols per kg of dry biomass (Savangikar and Savangikar, 2010).

Liquid–liquid extraction is also a potenital process for extracting valuable chemicals from aqueous biorfinery effluents. Malmary *et al.* (2000) explored the recovery of aconitic and lactic acids from simulated effluents of the sugarcane biorefinery with solvent extraction. The two carboxylic acids were extracted by tributylphosphate. As the viscosity of tributylphosphate is relatively high, dodecane was added as a diluent. The extraction of carboxylic acids with tributylphosphate is based on the undissociation–dissociation mechanism. In addition to tributylphosphate, trioctylamine can be used as the extractant.

The criteria for solvent selection are: high distribution coefficient and high selectivity for desired solutes; high stability and low solubility in aqueous phase; significant density difference from that of the solution to ensure phase separation; low viscosity; large interfacial tension; non-toxic; and low cost.

3.4.2 Membrane Separation

Membrane technology has been commercially applied in the chemical, biochemical, and food industries for many years; for example, concentration of fruit juice using membrane filtration. It can also be used for the separation and concentration of phytochemicals from mixtures. For instance, polyphenols have been separated and concentrated from fruit solutions with membrane technologies (Borneman, Gökmen, and Nijhuis, 2001; Alper and Acar, 2004; Nawaz *et al.*, 2006). As oxidants, polyphenols can prevent oxidation of high-density lipids and remove low-density lipids, as well as absorb free radicals, so as to reduce ulceration (Nawaz *et al.*, 2006). Wei, Hossain, and Saleh (2010) investigated the concentration and separation of polyphenolics from sugar solution with membrane ultra-filtration. Rutin model solutions and a Pellicon-2 regenerated cellulose ultrafiltration membrane with a nominal molecular weight cut-off (MWCO) of 1000 Da were used in their study. During the membrane separation, glucose passed through the membrane pores while rutin was retained in the retentate. The effects of the operating variables, including initial feed concentration, free flow rate, pH, and temperature on the concentration factors and flux through the membrane were observed. It was shown that rutin concentrations could be enriched by a factor of 2.9 and a recovery of 72.9% was obtained. Membrane systems also can be used to separate phytochemicals from plant extracts. Krishnaiah, Sarbatly, and Nah (2007) used a nanomembrane with an MWCO of 5 kDa to recover antioxidants from the methanolic extracts of various parts, leaf, fruit and roots, of *Morinda Citrifolia*. The effect of transmembrane pressure on separation was observed. Results showed that membrane separation was possible for antioxidant recovery from plant extracts.

Compared to conventional separation methods, membrane separation has many advantages (Wei, Hossain, and Saleh, 2010): no requirement for solvents or additives, easy automation and scale-up, lower operating costs, shorter processing time, mild operating conditions, and less waste. The major disavantage of membrane filtration is the potential membrane fouling during the separation; food industry processing is a good case-in-point for this observation.

3.4.3 Molecular Distillation

Molecular distillation (MD) is a process operated under high vacuum conditions that makes the mean free path of the molecules to be separated longer than the distance from the evaporation surface to the condenser surface. Thus, theoretically, the molecules in the vapour phase will not return to the liquid phase and the evaporation rate is only governed by the rate of the molecules that escape from the liquid surface. However, in industrial plants, the mean free path of the molecules is shorter than the distance between the evaporating and condensing surfaces. In this case, the process is called short path distillation (Chen *et al.*, 2005). At high vacuum conditions, the boiling points of the components to be separated are 200–300 °C lower than normal, allowing for efficient separation of high molecular weight compounds without degradation (Meloan, 1999).

MD can be used for distillation of heat-sensitive materials; for instance, separating naturally occurring phytochemicals such as omega-3 fatty acids (eicosapentaenoic acid (EPA) and docosahexaenoic acid (DHA)) in fish oil, free fatty acids from vegetable oil deodorizer distillate (Martins *et al.*, 2006), tocotrienols from palm oil fatty acid distillates (Posada *et al.*, 2007), tocopherol and fatty acid methyl esters from rapeseed oil deodorizer distillate (Jiang *et al.*, 2006), and purification of crude lipids from algae oil, palm oil, and so on. Currently, MD has been used both at the lab and at industrial scales. Posada *et al.* (2007) extracted tocotrienols from palm oil fatty acid distillates using MD. Rada *et al.* (2007) used MD to extract, separate and purify hydroxytyrosol, from olive tree leaf extracts. The whole process included two steps: (1) extraction of olive tree leaf biomass with ethanol and (2) incorporation of the ethanolic extract into glycerine and then separation of terpenic and phenolic compounds from the glycerine extract mixture by MD. Chen *et al.* (2007) used MD to concentrate the octacosanol contained in rice-bran-wax-derived policosanol extracts up to 25.4%. Below a temperature of 150 °C and a pressure of 66.7 Pa, distillates with 37.6% octacosanol were obtained, resulting in a 48.0% concentration factor. Results showed that the increase in distillation temperature or the decrease in distillation pressure could increase the mean free path of molecules; therefore, components with larger molecular weight could be separated and concentrated onto the condenser.

Basically, there are two kinds of molecular distillators: falling film and centrifugal distillators. In both models, the separation occurs under vacuum conditions, enabling molecules to evaporate from the evaporator to the condenser, and the formation of a thin liquid film, which promotes effective heat and mass transfers. Falling film distillators use gravity to promote a thin film on the evaporating cylinder (evaporator), usually with a wiping element that mixes and distributes the liquid over the whole evaporator surface (Cvengros *et al.*, 2001), while centrifugal distillators use centrifugal force to create this thin film. Two product streams are generated: a distillate rich in the light molecules that escape from the evaporator and reach the condenser, and a residue, rich in heavier molecules from the evaporator.

MD has many advantages: no requirement for solvent; less or no degradation of products due to high vacuum or low temperature operation; high product purity or less contamination (far below industry standards) in the product due to the separation being based on molecular weight. Its disadvantages include relatively high costs and loss of solute, as illustrated by the loss of initial triglyceride concentrations in a fatty acid extraction process (Bioriginal Food & Science Corp., 2011).

3.5 Summary

Separation and purification of phytochemicals or bioactive compounds can bring value-added co-products, which could transform the overall biorefinery economics. There are many separation approaches for extraction, concentration, and purification of phytochemicals or bioactive compounds from biomass prior to biofuels production, each having different advantages and disavantages.

Steam distillation is a simple, clean, and commercially used method without product contamination. Heavy components whose vapour pressures are less than 0.67 kPa (5 mmHg), however, are difficult to separate by this method due to higher boiling points. Conventional solid–liquid extraction with organic solvents is an effective process, but it is not suitable for the extraction of nutraceuticals or pharmaceuticals for human use because of contamination of the product by the solvents used. Edible oil and water are environmentally friendly solvents. The former can be used for extracting hydrophobic active compounds, with the latter used for extracting hydrophilic compounds. However, oil extraction is less commercially used due to its large consumption of oil and the difficulty in further separation of phytochemicals from oil. Water extraction is a green process, but it may lead to a low extraction yield under mild conditions, and hot water extraction may cause potential degradation of phytochemicals at high temperatures due to hydrolysis.

Compared to conventional solid–liquid extraction, in general, under the same operating conditions, such as temperature, extraction time and solvent to solid ratio, UAE has a higher extraction yield, and can be operated under milder conditions for comparable yields, which can avoid or lower the degradation of heat-sensitive extracts, resulting in the production of a less toxic extract in a shorter time. MAE has even higher extraction yields, less solvent consumption, and reduced extraction times with equivalent or higher extraction yields under similar operating conditions, compared to conventional solid–liquid extraction (e.g., Soxhlet extraction) and UAE. The major disadvantages of MAE are the requirement for both polar and non-polar solvents for extracting non-polar phytochemicals. Compared to supercritical fluid extraction, MAE needs additional solid–liquid separation steps to remove the solid residue from the liquid phase.

PLE can achieve a high extraction performance due to its elevated temperature and hence the increase in the solute solubility and the diffusivity of solutes in the solvent. However, its high temperature conditions favour the degradation of solutes, especially in aqueous solution. Hence, PLE operating at high temperatures is not suitable for extracting heat-sensitive active compounds from biomass.

Compared to conventional solvent extraction, $ScCO_2$ has many advantages: higher extraction yield and higher selectivity of desired solutes; milder temperature conditions and hence less degradation of heat-sensitive active compounds; shorter time for equivalent or better extraction performance; tunable physico-chemical properties of $ScCO_2$; no residue or solvent contamination in the products. Also, CO_2 is non-toxic and non-flammable (and hence environmentally friendly), inexpensive, and recyclable.

Liquid–liquid extraction and membrane separation are two potential methods for separation and concentration of phytochemicals from dilute solutions that are produced within the cadre of a biorefinery. Compared to conventional solution separation methods such as liquid–liquid extraction and disillation processes, membrane separation has many advantages: no requirement for solvents or additives; easy automation and scale-up; lower operating costs; shorter processing times; mild operating conditions, and less waste. Its major disadvantage is the potential for membrane fouling during the separation.

Molecular distillation is an efficient means for separation and purification of phytochemicals from plant extracts obtained by conventional solid–liquid extraction or enhanced solid–liquid extraction processes (UAE, MAE, and PLE). MD has many advantages:

no requirement for solvent; less or no degradation of products due to high vacuum or low temperature operation; high product purity or far less contamination in the product. Its disadvantages include a relatively high cost and the loss of the starting material to concentrate.

It is clear that value-added co-products integrated with a biomass-based biorefiney offers tremendous opportunities. While there are significant challenges that need to be overcome and better understood in the separation and purification of co-products from diverse biomass species in future biorefineries, experience in co-product manufacture from current food, pharmaceutical, and nutraceutial industries can be brought to bear in research, development, and successful commercialization in biorefineries.

References

Aalford, S.N. and Morel, duB.P.G. (2006) A survey of value addition in the sugar industry. *Proceedings. Congress of the South African Sugar Technologists Association*, **80**, 39–61.

Alper, N. and Acar, J. (2004) Removal of phenolic compounds in pomegranate juice using ultrafiltration and laccase-ultrafiltration combinations. *Nahrung*, **48**, 184–187.

Bioriginal Food & Science Corp. (access 2011) http://www.bioriginal.com/services/files/moleculardistille-doroilrefining.pdf.

Badami, S., Manohara Reddy, S.A., Kumar, E.P. *et al.* (2003) Antitumor activity of total alkaloid fraction of solanum pseudocapsicum leaves. *Phytother Research*, **17** (9), 1001–1004.

Borneman, Z.V., Gökmen, V., and Nijhuis, H.H. (2001) Selective removal of polyphenols and brown colour in apple juices using PES/PVP membranes in a single ultrafiltration process. *Separation and Purification Technology*, **22–23**, 53–61.

Chen, F., Cai, T., Zhao, G. *et al.* (2005) Optimizing conditions for the purification of crude octacosanol extract from rice bran wax by molecular distillation analyzed using response surface methodology. *Journal of Food Engineering*, **70**, 47–53.

Chen, F., Wang, Z., Zhao, G. *et al.* (2007) Purification process of octacosanol extracts from rice bran wax by molecular distillation. *Journal of Food Engineering*, **79**, 63–68.

Cravotto, G., Boffa, L., Mantegna, S. *et al.* (2008) Improved extraction of vegetable oils under high-intensity ultrasound and/or microwaves. *Ultrason Sonochemistry*, **15**, 898–902.

Cvengros, J., Pollák, S., Mikov, M., and Lutisan, J. (2001) Film wiping in the molecular evaporator, *Chemical Engineering Journal*, **81**, 9–14.

Cruz, J.M., Dominguez, J.M., Dominguez, H., and Parajo, J.C. (1999) Solvent extraction of hemicellulosic wood hydrolyzates: a procedure useful for obtaining both detoxified fermentation media and polyphenols with antioxidant activity. *Food Chemistry*, **67**, 147–153.

Diouf, P.N., Stevanovic, T., and Boutin, Y. (2009) The effect of extraction process on polyphenol content, triterpene composition and bioactivity of yellow birch (Betula alleghaniensis Britton) extracts. *Industrial Crops and Products*, **30**, 297–303.

Duan, L., Wallace, S.N., Engelberth, A. *et al.* (2009) Extraction of co-products from biomass: example of thermal degradation of silymarin compounds in subcritical water. *Applied Biochemistry and Biotechnology*, **158**, 362–373.

FERA (The Food and Environment Research Agency) (2011) A review of current knowledge on the economic potential of chemical products from the main commercial UK tree species. Available at https://secure.fera.defra.gov.uk/treechemicals/review/.

Krishnaiah, D., Sarbatly, R., and Nah, N.L. (2007) Recovery of phytochemical components from various parts of Morinda citrifolia extracts by using membrane separator. *Journal of Applied Science*, **7** (15), 2093–2098.

Gonzalez, J., Cruz, J.M., Dominguez, H., and Parajo, J.C. (2004) Production of antioxidants from Eucalyptus globulus wood by solvent extraction of hemicellulose hydrolyzates. *Food Chemistry*, **84**, 243–251.

Halim, R., Gladman, B., Danquah, M.K., and Webley., P.A. (2011) Oil extraction from microalgae for biodiesel production. *Bioresource Technology*, **102** (1), 178–185.

Ibañez, E., Kubátová, A., Señoráns, F. *et al.* (2003) Subcritical water extraction of antioxidant compounds from rosemary plants. *Journal of Agricultural and Food Chemistry*, **51**, 375–382.

Kawamura, F., Kikuchi, Y., Ohira, T., and Yatagai, M. (1999) ASE of paclitaxel and related compounds from the bark of taxus cuspidata. *Journal of Natural Products*, **62**, 244–247.

Jiang, S.T., Shao, P., Pan, L.J., and Zhao, Y.Y. (2006) Molecular distillation for recovering tocopherol and fatty acid methyl esters from rapeseed oil deodoriser distillate. *Biosystems Engineering*, **93** (4), 383–391.

Kim, W.-J., Kim, J.-D., Kim, J. *et al.* (2008) Selective caffeine removal from green tea using supercritical carbon dioxide extraction. *Journal of Food Engineering*, **89**, 303–309.

Klejdus, B., Kopecky, J., Benesová, L., and Vacek, J. (2009) Solid-phase/supercritical-fluid extraction for liquid chromatography of phenolic compounds in fresh water microalgae and selected cyanobacterial species. *Journal of Chromatography A*, **1216**, 763–771.

Lee, J.-Y., Yoo, C., Jun, S.-Y. *et al.* (2010) Comparison of several methods for effective lipid extraction from microalgae. *Bioresource Technology*, **101**, S75–S77.

Malmary, G., Albet, J., Putranto, A. *et al.* (2000) Recovery of aconitic and lactic acids from simulated aqueous effluents of the sugar-cane industry through liquid-liquid extraction. *Journal of Chemical Technology and Biotechnology*, **75**, 1169–1173.

Martins, P.F., Ito, V.M., Batistella, C.B., and Maciel, M.R.W. (2006) Free fatty acid separation from vegetable oil deodorizer distillate using molecular distillation process. *Separation and Purification Technology*, **48** (1), 78–84.

Lavoie, J.-M. and Stevanovic, T. (2007) Selective ultrasound-assisted extractions of lipophilic constituents from betula alleghaniensis and *B. papyrifera* wood at low temperatures. *Phytochemical Analysis*, **18**, 291–299.

Macías-Sánchez, M.D., Serrano, C.M., Rodríguez, M.R., and Martínez de la Ossa, E. (2009a) Kinetics of the supercritical fluid extraction of carotenoids from microalgae with CO2 and ethanol as cosolvent. *Chemical Engineer Journal*, **150**, 104–113.

Macías-Sánchez, M.D., Mantell, C., Rodríguez, M. *et al.* (2009b) Comparison of supercritical fluid and ultrasound-assisted extraction of carotenoids and chl a from Dunaliella salina. *Talanta*, **77**, 948–952.

Masango, P. (2005) Cleaner production of essential oils by steam distillation. *Journal of Cleaner Production*, **13**, 833–839.

Meloan, C.E. (1999) *Chemical Separations: Principles, Techniques, and Experiments*, John Wiley & Sons, Inc., New York.

Mendes, R.L., Nobre, B.P., Cardoso, M.T. *et al.* (2003) Supercritical carbon dioxide extraction of compounds with pharmaceutical importance from microalgae. *Inorganica Chimica Acta*, **356**, 328–334.

Mendes, R.L. (2008) Supercritical fluid extraction of active compounds from algae, in *Supercritical Fluid Extraction of Nutraceuticals and Bioactive Compounds* (ed. J.L. Martinez), CRC Press, Taylor & Francis Group, LLC, New York.

Manzan, A.C.C.M., Toniolo, F.S., Bredow, E., and Povh, N.P. (2003) Extraction of essential oil andpigments from Curcuma longa (L.) by steam distillation and extraction with volatile solvents. *Journal of Agricultural and Food Chemistry*, **51**, 6802–6807.

Mata, T.M., Martins, A.A., and Caetano, N.S. (2010) Microalgae for biodiesel production and other applications: A review. *Renewable and Sustainable Energy Reviews*, **14**, 217–232.

Middleton Jr., E., Kandaswami, C., and Theoharides, T.C. (2000) The effects of plant flavonoids on mammalian cells: implications for inflammation, heart disease, and cancer. *Pharmacological Reviews*, **52**, 673–751.

Nawaz, H., Shi, J., Mittal, G.S., and Kakuda, Y. (2006) Extraction of polyphenols from grape seeds and concentration by ultrafiltration. *Separation and Purification and Technology*, **48**, 176–181.

National Alfalfa & Forage Alliance http://www.alfalfa.org/pdf/2007SGWhitepaper.pdf.

Pasquet, V., Chérouvrier, J.-R., Farhat, F. *et al.* (2011) Study on the microalgal pigments extraction process: Performance of microwave assisted extraction. *Process Biochemistry*, **46**, 59–67.

Patrick, H.R., Griffith, K., Liotta, C.L., and Eckert, C.A. (2001) Near-critical water: a benign medium for catalytic reactions. *Industrial & Engineering Chemistry Research*, **40**, 6063–6067.

Pawar, C.R. and Surana, S.J. (2010) Optimizing conditions for gallic acid extraction from *Caesalpinia Decapetala* Wood. *Pakistan Journal of Pharmaceutical Sciences*, **23** (4), 423–425.

Pereda, S., Bottini, S.B., and Brignole, E.A. (2008) Fundamentals of supercritical fluid technology, in *Supercritical Fluid Extraction of Nutraceuticals and Bioactive Compounds* (ed. J.L. Martinez), CRC Press, Taylor & Francis Group, LLC, New York.

Posada, L.R., Shi, J., Kakuda, Y., and Xue, S.J. (2007) Extraction of tocotrienols from palm fatty acid distillates using molecular distillation Original Research Article. *Separation and Purification Technology*, **57** (2), 220–229.

Pronyk, C. and Mazza, G. (2009) Design and scale-up of pressurized fluid extractors for food and bioproducts. *Journal of Food Engineering*, **95**, 215–226.

Rada, M., Guinda, A., and Cayuela, J. (2007) Solid/liquid extraction and isolation by molecular distillation of hydroxytyrosol from Olea europaea L. leaves. *European Journal of Lipid Science and Technology*, **109**, 1071–1076.

Renuka Devi, R. and Arumughan, C. (2007) Phytochemical characterization of defatted rice bran and optimization of a process for their extraction and enrichment. *Bioresource Technology*, **98**, 3037–3043.

Savangikar, C.V. and Savangikar, V.A. (2010) Integrated production of phytochemical rich plant products or isolates from green vegetation, US Patent 20100040758.

Sousa, A.F., Pinto, P.C.R.O., Silvestre, A.J.D., and Neto, C.P. (2006) Triterpenic and other lipophilic components from industrial cork byproducts. *Journal of Agricultural and Food Chemistry*, **54**, 6888–6893.

Temelli, F., Saldana, M.D.A., Moquin, P.H.L., and Sun, M. (2008) Supercritical fluid extraction of specialty oils, in *Supercritical Fluid Extraction of Nutraceuticals and Bioactive Compounds* (ed. J.L. Martinez), CRC Press, Taylor & Francis Group, LLC, New York.

Turley, D.B., Chaudhry, Q., Watkins, R.W. *et al.* (2006) Chemical products from temperate forest tree species – developing strategies for exploitation. *Industrial Crops and Products*, **24**, 238–243.

Wang, L., Kumar, A., Weller, C.L., Jones, D.D., and Hanna, M.A. (2007) Co-production of chemical and energy products from distillers grains using supercritical fluid extraction and thermochemical conversion technologies. 2007 ASABE Annual International Meeting, Minneapolis, Minnesota, 17–20, June 2007.

Wang, L. and Weller, C.L. (2006) Recent advances in extraction of nutraceuticals from plants. *Trends in Food Science & Technology*, **17**, 300–312.

Uppugundla, N., Engelberth, A., Ravindranath, S.V. *et al.* (2009) Switchgrass water extracts: extraction, separation and biological activity of rutin and quercitrin. *Journal of Agricultural and Food Chemistry*, **57**, 7763–7770.

Vaughn, K., Mcclain, C., Carrier, D.J. *et al.* (2007) Effect of albizia julibrissin water extracts on low-density lipoprotein oxidization. *Journal of Agricultural and Food Chemistry*, **55**, 4704–4709.

Vilkhu, K., Mawson, R., Simons, L., and Bates, D. (2008) Applications and opportunities for ultrasound assisted extraction in the food industry - A review. *Innovative Food Science & Emerging Technologies*, **9** (2), 161–169.

Vinson, J.A., Jang, J., Dabbagh, Y.A. *et al.* (1995) Plant polyphenols exhibit lipoprotein-bound antioxidant activity using an in vitro oxidation model for heart disease. *Journal of Agricultural and Food Chemistry*, **43**, 2798–2799.

Wei, S., Hossain, M.M., and Saleh, Z.S. (2010) Concentration of rutin model solutions from their mixtures with glucose using ultrafiltration. *International Journal of Molecular Sciences*, **11**, 672–690.

Xiao, X.-H., Wang, J.-X., Wang, G. *et al.* (2009) Evaluation of vacuum microwave-assisted extraction technique for the extraction of antioxidants from plant samples. *Journal of Chromatography A*, **1216**, 8867–8873.

4

Phytochemicals from Corn: a Processing Perspective

Kent Rausch

Department of Agricultural and Biological Engineering,
University of Illinois, Urbana, Illinois, USA

4.1 Introduction: Corn Processes

Corn (zea maize) is a high yielding, genetically diverse and widely accepted grain crop in the US. Historically, corn was grown for consumption in animal diets. Use of corn for processing at an industrial scale, which began approximately 150 years ago, has continued to increase and has accelerated in recent years. The primary components of the corn kernel, from a processing perspective, consist of the pericarp, germ, tip cap and endosperm (Figure 4.1). The pericarp provides a protective layer around the kernel and is a barrier to moisture penetration. The germ contains the embryo plant and contains most of the kernel oil. The tip cap is the point of attachment to the cob and is the primary point of entry for moisture during processing. The corn kernel is unique amongst grains because its endosperm contains both regions of hard and soft endosperm within the same kernel. The endosperm is a large percentage of the kernel and contains most of the starch, along with storage proteins. As a result, the corn wet milling process is highly optimized to recovery of starch contained in the endosperm.

There are three primary commercial processes that consume significant amounts of corn: dry milling, wet milling and dry grind. Research has developed new processes that have advantages for general process efficiency, but also benefits for producing phytochemicals from corn.

Biorefinery Co-Products: Phytochemicals, Primary Metabolites and Value-Added Biomass Processing, First Edition.
Edited by Chantal Bergeron, Danielle Julie Carrier and Shri Ramaswamy.
© 2012 John Wiley & Sons, Ltd. Published 2012 by John Wiley & Sons, Ltd.

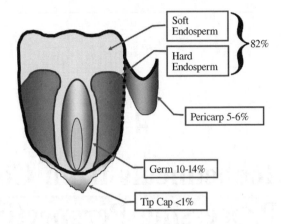

Figure 4.1   *A schematic of the corn kernel showing major components important in processing.*

4.1.1   Dry Milling

The primary objective of a dry milling process is to produce large pieces of endosperm, called grits, which are used in an array of human food products: breakfast cereals, extruded and baked snacks, breads and brewing (Brekke, 1970; Duensing, Roskens, and Alexander, 2003; Rausch and Belyea, 2006). Dry milling involves a brief tempering period where moisture is added to increase kernel moisture from 15 to 22% (Figure 4.2). Many dry millers add the moisture in a single stage. This sudden increase in moisture hydrates the pericarp and germ of the kernel. Before moisture can diffuse more deeply into the endosperm region, the tempered kernels are sent to a degermination unit operation, where the kernel is fractured, ideally, into germ,

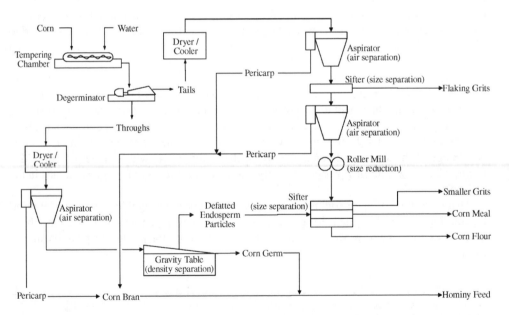

Figure 4.2   *Corn dry milling process (Rausch and Belyea, 2006). Reprinted with kind permission from Rausch et al., 2006 © Springer Science + Business Media (2006).*

pericarp and endosperm particles. If tempering and other parameters are correct, the endosperm is cleanly separated from the germ and pericarp, with large endosperm pieces resulting. Ideally, germ and pericarp should not have large amounts of endosperm attached. The germ, due to moisture from tempering, should be resilient enough to not fracture and release oil. In reality, separation of germ, pericarp and endosperm is incomplete so that small amounts of endosperm are attached to pericarp and germ, while small germ fragments attach to endosperm and pericarp. Because of attachment of residual endosperm and pericarp particles to the germ, oil concentration in the germ is reduced; typical fat contents of germ are 18 to 21% db (Johnston *et al.*, 2005). Similarly, because of endosperm and germ attachment to pericarp, components of interest within the pericarp region, such as phytosterols, will have lower concentrations. While the germ has higher fat content relative to whole kernels, very few dry mills expel or extract germ oil due to the economy of scale needed for oil extraction facilities. In a typical dry mill, germ, pericarp and low-valued endosperm products (e.g., low-grade corn flour) are combined to form a co-product, hominy feed. Because no chemicals are added during the dry milling process and process streams are subject to small amounts of drying, off odors and flavors are minimized, allowing the ready addition of dry milling fractions into human foods.

4.1.2   Wet Milling

Corn wet milling is designed to recover starch from the kernel at a purity of at least 99.5%. Starch products are found in numerous human and animal foods, as well as industrial products. Much of the recovered starch from wet milling is transformed into corn sweeteners or fermented into ethanol. In wet milling, corn is steeped for 24 to 36 h in a solution of sulfurous and lactic acid, using a complex semicontinuous system of batch tanks which facilitates kernel hydration, lactic acid fermentation and sulfur dioxide reactions within the endosperm (Figure 4.3). The steeping process is critical to the economical recovery of pure starch. Steeping hydrates the kernel and weakens protein bonds in the endosperm so that subsequent milling steps can separate the starch from the other corn components. Steeping also removes soluble materials from germ, allowing steeped germ to be recovered at high purity compared to dry milling and increasing oil content of the recovered germ fraction to 40–45% db. High oil content of steeped germ allows economical extraction of oil from the corn germ. Corn fiber and steeping solubles are combined to form corn gluten feed (21% wb protein minimum). Endosperm proteins are recovered to form corn gluten meal (60% wb protein minimum).

Wet milling facilities are more expensive to construct than dry milling or dry grind facilities, but some of their co-products (i.e., corn germ, corn gluten meal) have higher market value than the DDGS (distillers' dried grains with solubles) co-product from dry grind. The wet milling industry grew during the 1960s, with larger growth during the 1970s and 1980s as high fructose corn syrup became a commodity for food and beverage industries. However, demand for starch products and sweeteners has remained relatively stable during the period from about 1990 to 2010, with nominal growth in wet milling capacity.

4.1.3   Alternative Wet Milling Processes

An alkali wet milling method (Figure 4.4) was developed which avoided the long steep times needed for sulfurous acid steeping (Du *et al.*, 1999; Eckhoff *et al.*, 1999). Corn was soaked

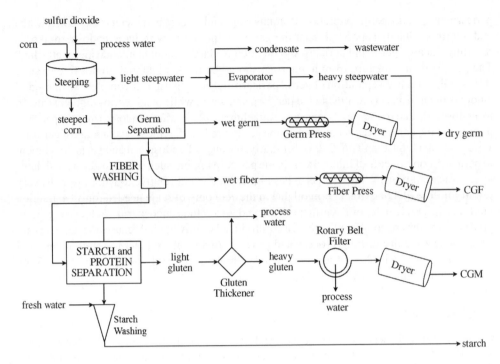

Figure 4.3   *A simplified schematic of the conventional corn wet milling process (Rausch and Belyea, 2006; Rausch et al., 2005; Rausch et al., 2007). Reprinted with permission from Rausch et al., 2005 © AACC International (2005).*

briefly in a solution containing sodium hydroxide at a relatively high temperature (5 min, 2% NaOH, 90 °C). Because the soak time was short, starch in the endosperm was not affected but the pericarp was loosened from the kernel. The pericarp was detached using gentle abrasion and separated using a screen. The debranned kernels were broken open using a roller mill, then steeped with sodium hydroxide (60 min, 0.5% NaOH, 45 °C). The kernel was then subjected to conventional wet milling steps to recover germ, fiber, starch and protein (gluten).

A modified wet milling process (Figure 4.5) was developed that reduced steeping times by first conducting a short (2 h) water soaking step followed by a coarse grind (Lopes-Filho *et al.*, 1997). The fractured kernels were then subjected to steeping (6 h, 2000 ppm SO$_2$, 0.55% lactic acid) in an agitated tank which intermittently recirculated the coarsely ground kernels through a mill to continue to break apart the kernel. Using these initial steps, germ was properly hydrated and not damaged, which allowed recovery of high-quality germ. Because the kernels were broken open during the coarse grinding, the diffusion pathways for SO$_2$ were shortened. Following the agitated, or dynamic, steeping step, germ was recovered by flotation. The subsequent process steps followed conventional wet milling. Process yields from the intermittent and dynamic steeping process were similar to conventional wet milling. Protein contents ($\sim$0.35%) of the starch fractions and oil contents (42–47%) of the germ fractions were similar. The method greatly reduced the time required to prepare the kernel for germ and starch recovery.

In another method, rather than an aqueous medium being used to convey SO$_2$ into the corn endosperm during steeping, a gaseous treatment of SO$_2$ was applied to corn kernels

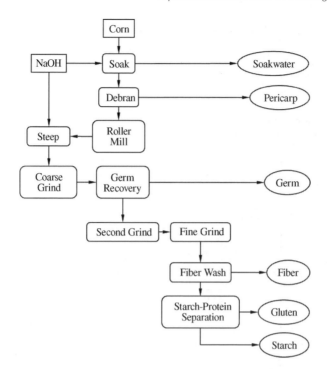

Figure 4.4 *Method for alkali wet milling (Du et al., 1999; Eckhoff et al., 1999). Reprinted with permission from Du et al., 1999 © AACC International (1999).*

(Eckhoff and Tso, 1991). Gaseous SO_2 was injected into sealed containers holding dry kernels for 15 min at room temperature. The treated corn was hydrated in steep tanks for up to 24 h using plain water or water with 0.55% lactic acid added. The steeped corn was processed following conventional wet milling methods. Lactic acid addition was found to increase starch yields compared to no addition. Using a gaseous treatment, a 12 h steep time had similar starch yields to conventionally steeped corn (48 h, 2000 ppm SO_2, 0.55% lactic acid).

4.1.4 Dry Grind

The corn dry grind process for fuel ethanol production is based on the ancient process of fermenting the kernel starch for production of ethanol beverages. In a typical dry grind process, whole corn kernels are ground in a mill, mixed with water, cooked with enzymes and fermented with enzymes and yeast (Figure 4.6). Enzymes are added during the cooking process to control the viscosity of the gelatinized starch and begin the process of degrading the starch granules into shorter-chain compounds called dextrins. Enzymes added during fermentation further degrade the dextrins into glucose, which can be consumed by the yeast to make ethanol. A distillation process recovers ethanol for use as a biofuel. Material in the corn kernel that does not ferment (fiber, protein, fat, ash) is recovered and dried to form distillers' dried grains with solubles (DDGS). The marketing of this co-product has played an important role in the sustainability of dry grind plants. For every three units of corn processed by a dry grind plant, one unit of DDGS is produced. Ethanol production by the dry grind process saw

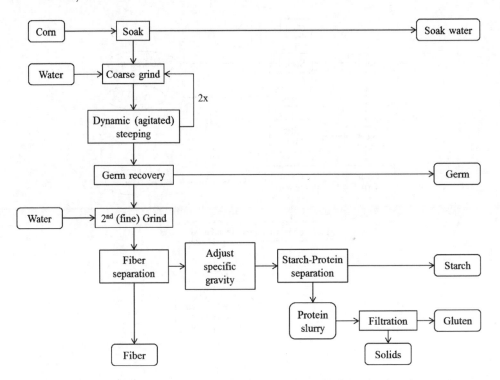

Figure 4.5   *Intermittent milling and dynamic steeping process for wet milling. Reprinted with permission from Lopes-Filho et al., 1999 © AACC International (1999).*

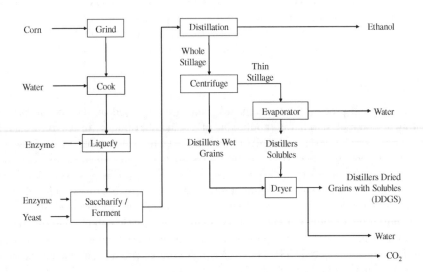

Figure 4.6   *Conventional dry grind process for production of fuel ethanol.*

rapid growth during the period from 1990 to 2008, as ethanol was added to gasoline as a fuel oxygenate.

During the period of rapid growth in the dry grind industry (approximately 1998 to 2008), consumption of DDGS became a challenge, especially since the value of DDGS began approaching the value of raw corn. The low value of DDGS relative to co-products from wet milling, corn gluten feed and corn gluten meal, is due in part to DDGS being a mixture of protein, fat, fiber and ash. This is in contrast to corn gluten meal and corn germ from the wet milling process, which have high protein and high oil contents, respectively. Because of relatively high fiber content, most DDGS is fed to ruminant animals; this further depresses the market value of DDGS since it competes with ruminant diet ingredients such as forage products.

4.1.5   Alternative Dry Grind Processes

To address the situation of low co-product value from the dry grind process, several process modifications have been developed. Each of these modifications affected phytochemical contents of the process streams and eventual co-products, which are discussed in following sections. The first modification to the dry grind process, quick germ, was to recover germ prior to the fermentation step (Singh and Eckhoff, 1996, 1997). The corn was hydrated for 3 to 12 h in a simple water soaking step, followed by coarse grind and germ separation operations that used wet milling technology. As a result, germ with fat contents similar to wet milled germ was recovered (Johnston *et al.*, 2005). Most of the pericarp and endosperm fiber components continued through the process, were placed in the fermenter and became part of the DDGS fraction. Since most of the germ components do not ferment, and since germ contains nearly all of the oil in the kernel, this preseparation had several advantages. Recovery of the germ fraction provided dry grind processors with the option of marketing germ, which was typically three to six times more valuable than DDGS. Furthermore, recovered germ was removed from the main process flow and did not take up fermenter capacity (Taylor *et al.*, 2001).

An alternative to the quick germ method, called dry fractionation, used tempering and degerminating processes from the dry milling industry prior to liquefaction, saccharification and fermentation steps. While this approach required less capital, its drawbacks included low germ fat content (18–21%) and ethanol yield reduction due to the high starch content of the germ fraction (20%, Johnston *et al.*, 2005). Due to germ recovery from the dry grind process, the resulting material after fermentation had slightly higher protein content. Because of the economics involved with oil extraction from germ, most of the germ fraction recovered by this method was typically combined with DDGS or marketed separately as a higher fat ruminant diet ingredient. Due to its composition, germ recovered by dry fractionation would compete on the market with hominy feed, which historically has lower value than DDGS or corn gluten feed.

A refinement to the quick germ process, called quick fiber, recovered germ and pericarp fiber prior to fermentation (Wahjudi *et al.*, 2000). Kernels were soaked 3 to 12 hours in water and coarsely ground (Figure 4.7). Initial versions of the process used recycled process slurry to adjust the specific gravity so that both germ and pericarp fiber could be recovered using hydrocyclones (Singh *et al.*, 1999). With the development of proteases and granular starch hydrolyzing enzymes, slurry specific gravities could be increased without recycling (Wang *et al.*, 2009; Wang *et al.*, 2005; Wang *et al.*, 2007). With the removal of germ and

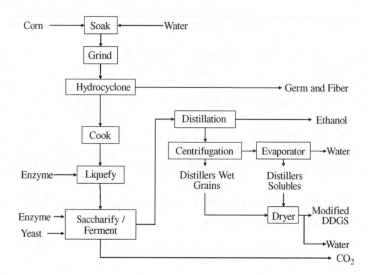

Figure 4.7 *Modified dry grind process (quick germ quick fiber) for production of fuel ethanol, germ, fiber and modified DDGS.*

pericarp fiber, the material remaining after fermentation had increased protein contents as well as reduced fiber levels, making the material more suitable for non-ruminant diets (Singh *et al.*, 2005). Because of the use of a soaking step, quick germ and quick fiber processes are called wet fractionation.

An additional modification to wet fractionation used sieving of the mash (prefermentation) or whole stillage (postfermentation) to recover endosperm fiber. This process, called enzymatic dry grind or e-milling, resulted in several unique and high valued co-products: high quality germ, pericarp fiber, endosperm fiber and a high-protein, low-fiber "DDGS." These changes to the dry grind process resulted in a DDGS co-product with 58% db protein, compared to conventional DDGS with a minimum of 26% db protein (Singh *et al.*, 2005).

4.1.6 Nixtamalization

The nixtamalization, or lime cooking, process is an ancient method used to produce a unique dough, masa (Serna-Saldivar, 2010). This dough is used to make a variety of corn-based products, including tortillas, snack chips and other food items. The process parameters used to produce masa vary widely and depend on the product being produced, the size of the operation, regional influences and operator preferences. The initial step is lime cooking, which involves the addition of approximately 1% lime (based on corn weight) to kettles containing water and corn at a ratio of about 3:1 (Figure 4.8). The cooking occurs at 85 to 100 °C for 15 to 60 min. Steeping follows for 8 to 16 h at temperatures above 68 °C, but below 100 °C. The combination of cooking and steeping serves several functions, which also affect the separation of nutrients in the following steps. During cooking and steeping, pericarp is loosened from the kernel and a portion of the starch is gelatinized. The washing step removes pericarp and solids that have become solubilized during cooking and steeping, producing a process stream called nejayote. Unfortunately, nutrients contained in the nejayote are not incorporated into the human food

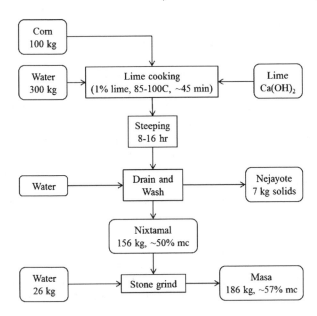

Figure 4.8 *Typical process for the production of masa (adapted from Serna-Saldivar, 2010). Adapted from Serna-Saldivar., 2012 © Taylor & Francis Group LLC (2010).*

product, but instead typically are treated as a waste stream. The remaining material is called nixtamal, which is further ground with the addition of water to form masa dough. The grinding is typically done using stone plates to give masa dough the desired texture. Masa can be dried and ground to make a flour product, or further processed into tortillas.

Nixtamalization has beneficial effects on the nutritional value of the proteins in corn by improving the essential amino acid profile and lead to assimilation of calcium ions (originating from the lime during the cooking step) into the masa material. For Latin Americans, tortillas are an important source of protein, calcium and calories (Maya-Cortes *et al.*, 2010). However, the nejayote that is drained away after cooking poses disposal issues due the high pH and the 5 to 12% soluble solids content. Valuable nutrients are lost with the nejayote, including fat, protein, dietary fiber, vitamins and minerals (Maya-Cortes *et al.*, 2010). The nejayote process stream thus creates dual issues for the processor: an environmental liability and a loss of nutrients needed by the consumer. Process improvements addressing these challenges would likely improve the economic sustainability for the processor, as well as result in more nutritious nixtamal products.

4.2 Phytochemicals Found in Corn

4.2.1 Introduction

While phytochemicals are present in sweet corn and waxy corn consumed as a fresh vegetable (Hu and Xu, 2011; Song *et al.*, 2010), this represents a small percentage of total US corn production (USDA, 2011). Plant breeding during the twentieth century improved the composition of horticultural crops, particularly the compositions of macronutrients (protein, fat and carbohydrate) and fiber (Farnham, Simon, and Stommel, 1999). Attempts to improve the amino acid profile of zein protein in corn through breeding and genetic improvements is an example. Relatively more current research has focused on micronutrients (minerals, vitamins). Farnham, Simon, and Stommel (1999) postulated that successful improvement of phytonutrient contents of crops will be realized through multidisciplinary research, advances in biotechnology and analytical techniques and education. This would lead to improved phytochemical content of crops of all types.

Liu (2007) summarized phytochemicals found in whole grains and compared these compounds to those in fruits and vegetables. Whole grain phytochemicals complement those found in fruits and vegetables, such as ferulic acid. Whole grains also contain tocotrienols, tocopherols and oryzanols, which have structural and plant defense roles in grains (Liu, 2007). The most important phytochemical groups are phenolics, carotenoids, vitamin E compounds, lignans, β-glucans and inulin. Liu (2007) provided a detailed review of these groups. Phenolic compounds contain one or more aromatic rings. Phenolics derived from plants have been implicated in reduced risk of chronic diseases. The most common phenolic compounds found in whole grains are phenolic acids and flavonoids. Phenolics have been measured in several grains, but most research has focused on wheat. Sosulski, Krygier, and Hogge (1982) measured total phenolic acid content (71 to 309 ppm) of wheat, rice, oat and corn flours which had been debranned by pearling or roller milling to remove germ and bran or pericarp layers. Of the grain flours tested, corn had the highest phenolic acid content (Sosulski, Krygier, and Hogge, 1982).

Liu (2007) reported that total phenolic content of whole grains was nearly as high or higher than some fruits and vegetables that are well regarded for their health benefits. Total phenolics for whole grains ranged 100 to 280 mg gallic acid equiv./100 g sample. Corn had the highest total phenolic content of the grains reported. Common fruits and vegetables were similar or higher: 110, 160, 300 and 530 mg gallic acid equiv./100 g sample for broccoli, strawberry, apple and cranberry, respectively. Total antioxidant activity was highest for corn (180 µmol vitamin C equiv./g grain) amongst cereal grains, and comparable to that of some fruits and vegetables (cranberry, apple and strawberry).

Carotenoids are widespread naturally occurring pigments. Carotenoids are divided into two classifications based on oxygen content (or lack thereof); xanthophylls are oxygenated derivatives of carotenes. Carotenoids consist of a 40-carbon skeleton of isoprene units (Liu, 2007). Carotenoids provide colour in plant materials and milling products (yellow, orange and red) as well as antioxidant activity.

Liu (2007) summarized research (Hendriks *et al.*, 1999; Miettinen *et al.*, 1995) that found intake of plant sterols or stanols can lower serum total and LDL cholesterol in humans. While not fully understood, phytosterols appear to compete with cholesterol in the human intestine and inhibit cholesterol absorption (Nissinen *et al.*, 2002), probably due in part to their structural similarities.

In quantifying phytochemical content of processing and other plant materials, Liu (2007) reported that many earlier studies (Bryngelsson, Dimberg, and Kamal-Eldin, 2002; Dietrych-Szostak and Oleszek, 1999; Miller *et al.*, 2000; Velioglu *et al.*, 1998; Zielinski and Kozlowska, 2000) presumed that analytical extraction techniques were adequate in recovering all phenolic compounds from grains. It appears that these earlier methods extracted only the loosely attached or soluble phenolic compounds and did not recover the phenolics tightly bound to the cell wall materials. Extraction techniques that used digestion were thought to release bound phytochemicals from the cell wall materials, thus more accurately reporting composition (Adom and Liu, 2002; Adom, Sorrells, and Liu, 2003; Adom, Sorrells, and Liu, 2005). Results using more extensive extraction methods have found that most grain phenolic compounds were in a bound form. Free phenolics in corn consisted of 15% of total phenolics. From a processing standpoint, Liu (2007) reported phytochemical compositions in milled fractions found in wheat only.

Inglett and Chen (2011) investigated free and bound phenolic compounds and antioxidant activities of corn bran that had been ground and air classified (Inglett and Chen, 2011). They reported on several extraction methods and their effects on quantification of free and bound phenolics. As particle size increased, free phenolic contents and free antioxidant activities decreased. However, as particle size increased, the bound phenolic contents and antioxidant activities also increased. They hypothesized that smaller particles, with increased surface area, may facilitate solvent extraction and increase values observed for free antioxidant activities. Certain measurement methods, such as use of alkaline treatment, may have changed the size characteristics of the particles and caused release of phenolic compounds that were not released by other methods. This work identifies the importance of analytical technique when determining composition, and when comparing concentrations amongst milling fractions. As a result, this work was more about analytical methods to measure phytochemicals than a method to concentrate phytochemicals by air classification.

4.2.2 Phytosterols

Phytosterols include more than 250 compounds, with the most common being sitosterol, stigmasterol and campesterol (4-desmethyl sterols; Piironen *et al.*, 2000). The beneficial effects of phytosterols have been described in several reviews (Ling and Jones, 1995; Moreau, Whitaker, and Hicks, 2002; Piironen *et al.*, 2000). Sitosterol was used and marketed in the 1950s as a method to lower serum cholesterol levels in patients with hypercholesterolemia (Moreau, Whitaker, and Hicks, 2002). Serum levels of β-sitosterol and campesterol were positively associated with absorption of dietary cholesterol and negatively with cholesterol synthesis (Miettinen, Tilvis, and Kesaniemi, 1990). In the 1970s, a mixture of sitostanol and campesterol was shown to be more effective in lowering cholesterol levels than previously used mixtures of phytosterols. The preparation of fat-soluble stanyl esters showed consistent reductions in serum total cholesterol when consumed as a mayonnaise or margarine spread (Piironen *et al.*, 2000). Free phytosterols were poorly soluble and had low bioavailability, but phytostanols esterified with fatty acids to form stanyl esters allowed the first commercial production of phytosterol fatty acid esters in the early 1970s. It was also discovered that fatty acid esters of sterols (steryl esters) could be added to food products. Moreau, Whitaker, and Hicks (2002) and Piironen *et al.* (2000) provide detailed reviews of the histories and apparent mechanisms of these compounds in lowering serum cholesterol.

Norton (1994, 1995) and Seitz (1989) reported that hexane-extracted corn bran had high levels of ferulate esters, which were similar in composition to γ-oryzanol found in rice bran and rice bran oil (Norton, 1994, 1995; Seitz, 1989). These studies on corn bran were originally conducted to gain understanding of how these compounds were involved in production of mycotoxins prior to grain harvest. The exact role of γ-oryzanol and other compounds in the plant were not determined, but because oryzanol compounds from rice had a long history of being linked with beneficial effects on hyperlipidemia (cholesterol levels), additional study on corn oryzanol was recommended. Moreau and coworkers (1996) used hexane extraction on corn fiber co-products obtained from commercial wet milling and dry milling processes (Moreau, Powell, and Hicks, 1996). The study revealed that commercially obtained corn fiber fractions had relatively low oil contents (0.54–3.5%) compared to rice bran (18%) that had been observed in a previous study (Kahlon *et al.*, 1992), but the corn fiber oil had higher concentrations of ferulate esters (up to 6.75%) than rice bran oil (0.1–0.8%). The extracts obtained from commercial wet milling sources that originated from the pericarp region of the kernel had the highest ferulate esters (2.7–6.75% db of fiber fraction), whereas extracts from dry milling fractions (bran) had concentrations of 1.5% db.

Corn fiber recovered from the wet milling process was found to contain a unique oil (Moreau, Powell, and Hicks, 1996) and was proposed as a natural cholesterol-lowering oil. The corn fiber oil, called Amaizing Oil, was shown to lower serum cholesterol levels in animal feeding trials at the University of Massachusetts and was patented as a joint invention between the Agricultural Research Service of USDA and the university (Moreau *et al.*, 1998). Additional animal studies (Moreau, Norton, and Hicks, 1999b; Ramjiganesh *et al.*, 2000; Wilson *et al.*, 2000) have confirmed its cholesterol-lowering effects.

Moreau, Singh, and Hicks (2001b) compared corn kernels to other potential sources of phytosterols, Job's tears and tesosinte (Moreau, Singh, and Hicks, 2001b). Whole seeds were ground and extracted; seeds were not fractionated. Total oil in seeds were higher in Job's tears than in the other seed groups, corn and teosinte, but corn had higher concentrations of ferulate-phytosterol esters (FPEs), free phytosterols and total sterols than teosinte (nine accessions) and Job's tears (three accessions). While there was significant variation in phytosterol concentrations in the corn kernels, Moreau, Singh, and Hicks (2001b) concluded that corn hybrids are a good source of these phytochemicals relative to other potential plant groups. In work with whole kernels from cereal grains (Moreau, Powell, and Singh, 2003), corn was found to be higher in total phytosterols (0.6–2.1wt%) than oats (0.1wt%), even though oats were higher in total lipid content (5.5–6.7%) than corn kernels (2.9–5.9%).

The corn fiber obtained from large-scale, commercial processes consists primarily of the outer layers of the corn kernel with smaller amounts of other kernel tissues originating from the endosperm and germ regions. Actually, the outer layer of the corn kernel contains two major tissues: pericarp (outer layer) and aleurone (inner layer). The pericarp consists of non-living cell walls, while the aleurone consists of a single layer of living cells surrounded by thick cell walls (Moreau *et al.*, 2000). Kernels were dissected by hand to separate pericarp and aleurone tissues and the fractions extracted with hexane. High levels of FPE and sitostanol were found to originate in the aleurone cells. Therefore, the aleurone layer is the major source of these compounds. Additional kernel fractionation research would determine whether fiber fractions would be obtained with the aleurone layer attached or if aleurone would be recovered with the endosperm.

Corn fiber oil has both phytosterol and phytostanyl esters and hydroxycinnamate esters, of which the latter may have unique functional properties unique absent in other phytosterol products. Moreau *et al.* (1998) reported that most of the phytosterols in corn fiber oil are esterified naturally with either fatty acids or phenolic acids, such as ferulic acid, a potent antioxidant, in contrast to phytosterols in soy or tall oil (Moreau *et al.*, 1998). Corn fiber oil contains high levels of sitostanol in the ferulic acid ester fraction and appears to be the richest source of natural stanols (and stanyl esters) reported (Moreau, Whitaker, and Hicks, 2002). Corn fiber oil also contains γ-tocopherol and various carotenoids, both with important antioxidant properties. The levels of total phytosterols in corn fiber oil range from about 15% to more than 50%, depending on extraction and fiber pretreatment conditions (Moreau, Powell, and Hicks, 1996). Levels of γ-tocopherol also vary with fiber pretreatment conditions (Moreau, Hicks, and Powell, 1999a) and range from about 0.3 to 3%. A combination of natural cholesterol-lowering components and antioxidants, which are thought to prevent oxidation of LDL-cholesterol, could make corn fiber oil an effective product to reduce heart disease.

4.2.3 Carotenoids

Certain carotenoids (more specifically, xanthophylls), lutein and zeaxanthin, have been reported to affect age-related macular degeneration, although standards for specific health claims have not yet been attained (Mozaffarieh, Sacu, and Wedrich, 2003; Trumbo and Ellwood, 2006). Corn is known to contain the xanthophylls lutein, zeaxanthin and β-cryptoxanthin. Although a body of research had studied lutein and zeaxanthin levels in whole kernels, there were no existing publications using corn oil (Hao *et al.*, 2005; Humphries and Khachik, 2003; Kurilich and Juvik, 1999a, 1999b; Moreau, Johnston, and Hicks, 2007; Moros *et al.*, 2002; Panfili, Fratianni, and Irano, 2004). Daily dosage currently believed to reduce age related macular degeneration is 6 mg lutein + zeaxanthin. Moreau, Johnston, and Hicks (2007) found the combined levels of lutein and zeaxanthin from either corn germ or corn fiber were similar ($1.4 \, \text{mg kg}^{-1}$). These levels made it practical for including the recommended levels in the diet. For example, 15 ml (1 tablespoon) of unrefined corn kernel oil contained about the same xanthophylls as 450 g (1 lb) of whole corn meal (3.3 mg lutein + zeaxanthin). Lutein + zeaxanthin levels ranged from $2.3 \, \mu\text{g g}^{-1}$ for hexane extracted corn germ oil to $221 \, \mu\text{g g}^{-1}$ for ethanol-extracted ground corn (Moreau, Johnston, and Hicks, 2007). Ethanol extracted more xanthophylls and carotenoids than hexane, but ethanol extraction also recovered compounds that had unmeasured or unknown toxicity, that is, diferuloylputrescine and *p*-coumaroylferululoylputrescine (Moreau and Hicks, 2005). As a result, ethanol extraction needs additional study to verify that these compounds would be removed by the oil refining process. Unrefined corn kernel oil extracted using ethanol was found to have high levels of lutein and zeaxanthin, but these were removed by the conventional refining, bleaching and deodorizing (RBD) steps used to produce refined corn oil. Nearly all carotenoids are removed by conventional RBD steps (Moreau and Hicks, 2005). Moreau, Johnston, and Hicks (2007) suggested that xanthophyll levels may provide incentive to use ethanol extraction to produce a healthy edible oil.

4.2.4 Polyamine Conjugates

Many vegetable oils are extracted by use of hexane, which originates from the petroleum industry and has properties that make it difficult to handle safely. An effective solvent that originates from a renewable source and has fewer health and safety risks has long been sought by the grain-processing industry. For example, ethanol produced at a corn-based facility would be a locally obtained and economical source for alternative solvent (Kwiatkowski and Cheryan, 2002). Moreau *et al.* (2001) investigated the effects of alternative solvents on oils obtained from various corn components (Moreau, Nunez, and Singh, 2001a). Corn fiber and bran were obtained from commercial wet milling or dry milling samples, respectively.

Extraction solvents included conventional hexane, ethanol, methylene chloride and chloro-form/methanol solvents. Extraction of corn bran with hexane yielded oils that contained triacylglycerols, phytosterols and other minor components. Extraction of corn bran with solvents such as methylene chloride or ethanol yielded two unknown peaks, later to be determined as *p*-coumaroyl-feruloylputrescine (CFP) and diferuloylputrescine (DFP). Extraction methodology and technique played a significant role in extraction of CFP and DFP. Until this study, CFP was previously unknown to occur in plants. DFP was previously reported to occur in tobacco, corn kernels and the male reproductive organs of the corn plant (tassel) (Moreau, Nunez, and Singh, 2001a).

Corn bran (from dry milling) had higher levels of CFP and DFP than corn fiber (from wet milling), suggesting that these compounds are concentrated in the pericarp tissue and not in the aleurone layer or in endosperm fiber (cellular fiber material). Previous work had determined that tempering during the dry milling process caused the outer kernel layer to fracture between the pericarp and aleurone, while conventional steeping during the wet milling process caused outer layers to fracture between the aleurone and the endosperm tissue (Moreau, Powell, and Hicks, 1996; Moreau *et al.*, 1999c; Singh, Moreau, and Cooke, 2001a; Singh *et al.*, 1999; Wolf *et al.*, 1952). The changes in the line of fracture, as well as the changes in compositions of oils obtained from milling fractions is discussed in a separate section below. Steeping of corn kernels with SO_2 and lactic acid, following by hand dissection, allowed recovery of pericarp and aleurone tissues separately. These tissues were then subject to extraction with ethanol. Aleurone had 12 times more extractable oil than pericarp (7.0 vs. 0.6 wt%, respectively), but pericarp had much higher levels than aleurone of CFP (2.93 vs. 0.01 wt%, respectively) and DFP (11.95 vs. 0.04 wt%, respectively).

Additional work by Moreau and Hicks (2005) sought to quantify the composition of oils obtained by ethanol extraction (kernel, germ), especially the levels of polyamine conjugates. Previous work (Moreau, Nunez, and Singh, 2001a) found that fiber oil and bran oil contained up to 10% polyamine conjugates, while germ oil (crude and refined) contained very low levels of these compounds. Corn bran was obtained from a commercial dry mill, corn germ obtained from commercial wet mill and whole kernels originated from a single hybrid sample. Ground kernels extracted with ethanol had components detected in significant amounts were hydro-xycinnamate sterol ester (HSE, 0.23%) and the polyamine conjugatesCFP (0.18%)and DFP (0.66%) (Table 4.1; Moreau and Hicks, 2005). High levels of HSE were reported in hexane extracted corn fiber oil (4–6%) and low levels in corn germ oil ($<0.01\%$) in previous work (Moreau, Powell, and Hicks, 1996).

Moreau and Hicks (2005) also observed the effects of ethanol concentration (70 to 100%) on extraction of polyamine conjugates. Similar levels of CFP ($\sim0.2\%$) and DFP ($\sim0.6\%$) were extracted as ethanol concentration was varied, in contrast to results found by Kwiatkowski and Cheryan with corn oil extraction (100% ethanol was optimal) and corn zein (70% ethanol was optimal) (Kwiatkowski and Cheryan, 2002). Moreau and Hicks (2005) studied the effects of extraction on tocopherols and tocotrienols and confirmed the presence of α- and γ-tocopherols and α- and γ-tocotrienols and confirmed low levels of δ-tocopherol (Table 4.2; Moreau and Hicks, 2005). High levels of total tocopherols and tocotrienols (350 to 550 mg kg^{-1} oil, respectively) in oil extracted by ethanol from ground kernels indicated this type of oil may have value as a health-promoting edible oil, since other commercially available oils marketed as being high in tocotrienols have similar levels (530 and 770 mg kg^{-1} oil for palm and rice bran oils, respectively).

Table 4.1 *Sterols (CFP) and polyamine conjugates (DFP) in oil extracted at 50 and 100°C from ground corn, corn bran, wet milled (WM) corn germ and dry milled (DM) corn germ (from Moreau and Hicks (2005)).*

Sample, solvent	°C	Oil extracted (%)	wt % of oil				
			SE	FS	HSE	CFP	DFP
Corn kernels							
Hexane	50	2.70 ± 0.06	1.03 ± 0.02	0.74 ± 9.92	0.38 ± 0.03	0	0
	100	3.28 ± 0.04	1.01 ± 0.02	0.76 ± 0.04	0.30 ± 0.00	0	0
Isopropanol	50	3.36 ± 0.06	1.33 ± 0.29	0.59 ± 0.07	0.33 ± 0.02	0.081 ± 0.007	0.290 ± 0.023
	100	5.03 ± 0.24	0.86 ± 0.14	0.52 ± 0.03	0.27 ± 0.02	0.126 ± 0.003	0.459 ± 0.056
Ethanol	50	3.38 ± 0.04	0.69 ± 0.03	0.81 ± 0.04	0.31 ± 0.02	0.194 ± 0.024	0.568 ± 0.083
	100	5.53 ± 0.20	1.04 ± 0.25	0.52 ± 0.02	0.25 ± 0.01	0.120 ± 0.002	0.463 ± 0.022
Corn bran							
Hexane	50	1.92 ± 0.07	5.70 ± 0.25	1.49 ± 0.08	0.44 ± 0.01	0	0
	100	2.11 ± 0.00	5.01 ± 0.05	1.65 ± 0.05	0.42 ± 0.01	0	0
Isopropanol	50	2.46 ± 0.01	3.82 ± 0.08	0.93 ± 0.01	0.23 ± 0.04	0.99 ± 0.15	2.58 ± 0.32
	100	6.41 ± 0.28	2.80 ± 0.06	1.10 ± 0.09	0.28 ± 0.01	1.18 ± 0.11	4.36 ± 0.38
Ethanol	50	3.47 ± 0.61	3.03 ± 0.14	0.80 ± 0.04	0.23 ± 0.02	1.12 ± 0.07	3.24 ± 0.25
	100	5.08 ± 0.43	2.33 ± 0.08	2.20 ± 0.03	0.25 ± 0.01	0.88 ± 0.04	3.34 ± 0.21
WM gem							
Hexane	50	32.18 ± 0.22	0.51 ± 0.15	0.25 ± 0.04	0.03 ± 0.01	0	0
	100	34.21 ± 0.59	0.63 ± 0.01	0.34 ± 0.01	0.03 ± 0.01	0	0
Isopropanol	50	27.91 ± 0.29	0.49 ± 0.02	0.27 ± 0.02	0.02 ± 0.00	0	0
	100	33.43 ± 0.02	0.63 ± 0.01	0.37 ± 0.02	0.02 ± 0.00	0.02 ± 0.00	0.04 ± 0.00
Ethanol	50	27.18 ± 0.71	0.45 ± 0.06	0.26 ± 0.00	0.02 ± 0.00	0	0
	100	41.57 ± 0.03	0.59 ± 0.04	0.35 ± 0.02	0.03 ± 0.01	0.02 ± 0.00	0.06 ± 0.01
DM germ							
Hexane	50	14.31 ± 0.15	0.53 ± 0.06	0.48 ± 0.07	0.04 ± 0.00	0	0
	100	17.42 ± 0.30	0.44 ± 0.05	0.45 ± 0.05	0.04 ± 0.01	0	0
Isopropanol	50	16.26 ± 0.19	0.48 ± 0.04	0.31 ± 0.03	0.04 ± 0.01	0.07 ± 0.00	0.27 ± 0.00
	100	22.27 ± 0.50	0.97 ± 0.14	0.31 ± 0.03	0.03 ± 0.00	0.10 ± 0.01	0.31 ± 0.02
Ethanol	50	17.15 ± 0.59	0.72 ± 0.11	0.47 ± 0.04	0.05 ± 0.01	0.09 ± 0.01	0.33 ± 0.01
	100	23.57 ± 1.10	1.42 ± 0.55	0.47 ± 0.06	0.05 ± 0.01	0.11 ± 0.01	0.31 ± 0.03

Experiments demonstrated that composition of oils extracted using ethanol are different than oil extracted from corn germ by conventional hexane. Ethanol extracted corn kernel oil had higher levels of free phytosterols, phytosterol fatty acyl esters and tocopherols and tocotrienols (Table 4.3, Moreau and Hicks, 2005). Little is known about the physiological role of polyamine conjugates in plants or effects on humans and animals if included in the diet. The biological activities of HSE, CFP and DFP need additional study if they are to be included in an edible oil product.

Because of the interest in alternative solvents and their effects on oil composition, Moreau, Lampi, and Hicks (2009) investigated effects of extraction and refinement methods on oils extracted from corn kernel, germ and fiber. Crude corn oil, originating from the germ, is typically refined, bleached and deodorized before marketing to the consumer (Moreau, Lampi, and Hicks, 2009). It was hypothesized that these refinement steps removed oil components that may have health benefits and changed the value of the final corn oil product. Corn germ oil and fiber oil was extracted by use of hexane, while corn kernel oil was extracted using 100% ethanol. Crude oil extracts were refined, bleached and deodorized (RBD). Corn fiber

Table 4.2 Individual tocopherols (T) and tocotrienols (T3) in oil extracted at 50 and 100°C from ground corn, corn bran, WM corn germ and DM corn germ (from Moreau and Hicks (2005)).

Sample, solvent	°C	% oil extracted	mg/kg					
			α-T	γ-T	δ-T	α-T3	γ-T3	δ-T3
Corn kernels								
Hexane	50	2.70 ± 0.06	386.7 ± 2.3	1066.7 ± 26.7	72.2 ± 3.37	138.0 ± 3.5	214.8 ± 2.8	0
	100	3.28 ± 0.04	425.6 ± 11.0	1034.7 ± 5.2	112.4 ± 0.8	173.6 ± 0.4	324.2 ± 0.5	49.5 ± 2.8
Isopropanol	50	3.36 ± 0.06	330.9 ± 26.7	911.5 ± 61.5	60.7 ± 05.4	133.0 ± 12.7	239.1 ± 19.4	0
	100	5.03 ± 0.24	228.7 ± 9.7	616.0 ± 7.8	74.8 ± 01.1	123.2 ± 0.2	277.4 ± 9.8	39.4 ± 5.5
Ethanol	50	3.38 ± 0.04	284.2 ± 30.2	865.8 ± 148.0	61.1 ± 10.3	132.2 ± 14.6	257.3 ± 38.9	0
	100	5.53 ± 0.20	201.2 ± 5.4	548.0 ± 5.7	101.8 ± 0.4	113.2 ± 0.5	256.8 ± 5.3	35.6 ± 3.2
Corn bran								
Hexane	50	1.92 ± 0.07	207.5 ± 9.2	473.4 ± 19.2	77.0 ± 4.2	62.9 ± 1.3	108.3 ± 8.8	0
	100	2.11 ± 0.00	174.9 ± 6.5	578.9 ± 15.4	113.0 ± 2.8	58.0 ± 2.8	168.5 ± 2.1	0
Isopropanol	50	2.46 ± 0.01	93.4 ± 27.4	426.8 ± 0.7	57.3 ± 2.1	0	110.0 ± 6.4	0
	100	6.41 ± 0.28	131.2 ± 3.7	385.7 ± 5.9	68.5 ± 1.4	47.6 ± 1.8	126.8 ± 1.1	0
Ethanol	50	3.47 ± 0.61	62.4 ± 12.2	335.3 ± 6.4	49.4 ± 2.7	0	90.3 ± 0.4	0
	100	5.08 ± 0.43	113.0 ± 3.5	330.4 ± 11.9	59.8 ± 1.1	40.4 ± 0.1	106.3 ± 3.2	0
WM germ								
Hexane	50	32.18 ± 0.22	181.3 ± 8.5	827.1 ± 39.2	91.7 ± 8.7	0	71.1 ± 37.8	0
	100	34.21 ± 0.59	166.0 ± 1.4	993.3 ± 12.7	133.9 ± 3.7	0	67.7 ± 0.5	0
Isopropanol	50	27.91 ± 0.29	191.9 ± 10.7	880.3 ± 62.6	94.4 ± 7.2	0	48.7 ± 2.6	0
	100	33.43 ± 0.02	216.4 ± 11.9	1116.9 ± 51.5	138.9 ± 2.0	0	72.4 ± 3.4	0
Ethanol	50	27.18 ± 0.71	157.9 ± 6.5	889.7 ± 3.3	90.4 ± 5.5	0	52.5 ± 4.9	0
	100	41.57 ± 0.03	194.2 ± 1.9	1045.7 ± 25.7	134.4 ± 4.1	0	75.7 ± 1.2	0
DM germ								
Hexane	50	14.31 ± 0.15	0	268.0 ± 3.5	57.8 ± 12.4	0	0	0
	100	17.42 ± 0.30	0	201.4 ± 3.0	90.6 ± 1.1	0	0	0
Isopropanol	50	16.26 ± 0.19	0	323.7 ± 10.4	68.7 ± 2.3	0	0	0
	100	22.27 ± 0.50	31.9 ± 0.8	348.5 ± 4.2	93.4 ± 5.5	0	38.0 ± 1.4	0
Ethanol	50	17.15 ± 0.59	0	291.2 ± 0.2	68.3 ± 2.8	0	0	0
	100	23.57 ± 1.10	31.1 ± 0.4	309.0 ± 3.5	84.3 ± 2.5	0	33.8 ± 3.9	0

Table 4.3 *Lipid compositions of corn oil obtained by extracting corn germ with hexane and by extracting ground whole corn with 100% ethanol at 50°C and 1000 psi. Moreau and Hicks (2005).*[a]

Lipid class	Corn oil (wt %)	
	From hexane-extracted wet-milled corn germ	From ethanol-extracted ground corn kernels
TAG	97.70 ± 0.78	95.78 ± 2.33
DAG	0.07 ± 0.00	0.05 ± 0.01
FFA	1.26 ± 0.03	0.68 ± 0.09
Free phytosterols (FS)	0.48 ± 0.01	0.76 ± 0.07
Phytosterol fatty acyl esters (SE)	0.47 ± 0.00	0.50 ± 0.04
Hydroxycinnamate sterol esters (HSE)	0.02 ± 0.00	0.23 ± 0.03
Diferuloylputrescine (DFP)	0	0.66 ± 0.05
p-Coumaroyl feruloylputrecine (CFP)	0	0.18 ± 0.01

[a]Both oil samples were unrefined.

originated from a commercial wet mill; corn germ oil was obtained commercially. Total phytosterols were 10 times higher in crude corn fiber oil than in commercial corn (germ) oil; total phytosterols in crude corn kernel oil were two times higher than in conventional corn (germ) oil. When subjected to conventional RBD, crude kernel oil had about half of total phytosterols removed. Crude kernel oil had higher total sterols than RBD kernel oil (2.4% vs. 1.1% oil, respectively). Crude corn fiber oil had higher total phytosterols than RBD fiber oil (8.7% vs. 7.9% oil, respectively) that used "gentle" refining processes. Commercial refined germ oil had relatively low total phytosterols (0.8% oil).

Moreau, Lampi, and Hicks (2009) confirmed that no polyamine conjugates (CFP and DFP) were detected in hexane-extracted products (commercial corn oil and corn fiber oil). However, high levels of polyamine conjugates were found in kernel oil extracted with ethanol. In the RBD-processed kernel oil, no polyamine conjugates were detected, demonstrating that the RBD processes removed the conjugates. This study reported that the safety of polyamine conjugates was unknown, but would need to be determined if present in an edible oil used for human consumption.

4.3 Corn Processing Effects on Phytochemical Recovery

At a commercial scale, the most concentrated forms of corn fiber originate from wet milling and dry milling processes. During wet milling, fiber is recovered after steepwater is drained and germ is recovered. The remaining solids are finely ground and passed over sieves to separate fiber (larger particles) from the protein and starch (smaller particles). As a result of these processing steps, corn fiber recovered from wet milling contains both pericarp and endosperm fiber constituents. Wet milling typically combines the fiber recovered with concentrated steepwater to form the corn gluten feed co-product. Because of the steeping process, fiber recovered from wet milling has lower starch and protein contents than fiber recovered from dry milling. During dry milling, the kernel is briefly soaked (tempered) with plain water or steam, passed through a degerminator which fractures the kernel and passed through an aspirator which separates pericarp from other kernel components. As a result, corn fiber (bran) from dry milling will contain primarily pericarp and relatively low concentrations of endosperm fiber.

4.3.1 Research with Corn Fiber Obtained from Wet Milling and Dry-Grind-Based Processes

The earliest research with corn fiber and corn fiber oil used fractions obtained from these two processes, wet milling and dry milling, since they were representative of corn fiber fractions likely to be available. As new processes were developed to recover fiber from corn, additional research was conducted to determine the concentrations and compositions of corn fiber oil obtained by these processes.

Moreau, Powell, and Hicks (1996) studied corn fiber obtained from several sources, including commercial wet mills and dry mills (Moreau, Powell, and Hicks, 1996). It was not known whether the different processing steps involved in the production of corn fiber fractions would result in products with oils of different yields and chemical compositions. Wet milled fiber fractions were obtained from several corn sources: common dent, yellow waxy, white waxy and high amylose, as well as from the corn gluten feed (fiber and steepwater) process stream. Extractable oil contents ranged from 1.29 to 3.68% (gluten feed and high-amylose fiber, respectively, Table 4.4). Ferulate ester contents of the extracted oils ranged from 1.08 to 6.75% (gluten feed and common fiber, respectively). Grinding the fiber fractions as obtained from commercial sources resulted in large increases in the amounts of extractable oil, an increase of sixfold, from 0.31% for unground material to 1.7% for material ground to a 20 mesh before extraction. Using supercritical fluid extraction resulted in about 50% more extractable oil than with hexane, using unground fiber. However, extractions of 20 and 80 mesh fiber fractions yielded slightly less oil than hexane extraction. Gluten feed had the lowest concentrations of extractable oil (1.29–2.37%) and low concentrations of ferulate esters in oil

Table 4.4 Extractable oil and ferulate esters in corn fiber and other materials obtained from various sources (adapted from Moreau et al (1996)).

Source	Sample[a]	Extractable Oil (wt %)	Ferulate Esters in Oil (wt %)
American Maize	yellow dent no. 2 fiber	1.72	6.75
	yellow waxy fiber	2.12	5.05
	white waxy fiber	2.82	4.54
	high amylose fiber	3.68	2.00
	gluten feed	1.29	3.64
	whole yellow waxy kernels	2.81	0.23
Cargill	yellow dent no. 2 fiber	2.26	2.70
	gluten feed	2.37	1.08
CPC	pericarp fiber	1.09	5.65
	endosperm fiber	0.54	3.37
	corn cleanings	1.24	1.62
	spent flake	1.75	0.25
	food grade fiber	1.59	1.83
Lauhoff	corn bran (pericarp)	1.32	1.50
Sigma Chemical	DDGS	2.53	0.79
purchased locally	corn cobs	0.20	0
	wheat bran	3.36	0.80
	oat bran	6.75	0.40

[a] Samples ground to 20 mesh, and each 4 g sample extracted with 40 mL of hexane for 1 h at 25 °C.

(1.08–3.64%), probably due to the addition (dilution) of concentrated steepwater to the gluten feed. Fiber fractions obtained from different wet milling process streams were obtained: pericarp, endosperm, spent germ flakes and a food-grade fiber product. Endosperm fiber contained relatively low oil levels and low levels of ferulate esters in the oil. Spent germ flake contained 1.75% extractable oil and only 0.25% ferulate esters. Purified fiber (food grade) contained 1.59% extractable oil and 1.83% ferulate esters. Bran obtained from a dry mill had low levels of extractable oil and ferulate esters. A DDGS sample contained 2.53% ext. oil and 0.79% ferulate esters. It was not clear if the differences in corn fiber sources were due to the milling processes used or due to hybrid differences, since hybrid type also was observed to have an effect on extractable oil and ferulate ester content. This study revealed that commercial corn fibers contain relatively low levels of oil (0.54–3.5 wt%) compared to fiber oil contents previously reported for rice bran (~18 wt%; Kahlon *et al.*, 1992), the oil obtained from corn fiber is richer in ferulate esters (up to 6.75 wt%) than rice bran oil (0.1 to 0.8 wt%) which was reported in an earlier study (Rogers *et al.*, 1993). However, due to higher concentrations of ferulate esters in corn fiber fractions (0.12%) compared to rice bran (estimated as 0.018–0.14%), corn fiber may contain more ferulate esters than rice bran.

A study conducted by Moreau *et al.* (1999c) processed three corn hybrids using laboratory conditions. Laboratory wet milling followed a 100 g scale procedure which recovers a pericarp (coarse) fiber fraction separately from an endosperm (fine) fiber fraction (Eckhoff *et al.*, 1996). The dry milling procedure used 2 kg batches and recovered germ by aspiration and sieving (Brekke *et al.*, 1972). The levels of ferulate-phytosterol esters (FPE), phytosterol fatty acyl esters (St:E) and free phytosterols (St) were measured in the milling fractions.

As had been previously reported (Moreau, Powell, and Hicks, 1996), FPE levels were higher in wet milling fiber fractions than in dry milling fiber fractions. FPE levels in the fiber fractions recovered from laboratory milling procedures were similar to those previously obtained from commercial corn fiber products (Moreau, Powell, and Hicks, 1996). Pericarp (coarse) fiber fractions contained 59–74%, endosperm (fine) fiber fractions contained 35–41% of FPE present in the kernels. From dry milling, the pericarp (bran) fraction contained only 10–16% of FPE present and about 75% of FPE was recovered in the grit products. In dry milling fractions, the FPE level in oil from the grits was higher than the FPE levels in oil obtained from the bran. During wet milling, 94% of the FPE in original whole kernels was recovered in the fiber fractions; in dry milling, 17% was recovered in the pericarp fraction and the rest recovered in the grits (75%). The other phytosterols, St:E and St, were distributed over several milling fractions. In wet milling, the highest levels of St:E and St were observed in the germ, but appreciable quantities were also observed in gluten, pericarp and endosperm fiber fractions. In dry milling, St:E and St were evenly distributed between grits and germ fractions. Dry milled and wet milled germ fractions had the highest levels of γ-tocopherol. Commercial wet milling fiber appeared to be the best industrial source material for FPE. Wet milling fiber contained the highest levels of natural sitostanol of any other plant material yet reported. Since Seitz (1989) found that these ester compounds were associated with the inner pericarp of the corn kernel, Moreau and co-workers theorized that the inner pericarp adheres to pericarp during wet milling, but not during dry milling.

A modified dry grind process was developed in the 1990s that was designed to improve dry grind processing efficiency and increase the value of co-products recovered during ethanol recovery. Details of the modified dry grind process (Singh and Eckhoff, 1996, 1997; Wahjudi *et al.*, 2000) were summarized in an earlier section (Figure 4.7). Because the modified process recovered pericarp fiber as a separate process stream, it was desired to measure the fiber stream's phytochemical content (Singh *et al.*, 1999). Two yellow dent hybrids (3.4 and 6.8% total oil content) were processed using the modified dry grind and wet milling processes and fiber fractions recovered. The modified dry grind process followed soaking parameters (12 h, 59 °C) in water based on earlier work (Singh and Eckhoff, 1996). Treatments included soaking in water and with steeping chemicals (12 h, 59 °C, 0.2% SO_2, 0.55% lactic acid) added. Germ and fiber (quick fiber) were recovered. For wet milling treatments, corn was steeped with no steeping chemicals (24 h, 52 °C, 0.2% SO_2, 0.55% lactic acid) and processed using a laboratory method (Eckhoff *et al.*, 1996) to recover germ and pericarp (coarse) and endosperm (fine) fiber fractions. It was hypothesized that quick fiber would differ from wet milled pericarp fiber because wet milled fiber was exposed to steeping chemicals for extended periods, while quick fiber has not.

The modified dry grind process had quick fiber yields of 6.2 to 7.0% of corn, or 46 to 60% of total wet milling fiber yields (pericarp and endosperm fiber yields). Adding steep chemicals (SO_2 and lactic acid) during the soak period increased quick fiber yields (11.3–12.1%), percent FPE recovered (77.7–92.3%) and total phytosterol recoveries. With addition of steep chemicals, FPE and phytosterol recoveries were similar to or higher than those from total fiber from wet milling. For wet milling without steeping chemicals, recoveries of FPE from pericarp fiber decreased by half, while the endosperm fiber FPE recovery increased more than twofold. Yields of pericarp and endosperm fiber fractions also shifted rather dramatically. Pericarp fiber yields decreased while endosperm fiber yields increased.

In modified dry grind and wet milling processes, corn is coarsely ground using disk tooth mills following soaking or steeping steps, respectively. Singh and co-workers (1999) speculated that without the aid of steeping chemicals, fiber recovered as part of the endosperm fiber fraction increased, apparently due to changes in locations where the kernel fractured during grinding. They cited previous research that had determined steeping with SO_2 and lactic acid increased water uptake (Cox, Macmasters, and Hilbert, 1944; Roushdi and Fahmy, 1981; Ruan, Litchfield, and Eckhoff, 1992) and softened the kernel (Cox, Macmasters, and Hilbert, 1944; Roushdi and Fahmy, 1981; Shandera, Parkhurst, and Jackson, 1995). The wet milling process separated pericarp from endosperm along the tube and cross cells, a narrow region between pericarp and endosperm tissues (Wolf *et al.*, 1953). Singh and co-workers (1999) attributed the change in fiber fraction yields and compositions to a change in the line of fracture between pericarp and endosperm, due to the presence or absence of steeping chemicals. Because Seitz (1989) determined FPE was concentrated in the inner pericarp, FPE concentrations of recovered fiber fractions would vary depending on which fraction of the inner pericarp was recovered.

To help verify the theory put forth in Singh *et al.* (1999), experiments were designed to observe the line of fracture between pericarp and endosperm regions during various milling techniques (Singh, Moreau, and Cooke, 2001a). Wolf *et al.* (1952) had previously observed during dissection and microscopy work that the line of fracture between the pericarp and the endosperm was at the cross and tube cells, if there is absorption of water by the corn kernel. This suggested that the line of fracture should be located at the cross and tube cell region if water absorption had previously occurred (Wolf *et al.*, 1952). Additional work theorized that FPE in corn kernels were probably located in the inner pericarp fraction, probably near the cross and tube cells (Seitz, 1989). Based only on the observation of Seitz (1989), Singh,

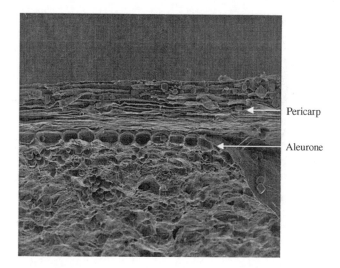

Figure 4.9 *Wet milled pericarp (coarse) fiber cross section. Sample obtained using steeping chemicals according to method of Eckhoff et al. (1996). Aleurone is shown attached to pericarp (left portion) and separating from pericarp on right side (250 ×) (Singh, Moreau, and Cooke, 2001a).*

Moreau, and Cooke (2001a) reasoned that FPE recovered from pericarp with and without chemicals during steeping would be about equal. However, this was in contrast to results obtained in work by Singh *et al.* (1999). To clarify where the lines of fracture occurred and the associated changes in phytochemical recovery during processing, modified dry grind (quick fiber) and wet milling procedures were carried out under laboratory conditions (Singh, Moreau, and Cooke, 2001a). Coarse and fine fiber fractions were obtained by wet milling with and without SO_2 and lactic acid during steeping.

Singh, Moreau, and Cooke (2001a) found that wet milling, with and without chemicals, changed the line of fracture between pericarp and endosperm, as well as the amount of aleurone recovered as coarse (pericarp) and fine (endosperm) fiber (Figure 4.9). Microscopic examination showed that pericarp fiber fractions steeped with conventional steeping chemicals (24 h, 52 °C, 0.2% SO_2, 0.55% lactic acid) contained pericarp and aleurone layers (Figure 4.10). Fiber fractions obtained by steeping without chemicals appeared to contain only the pericarp component and no evidence of the aleurone layer. Microscopic examination of the endosperm (fine) fiber fraction obtained without use of steeping chemicals showed greater presence of aleurone cells than in endosperm fiber fractions obtained from use of steeping chemicals.

Using hand dissection experiments with steeped corn, it was found that a single aleurone layer was similar to the innermost layer of pericarp fiber obtained with use of steeping chemicals (Singh, Moreau, and Cooke, 2001a). This work confirmed that the line of fracture changed as a result of chemical addition (SO_2 and lactic acid) during the steeping period. These observations support observations reported by Wolf *et al.* (1952) that fracture would occur at the cross and tube cell region with a simple water soak, and by other work that reported higher FPE and other phytosterol concentrations in fiber fractions obtained by wet milling and dry milling methods (Moreau, Powell, and Hicks, 1996; Moreau *et al.*, 1999c; Singh *et al.*, 1999). Since commercial dry milling (Figure 4.2) uses only water during tempering, the line of fracture would be at the

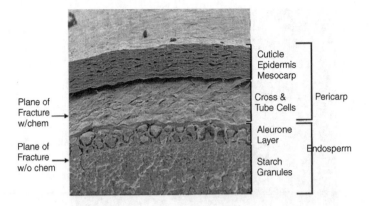

Figure 4.10   *Cross section of steeped kernel (Singh, Moreau, and Cooke, 2001a).*

cross and tube cells, resulting in the observed decrease in phytosterol concentrations in the pericarp fiber fractions, while the aleurone layer remained attached to the endosperm (grit) fractions (Moreau, Powell, and Hicks, 1996; Moreau *et al.*, 1999c).

To determine whether phytosterols were located in the aleurone region, phytosterol levels were measured. For fiber fractions steeped with chemicals, 63% of total FPE recovered in the fiber fraction was recovered in the pericarp (coarse) fraction and 37% recovered in the endosperm (fine) fiber fraction. For fractions steeped without chemicals, 21% was recovered from the pericarp fiber fraction and 79% recovered from the endosperm fiber fraction. These results were consistent with previous work (Singh *et al.*, 1999). Similar shifts were observed in free phytosterols (St) and phytosterol fatty acyl esters (St:E). Total phytosterol yields from the recovered fiber showed 60% from pericarp fiber and 40% endosperm fiber when the corn was first steeped with chemicals. The distribution shifted to 22 and 79% for pericarp and endosperm fiber fractions, respectively, when steeping chemicals were not added. Hand dissection studies found that aleurone contained 30 times more oil and eight times more phytosterols than pericarp.

To determine the effects of wet milling and other milling processes on germ and fiber yields and compositions of oil recovered in the germ and fiber fractions, a laboratory milling study was conducted using a single yellow dent corn hybrid processed by five milling techniques (Singh *et al.*, 2001b). The control milling treatment was a conventional wet milling procedure that used a 100 g sample (Eckhoff *et al.*, 1996). Gaseous SO_2 wet milling followed the same 100 g wet milling procedure as Eckhoff *et al.* (1996), but used a gaseous SO_2 treatment followed by a 14-h steep period (Eckhoff and Tso, 1991). The intermittent milling and dynamic steeping (IMDS) process used 1 kg samples (Lopes-Filho *et al.*, 1997). An alkali wet milling process used 100 g samples (Eckhoff *et al.*, 1999) through the cracking and steeping steps and then was modified to follow the conventional wet milling procedure of Eckhoff *et al.* (1996) using steepwater as the process water. For modified dry grind processing, soaking parameters followed the method of Singh and Eckhoff (1996) but included chemicals added during the soaking step (0.2% SO_2, 0.55% lactic acid) based on improved phytosterol yields in recovered quick fiber previously observed (Singh *et al.*, 1999). Germ and fiber fractions were analyzed for oil and phytosterol composition (FPE, St and St:E contents).

Table 4.5 Yields of germ and germ oil and recoveries of phytosterol compounds in corn germ oil (Singh et al., 2001b).[*]

Process	Germ Yield	Germ Oil Yield	FPE	St	St:E	Total Phytosterol
Conventional wet milling	5.01d[a]	44.94[a]	0.17[a]	12.21[a]	9.85[a]	22.22
Gaseous SO₂	5.36[cd]	43.04[a]	0.16[a]	13.79[ab]	12.88[a]	26.83
Alkali wet milling	3.22[e]	36.49[a]	0.08[d]	7.27[a]	4.99[d]	12.35
IMDS[a]	5.75[a]	35.12[a]	0.47[a]	11.82[a]	9.65[a]	21.95
Modified dry grind w/chem	6.72[a]	41.77[a]	0.32[a]	17.29[a]	15.08[a]	32.70

[*] Germ yield and germ oil yield are on wt% basis. Ferulate phytosterol esters (FPE), free phytosterols (St), and fatty acyl phytosterol esters (St:E) have units of (mg/100 g of corn).
[a] All yields are averages of two values.
[b] Values followed by the same letter in the same column are not significantly different ($P < 0.05$).
[c] Intermittent milling and dynamic steeping.

Singh *et al.* (2001b) reported germ yields for the five processes ranged from 3.2% (alkali) to 6.7% (modified dry grind), with total oil contents ranging from 35.1 to 44.9 wt% for IMDS and conventional wet milling processes, respectively (Table 4.5). In determining process economics, total germ oil yield is an important consideration; conventional and gaseous SO₂ wet milling and modified dry grind processes had the highest total oil recoveries, ranging from 41.8 to 44.9 wt%. In the germ fractions recovered by the various processes, FPE recoveries in germ oil were highest for the IMDS and quick fiber processes (0.32 to 0.47 mg/100 g corn) and lowest for the alkali process (0.8 mg/100 g corn). St and St:E recoveries in germ oil were highest for the modified dry grind process (17.3 and 15.1 mg/100 g corn, respectively). Total phytosterol recoveries ranged from 12.3 (alkali) to 32.7 (modified dry grind) mg/100 g corn (Table 4.6).

Fiber yields ranged from 11.4% (alkali) to 13.4% (conventional wet milling) and were found to not differ amongst milling processes; total fiber oil yields ranged 1.5% (modified dry grind) to 7.5% (alkali) (Table 4.7). It was observed that the alkali fiber fraction contained a large amount of unrecovered germ, which affected total oil yield from the alkali fiber fraction. Although the fiber fraction from modified dry grind had one of the lowest oil yields, it had one of the highest FPE recoveries (8.6 mg/100g corn). FPE recoveries in fiber fractions were rather bimodal: the alkali process had FPE recovery of 3.5 mg/100 g corn while the other processes

Table 4.6 Compositions of germ oil recovered by various milling processes (adapted from Singh et al., 2001b).[*] Reprinted with kind permission form Dien et al., 2004 © Springer Science + Business Media (2004).

Process	%wt of Germ Oil[a]			
	FPE	St	St:E	Total
Conventional wetmilling	0.01	0.54	0.44	0.99
Gaseous SO₂ wet milling	0.01	0.60	0.56	1.16
Alkali wetmilling	0.01	0.62	0.43	1.05
IMDS[b]	0.02	0.59	0.48	1.09
Modified dry grind w/chem	0.01	0.62	0.54	1.17

[*] Ferulate phytosterol esters (FPE), free phytosterols (St), fatty acyl phytosterol esters (St:E) and total phytosterol compounds.
[a] Values are means of two observations; individual standard deviations were ≤9% of the means (Singh et al., 2001b).
[b] Intermittent milling and dynamic steeping.

Table 4.7   *Yields and compositions of fiber and fiber oil recovered from various processes (reproduced from Singh et al. © (2001b), AACC International Press).* Reprinted with kind permission form Dien et al., 2004 © Springer Science + Business Media (2004).*

Process	Fiber Yield	Fiber Oil Yield	FPE	St	St:E	Total Phytosterol
Conventional wet milling	13.36[ab]	2.09[b]	7.75[ab]	3.19[c]	10.55[c]	21.48
Gaseous SO_2	13.19[a]	1.98[bc]	7.39[bc]	3.21[c]	10.91[c]	21.52
Alkali wet milling	11.44[a]	7.51[a]	3.45[d]	7.65[a]	15.40[a]	26.49
IMDS[c]	11.85[a]	1.82[bc]	8.03[ab]	3.70[b]	7.14[d]	18.87
Modified dry grind w/chem	11.90[a]	1.50[cd]	8.60[a]	2.88[d]	12.81[b]	24.30

* Fiber yield and fiber oil yield are in wt% basis. Ferulate phytosterol esters (FPE), free phytosterols (St), and fatty acyl phytosterol esters (St:E) have units of (mg/100 g of corn).
[a] All yields are means of two values.
[b] Values followed by the same letter in the same column are not significantly different ($P < 0.05$).
[c] Intermittent milling and dynamic steeping.

had recoveries of 7.4 to 8.6 mg/100 g corn (Table 4.7). Total phytosterol recoveries in fiber fractions ranged from 18.9 (IMDS) to 26.5 (alkali) mg/100 g corn (Table 4.8, original reference). Although the germ recovered using the alkali process had the lowest total phytosterol recovery, it had the highest recovery of total phytosterols in the fiber fraction. Of total phytosterols recovered from fiber in conventional wet milling, 49 and 36% were from the St:E and FPE fractions, respectively. The total percentage weight of phytosterols in fiber oil for conventional wet milling was 16%; for alternative milling processes, the total percentage phytosterols in fiber oil ranged from 8.8% (IMDS) to 20.1% (alkali). In germ and fiber fractions, the alkali process recovered higher levels of free phytosterols (St) than the other processes; this was attributed to alkaline hydrolysis of phytosterol esters to form free phytosterols, fatty acids and ferulic acid. FPE recoveries in fiber fractions (3.5 to 8.6 mg/100 g corn) were an order of magnitude higher than FPE recoveries in germ fractions (0.08 to 0.32 mg/100 g corn). Gaseous SO_2, IMDS and quick fiber processes had FPE recoveries similar to the conventional wet milling process.

Alternative milling processes had significant effects on yield and composition of germ and fiber oil compared to a conventional wet milling process (Singh *et al.*, 2001b). Alkali wet milling negatively affected phytosterol recovery from germ. The modified dry grind processes,

Table 4.8   *Compositions of fiber oil recovered by various processes (reproduced from Singh et al. © (2001b), AACC International Press).*

Process	%wt of Oil[a]			
	FPE	St	St:E	Total
Conventional wet milling	5.70	2.29	7.78	15.78
Gaseous SO_2	5.89	2.47	8.77	17.13
Alkali wet milling	2.21	7.44	10.48	20.12
IMDS[b]	3.74	1.72	3.32	8.77
Modified dry grind w/chem	5.00	1.68	7.44	14.11

* Ferulate phytosterol esters (FPE), free phytosterols (St), fatty acyl phytosterol esters (St:E) and total phytosterol compounds.
[a] Values are means of two observations; individual standard deviations were ≤12% of the means (Singh et al., 2001b).
[b] Intermittent milling and dynamic steeping.

quick germ and quick fiber, were either comparable or significantly higher in phytosterol recovery than conventional wet milling. The study suggested significant levels of phytosterols are present in crude germ oil. Current oil refining processes significantly reduce the levels of these compounds. Refining techniques could be modified to include phytosterols, resulting in germ oil that could be marketed as a healthy alternative to conventional corn oil.

Effects due to hybrid variability have been found to affect yields of primary products and co-products for wet milling, dry milling and dry grind processes. Therefore, it would logically follow that the hybrid would also affect recovery of corn fiber, corn fiber oil and fiber oil composition. To observe hybrid effects, Singh and co-workers (2000) measured variability in corn fiber yields, extractable fiber oil and levels of total phytosterol components in fiber oil. The laboratory wet milling procedure of Eckhoff *et al.* (1996) was used to obtain pericarp and endosperm fiber fractions. One replicate milling run was performed on each commercially available hybrid. In one experiment, 16 hybrids were grown in a single location. Variation was observed amongst hybrids for fiber recovered from corn (13.2 to 16.6% db) and fiber oil recovered from the fiber fraction (0.9 to 2.4% db). No significant correlations between fiber and fiber oil yields were observed. Amongst hybrids, variations in individual phytosterol components were detected (2.9–9.2, 1.9–4.3, 6.5–9.5 12.0–21.6% db of fiber oil, for FPE, St, St:E and total yield of phytosterols from fiber oil, respectively; Table 4.9).

In a second experiment, two hybrids were grown in thirteen locations. Amongst growing locations, effects were significant. The same hybrids grown at varying locations showed a range of 4.0–17.5, 4.9–12.2, 1.95–4.45% db of fiber oil for FBE, St:E and St, respectively. Ranking of hybrids with respect to their total phytosterols contents was consistent for almost all growing locations.

Wang *et al.* (1998) investigated effects of steeping conditions and steep chemical addition on recovery of tocopherols and tocotrienols. They found that steeping conditions had little effect on concentrations of α-tocopherol and α-tocotrienol, but at higher SO_2 concentrations, a shorter steep time gave slightly higher γ-tocotrienol and lower γ-tocopherol contents (Wang *et al.*, 1998). This work did not fractionate the kernel following steeping, but simply ground the whole kernel prior to performing extractions.

Because Wang *et al.* (1998) found effects of steeping conditions on tocopherol and tocotrienols recovery, Singh *et al.* (2000b) determined the effects of steep acid and sulfite compounds using during the wet milling steeping step on phytosterol compounds. Several steeping treatments were investigated and measured the effects on corn fiber yield, fiber oil yield and concentrations of FPE, St and St:E in fiber oil. To observe the effects of acid addition, each of five acids (organic and inorganic) were added at equivalent rates in place of the 0.55% lactic acid typically used plus conventional 0.2% SO_2. To observe the effects of sulfite compounds, potassium sulfite, sodium metabisulfite and ammonium bisulfite were added at equivalent 0.2% SO_2 rates with 0.55% lactic acid also added. There were no effects detected

Table 4.9 *Range of variations in phytosterol compound concentrations observed due to hybrid and growing location effects (Singh et al., 2000).*

Effect	Range of variation observed (% db phytosterol compound in fiber oil)			
	FPE	St:E	St	Total Phytosterols
Hybrid	2.9–9.2	6.5–9.5	1.9–4.3	12.0–21.6
Location	4.0–17.5	4.9–12.2	1.95–4.45	n/r

due to steep acid or sulfite compound on fiber yields. Effects were observed on fiber oil yield and composition of fiber oil. For one hybrid, ammonium sulfite treatment had higher FPE and St:E yields; for the second hybrid, it increased only the FPE yield. This indicated sulfite treatment effects were probably hybrid dependent. The maximum increase in phytochemicals (FPE, St, St:E) measured was an 8.4% increase in St:E using citric acid. Ammonium sulfite, acetic acid, citric acid and hydrochloric acid gave the largest overall positive effects of phytosterol increases in corn fiber oil.

Pretreatment methods were used to determine if fiber oil concentrations could be increased for more economical extraction (Singh *et al.*, 2003). Corn fiber fractions were pretreated with dilute sulfuric acid, enzymes or a combination of acid and enzyme to remove non-lipid components from the fiber fraction. Several enzyme combinations were used without acid addition, or sulfuric acid followed by cellulase pretreatment. These pretreatment methods hydrolyzed the starch and cell wall polysaccharides of the fiber fraction. Oil concentrations in the residual fiber solids increased from initial concentrations of 1.4% in untreated fiber to 12.2 and 7.9% for fiber pretreated with acid or acid plus enzyme, respectively. Total phytosterol concentrations increased from 177 mg/100 g fiber to 1433 and 1027 mg/100 g fiber for acid and acid plus enzyme pretreatments, respectively. Based on this study, it appears that phytosterols should be extracted following fiber recovery and pretreatment, but before SSF process steps.

Oil concentrations in fiber extracted by wet milling are rather low, 1.5 to 3.0%. Previous work (Singh, Moreau, and Cooke, 2001a) found that more than 90% of corn fiber oil is contained in the aleurone layer, which is a single layer of cells located beneath the pericarp tissue (Watson, 1984). Oil in the aleurone layer is contained in individual oil body cells. During size reduction, oil-containing cells may be released into an aqueous media, allowing them to float (Singh and Moreau, 2003). To determine whether particle size reduction and floatation could increase fiber oil concentrations, fiber fractions were ground to reduce particle size and subject to floatation and skimming. The floating fraction contained higher concentrations of total oil and total phytosterols, but significant amounts of phytosterols were lost to the higher density fraction that did not float. It appeared that a portion of the phytosterols remained attached to higher density material. The authors noted that reducing particle size further may be needed to affect higher phytosterol recovery, but this may prove impractical in a commercial setting.

Use of modified dry grind processes allows recovery of kernel fiber to be used for phytochemical production and also affords the opportunity for the processor to produce cellulosic ethanol in addition to ethanol resulting from starch conversion. Understanding how phytochemicals are affected during processing of fiber into ethanol is important, as it is possible that both phytochemical recovery and cellulosic ethanol production will be carried out in tandem to improve process economics. Previous work (Singh *et al.*, 2003) had not conducted simultaneous saccharification and fermentation on pretreated fiber fractions or determined phytosterol concentrations in the residual fermented materials. Dien *et al.* (2004, 2005) reported work that recovered fiber during modified dry grind processing and converted the cellulose in the fiber fraction to ethanol. Fiber compositions prior to and following fermentation were measured. The fiber recovered contained 15% w/w starch and 17% cellulose, approximately the same starch and fiber contents obtained by the more involved steeping process of wet milling (Table 4.10).

Fiber pretreatment used a dilute acid method with H_2SO_4 (0–12 g H_2SO_4/100 g db biomass or 0.7 to 4.8% based on g H_2SO_4/g biomass). Simultaneous saccharification and fermentation

Table 4.10 *Compositions of corn fiber fractions adapted from other sources (Dien et al., 2004; Dien et al., 2005).*

Component	Corn Fiber from Wet Milling (%w/w db)	Quick Fiber (%w/w db)
Starch	11–23	15
Cellulose	12–18	17
Xylan	18–28	22
Arabinan	11–19	11
Total Carbohydrate*	70	65
Protein	11–12	11
Oil	2	1

*Sum of the midpoints of ranges for fiber from wet milling.

of fiber included addition of enzymes (cellulose, β-glucosidase and glucoamylase) and yeast (*S. cerevisiae*). Following pretreatment and fermentation, 1.12% oils present in residual solids (post-SSF), comparable to untreated QF (1.24–3.49% oil). Low oil concentrations may be due to dilution from addition of materials during the SSF step (Dien *et al.*, 2004; Dien *et al.*, 2005). The fiber fraction prior to neutralizing of pretreatment acids and SSF contained 7× more oil and 4.8× more total phytosterols than solids recovered post SSF (Table 4.11). They concluded that pretreated fiber is a valuable source of phytosterols, provided they are recovered prior to SSF.

As described earlier, distillers dried grains with solubles (DDGS) is a co-product from the dry grind process consisting of unfermented materials from the production of ethanol. DDGS has value relatively similar to that of corn, due to its suitability mainly in ruminant animal diets (Rausch and Belyea, 2006). Removal of fiber from the DDGS could increase the percentage protein, thus increasing the value of the higher protein portion. Furthermore, isolation of the DDGS fiber would create a potential source of phytochemicals. A method was developed that combined size separation (sieving) followed by air classification (elutriation) process steps to increase protein concentrations in one fraction while increasing fiber fractions in another (Srinivasan *et al.*, 2005; Srinivasan *et al.*, 2006). This method resulted in one fraction (elusieve fiber) that was higher in neutral detergent fiber (NDF) contents and another fraction (enhanced DDGS) that had increased protein and fat contents. Based on this work, additional research investigated the distribution of phytochemicals in separated DDGS fractions, sieving and elutriation were conducted on commercial samples of DDGS (Srinivasan *et al.*, 2007). There were no differences in phytosterol distributions amongst sieve size categories, but phytosterol contents of oil from the elusieve fiber fraction were higher than or the same as the enhanced DDGS. Phytosterol contents of elusieve fiber (112–142 mg/100 g) was higher than wet milling

Table 4.11 *Oil concentrations and compositions of corn fiber fractions from a fermentation stream.**

Fiber Source	Total Oil (% w/w)	Free Sterol	FPE**	St:E**	Total Sterols
Pre SSF[+]	8.15 ± 0.21	4.43 ± 0.19	3.27 ± 0.04	7.9 ± 0.1	15.6
Post-SSF[++]	1.12 ± 0.14	6.03 ± 3.74	5.82 ± 3.66	11.8 ± 5.7	23.6
Post FBR5	8.28 ± 0.14	5.8 ± 0.79	4.29 ± 0.69	12.2 ± 1.8	22.3

*Data taken directly from (Dien *et al.*, 2005).
**FPE – ferulate phystosterol esters; St:E – phytosterol fatty acyl esters.
[+] Pretreated with dilute acid.
[++] Residual from SSF.

endosperm fiber (71 mg/100 g) obtained in an earlier study (Singh *et al.*, 1999). Phytosterol content of elusieve fiber (112–142 mg/100 g) was higher than that of wet-milling endosperm fiber (71 mg/100 g; Singh *et al.*, 1999). Based on phytosterol concentrations observed in previous work, Srinivasan *et al.* (2007) concluded that enhanced DDGS (226–232 mg/100 g) would be a more concentrated feedstock than dry milling fiber (152 mg/100 g; Moreau *et al.*, 1999c), wet milling endosperm fiber (71 mg/100 g; Singh *et al.*, 1999), quick fiber obtained from steeping without chemicals (215 mg/100 g; Singh *et al.*, 1999) and DDGS (216 mg/100 g; Srinivasan *et al.*, 2007). Because sieving and elutriation process steps would be a straightforward addition to convention dry grind processing, this process modification could be used to fractionate all or a portion of the DDGS co-product at a dry grind plant.

4.3.2 Research on Phytochemicals and the Nixtamalization Process

As described in an earlier section, nixtamalization involves the thermal-alkaline treatment of corn, resulting in changes in composition, structure and nutritional value of its products. The nixtamalization process partially removes the pericarp due to the alkali treatment, nixtamal washing and handling (Serna-Saldivar *et al.*, 1993). De la Parra, Saldivar, and Liu (2007) state that the phytochemicals in whole grain are mainly in bound form based on several studies that found that phytochemical compounds appeared to form bonds to pericarp tissues (Adom and Liu, 2002; Adom, Sorrells, and Liu, 2003; Adom, Sorrells, and Liu, 2005). In work by Adom and Liu (2002), corn was found to have the highest total phenolic content of the cereal grains tested (wheat, oats and rice); 85% of phenolics were in bound form. In work with other cereal grains and with sweet corn, ferulic acid was found in bound form, but these works focused more on other cereal grains and sweet corn (Adom and Liu, 2002; Adom, Sorrells, and Liu, 2003; Adom, Sorrells, and Liu, 2005; Dewanto, Wu, and Liu, 2002; Hernanz *et al.*, 2001) rather than commercial corn harvested at maturity. It is thought that bound phytochemicals may survive stomach and intestinal digestion, leaving them intact to function in the colon where they have been positively linked with human health (Adom and Liu, 2002).

Nejayote, the cooking liquor that is drained off during nixtamalization, has been found to include high levels of phytochemicals (Velasco-Martinez *et al.*, 1997). Typically, nejayote is not recycled or only partially recycled in the process, thus resulting in the loss of the phytochemical compounds to waste treatment. If bound forms of phytochemicals are attached to the pericarp, it is apparent that nejayote has phytochemical content due to the pericarp and perhaps the aleurone layer. However, lime cooking increases calcium content, increases niacinbioavailability, removes most of the pericarp and reduces mycotoxins in raw kernels (Serna-Saldivar, Gomez, and Rooney, 1990). In Mexico, most dietary calcium originates from tortillas and related products (Gonzalez *et al.*, 2005; Serna-Saldivar, Gomez, and Rooney, 1990). Prior to the De la Parra, Saldivar, and Liu (2007) study, relatively little was published regarding phytochemicals and antioxidant activity of nixtamalized corn products (Kern *et al.*, 2003; Pflugfelder, Rooney, and Waniska, 1988; Scott and Eldridge, 2005; Serna-Saldivar, Gomez, and Rooney, 1990; Sosulski, Krygier, and Hogge, 1982). Since corn has high concentrations of these compounds in bound form, nixtamalized corn products may prove to be a good source of these phytonutrients, if a method can be developed for recovering them from the nejayote.

De la Parra, Saldivar, and Liu (2007) measured total phenolics, anthocyanins, ferulic acid, carotenoids and antioxidant activities in five types of corn (white, yellow, high carotenoid, blue, red). The nixtamalization process (cooking step) was adjusted for optimum conditions for each corn type (De la Parra, Saldivar, and Liu, 2007). Cook times were optimized for each corn type, determined by a 100 g nixtamalization procedure which varied cooking times from 0 to 45 min. Procedures were developed from previous work (Gonzalez *et al.*, 2005; Serna-Saldivar *et al.*, 1993). Once cook time was decided for each corn type, additional nixtamalization experiments used a 3 kg sample size. The cooked and steeped samples were processed into masa, tortillas and tortilla chips.

Nixtamalization reduced total phenolics and antioxidant activities when compared to raw grains, while increasing concentrations of free phenolics and soluble conjugated ferulic acid, and reducing concentrations of bound phenolics and ferulic acid. Results from De la Parra, Saldivar, and Liu (2007) confirmed previous work that found >80% of total phenolic compounds were in bound form and appeared attached to cell wall structures (Adom and Liu, 2002; Adom, Sorrells, and Liu, 2003; Adom, Sorrells, and Liu, 2005; Hernanz *et al.*, 2001; Smith and Hartley, 1983; Sosulski, Krygier, and Hogge, 1982). Lime cooking resulted in phenolics being recovered in the steep liquor or nejayote (Table 4.12, data from De la Parra, Saldivar, and Liu (2007)). Most of the ferulic acid was associated with corn bran or pericarp. Gonzalez *et al.* (2004) determined alkaline treatment partially hydrolyzed hemicellulose fractions rich in phenolics (Gonzalez *et al.*, 2004). Ferulic acid had been previously found to be associated with the cell wall, linked by ester bonds. Ferulic acid has been found to be bound to cell wall material, such as aleurone layer and pericarp tissue; phenolic acids were esterified to cell wall components (Cortes *et al.*, 2006; Lozovaya *et al.*, 2000). De la Parra, Saldivar, and Liu (2007) found that 97% of ferulic acid was bound to cell wall material, which agreed with previous findings (Table 4.13 data from De la Parra, Saldivar, and Liu, 2007)

Table 4.12   *Total phenolic content of raw corn and their nixtamalized corn products (masa, tortilla, and chips) (mean ± sd, n = 3) (De la Parra, Saldivar, and Liu, 2007).*

type of corn	sample/product	total phenolics[a] (mg of gallic acid equiv/100 g of dry wt)		
		free	bound	total
white	corn	34.7 ± 0.4 b	226.0 ± 6.3 b	260.7 ± 6.1 c
	masa	39.1 ± 09 ab	126.1 ± 3.1 d	165.2 ± 4.0 f
	tortilla	47.2 ± 1.8 a	119.0 ± 6.2 d	166.2 ± 6.2 f
	chips	46.3 ± 2.7 a	97.3 ± 3.3 e	143.6 ± 2.1 g
yellow	corn	43.6 ± 1 8 ab	242.2 ± 13.1 b	285.8 ± 14.0 b
	masa	41.5 ± 0.6 ab	140.7 ± 2.6 d	182.2 ± 2.4 f
	tortilla	51.1 ± 1.3 a	132.2 ± 1.6 d	183.3 ± 1.4 f
	chips	43.1 ± 09 ab	102.1 ± 2.0 e	145.2 ± 1 2 g
red	corn	38.2 ± 0.4 ab	205.6 ± 4.5 c	243.8 ± 4.6 d
	masa	28.0 ± 0.8 c	97.2 ± 3 6 e	125.3 ± 2.8 h
	tortilla	30.5 ± 0 7 bc	106.0 ± 3.6 e	136.5 ± 2 9 g
	chips	26.4 ± 0.8 c	85.3 ± 1.8 f	111.7 ± 1.9 i
blue	corn	45.5 ± 0.5 a	220.7 ± 0.5 b	266.2 ± 0 7 c
	masa	30.3 ± 1.1 bc	128.2 ± 1.7 d	158.5 ± 1.2 g
	tortilla	39.1 ± 1.5 ab	122.7 ± 0.6 de	161.8 ± 2.1 g
	chips	41.4 ± 1.6 ab	95.5 ± 1.1 e	136.9 ± 1.2 g
high carotenoid	corn	50.0 ± 2.5 a	270.1 ± 9.4 a	320.1 ± 7.6 a
	masa	40.3 ± 1.0 ab	158.0 ± 0.8 d	198.3 ± 1.4 e
	tortilla	53.0 ± 1.3 a	154.5 ± 0.6 d	207.5 ± 1.3 e
	chips	46.4 ± 0.7 a	108.6 ± 2.2 e	155.0 ± 1.9 g

[a]Values with no letters in common are significantly different (p <0.05).

Table 4.13 Ferulic acid content and percentage contribution of raw corns and their nixtamalized corn products (masa, tortilla, and chips) (mean ± sd, n = 3) (De la Parra, Saldivar, and Liu, 2007).

type of corn	sample/product	ferulic acid content[a] (µg of ferulic acid /100 g of dry wt of sample)			total
		free	soluble conjugated	bound	
white	corn	495 ± 18 f (0.41)[b]	756 ± 73 d (0.63)	119201 ± 11173 d (98.96)	120453 b
	masa	7558 ± 757 e (17.90)	14201 ± 1305 c (33.63)	20470 ± 1617 d (48.47)	42229 d
	totila	9988 ± 495 d (11.73)	23267 ± 1832 b (27.32)	51905 ± 3582 c (60.95)	85160 c
	chips	6643 ± 832 e (10.94)	15989 ± 100 c (26.34)	38070 ± 4772 cb (62.72)	60702 d
yellow	corn	645 ± 14 f (0.63)	1474 ± 102 d (1.43)	100849 ± 5034 b (97.94)	102968 c
	masa	12596 ± 964 c (16.17)	20747 ± 1331 cb (26.64)	44527 ± 3083 cd (57.18)	77870 cd
	tortilla	17462 ± 668 b (15.28)	27827 ± 2500 b (24.34)	69025 ± 3850 c (60.38)	114314 bc
	chips	13237 ± 551 c (14.67)	25860 ± 1861 b (28.66)	51142 ± 2377 c (56.67)	90239 c
red	corn	588 ± 21 f (0.45)	1259 ± 48 d (0.97)	128450 ± 11553 b (98.58)	130297 b
	masa	6224 ± 364 e (12.49)	20747 ± 1331 cb (41.63)	22865 ± 3303 d (45.88)	49836 d
	tortilla	8202 ± 455 ed (11.11)	15588 ± 720 c (21.11)	50036 ± 7934 c (67.78)	73826 cd
	chips	8083 ± 572 ed (10.72)	29391 ± 2084 b (39.0)	37890 ± 4177 cd (50.28)	75364 cd
blue	corn	683 ± 47 f (0.53)	1451 ± 27 d (1.12)	127851 ± 8094 b (98.35)	129985 b
	masa	10220 ± 1370 ed (19.17)	19760 ± 1145 cb (37.07)	23330 ± 1917 d (43.76)	53310 d
	tortilla	17587 ± 1259 b (17.35)	31746 ± 2519 b (31.32)	52030 ± 1499 c (51.33)	101363 c
	chips	14360 ± 965 c (16.77)	28163 ± 2548 b (32.88)	43129 ± 2172 cd (50.35)	85652 c
high carotenoid	corn	970 ± 82 f (0.64)	1965 ± 33 d (1.28)	150077 ± 12419 a (98.08)	153012 a
	masa	11170 ± 267 ed (14.65)	24450 ± 654 b (32.08)	40610 ± 3798 cb (53.27)	76230 cd
	tortilla	21566 ± 505 a (15.79)	37743 ± 1295 a (27.63)	77307 ± 6570 c (56.58)	136616 b
	chips	13963 ± 747 c (13.10)	29931 ± 2169 c (28.08)	62714 ± 3322 c (58.82)	106608 c

[a] Values with no letters in common are significantly different (p <0.05).
[b] Percent contribution o total.

Table 4.14 *Anthocyanin content of white, yellow, red, blue, and high-carotenoid corns and their nixtamalized corn products (masa, tortilla, and chips) (mean ± sd, n = 3) (De la Parra, Saldivar, and Liu, 2007).*

type of corn	anthocyanin content[a] (mg of cyanidin-3-glucoside equiv/100 g of dry wt of sample)			
	corn	masa	tortilla	chips
white	1.33 ± 0.02 e	0.28 ± 0.01 f	0.48 ± 0.01 f	0.51 ± 0.01 f
yellow	0.57 ± 0.01 f	0.31 ± 0.01 f	0.29 ± 0.00 f	0.36 ± 0.01 f
red	9.75 + 0.44 b	2.21 ± 0.07 e	2.08 + 0.05 e	2.41 ± 0.04 e
blue	36.87 ± 0.71 a	2.63 ± 0.12 e	3.81 ± 0.11 d	3.29 ± 0.10 d
high carotenoid	4.63 ± 0.06 c	0.56 ± 0.01 f	0.68 ± 0.02 f	0.97 ± 0.02 f

[a] Values with no letters in common are significantly different ($p < 0.05$).

(Adom and Liu, 2002; Adom, Sorrells, and Liu, 2005; Lozovaya *et al.*, 2000; Rouau *et al.*, 2003; Sosulski, Krygier, and Hogge, 1982). Lime cooking and steeping caused losses of ferulic acid. Higher concentrations of ferulic acid in tortillas and tortilla chips compared to masa indicated that baking and frying operations complexed ferulic acid.

Anthocyanins are located mainly in the aleurone and pericarp regions of the kernel (Salinas-Moreno *et al.*, 2003). Other work (Cortes *et al.*, 2006) found that corn cooked with increasing amounts of lime had increased anthocyanin losses. Work by De la Parra, Saldivar, and Liu (2007) found higher anthocyanin losses in blue and red corn types (Table 4.14, data from De la Parra, Saldivar, and Liu, 2007). Anthocyanin losses ranged from 70 to 80%, which compared to work from Salinas-Moreno *et al.* (2003) which observed 70 to 100% losses. Lime cooking had the largest negative effect on anthocyanin losses; processing into tortillas and tortilla chips did not affect anthocyanin concentrations. Amongst corn types tested by De la Parra, Saldivar, and Liu (2007), the highest concentrations of total phenolics, ferulic acid and antioxidant activity were observed in the high carotenoid genotype followed by the regular yellow corn (Table 4.15, data from De la Parra, Saldivar, and Liu, 2007). White corn contained the lowest concentrations of total phenolics and antioxidant activity, while blue corn had the highest anthocyanin concentration, followed by red corn. This points out that hybrid selection plays a major role in phytonutrient concentrations, and should be considered in the overall economic feasibility of recovering phytochemicals.

Gutierrez-Uribe *et al.* (2010) investigated the distribution of phytochemicals in six types of corn that were lime cooked. The lime cooking procedure of Serna-Saldivar *et al.* (1993) was used to produce masa and nejayote samples from whole corn (Gutierrez-Uribe *et al.*, 2010). Kernels, masa and nejayote samples were analyzed for free and bound phenolics, ferulic acid and antioxidant activity. A majority (82 to 91%) of the phenolic compounds were present in bound form. Phenolics concentrations in nejayote were approximately two times higher than those in either whole kernels or masa. Bound ferulic acid was three times lower compared to original kernel concentrations. Nejayote contained 125 and 10 times more free ferulic acid and 15 and 50 times more bound ferulic acid compared to kernels and masa, respectively. Nejayote antioxidant activities were 40 and eight times higher in free form and 190 and 60 times higher in bound form compared to kernels and masa, respectively. Results from Gutierrez-Uribe *et al.* (2010) provided conclusive evidence that nixtamalization created high concentrations of phytochemicals in nejayote, but unfortunately these solids are not consumed in the diet, but must be disposed of properly.

Table 4.15 Carotenoid content of white, yellow, red, blue, and high-carotenoid corns and their nixtamalized products (De la Parra, Saldivar, and Liu, 2007).

type of corn	sample/product	carotenoid content[a] (µg/100 g of dry wt of sample)			
		lutein	zeaxanthin	β-cryploxanthin	β-carotene
white	com	5.73 ± 0.18 f	6.01 ± 0.06 f	1.27 ± 0.06 d	4.92 ± 0.18 e
	masa	8.16 ± 0.15 f	7.76 ± 0.23 f	1.48 ± 0.01 d	nd
	tortilla	8.61 ± 0.27 f	7.50 ± 0.43 f	1.57 ± 0.03 d	nd
	chips	0.82 ± 0.07 f	9.80 ± 0.31 f	0.08 ± 0.00 d	nd
yellow	com	406.2 ± 4.9 a	353.2 ± 23.1 a	19.1 ± 12 b	33.6 ± 1.2 b
	masa	129.2 ± 14.3 c	132.4 ± 11.3 b	3.00 ± 0.18 d	20.7 ± 0.9 c
	tortilla	108.2 ± 7.2 d	101.6 ± 10.1c	1.81 ± 0.14 d	11.1 ± 0.8 d
	chips	77.3 ± 7.2d	64.9 ± 5.5 d	0.11 ± 0.01 d	nd
red	com	121.7 ± 12.1c	111.9 ± 9.2 c	13.1 ± 1.8 c	20.2 ± 1.9 c
	masa	42.6 ± 3.2 e	44.9 ± 3.9 d	1.42 ± 0.14 d	7.02 ± 0.91 e
	tortilla	41.4 ± 4.1 e	42.3 ± 1.1 e	1.32 ± 0.10 d	6.39 ± 0.11 e
	chips	40.9 ± 2.5 e	30.2 ± 2.1 e	1.78 ± 0.09 d	nd
blue	com	5.17 ± 0.49 f	14.3 ± 1.0 f	3.41 ± 0.39 d	23.1 ± 2.1 c
	masa	4.71 ± 0.35 f	8.71 ± 0.10 f	1.00 ± 0.06 d	12.5 ± 1.2 d
	tortilla	4.53 ± 0.02 f	8.81 ± 0.90 f	1.22 ± 0.09 d	12.1 ± 0.4 d
	chips	4.48 ± 0.28 f	8.78 ± 0.71 f	0.96 ± 0.08 d	nd
high carolenoid	com	245.6 ± 9.4 a	322.3 ± 10.7 a	23.1 ± 1.0 a	45.8 ± 3.9 a
	masa	74.9 ± 7.2 d	112.6 ± 14.4 c	11.5 ± 1.0 c	25.9 ± 1.7 c
	tortilla	72.5 ± 4.8 d	105.3 ± 6.6 c	12.4 ± 0.4 c	14.6 ± 1.0 d
	chips	61.1 ± 3.4 d	92.5 ± 1.4 c	4.71 ± 0.20 d	8.39 ± 0.21 e

[a] Values with no letters in common are significantly different ($p < 0.05$), nd, not determined.

4.4 Conclusions

Phytochemical recovery from corn kernels has progressed from initial discovery and characterization in whole kernels, to kernel fractionation using conventional processes, to improved phytochemical recovery using modified processes. Phytochemicals are obtained commercially from the wet milling fiber fraction. In the case of phytochemical recovery from corn kernel fiber, conventional wet milling and dry milling processes were natural starting points for extraction from fiber/bran fractions. In many cases, validation of phytochemical compositions using modified or larger-scale research processes is still needed. In the dry grind industry, recovery of phytochemicals from DDGS may prove to be profitable as this industry matures and additional value from co-products is needed. Additional modifications to the dry grind process, if adopted, may become attractive since recovery of phytochemicals, as well as other co-products with improved market value (e.g., germ), are possible.

In the case of nixtamalization, nejayote is typically sent to a waste treatment facility where the nutritional benefits of this process stream are lost. The phytochemical content of the nejayote may add economic robustness to the process if the compounds can be recovered economically and added to the nixtamalized food products. This is an important opportunity for cultures and regions where corn tortillas are a dietary staple; diverting phytonutrients into the tortilla may improve the overall health of the population. In countries with higher per capita income and readily available food choices, such as the US, diverting phytonutrients from low-value nejayote to nixtamalized snack foods (e.g., corn chips) will improve overall human nutrition because snack foods contribute a significant portion of the overall diet. The value of recovering the phytochemicals in the nejayote must be determined. The scale of the nixtamalization process where economical recovery is possible should be determined.

As efficacy of specific phytochemicals in the corn kernel are proven clinically, additional process research will be needed to verify efficacy when produced at larger scales or at a commercial scale. The corn processing industry, in general, is a relatively conservative segment of the food industry. From a commercial scale perspective, fractionation of the corn kernel to concentrate phytochemicals is in the early stages and perceived as a high-risk enterprise. Most research has been done at a scale that collects less than 1 kg of the desired phytochemicals. As process scale is increased, composition, efficacy and cost effectiveness must be tested. Once health claims can be legitimately stated for the corn-based phytochemical in question, commercial scale processing will become attractive to manufacturers.

References

Adom, K.K. and Liu, R.H. (2002) Antioxidant activity of grains. *Journal of Agricultural and Food Chemistry*, **50**, 6182–6187.

Adom, K.K., Sorrells, M.E., and Liu, R.H. (2003) Phytochemical profiles and antioxidant activity of wheat varieties. *Journal of Agricultural and Food Chemistry*, **51**, 7825–7834.

Adom, K.K., Sorrells, M.E., and Liu, R.H. (2005) Phytochemicals and antioxidant activity of milled fractions of different wheat varieties. *Journal of Agricultural and Food Chemistry*, **53**, 2297–2306.

Brekke, O.L. (1970) Corn dry milling industry, in *Corn: Culture, Processing, and Products* (ed. G.E. Inglett), AVI Publishing Company, Westport, CT, pp. 265–291.

Brekke, O.L., Peplinski, A.J., Griffin, E.L., and Ellis, J.J. (1972) Dry milling of corn attacked by southern leaf blight. *Cereal Chemistry*, **49**, 466.

Bryngelsson, S., Dimberg, L.H., and Kamal-Eldin, A. (2002) Effects of commercial processing on levels of antioxidants in oats (avena sativa l.). *Journal of Agricultural and Food Chemistry*, **50**, 1890–1896.

Cortes, G.A., Salinas, M.Y., San Martin-Martinez, E., and Martinez-Bustos, F. (2006) Stability of anthocyanins of blue maize (zea mays l.) after nixtamalization of seperated pericarp-germ tip cap and endosperm fractions. *Journal of Cereal Science*, **43**, 57–62.

Cox, M.J., Macmasters, M.M., and Hilbert, G.E. (1944) Effect of sulfurous acid steep in corn wet milling. *Cereal Chemistry*, **21**, 447–465.

De la Parra, C., Saldivar, S.O.S., and Liu, R.H. (2007) Effect of processing on the phytochemical profiles and antioxidant activity of corn for production of masa, tortillas, and tortilla chips. *Journal of Agricultural and Food Chemistry*, **55**, 4177–4183.

Dewanto, V., Wu, X.Z., and Liu, R.H. (2002) Processed sweet corn has higher antioxidant activity. *Journal of Agricultural and Food Chemistry*, **50**, 4959–4964.

Dien, B.S., Nagle, N., Hicks, K.B. *et al.* (2004) Fermentation of "quick fiber" produced from a modified corn-milling process into ethanol and recovery of corn fiber oil. *Applied Biochemistry and Biotechnology*, **113**, 937–949.

Dien, B.S., Nagle, N., Singh, V. *et al.* (2005) Review of process for producing corn fiber oil and ethanol from "quick fiber". *International Sugar Journal*, **107**, 187–191.

Dietrych-Szostak, D. and Oleszek, W. (1999) Effect of processing on the flavonoid content in buckwheat (fagopyrum esculentum moench) grain. *Journal of Agricultural and Food Chemistry*, **47**, 4384–4387.

Du, L., Rausch, K.D., Yang, P. *et al.* (1999) Comparison of alkali and conventional corn wet-milling: 1-kg procedures. *Cereal Chemistry*, **76**, 811–815.

Duensing, W.J., Roskens, A.B., and Alexander, R.J. (2003) Corn dry milling: Processes, products, and applications, in *Corn: Chemistry and Technology* (eds P.J. White and L.A. Johnson), American Association of Cereal Chemists, Inc., St. Paul, MN, pp. 407–447.

Eckhoff S.R., Du, L., Yang, P. *et al.* (1999) Comparison between alkali and conventional corn wet-milling: 100-g procedures. *Cereal Chemistry*, **76**, 96–99.

Eckhoff, S.R., Singh, S.K., Zehr, B.E. *et al.* (1996) A 100-g laboratory corn wet-milling procedure. *Cereal Chemistry*, **73**, 54–57.

Eckhoff, S.R. and Tso, C.C. (1991) Wet milling of corn using gaseous so2 addition before steeping and the effect of lactic-acid on steeping. *Cereal Chemistry*, **68**, 248–251.

Farnham, M.W., Simon, P.W., and Stommel, J.R. (1999) Improved phytonutrient content through plant genetic improvement. *Nutrition Reviews*, **57**, S19–S26.

Gonzalez, R., Reguera, E., Figueroa, J.M., and Sanchez-Sinencio, F. (2005) On the nature of the ca binding to the hull of nixtamalized corn grains. *Lwt-Food Science and Technology*, **38**, 119–124.

Gonzalez, R., Reguera, E., Mendoza, L. *et al.* (2004) Physicochemical changes in the hull of corn grains during their alkaline cooking. *Journal of Agricultural and Food Chemistry*, **52**, 3831–3837.

Gutierrez-Uribe, J.A., Rojas-Garcia, C., Garcia-Lara, S., and Serna-Saldivar, S.O. (2010) Phytochemical analysis of wastewater (nejayote) obtained after lime-cooking of different types of maize kernels processed into masa for tortillas. *Journal of Cereal Science*, **52**, 410–416.

Hao, Z.G., Parker, B., Knapp, M., and Yu, L.L. (2005) Simultaneous quantification of alpha-tocopherol and four major carotenoids in botanical materials by normal phase liquid chromatography-atmospheric pressure chemical ionization-tandem mass spectrometry. *Journal of Chromatography A*, **1094**, 83–90.

Hendriks, H.F.J., Weststrate, J.A., van Vliet, T., and Meijer, G.W. (1999) Spreads enriched with three different levels of vegetable oil sterols and the degree of cholesterol lowering in normocholesterolaemic and mildly hypercholesterolaemic subjects. *European Journal of Clinical Nutrition*, **53**, 319–327.

Hernanz, D., Nunez, V., Sancho, A.I. *et al.* (2001) Hydroxycinnamic acids and ferulic acid dehydrodimers in barley and processed barley. *Journal of Agricultural and Food Chemistry*, **49**, 4884–4888.

Hu, Q.P. and Xu, J.G. (2011) Profiles of carotenoids, anthocyanins, phenolics, and antioxidant activity of selected color waxy corn grains during maturation. *Journal of Agricultural and Food Chemistry*, **59**, 2026–2033.

Humphries, J.M. and Khachik, F. (2003) Distribution of lutein, zeaxanthin, and related geometrical isomers in fruit, vegetables, wheat, and pasta products. *Journal of Agricultural and Food Chemistry*, **51**, 1322–1327.

Inglett, G.E. and Chen, D.J. (2011) Antioxidant activity and phenolic content of air-classified corn bran. *Cereal Chemistry*, **88**, 36–40.

Johnston, D.B., McAloon, A.J., Moreau, R.A. *et al.* (2005) Composition and economic comparison of germ fractions from modified corn processing technologies. *Journal of the American Oil Chemists Society*, **82**, 603–608.

Kahlon, T.S., Saunders, R.M., Sayre, R.N. *et al.* (1992) Cholesterol-lowering effects of rice bran and rice bran oil fractions in hypercholesterolemic hamsters. *Cereal Chemistry*, **69**, 485–489.

Kern, S.M., Bennett, R.N., Mellon, F.A. *et al.* (2003) Absorption of hydroxycinnamates in humans after high-bran cereal consumption. *Journal of Agricultural and Food Chemistry*, **51**, 6050–6055.

Kurilich, A.C. and Juvik, J.A. (1999a) Quantification of carotenoid and tocopherol antioxidants in zea mays. *Journal of Agricultural and Food Chemistry*, **47**, 1948–1955.

Kurilich, A.C. and Juvik, J.A. (1999b) Simultaneous quantification of carotenoids and tocopherols in corn kernel extracts by HPLC. *Journal of Liquid Chromatography & Related Technologies*, **22**, 2925–2934.

Kwiatkowski, J.R. and Cheryan, M. (2002) Extraction of oil from ground corn using ethanol. *Journal of the American Oil Chemists Society*, **79**, 825–830.

Ling, W.H. and Jones, P.J.H. (1995) Dietary phytosterols - a review of metabolism, benefits and side-effects. *Life Sciences*, **57**, 195–206.

Liu, R.H. (2007) Whole grain phytochemicals and health. *Journal of Cereal Science*, **46**, 207–219.

Lopes-Filho, J.F., Buriak, P., Tumbleson, M.E., and Eckhoff, S.R. (1997) Intermittent milling and dynamic steeping process for corn starch recovery. *Cereal Chemistry*, **74**, 633–638.

Lozovaya, V.V., Gorshkova, T.A., Rumyantseva, N.I. *et al.* (2000) Cell wall-bound phenolics in cells of maize (zea mays, gramineae) and buckwheat (fagopyrum tataricum, polygonaceae) with different plant regeneration abilities. *Plant Science*, **152**, 79–85.

Maya-Cortes, D.C., Cardenas, J.D.F., Garnica-Romo, M.G. *et al.* (2010) Whole-grain corn tortilla prepared using an ecological nixtamalisation process and its impact on the nutritional value. *International Journal of Food Science and Technology*, **45**, 23–28.

Miettinen, T.A., Puska, P., Gylling, H. *et al.* (1995) Reduction of serum-cholesterol with sitostanol-ester margarine in a mildly hypercholesterolemic population. *New England Journal of Medicine*, **333**, 1308–1312.

Miettinen, T.A., Tilvis, R.S., and Kesaniemi, Y.A. (1990) Serum plant sterols and cholesterol precursors reflect cholesterol absorption and synthesis in volunteers of a randomly selected male-population. *American Journal of Epidemiology*, **131**, 20–31.

Miller, H.E., Rigelhof, F., Marquart, L. *et al.* (2000) Antioxidant content of whole grain breakfast cereals, fruits and vegetables. *Journal of the American College of Nutrition*, **19**, 312S–319S.

Moreau, R., Hicks, K., Nicolosi, R., and Norton, R. (1998) Corn fiber oil: Its preparation, composition, and use. Patent No. 5,843,499:US.

Moreau, R.A. and Hicks, K.B. (2005) The composition of corn oil obtained by the alcohol extraction of ground corn. *Journal of the American Oil Chemists Society*, **82**, 809–815.

Moreau, R.A., Hicks, K.B., and Powell, M.J. (1999a) Effect of heat pretreatment on the yield and composition of oil extracted from corn fiber. *Journal of Agricultural and Food Chemistry*, **47**, 2869–2871.

Moreau, R.A., Johnston, D.B., and Hicks, K.B. (2007) A comparison of the levels of lutein and zeaxanthin in corn germ oil, corn fiber oil and corn kernel oil. *Journal of the American Oil Chemists Society*, **84**, 1039–1044.

Moreau, R.A., Lampi, A.-M., and Hicks, K.B. (2009) Fatty acid, phytosterol, and polyamine conjugate profiles of edible oils extracted from corn germ, corn fiber, and corn kernels. *Journal of the American Oil Chemists' Society*, **86**, 1209–1214.

Moreau, R.A., Norton, R.A., and Hicks, K.B. (1999b) Phytosterols and phytostanols lower cholesterol. *INFORM - International News on Fats, Oils and Related Materials*, **10**, 572–577.

Moreau, R.A., Nunez, A., and Singh, V. (2001a) Diferuloylputrescine and p-coumaroyl-feruloylputrescine, abundant polyamine conjugates in lipid extracts of maize kernels. *Lipids*, **36**, 839–844.

Moreau, R.A., Powell, M.J., and Hicks, K.B. (1996) Extraction and quantitative analysis of oil from commercial corn fiber. *Journal of Agricultural and Food Chemistry*, **44**, 2149–2154.

Moreau, R.A., Powell, M.J., and Singh, V. (2003) Pressurized liquid extraction of polar and nonpolar lipids in corn and oats with hexane, methylene chloride, isopropanol, and ethanol. *Journal of the American Oil Chemists Society*, **80**, 1063–1067.

Moreau, R.A., Singh, V., Eckhoff, S.R. *et al.* (1999c) Comparison of yield and composition of oil extracted from corn fiber and corn bran. *Cereal Chemistry*, **76**, 449–451.

Moreau, R.A., Singh, V., and Hicks, K.B. (2001b) Comparison of oil and phytosterol levels in germplasm accessions of corn, teosinte, and Job's tears. *Journal of Agricultural and Food Chemistry*, **49**, 3793–3795.

Moreau, R.A., Singh, V., Nunez, A., and Hicks, K.B. (2000) Phytosterols in the aleurone layer of corn kernels. *Biochemical Society Transactions*, **28**, 803–806.

Moreau, R.A., Whitaker, B.D., and Hicks, K.B. (2002) Phytosterols, phytostanols, and their conjugates in foods: Structural diversity, quantitative analysis, and health-promoting uses. *Progress in Lipid Research*, **41**, 457–500.

Moros, E.E., Darnoko, D., Cheryan, M. *et al.* (2002) Analysis of xanthophylls in corn by HPLC. *Journal of Agricultural and Food Chemistry*, **50**, 5787–5790.

Mozaffarieh, M., Sacu, S., and Wedrich, A. (2003) The role of the carotenoids, lutein and zeaxanthin, in protecting against age-related macular degeneration: A review based on controversial evidence. *Nutrition Journal*, **2**, 20.

Nissinen, M., Gylling, H., Vuoristo, M., and Miettinen, T.A. (2002) Micellar distribution of cholesterol and phytosterols after duodenal plant stanol ester infusion. *American Journal of Physiology-Gastrointestinal and Liver Physiology*, **282**, G1009–G1015.

Norton, R.A. (1994) Isolation and identification of steryl cinnamic acid-derivatives from corn bran. *Cereal Chemistry*, **71**, 111–117.

Norton, R.A. (1995) Quantitation of steryl ferulate and p-coumarate esters from corn and rice. *Lipids*, **30**, 269–274.

Panfili, G., Fratianni, A., and Irano, M. (2004) Improved normal-phase high-performance liquid chromatography procedure for the determination of carotenoids in cereals. *Journal of Agricultural and Food Chemistry*, **52**, 6373–6377.

Pflugfelder, R.L., Rooney, L.W., and Waniska, R.D. (1988) Dry-matter losses in commercial corn masa production. *Cereal Chemistry*, **65**, 127–132.

Piironen, V., Lindsay, D.G., Miettinen, T.A. *et al.* (2000) Plant sterols: Biosynthesis, biological function and their importance to human nutrition. *Journal of the Science of Food and Agriculture*, **80**, 939–966.

Ramjiganesh, T., Roy, S., Nicolosi, R.J. *et al.* (2000) Corn husk oil lowers plasma LDL cholesterol concentrations by decreasing cholesterol absorption and altering hepatic cholesterol metabolism in guinea pigs. *Journal of Nutritional Biochemistry*, **11**, 358–366.

Rausch, K.D. and Belyea, R.L. (2006) The future of co-products from corn processing. *Applied Biochemistry and Biotechnology*, **128**, 47–86.

Rausch, K.D., Raskin, L.M., Belyea, R.L. *et al.* (2005) Phosphorus concentrations and flow in maize wet-milling streams. *Cereal Chemistry*, **82**, 431–435.

Rausch, K.D., Raskin, L.M., Belyea, R.L. *et al.* (2007) Nitrogen and sulfur concentrations and flow rates of corn wet-milling streams. *Cereal Chemistry*, **84**, 260–264.

Rogers, E.J., Rice, S.M., Nicolosi, R.J. *et al.* (1993) Identification and quantitation of gamma-oryzanol components and simultaneous assessment of tocols in rice bran oil. *Journal of the American Oil Chemists Society*, **70**, 301–307.

Rouau, X., Cheynier, V., Surget, A. *et al.* (2003) A dehydrotrimer of ferulic acid from maize bran. *Phytochemistry*, **63**, 899–903.

Roushdi, M. and Fahmy, A.S. (1981) Role of lactic-acid in corn steeping and its relation with starch isolation. *Starke*, **33**, 426–428.

Ruan, R.S., Litchfield, J.B., and Eckhoff, S.R. (1992) Simultaneous and nondestructive measurement of transient moisture profiles and structural-changes in corn kernels during steeping using microscopic nuclear-magnetic-resonance imaging. *Cereal Chemistry*, **69**, 600–606.

Salinas-Moreno, Y., Martinez-Bustos, F., Soto-Hernandez, M. *et al.* (2003) Effect of alkaline cooking process on anthocyanins in pigmented maize grain. *Agrociencia*, **37**, 617–628.

Scott, C.E. and Eldridge, A.L. (2005) Comparison of carotenoid content in fresh, frozen and canned corn. *Journal of Food Composition and Analysis*, **18**, 551–559.

Seitz, L.M. (1989) Stanol and sterol esters of ferulic and p-coumaric acids in wheat, corn, rye, and triticale. *Journal of Agricultural and Food Chemistry*, **37**, 662–667.

Serna-Saldivar, S.O. (2010) Milling of maize into lime-cooked products, in *Cereal grains: Properties, Processing, and Nutritional Attributes*, CRC Press, Boca Raton, FL, pp. 239–257.

Serna-Saldivar, S.O., Gomez, M.H., Almeida-Dominguez, H.D. *et al.* (1993) A method to evaluate the lime-cooking properties of corn (zea-mays). *Cereal Chemistry*, **70**, 762–764.

Serna-Saldivar, S.O., Gomez, M.H., and Rooney, L.W. (1990) Technology, chemistry and nutritive value of alkaline cooked corn products, in *Advances in Cereal Science and Technology* (ed. Y. Pomeranz), American Association of Cereal Chemists, St. Paul, MN, pp. 243–295.

Shandera, D.L., Parkhurst, A.M., and Jackson, D.S. (1995) Interactions of sulfur-dioxide, lactic-acid, and temperature during simulated corn wet-milling. *Cereal Chemistry*, **72**, 371–378.

Singh, V. and Eckhoff, S.R. (1996) Effect of soak time, soak temperature, and lactic acid on germ recovery parameters. *Cereal Chemistry*, **73**, 716–720.

Singh, V. and Eckhoff, S.R. (1997) Economics of germ preseparation for dry-grind ethanol facilities. *Cereal Chemistry*, **74**, 462–466.

Singh, V., Johnston, D.B., Moreau, R.A. *et al.* (2003) Pretreatment of wet-milled corn fiber to improve recovery of corn fiber oil and phytosterols. *Cereal Chemistry*, **80**, 118–122.

Singh, V., Johnston, D.B., Naidu, K. *et al.* (2005) Comparison of modified dry-grind corn processes for fermentation characteristics and DDGS composition. *Cereal Chemistry*, **82**, 187–190.

Singh, V. and Moreau, R.A. (2003) Enrichment of oil in corn fiber by size reduction and floatation of aleurone cells. *Cereal Chemistry*, **80**, 123–125.

Singh, V., Moreau, R.A., and Cooke, P.H. (2001a) Effect of corn milling practices on aleurone layer cells and their unique phytosterols. *Cereal Chemistry*, **78**, 436–441.

Singh, V., Moreau, R.A., Doner, L.W. *et al.* (1999) Recovery of fiber in the corn dry-grind ethanol process: A feedstock for valuable co-products. *Cereal Chemistry*, **76**, 868–872.

Singh, V., Moreau, R.A., Haken, A.E. *et al.* (2000) Hybrid variability and effect of growth location on corn fiber yields and corn fiber oil composition. *Cereal Chemistry*, **77**, 692–695.

Singh, V., Moreau, R.A., Haken, A.E., Hicks, K.B., and Eckhoff, S.R. (2000b) Effect of various acids and sulfites in steep solution on yields and composition of corn fiber and corn fiber oil. *Cereal Chemistry*, **77**, 665–668.

Singh, V., Moreau, R.A., Hicks, K.B., and Eckhoff, S.R. (2001b) Effect of alternative milling techniques on the yield and composition of corn germ oil and corn fiber oil. *Cereal Chemistry*, **78**, 46–49.

Smith, M.M. and Hartley, R.D. (1983) Occurrence and nature of ferulic acid substitution of cell-wall polysaccharides in graminaceous plants. *Carbohydrate Research*, **118**, 65–80.

Song, W., Derito, C.M., Liu, M.K. *et al.* (2010) Cellular antioxidant activity of common vegetables. *Journal of Agricultural and Food Chemistry*, **58**, 6621–6629.

Sosulski, F., Krygier, K., and Hogge, L. (1982) Free, esterified, and insoluble-bound phenolic-acids .3. Composition of phenolic-acids in cereal and potato flours. *Journal of Agricultural and Food Chemistry*, **30**, 337–340.

Srinivasan, R., Moreau, R.A., Rausch, K.D. *et al.* (2005) Separation of fiber from distillers dried grains with solubles (DDGS) using sieving and elutriation. *Cereal Chemistry*, **82**, 528–533.

Srinivasan, R., Moreau, R.A., Rausch, K.D. *et al.* (2007) Phytosterol distribution in fractions obtained from processing of distillers dried grains with solubles using sieving and elutriation. *Cereal Chemistry*, **84**, 626–630.

Srinivasan, R., Vijay, S., Belyea, R.L. *et al.* (2006) Economics of fiber separation from distillers dried grains with solubles (DDGS) using sieving and elutriation. *Cereal Chemistry*, **83** (4), 324–330.

Taylor, F., McAloon, A.J., Craig, J.C. *et al.* (2001) Fermentation and costs of fuel ethanol from corn with quick-germ process. *Applied Biochemistry and Biotechnology*, **94**, 41–49.

Trumbo, P.R. and Ellwood, K.C. (2006) Lutein and zeaxanthin intakes and risk of age-related macular degeneration and cataracts: An evaluation using the food and drug administration's evidence-based review system for health claims. *The American Journal of Clinical Nutrition*, **84**, 971–974.

USDA (2011) Data and statistics database. Quick stats 1.0 http://www.nass.usda.gov/Data_and_Statistics/ Quick_Stats_1.0/index.asp. USDA National Agricultural Statistics Service. Washington, DC.

Velasco-Martinez, M., Angulo, O., VazquezCouturier, D.L. *et al.* (1997) Effect of dried solids of nejayote on broiler growth. *Poultry Science*, **76**, 1531–1534.

Velioglu, Y.S., Mazza, G., Gao, L., and Oomah, B.D. (1998) Antioxidant activity and total phenolics in selected fruits, vegetables, and grain products. *Journal of Agricultural and Food Chemistry*, **46**, 4113–4117.

Wahjudi, J., Xu, L., Wang, P. *et al.* (2000) Quick fiber process: Effect of mash temperature, dry solids, and residual germ on fiber yield and purity. *Cereal Chemistry*, **77**, 640–644.

Wang, C.Y., Ning, J., Krishnan, P.G., and Matthees, D.P. (1998) Effects of steeping conditions during wet-milling on the retentions of tocopherols and tocotrienols in corn. *Journal of the American Oil Chemists Society*, **75**, 609–613.

Wang, P., Johnston, D.B., Rausch, K.D. *et al.* (2009) Effects of protease and urea on a granular starch hydrolyzing process for corn ethanol production. *Cereal Chemistry*, **86**, 319–322.

Wang, P., Singh, V., Xu, L. *et al.* (2005) Comparison of enzymatic (e-mill) and conventional dry-grind corn processes using a granular starch hydrolyzing enzyme. *Cereal Chemistry*, **82**, 734–738.

Wang, P., Singh, V., Xue, H. *et al.* (2007) Comparison of raw starch hydrolyzing enzyme with conventional liquefaction and saccharification enzymes in dry-grind corn processing. *Cereal Chemistry*, **84**, 10–14.

Watson, S.A. (1984) Corn and sorghum starches: Production, in *Starch: Chemistry and Technology* (eds R.L. Whistler, J.N. BeMiller, and E.F. Paschall), Academic Press, Orlando, FL, pp. 417–468.

Wilson, T.A., DeSimone, A.P., Romano, C.A., and Nicolosi, R.J. (2000) Corn fiber oil lowers plasma cholesterol levels and increases cholesterol excretion greater than corn oil and similar to diets containing soy sterols and soy stanols in hamsters. *Journal of Nutritional Biochemistry*, **11**, 443–449.

Wolf, M.J., Buzan, C.L., Macmasters, M.M., and Rist, C.E. (1952) Structure of mature corn kernel. III. Microscopic structure of the endosperm of dent corn. *Cereal Chemistry*, **29**, 349–436.

Wolf, M.J., Macmasters, M.M., Cannon, J.A. *et al.* (1953) Preparation and some properties of hemicellulose from corn hulls. *Cereal Chemistry*, **30**, 451–470.

Zielinski, H. and Kozlowska, H. (2000) Antioxidant activity and total phenolics in selected cereal grains and their different morphological fractions. *Journal of Agricultural and Food Chemistry*, **48**, 2008–2016.

5

Co-Products from Cereal and Oilseed Biorefinery Systems

Nurhan Turgut Dunford

Department of Biosystems and Agricultural Engineering and Robert M. Kerr Food & Agricultural Products Center, Oklahoma State University, Stillwater, Oklahoma, USA

5.1 Introduction

The biorefinery concept is similar to the petroleum refinery except that it is based on biomass feedstock rather than crude petroleum oil. A cereal- or oilseed-based biorefinery would use various fractions of these crops to produce various products, including fuels, power, heat, food, chemicals and materials. There are several examples of successful biorefinery systems operating in the US (Paster, Pellegrino, and Carole, 2003). Most of the current biorefinery systems are operated by large companies and utilize biomass delivered from locations that are not necessarily close to the processing plant. The biorefinery concept can also be very effective when it is based on a processing plant that would refine local agricultural crops. The idea is that the final products, not the feedstock, can be economically transported to distant markets and some of the products could be used locally, eliminating long-distance transportation costs.

This chapter will cover current and potential co-products from cereal and oilseed biorefineries. Whole crops rather than just grains and seeds will be evaluated as feedstock. Although the main emphasis of this chapter will be on the biorefinery products, the basics of plant structure and chemical composition will also be discussed since these properties determine the type of products that can be derived from oilseeds and cereals. Proteins,

Biorefinery Co-Products: Phytochemicals, Primary Metabolites and Value-Added Biomass Processing, First Edition.
Edited by Chantal Bergeron, Danielle Julie Carrier and Shri Ramaswamy.
© 2012 John Wiley & Sons, Ltd. Published 2012 by John Wiley & Sons, Ltd.

carbohydrates (i.e., starch and cellulose), and lipids (both non-polar and polar, i.e., triacyl-glycerides (TAG) and phospholipids, respectively) are the major components or "building blocks" of cereals and oilseeds. In an effort to help the readers with no chemistry and biology knowledge the chemical structures of the "building block" molecules are given in Figure 5.1.

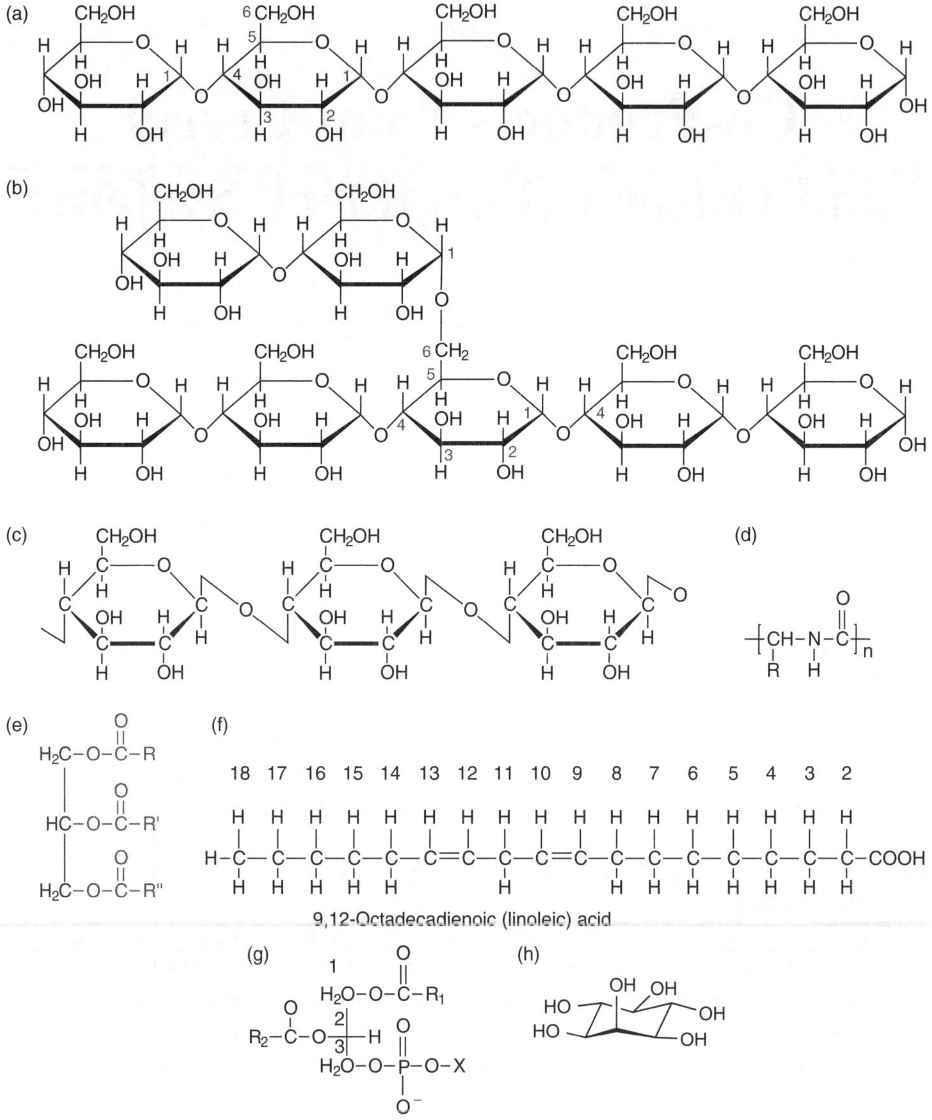

Figure 5.1   *Chemical structure of bioproduct building block molecules. (a) Amylose. (b) Amylopectin. (c) Cellulose. (d) Protein. (e) Triacylglyceride. R, R′ and R″ represent fatty acids. (f) Fatty acid (linoleic acid). (g) Phispholipids. R_1 and R_2 represent fatty acids, X = H (phosphatic acid), $CH_2CH_2NH_3^+$ (phosphatidylethanolamine), $CH_2CH_2(CH_3)_3^+$ (phosphatidylcholine), $CH_2CH(OH)CH_2OH$ (phosphatidylglycerol) and (h) phosphatidylinositol.*

5.2 Cereals

Cereals are the seeds or grains of plants which belong to the grass families Gramineae or Poaceae. They are by far the most important crops cultivated globally. In 2009 about 2.5 billion metric tons of cereals were produced worldwide (FAO, 2010). Wheat leads the global cereal production (685.6 million tons in 2009), followed by rice (685.2 million tons), barley (152.1 million tons) and sorghum (56.1 million tons). Cereals provide 95% of staple food globally. Even in developed countries 20–35% of the dietary energy comes from cereal grains (Dendy and Dobraszczyk, 2001).

Most cereal crops have thin stems, except maize, sorghum and pearl millet, and the height of the plants varies between 30 and 300 cm. Both stem thickness and height of the plant are important characteristics affecting biomass production, harvesting and processing. Harvest index (HI) which is defined as the proportion of grain yield to total above-ground biomass of a cereal crop is used for estimation of total straw, chaff and stubble yield from grain yield data. HIs of cereal crops vary between 50–80% (McCartney *et al.*, 2006). Hence, products from cereal crop residues are important for the economic feasibility of a whole crop cereal biorefinery system.

Seeds or grains, also botanically known as caryopsis, are the economically important parts of the cereal crops. Seeds are covered with a seed coat or testa. Some cereal grains also have a second layer of protection over the seed coat known as husk or hull which may stay with the grain after it is separated from the head or ears of the whole plant. A cereal grain consists of three parts, bran, germ and endosperm. Bran is the outer layer of the kernel and a byproduct of grain milling operations. Bran is rich in carbohydrates and proteins. Health beneficial phytochemicals are concentrated in the bran. Although lipids, vitamins and other bioactive compounds are minor components, they contribute significantly to the nutritional quality of cereal grains. Aleurone is the outermost cell layer in the endosperm. During conventional milling aleurone normally remains attached to the bran and is removed with it (Yu, 2008; Posner and Hibbs, 2005). Endosperm comprises the largest portion of a cereal grain and mainly contains protein and starch. The germ constitutes about 2–3% of the cereal grain and can be separated in a fairly pure form during the milling operation. Although lipids are present in relatively small quantities in whole grains, they play an important role during processing and the nature of the products by affecting the properties of protein and starch. Most of the oil is found in the germ fraction of the grain. For some cereals such as rice, corn and wheat, oil has economic significance. Oats contain a considerable amount of lipid in the endosperm. The lipid content of the germ varies with the grain type and it can be as high as 60%. Even though a small amount of germ oil is commercially extracted from other grains such as oats and barley, the most important commercial germ oils are obtained from rice, corn and wheat germ.

Conversion of cereal grain fractions and crop residues into high-value products within a biorefinery system would improve the economic feasibility of the entire system. A schematic of a typical cereal biorefinery system is shown in Figure 5.2.

5.2.1 Wheat

The wheat biorefinery concept has been evaluated by various groups, some focusing on economic feasibility and others emphasizing the types of products that can be manufactured within a biorefinery system (Annetts and Audsley, 2003; Koutinas, Wang, and Webb, 2004;

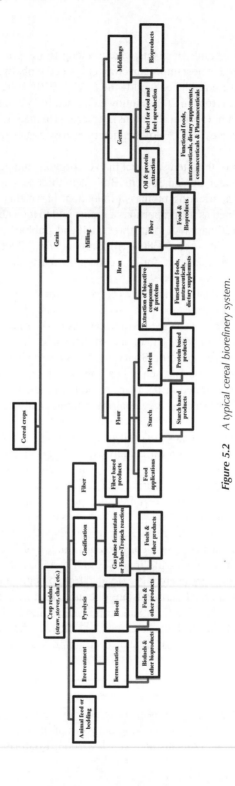

Figure 5.2   *A typical cereal biorefinery system.*

Deswarte *et al.*, 2007; Clark, Deswarte, and Farmer, 2009; Jenkins, 2008). Wheat is mainly used as a staple food and to make flour for leavened, flat and various other breads, cookies, cakes, pasta and noodles. The presence of gluten protein makes wheat the only grain suitable for making leavened bread. Wheat endosperm can be fractionated into starch and protein. Gluten is a very important protein that has unique structural and functional properties. Wheat gluten is used as an ingredient in diverse food and industrial products. Flours with low protein content are usually fortified with gluten to improve flour quality and desired performance in baking applications. A large portion of commercial gluten is used as a meat replacer in vegetarian foods and artificial forms of seafood analogues (Day *et al.*, 2006). Synthetic cheese, protein-fortified cereals and fruit fillings for nutritional bars are amongst the many other food applications of wheat gluten.

Non-food applications of wheat gluten include simulated meat for pet food and aquaculture feed. The thermoplasticity and good adhesive and film-forming properties make wheat gluten suitable for the production of natural adhesives and pressure-sensitive medical bandages. Peptides produced by gluten hydrolysis are beneficial in cosmetics, lotions, hair- and skin-care products and biodegradable resins. Numerous other industrial applications of wheat gluten, including edible films and biodegradable plastics, have been studied (Guilbert *et al.*, 2002).

Wheat flour is mainly comprised of starch (64–67%) (MacMasters, Hinton, and Bradbury, 1971). Starch is a polymer composed of glucose molecules which can be hydrolyzed into simple sugars. Once sugars have been depolymerized they can be used to produce a wide range of chemicals. Industrial starches have numerous applications in both food and non-food industrial product manufacturing. Potatoes, maize, wheat and tapioca are the most important sources of industrial starches. There are significant differences in properties of starches from different sources because of the relative proportions of amylose and amylopectin (Figure 5.1) and the presence or absence of non-starch components such as lipids, proteins and phosphate groups. Attributes such as physical dimensions of starch granules, their biodegradability, lubricity, texture, and the amount and nature of their non-carbohydrate components all affect industrial uses of starch. Native starches can be chemically or physically modified to improve their functionality. Non-food starch-based products include chemicals produced by fermentation, starch-containing plastics, starch-based adhesives, soil conditioners and extruded starch-based packaging materials. Starch is also used in the paper, binder and detergent industries. Excipients for tablet formation, binders for coatings and encapsulation material for the controlled release of active ingredients from a drug are some of the applications of starch and starch derivatives in the pharmaceutical industry (Ellis *et al.*, 1998).

In recent years bioethanol has been drawing attention as a transportation fuel additive. Although currently corn is the major feedstock, wheat flour and starch can also be used for bioethanol production. Utilization of pure starch and whole wheat flour for bioethanol production has been reported (Lang, Macdonald, and Hill, 2001). About 95% of the theoretical amount of ethanol could be produced by using either pure starch or whole wheat flour. An economic analysis of the fermentation process described in the latter publication indicated that ethanol production from wheat starch or flour would be economically feasible. In the same study it was assumed that a co-product, food-grade wheat gluten, would be produced along with bioethanol. However, wheat flour would not be the best feedstock for biofuel production because of the high demand for wheat for food applications.

A wheat-based fungal bioconversion process has been developed for the production of a generic substrate containing all the nutrients required for fermentation (Webb and Wang,

1997). A preliminary economic analysis of the process has shown that wheat can be an efficient and economically viable renewable feedstock for the production of many bioproducts (Koutinas, Wang, and Webb, 2004). Succinic acid, fumaric acid, malic acid, glutamic acid, itaconic acid and aspartic acid, as well as 3-hydroxypropionic acid can be produced by fermentation of a wheat-derived medium. A few of these products are discussed below.

Succinic acid is a platform chemical that has been traditionally manufactured from petrochemicals. It is also produced biochemically from glucose. US Department of Energy scientists engineered *Anaerobiospirillum succiniciproducens* and *Eschericia coli* for producing succinic acid and licensed these microorganisms to a small business (Werpy and Petersen, 2004). Economic feasibility of succinic acid production can be improved by using byproducts from food and agricultural material processing operations as feedstock. In a recent study wheat was fractionated into bran, middlings and flour (Dorado *et al.*, 2009). The bran fraction was used as the substrate for *Aspergillus awamori* and *A. oryzae* in solid-state fermentations to produce enzyme mixtures rich in amylolytic and proteolytic enzymes, respectively. The solids from the fermentation were then used as crude enzyme sources by direct addition into aqueous suspensions of bran and middlings to produce a hydrolysate containing glucose, maltose and free amino nitrogen. The hydrolysate was then used as the sole substrate for *Actinobacillus succinogenes* fermentation to produce succinic acid. Supplementation of the fermentation medium with yeast extract did not significantly increase succinic acid production, indicating that wheat-based fermentation medium contained all the nutrients needed for microbial growth and conversion.

Lactic acid is an important feedstock for production of biodegradable polymers, specifically polylactic acid. Sugars obtained from wheat starch hydrolysis can be converted to lactic acid by fermentation. However, this process faces several challenges: product inhibition at high lactic acid concentrations in the fermentation medium (Goncalves *et al.*, 1991); undesirable byproduct formation, that is, lactate and calcium sulfate formation (Datta *et al.*, 1995); and polymerization of lactic acid during product recovery by evaporation (Mellan, 1977). The commercial viability of lactic acid production from whole wheat flour has been examined (Akerberg and Zacchi, 2000). In the latter study processing-related issues were identified and alternative techniques were evaluated. Batch fermentation was economically better than continuous fermentation. The costs of feedstock, sodium hydroxide used for pH control during fermentation and the conversion of lactate to lactic acid by electrodialysis contributed substantially to the total production cost of lactic acid. It was suggested that lactic acid production costs could be reduced by lowering the pH and/or by recycling the sodium hydroxide produced by electrodialysis to the fermenter.

Polyhydroxyalkanoates (PHAs) are biodegradable plastics which can be produced from renewable carbon sources derived from agricultural materials or even industrial waste. Although production of a typical PHA, poly[(*R*)-3-hydroxybutyrate] (P(3HB) or PHB) in bacteria has been studied extensively (Braunegg, Lefebvre, and Genser, 1998), large-scale microbial production of PHAs has been hindered by the economic and technical difficulties involved in traditional fermentation practices. The potential of a novel wheat-based feedstock for PHB production by *Cupriavidus necator* has been examined (Koutinas *et al.*, 2007). The feedstock was developed at the Satake Centre for Grain Process Engineering. Initially the outer layers of wheat kernel were removed by pearling. Wheat gluten was extracted from pearled wheat flour as a value-added co-product. The enzymes required for hydrolysis of pearled wheat flour were produced by fungal fermentation. The remaining fungal mycelia and undigested wheat components were used to produce a nutrient-rich supplement (NRS) which had similar composition to yeast extract. The high cost of yeast extract, which is commonly used in commercial fermentation adversely affects the feasibility of the entire process. The bacterial PHB production was achieved in a medium formulated with wheat hydrolysate and NRS in the absence of yeast extracts (Koutinas *et al.*, 2007).

Currently wheat straw is baled for use as livestock bedding and/or low-grade animal feed providing minimal economic return. The green forage may be grazed by livestock or used as hay or silage. In the southern Great Plains, wheat serves a dual purpose by being grazed in the fall and early spring and then harvested as a grain crop in the early summer. Straw also represents a significant opportunity for fibre substitution in industrial products. For example, pulp from straw is partially substituted for wood fibre in some paper and paperboard products (Mckean and Jacobs, 1997). Wheat straw can also be used for energy generation or as a feedstock for the production of chemicals.

Pyrolysis and gasification are two thermochemical processes that have been examined extensively for fuel and bioproduct development. Production of methanol and hydrogen is viable by using gasification followed by the Fischer–Tropsch conversion process. However, production of simple alcohols, aldehydes and mixed alcohols via gasification route requires additional research and development (Werpy and Petersen, 2004). Advanced cellulose hydrolysis technologies are also being developed for converting straw to biofuels and other industrial bioproducts through fermentation (Saha *et al.*, 2005; Szczodrak, 1988; Ahring *et al.*, 1999; Ahring *et al.*, 1996).

Recovery and conversion of surface waxes and other bioactive compounds present in wheat straw into high-value products such as insecticides, nutraceuticals and cosmeceutical ingredients may improve the economic feasibility of other commodity or lower-value products generated in a biorefinery (Dunford and Edwards, 2010; Irmak, Dunford, and Milligan, 2005; Clark, 2007). Plant-based waxes have a growing market for the replacement of synthetic and animal-product-derived waxes and lanolin obtained from petroleum.

Wheat bran is rich in arabinoxylans which could be enzymatically hydrolysed into xylose and then converted to xylitol via fermentation. The US Department of Energy identified xylitol as one of the top 12 high-value platform chemicals that can be used to produce many other chemicals, including xylaric acid, propylene glycol, ethylene glycol, and mixtures of hydroxyl furans and polyesters (Koutinas *et al.*, 2008).

It is well established that wheat grain, particularly the milling industry byproducts, bran and germ, are rich in a number of health-beneficial bioactive compounds, including carotenoids, phytosterols, phenolic acids and tocopherols, which contribute to the antioxidant properties of wheat. Bran can be extracted with ethanol and water mixtures to obtain antioxidant-rich products that can be used in food, pharmaceutical and cosmetic formulations (Chen, 2011).

Wheat germ is a unique source of highly concentrated nutrients. Germ offers three times as much protein of high biological value (26%), seven times as much fat (11–15%), 15 times as much sugar (17%) and six times as much mineral content (4%) when compared with flour. Wheat germ oil (WGO) has a number of nutritional and health benefits, such as reducing plasma and liver cholesterol levels, improving physical endurance/fitness and possibly helping to delay the effects of ageing. These effects are attributed to the high concentration of bioactive compounds present in the oil. WGO is rich in unsaponifiable compounds, in particular phytosterols and tocopherols. It has been reported that WGO improves human physical fitness, an effect attributed to high policosanol, specifically high octacosanol content in WGO

(Cureton, 1972). In addition, wheat germ is the richest known source of α-tocopherols (natural vitamin E) of plant origin and also a good source of thiamine, riboflavin and niacin (Dunford, 2009).The total unsaturated and polyunsaturated fatty acid (PUFA) contents of WGO are about 81 and 64%, respectively.

Several extraction techniques, including pressurized solvent extraction, supercritical fluid extraction, aqueous extraction and enzymatic extraction have been evaluated for their efficiency to recover high quality oil from wheat germ (Dunford and Martinez, 2003; Dunford, 2009; Dunford and Zhang, 2003; Eisenmenger *et al.*, 2006; Xie and Dunford, 2010; Xie, 2009). Mechanical oil extraction rather than organic solvent extraction is preferred for recovering WGO that will be used in speciality products and health foods. Pressurized solvent and supercritical fluid extraction techniques result in similar oil yields as conventional hexane extraction, over 95% and produce higher quality residual meal The main disadvantage of high-pressure systems is their high capital cost. Yet, high-pressure extraction systems can be economically viable for production of high-value products. Although aqueous and enzymatic extraction yields are lower, about 65–70%, these techniques require lower capital cost, eliminate organic solvent use and facilitate recovery of oil and proteins simultaneously (Xie, 2009).

Currently WGO is produced for speciality markets in the US. It is used in dietary supplements, biological insect-control agents, pharmaceuticals and cosmetic formulations. However, growing consumer demand for healthy food products may change the market trends in the future and wheat germ may soon become not simply a byproduct of wheat milling, but a high-value speciality product itself and contribute to the sustainability and economic feasibility of a wheat biorefinery system.

5.2.2 Barley

Barley is amongst the four largest cereal crops (wheat, maize and rice) grown in the world (Newman and Newman, 2008). Barley kernel is covered with hull, which comprises about 13% of the kernel and consists mostly of cellulose, hemicelluloses (xylans), lignin and a small amount of protein (Andersson *et al.*, 1999). Hulls adhere to the caryopsis of hulled barley, while they are not attached or are loosely attached to the grain surface of hulless barley. Lower ash and dietary fibre, but higher starch, protein and oil contents of hulless barley are due to the absence of the hull on the grain. Carbohydrates comprise about 80% of the barley grain. Starch content of barley may be up to 65% (Song and Jane, 2000). The major non-starch carbohydrates in barley are (1,3)(1,4)-β-D-glucans and arabinoxylans. β-Glucans, which are mainly present in the endosperm cell walls (Oscarsson *et al.*, 1996) consist of high-molecular-weight linear chains of β-glucosyl residues polymerized through β-(1–3) and β-(1–4) linkages (Newman and Newman, 2008).

It is estimated that about 85% of the world barley production is utilized as feed (Mäkinen and Nuutila, 2004). Malting is the second largest application for barley grain. Only 2% of barley is used for food production in the US. However, in regions with extreme climates, such as Himalayan nations, Ethiopia and Morocco, barley remains an important food source (Baik and Ullrich, 2008). Renewed interest in barley for food uses in developed countries is due to its health benefits, including reduction of blood LDL (low density lipoprotein) cholesterol level, glycemic index and body mass which lead to control of heart disease and type-2 diabetes. Wholegrain barley foods appear to be associated with increased satiety and weight loss. The beneficial effects of barley are associated with bioactive compounds, such as β-glucans, tocopherols and tocotrienols in the grain (Baik and Ullrich, 2008). Indeed the US Food and Drug Administration (FDA) approved a health claim verifying that barley contains high levels of β-glucans which help to prevent coronary heart disease when consumed by humans (FDA, 2005). Interest in incorporating low-tannin barley in the human diet is also increasing because of its high nutritional value (Newman and Newman, 2008). Even though currently the main source of β-glucan in foods is oats, barley β-glucan has a potential for new formulations. Co-products from barley milling can be used to recover phenolic acids and other antioxidants such as flavonols and proanthocyanins, which can be incorporated into functional foods and nutraceuticals such as high-performance sports drinks and dietary supplements, anti-ageing cosmetics and sun-screen lotions (Griffey *et al.*, 2009; Newman and Newman, 2008).

Ground whole barley grain and flour have been successfully converted to bioethanol (Sohn *et al.*, 2007; Septiano *et al.*, 2010). Interest in using barley, particularly hulless winter barley, for bioproduct manufacturing and biofuel production is growing, specifically outside the Corn Belt area in the US (Griffey *et al.*, 2009; Septiano *et al.*, 2010). Production of valuable high-protein content distillers' dried grains with solubles (DDGS) as a byproduct improves the feasibility of barley based alcohol production for fuel and solvent uses (Ingledew *et al.*, 1995).

A substantial amount of residual hull would be generated when hulled barley is utilized as a feedstock for bioethanol production. Barley fractions with high hull content could be a source of thermal energy to replace natural gas used in ethanol production plants.

The high-temperature pyrolysis characteristics of high hull content barley fractions were examined (Boateng *et al.*, 2007). Although both synthetic gas and bio-oil yields could be maximized at high temperatures, substantial amounts of silica-like structures (char) were observed on the surfaces and interfaces of the pyrolized materials. It was suggested that low-temperature pyrolysis (500–600 °C) or steam reform gasification might be alternative processes to minimize char formation. The gross energy content of the high hull content barley fractions was similar to that of wood bark (19.7 MJ kg^{-1}), indicating that this byproduct produced during bioethanol production can be a good alternative fuel for direct-fired combustion in ethanol plants. NOx emissions would be minimized during the combustion because of the low nitrogen content of the hulls.

5.2.3 Sorghum

Sorghum is a member of the grass family Poaceae. It is a drought-tolerant crop that grows well with minimal input. The structure of sorghum grain is similar to that of other cereal grains. The endosperm, germ and pericarp account about 85, 9.55 and 6.5% of the whole grain, respectively. In terms of its utilization, sorghum is classified into three groups: grain, grass/forage and sweet sorghum (Monk, Miller, and McBee, 1984). Grain sorghums are grown for their seeds, which are rich in starch. Grassy sorghums grow fast and can reach 3.1 m in height, generating significant amounts of biomass. Sweet sorghums have sweet juicy stems and produce the highest amount of biomass amongst the sorghum varieties.

Food applications of sorghum have been reviewed extensively (Taylor and Dewar, 2001; Taylor, Schober, and Bean, 2006; O'Kennedy, Grootboom, and Shewry, 2006). White food sorghum grain is milled into flour and incorporated into snacks, cookies, bread and ethnic foods. In the US, food products made with white sorghum are marketed as substitutes for wheat-based products targeting people with coeliac disease who are allergic to wheat gluten

(Ciacci *et al.*, 2007). In some African cultures tannin sorghums are preferred because of their satiating properties. It is believed that pigmented sorghum varieties promote the health of unborn babies and have therapeutic affects against digestive-system-related diseases (Taylor, Schober, and Bean, 2006). Sorghum is suitable for use in fermentation because of its high starch content. A large proportion of grain sorghum is used for malting in Africa. It is estimated that about 200 000 tons of sorghum are malted and 3000 million litres of sorghum beer is produced annually in Southern Africa alone (Taylor and Dewar, 2001). Sweet sorghum is further classified as sugar- and syrup-types. Syrup-type sweet sorghum is rich in glucose and used for syrup, wine, lactic acid and alcohol production (Billa *et al.*, 1997). Sorghum syrup is made by pressing the juice from the stems with rollers and boiling it down to the desired consistency. Animal feed and silage can also be made from sweet sorghums. Sugar-type sweet sorghum, which mainly contains sucrose, can be used for crystal sugar production. Grain sorghum is used to produce industrial bioproducts such as starch, biopolymer films and coatings. Isolation of starch from sorghum is done by wet milling (Munck, 1995). However, pigments in tannin sorghum stain the starch, which makes the final product undesirable. Bleaching and low-cost abbreviated wet-milling have been used to improve sorghum starch colour (Yang and Seib, 1995; Beta *et al.*, 2000). It is anticipated that new developments in wet-milling techniques and sorghum breeding will benefit the industrial use of sorghum starch for bioethanol and other bioproduct manufacturing. Activated carbon production (Diao, Walawender, and Fan, 2002) and recovery of health-beneficial bioactive compounds such as phytosterols and policosanols from outer layers of sorghum grain have also been explored (Singh, Moreau, and Hicks, 2003).

Kafirin which is the most hydrophobic prolamin in sorghum and has good film-forming properties is a potential source for producing bio-plastics. When plasticized with glycerol and polyethylene glycol, kafirin films had similar tensile and water-vapour-barrier properties as the films made from commercial maize zein plasticized in the same way (Da Silva and Taylor, 2005). Coating fruits with kafirin to delay ripening, reduce stem-end withering and increase shelf-life has also been investigated (Taylor, Schober, and Bean, 2006). Although kafirin coating was successful with pears, similar studies carried out with litchis showed excessive darkening and formation of white deposits on the fruit. Hence, utilization of kafirin as a fruit coating needs to be tested for different fruits for successful applications.

High photosynthetic efficiency of sorghum makes it very attractive as biofuel feedstock. Currently sweet and energy sorghums are being developed as dedicated bioenergy crops. Recovery of sorghum wax from DDGS produced during bioethanol production from grain sorghum has been examined (Weller *et al.*, 1998). Sorghum wax can be utilized as edible coatings and its production can be easily incorporated in a biorefinery system.

5.3   Oilseed Biorefineries

Oilseed crops are major agricultural commodities grown globally. Total world oilseed production was about 450 million metric tons during the 2010–2011 crop year (USDA, 2011). It is expected that oilseed production will increase to 460 million metric tons in 2011–2012. Currently rapeseed, canola, soybean, cottonseed, peanuts and sunflower seed are the most important commercial oilseeds. About 376 million metric tons of oilseeds were crushed for oil last year.

All oilseeds contain lipids, proteins and carbohydrates. Plant oils mainly contain triacylglycerol (TAG) but some free fatty acids (FFAs), mono- and diacylglycerides and unsaponifiable compounds such as phytosterols, tocopherols and tocotrienols are also present in oils. Oil and oilseed processing is a well-established industry. A number of publications reviewing various aspects of oil and oilseed processing are available (Khan and Hanna, 1983; Dunford and Dunford, 2004; Gupta, 2007). In summary, oilseeds need to be cleaned to remove plant stems, sticks, leaves and foreign material, and separated from their outer husk or shell before oil extraction. Seed size reduction and flaking improve oil yields when solvent extraction is used for oil recovery. Oilseeds are cooked or tempered to denature proteins, release oil from the cells and inactivate enzymes. For example, rapeseed contains the enzyme myrosinase. This enzyme catalyses hydrolysis of glucosinolates which are naturally present in rapeseed. During the hydrolysis process undesirable isothiocyanates and nitriles are formed. Seed cooking and tempering minimize enzymatic reactions.

There are several techniques for extracting oil from oilseeds. Two common oilseed extraction processes are solvent extraction and mechanical extraction using a screw press (Abraham, Hron, and Koltun, 1988; Mrema and McNulty, 1985). Use of a screw press is preferred by small processors because of its low capital cost. Solvent extraction is the standard practice in today's modern oilseed-processing facilities. An organic solvent, usually a mixture of hexane isomers, is brought into contact with flaked seeds to dissolve the oil, then oil and solvent are separated through distillation. Supercritical fluid, water and enzyme-aided water extraction processes are of interest for speciality and gourmet oil production (Dunford, 1995; Dunford, 2004).

The crude oil produced through mechanical pressing contains nearly 2% solids which must be separated. A filter press is commonly used for crude oil filtration. Crude oil comprises both desirable and undesirable compounds. Desirable oil components are TAGs (neutral lipids) and health-beneficial compounds such as tocopherols and phytosterols. FFAs, phospholipids (PLs), also referred to as gums, and lipid oxidation products must be removed during oil refining due to their adverse effects on oil quality, flavour and functionality. There are several unit operations in a crude oil refining operation. Degumming, deacidification/refining, bleaching, deodorization and winterization are commonly used for edible oil production. PLs settle out of the oil during shipping and storage and have adverse effects on the colour and flavour of oil. Therefore, they are removed from the oil during the degumming step. This process produces degummed oil and crude lecithin as a co-product. As an industry standard, FFA content of refined oil is reduced to less than 0.1% during deacidification, also referred to as neutralization or refining, which produces neutral oil and soap stock as a byproduct. Originally bleaching was used to remove coloured compounds such as carotenoids and chlorophyll. Today, bleaching is designed to remove undesirable oil components, including peroxides, aldehydes, ketones, phosphatides, oxidative trace metals, soaps and other contaminants, such as pesticides and polycyclic aromatic hydrocarbons. Deodorization is a steam-distillation process during which volatile and odoriferous compounds are stripped off with steam. The objective is to produce a bland and stable product. Deodorization removes FFAs, aldehydes, ketones and peroxides from bleached oil and deodorizer distillate (DD) is produced as a byproduct. In some cases, for example, salad oil production, RBD (refined-bleached-deodorized oil) oil is winterized to separate out waxes, thus minimizing solidification of the product under refrigerated conditions.

As discussed above, oil and oilseed processing generates numerous co-products that could be further processed to produce high-value products within a biorefinery system. A schematic of a simplified oil-/oilseed-based biorefinery system is shown in Figure 5.3.

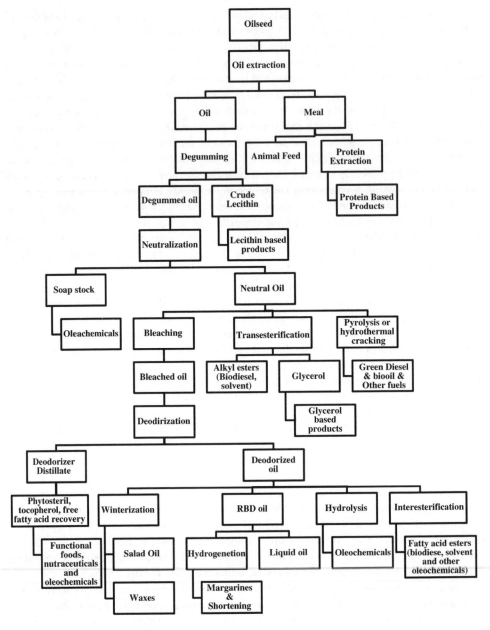

Figure 5.3   *A schematic of a typical oil- and oilseed-based biorefinery system.*

5.3.1   Oil- and Oilseed-Based Products

Edible Products

Oilseeds are grown particularly for oil and are commonly used for the production of edible oils for human consumption. In the US soybean oil is the major commodity product used in many food applications because of its availability and relatively low cost. Food applications of soybean oil have been reviewed by other authors (Liu, 1997; Hammond *et al.*, 2005).

Although fats and oils have an unhealthy connation to the public, they play an important role in human growth and health. In the past, edible oil quality was mainly defined by organoleptic parameters such as taste, odour and colour. Today, more emphasis is placed on the nutritional composition of oils. Vegetable oils are the major sources of the fat-soluble vitamins A, D, E, K, and other health-beneficial compounds such as PLs, glycolipids, phytosterols and policosanols. Oils are vital in the human diet for providing the essential fatty acids linoleic and linolenic acids. Some of these beneficial compounds may be removed during conventional oil refining processes (Dunford and Dunford, 2004; Dunford, Teel, and King, 2002). For example, a significant portion of the phytosterols present in crude oil is removed during the refining process and ends up in the soap stock and DD, byproducts of the edible-oil processing industry (Gutfinger and Letan, 1974). Commercial phytosterol-enriched products are made with phytosterols isolated from DD. The conventional sterol isolation process involves a series of energy-intensive and complex unit operations such as liquid/liquid extraction, trans- and inter-esterification, molecular distillation and crystallization. A patented process based on supercritical fluid fractionation technology produces oils enriched in phytosterol esters while refining, removing FFAs, from crude oil (Dunford and King, 2004). The same process can be used to recover phytosterols from DD (Dunford and King, 2000). Similar to phytosterols, tocopherols are removed during the oil refining process and are concentrated in the DD, which is an important feedstock for the industrial production of natural vitamin E (Nagao *et al.*, 2005; Mendes, Pessoa, and Uller, 2002). Phospholipids removed during oil degumming process are further refined to produce lecithins, which are important emulsifiers used in food, feed, pharmaceutical and other industrial products (Wendel, 2000; Van Nieuwenhuyzen, 1976).

RBD grade vegetable oils have been widely used in food applications. In the US liquid canola is preferred in salad dressings because of its low saturated fatty acid content. For frying applications canola oil is lightly hydrogenated to improve its oxidative stability. Soft (tub) and hard (stick) margarines, shortenings, and baking and pastry margarines are formulated with partially hydrogenated and liquid canola oil. In general high erucic acid rapeseed (HEAR) oil is not used for human consumption in the US and Canada. HEAR is only used in speciality food applications in Canada. Fully hydrogenated HEAR oil is added to peanut butter at 1–2% levels to prevent oiling (Przybylski *et al.*, 2005).

Cottonseed oil has always been a premium product for the food industry, specifically for frying applications because of its neutral flavour and extended fry life. A few years ago snack manufacturers moved away from cottonseed oil because of its relatively high saturated fatty acid content and consumer concerns over the adverse effects of saturated fats. Yet, since the announcement of the US Food and Drug Administration (FDA) trans fat labelling rule on food packaging (FDA, 2003a), a reversal has been observed and cottonseed oil is now back in high demand as a trans-free oil. Cottonseed oil forms beta-prime crystals that contribute a smooth consistency, fine texture and pleasant mouth feel to the products. Hence, cottonseed oil is suitable for food formulations that require solid fat. Zero-calorie fat substitutes have also been formulated with cottonseed oil (Akoh, 1998).

Traditionally calcium-rich sunflower seeds were roasted, ground into flour and incorporated into foods. Sunflower florets were extracted to obtain a yellow dye. Sunflower seed oil was also utilized in ceremonial body painting and residual plant parts were dried and used as building material (Putt, 1997). Sunflowers are classified in three groups; oilseed, confectionary and ornamentals, which are planted in gardens to attract birds. Some sunflower varieties with large seeds are roasted for snack foods or blended with other grains to make bird seed. Whole, large sunflower seeds contain about 40% oil and 25% protein, while small black sunflower seeds may contain up to 55% oil. In the 1980s sunflower became a major oilseed crop after the introduction of high-yielding varieties. Sunflower oil is rich in PUFAs. With the growing emphasis on the beneficial effects of PUFA in the human diet, sunflower oil has become a premium oil for many food applications (Carvalho, Miranda, and Pereira, 2006).

In 2003 the FDA approved the first qualified health claim for a food. The claim is allowed to be used on packaging labels stating that "Scientific evidence suggests but does not prove that eating 1.5 ounces per day of most nuts, as part of a diet low in saturated fat and cholesterol, may reduce the risk of heart disease" (FDA, 2003b). Numerous studies have shown that including nuts in the diet can reduce the risk of heart disease (Haumann, 1998; Kris-Etherton *et al.*, 2001; Higgs, 2002). Peanuts are rich in oil (40–50% oil) and a good source of a wide range of nutrients and bioactive compounds with health benefits. Many of the bioactive compounds that occur in nuts are associated with the oil fraction (Jonnala, Dunford, and Dashiell, 2005). Nuts contain tocopherols, tocotrienols, phytosterols and various types of flavonoids, including isoflavones and quercetin. About two-thirds of the total world peanut production is crushed for oil. Peanuts are found in a wide variety of products and are an integral part of many cuisines around the world. Unlike other countries, where most of the peanuts produced are used for oil, cake and meal production, the prime market in the US is peanut butter. About half of the US peanut crop is made into peanut butter, which is one of America's favourite foods (Anonymous, 2003). Other forms of peanut consumption in the US are confectionary (25%) and snack/salted peanuts (10%). Peanuts are even considered as a potential food for space travel (Haumann, 1998).

5.3.2 Industrial Products

An increasing proportion of vegetable oils is used for production of biofuel, high-value speciality chemicals and bioproducts. Oils from oilseeds are already very important feedstocks for the oleochemical industry. Oleochemicals are chemicals that are derived from animal or plant-based TAG and/or fatty acids. Basic oleochemicals are fatty acids, fatty alcohols, methyl esters and glycerine. Unsaturated fatty acids in oils can be readily polymerized or functional groups can be added to these renewable resources to produce a variety of value-added products and chemical intermediates.

Oil hydrolysis by steam splitting, saponification or enzyme action is the most important commercial modification of oils. Mono- and diacylglycerides which are produced by partial hydrolysis of TAG are excellent emulsifiers. They are widely used in food formulations. Sulfation of mono- and diacylglycerides improves their surface active properties and turns them into laundry detergents. Glycolipids present in plant oils are also very effective emulsifiers. Alcoholysis or aminolysis of fatty acids offer numerous opportunities for development of high-value multifunctional fatty acid derivatives. For example, fatty amides are used as antistatic or antiblock and mould-release agents. Due to their superior performances erucamides derived from high-erucic acid oils such as crambe and rapeseed receive a premium over the price of shorter-chain amides (Pryde and Rothfis, 1989; Murphy, 1992). Reactions of fatty alcohols with vinyl acetate or acetylene under pressure produce vinyl ethers which can be polymerized by strong acid to produce adhesives, surface coating materials and waterproofing agents or, depending on the type of alcohol, possibly polymeric plasticizers.

One of the largest non-food industrial uses of vegetable oils is the production of biodiesel. Sunflower seed, rapeseed, soybean or palm oils are the most common feedstocks for biodiesel production. Although TAG oils can be directly used as fuel in diesel engines they may lead to low engine efficiency, short life and high pollutant emissions. Additionally, crude oil contains compounds such as phosphorus and FFAs which may damage the engine. To overcome these problems oil is converted to biodiesel through a transesterification reaction. Crude vegetable oils to be used for biodiesel production must be at least degummed and neutralized prior to conversion to biodiesel (Dunford and Su, 2010). Methanol or ethanol is used to convert TAG to biodiesel (fatty acid methyl or ethyl esters).

Biodiesel contains a substantial amount of oxygen that reduces its heating value and results in higher NOx emissions when blended with petroleum diesel at high concentrations. Vegetable oils can be deoxygenated using existing hydrotreatment techniques to produce green diesel. This process involves conversion of vegetable oils to paraffins under high hydrogen pressure and at relatively high temperatures (Lappas, Bezergianni, and Vasalos, 2009). The main advantages of green or renewable diesel obtained from hydrotreatment processes over biodiesel obtained by transesterification are: superior cold weather properties; generation of a higher-value byproduct (propane vs. glycerol); higher energy content; greater cetane number and lower capital cost of the process. Feedstock quality requirements for hydrotreatment processes are more flexible than those for biodiesel (Stumborg, Wong, and Hogan, 1996). This leads to lower operating costs for green diesel production.

Pyrolysis is another process that can be used to produce vegetable-oil-based biofuels (Milne, Evans, and Nagle, 1990; Sharma and Bakhshi, 1993; Schwab, Bagby, and Freedman, 1987; Maher and Bressler, 2007). This process decomposes oils by heating them in the absence of air and produces a bio-oil which can be further upgraded to renewable diesel, gasoline or jet fuel. Pyrolysis of vegetable oils for fuel purposes is not new. Increasing product yield, cleaning and stabilizing bio-oil and developing improved catalysts for upgrading bio-oil to fuel are the main challenges affecting the economic feasibility of this process for commercial applications (Czernik and Bridgwater, 2004). Phenolics and cyclic ketones for resins and solvents, levoglucosan and levoglucosenon for polymers, and aromatic hydrocarbons for fuel and solvents are some of the potential renewable chemicals that can be produced from bio-oil (Elliot, 2004).

During biodiesel production by transesterification, glycerol is produced as a co-product. Glycerol supply has increased significantly due to increased biodiesel production during recent years. Although pharmaceutical-grade glycerol has numerous applications in food and industrial products, there is a need for new applications to utilize excess glycerol in the market. Development of innovative technologies for conversion of glycerol to high-value products could present new opportunities to biodiesel producers (Tyson, 2005). Glycerol can be burnt as a fuel or converted into higher-value chemicals by oxidation, reduction, halogenation, etherification and esterification reactions (Demirel *et al.*, 2007; Deckwer, 1995; Zhang *et al.*, 2011; Karinen and Krause, 2006). Dihydroxyacetone, glyceraldehyde, glyceric acid, glycolic acid, hydroxypyruvic acid, mesoxalic acid, oxalic acid and tartronic acid can be produced by glycerol oxidation. Dihydroxyacetone is a tanning agent also used as a synthon in organic synthesis. Dihydroxyacetone and hydroxypyruvic acid are used for D,L-serine synthesis. Recently, mesoxalic acid has been shown to have anti-HIV activity. Hydroxypyruvic acid is a cheese flavour and can be used to manage fruit maturation. Glyceryl carbonate can be converted to propanediol and ethylene glycol, which are used in polymer production and de-icing applications (Aresta *et al.*, 2009). Glycerol can also be a carbon source for micro-organisms. For example, the bacteria *Klebsiella planticola* and a mutant of *Enterobacter aerogenes* are able to use glycerol to produce ethanol (da Silva, Matthias, and Jonas, 2009; Amaral *et al.*, 2009).

Utilization of rapeseed-based methyl esters as solvents in paints and coatings is also increasing. Methyl esters have a number of advantages over petroleum-based solvents. They are biodegradable, have very low volatility and viscosity, high flash points and good water wetting and penetration properties. Plant oils have the right physico-chemical characteristics (inert, high viscosity index, low volatility) for production of lubricants, hydraulic fluids, engine oils, heat transfer fluids, inks and demoulding agents (Przybylski *et al.*, 2005). Brassylic and pelargonic acids derived from rapeseed oil are used for polymer production (nylon, polyester and melamine resin coatings).

Meal residue after oil extraction from oilseed is rich in proteins and generally used as feed in the livestock and aquaculture industries. Soybean proteins are widely used in many food and industrial applications. Canola proteins have high biological value due to their well-balanced amino acid composition and are potential high-quality protein sources for human consumption (Sosulski, 1983). Canola proteins have good functional properties, such as emulsifying, foaming and gelling abilities, which make them potential candidates for new product development in the food and bioproduct manufacturing industries (Yoshie-Stark, Wada, and Wasche, 2008; Aider and Barbana, 2011).

Utilization of oils from non-food oilseeds is critical for the long-term sustainability of a lipid-based biofuels and bioproduct manufacturing industry. Lesqueralla (about 20–25% oil), came-lina (30–40% oil), crambe (around 30–40% oil), vernonia (20–50% oil), castor (around 50% oil), jojoba (40–60% oil) and cuphea (about 30% oil) are some of the non-food oilseed crops that have been evaluated for their potential as feedstocks for bio-product and biofuel production. Furthermore, these oilseeds contain specific fatty acids that are not suitable for human consumption, but have unique chemical properties that are valuable for specific industrial applications. Some of the applications of specific fatty acids include; ricinoleic acid, an unusual hydroxy fatty acid in castor bean and lesquerella used for bio-plastics, erucic acid in rapeseed and crambe for lubricants, nylon and plasticizers, vernolic acid, an epoxy fatty acid in vernonia used in plasticizers and linolenic acid in flax for coatings and drying agents (Battey, Schmid, and Ohlrogge, 1989). Oils from non-food oilseed also have various functionalities as they are. For example when lesquerella oil was mixed with engine oil at low concentrations, 0.25%, it had superior performance in reducing wear and damage in diesel engines compared to castor, soybean and rapeseed methyl esters (Goodrum and Geller, 2004). Lesquerella contains unique molecules, estolides, which are rare in other seed oils. Estolides allow lesquerella oil to flow more easily than petroleum at cold temperatures. Lesquerella oil can also be used as an ingredient for a number of other bioproducts such as lubricants, motor oils, plastics, inks and adhesives. Lesquerella seed coat and seed meal contain a gum that is useful in coatings and food thickeners.

5.4   Conclusions

Cereal and oilseed crops are excellent feedstocks for the establishment of whole-crop biorefinery systems. Significant amounts of crop residues and several grain fractions are generated during harvest and grain milling. These co-products can readily be converted in to high-value foods and industrial bioproducts. Utilization of wheat for industrial product manufacturing needs to be minimized because of the high demand for this crop for food applications. However, cereal crop residues, including hulls, straw and stover have good potential for industrial bioproduct and biofuel production.

Oilseeds are very versatile crops, producing significant amounts of oil and protein. Vegetable oils have been very important renewable resources for the oleochemical industry. Currently there is a great interest in the production of vegetable-oil-based biofuels. The US Department of Energy (DOE) commissioned the National Renewable Energy Laboratory (NREL) to develop a long-term research plan for commercializing lipid fuels and co-products, thereby displacing petroleum and establishing a new bio-based industry. A detailed report, *Biomass Oil Analysis: Research Needs and Recommendations*, NREL/TP-510-34796, was published. I believe this report is an excellent document highlighting the current status of lipid-based biofuels. The findings of the report led to the conclusion that "The key issue is the current division of Office of the Biomass Program (OBP) budgets between ethanol and lipidfuels. DOE invests heavily in biomass ethanol technology that entails substantial risks and would not provide significant petroleum displacement benefits for at least another decade. Lipid fuels are produced today and face few barriers to use, particularly in low blends, and can displace over 1 billion gallons per year of petroleum diesel with existing supplies before 2014. In order to meet the goals set by the Office of Energy Efficiency and Renewable Energy to displace petroleum, budget allocations need to be revisited and Congressional support must be enlisted. This analysis found that modest production cost reductions and significant increases in lipid production could be achieved, but only if there are government incentives to promote the use of the high cost lipid fuels. Without the incentives, lipid fuel use will remain modest and there will be little need for research to increase supply" (Tyson, 2005).

There has been very important developments in the genetic modification of oilseeds, resulting in increased crop yields and oil content, incorporation of fatty acids of nutritional importance that are usually obtained from other sources and expression of industrial fatty acids that are currently sourced from petrochemicals or from low-yielding plants. Certainly biotechnology will continue to play a very important role in the economic viability and type of bioproducts that can be derived from oilseeds.

Utilization of non-food oilseeds for industrial bioproducts is critical for the sustainability of lipid-based biorefinery systems. Some of the non-food oilseeds contain unusual fatty acids with unique properties for specific applications. For these unusual fatty acids to be economically viable for bioproduct development, crops containing these fatty acids need to be genetically engineered so that only a single predominant unusual fatty acid is produced in the seeds. Considering the successful advancements in genetic modifications of oilseeds during the last two decades it is very likely that concentration of unusual fatty acids in non-food oilseeds will be increased significantly in the near future. The production of such new fatty acids in large quantities and a cost-effective way will definitely diversify the potential co-products that can be processed in a biorefinery system.

References

Abraham, G., Hron, R.J., and Koltun, S.P. (1988) Modeling the solvent extraction of oilseeds. *Journal of the American Oil Chemists' Society*, **65** (1), 125–135.

Ahring, B.K., Jensen, K., Nielsen, P. *et al.* (1996) Pretreatment of wheat straw and conversion of xylose and xylan to ethanol by thermophilic anaerobic bacteria. *Bioresource Technology*, **58**, 107–113.

Ahring, B.K., Licht, D., Schmidt, A.S. *et al.* (1999) Production of ethanol from wet oxidised wheat straw by thermoanaerobacter mathranii. *Bioresource Technology*, **68** (1), 3–9.

Aider, M. and Barbana, C. (2011) Canola proteins: composition, extraction, functional properties, bioactivity, applications as a food ingredient and allergenicity - a practical and critical review. *Trends in Food Science & Technology*, **22** (1), 21–39.

Akerberg, C. and Zacchi, G. (2000) An economic evaluation of the fermentative production of lactic acid from wheat flour. *Bioresource Technology*, **75**, 119–126.

Akoh, C.C. (1998) Fat replacers. *Food Technology*, **52** (3), 47–53.

Amaral, P.F.F., Ferreira, T.F., Fontes, G.C., and Coelho, M.A.Z. (2009) Glycerol valorization: new biotechnological routes. *Food and Bioproducts Processing*, **87** (3), 179–186.

Andersson, A.A.M., Elfverson, C., Andersson, R. *et al.* (1999) Chemical and physical characteristics of different barley samples. *Journal of the science of food and agriculture*, **79** (7), 979–986.

Annetts, J.E. and Audsley, E. (2003) Modelling the value of a rural biorefinery–Part Ii: analysis and implications. *Agricultural Systems*, **76** (1), 61–76.

Anonymous. American Peanut Council http://www.peanutsusa.com, accessed 7 July 2010.

Aresta, M., Dibenedetto, A., Nocito, F., and Ferragina, C. (2009) Valorization of bio-glycerol: new catalytic materials for the synthesis of glycerol carbonate via glycerolysis of urea. *Journal of Catalysis*, **268** (1), 106–114.

Baik, B.-K. and Ullrich, S.E. (2008) Barley for food: characteristics, improvement, and renewed interest. *Journal of Cereal Science*, **48** (2), 233–242.

Battey, J.F., Schmid, K.M., and Ohlrogge, J.B. (1989) Genetic engineering for plant oils: potential and limitations. *Trends in Biotechnology*, **7** (5), 122–126.

Beta, T., Rooney, L.W., Marovatsanga, L.T., and Taylor, J.R.N. (2000) Effect of chemical treatments on polyphenols and malt quality in sorghum. *Journal of Cereal Science*, **31**, 295–302.

Billa, E., Koullas, D.P., Monties, B., and Koukios, E.G. (1997) Structure and composition of sweet sorghum stalk components. *Industrial Crops and Products*, **6**, 297–302.

Boateng, A.A., Hicks, K.B., Flores, R.A., and Gutsol, A. (2007) Pyrolysis of hull-enriched byproducts from the scarification of hulled barley (Hordeum Vulgare L.). *Journal of Analytical and Applied Pyrolysis*, **78** (1), 95–103.

Braunegg, G., Lefebvre, G., and Genser, K.F. (1998) Polyhydroxyalkanoates, biopolyesters from renewable resources: physiological and engineering aspects. *Journal of Biotechnology*, **65** (2–3), 127–161.

Carvalho, I.S., Miranda, I., and Pereira, H. (2006) Evaluation of oil composition of some crops suitable for human nutrition. *Industrial Crops and Products*, **24** (1), 75–78.

Chen, Y. (2011) Efficiency of abrasive dehulling to produce wheat grain fractions enriched in antioxidants, in *Biosystems and Agricultural Engineering*, Oklahoma State University, Stillwater, OK, p. 131.

Ciacci, C., Maiuri, L., Caporaso, N. *et al.* (2007) Celiac disease: in vitro and in vivo safety and palatability of wheat-free sorghum food products. *Clinical Nutrition*, **26** (6), 799–805.

Clark, J.H. (2007) Perspective green chemistry for the second generation biorefinery – sustainable chemical manufacturing based on biomass. *Journal of Chemical Technology & Biotechnology*, **82**, 603–609.

Clark, J.H., Deswarte, F.E.I., and Farmer, T.J. (2009) The integration of green chemistry into future biorefineries. *Biofuels, Bioproducts and Biorefining*, **3** (1), 72–90.

Cureton, T.K. (1972) *The Physiological Effects of Wheat Germ Oil on Humans in Exercise, Forty-Two Physical Training Programs Utilizing 894 Humans*, Charles C. Thomas., Springfield, IL.

Czernik, S. and Bridgwater, A.V. (2004) Overview of applications of biomass fast pyrolysis oil. *Energy Fuels*, **18** (2), 590–598.

da Silva, G.P., Matthias, M., and Jonas, C. (2009) Glycerol: a promising and abundant carbon source for industrial microbiology. *Biotechnology Advances*, **27** (1), 30–39.

Da Silva, L.S. and Taylor, J.R.N. (2005) Physical, mechanical and barrier properties of kafirin films from red and white sorghum milling fractions. *Cereal Chemistry*, **82**, 9–14.

Datta, R., Tsai, S.-P., Bonsignore, P. *et al.* (1995) Technological and economical potential of poly(lactic) acid and lactic acid derivatives. *FEMS Microbiology Reviews*, **16**, 221–231.

Day, L., Augustin, M.A., Batey, I.L., and Wrigley, C.W. (2006) Wheat-gluten uses and industry needs. *Trends in Food Science & Technology*, **17** (2), 82–90.

Deckwer, W.-D. (1995) Microbial conversion of glycerol to 1,3-propanediol. *FEMS Microbiology Reviews*, **16** (2–3), 143–149.

Demirel, S., Lehnert, K., Lucas, M., and Claus, P. (2007) Use of renewables for the production of chemicals: glycerol oxidation over carbon supported gold catalysts. *Applied Catalysis B: Environmental*, **70** (1–4), 637–643.

Dendy, D.A.V. and Dobraszczyk, B.J. (2001) *Cereals and Cereal Products Chemistry and Technology*, Aspen Publishers, Inc., Gaithersbur, Maryland.

Deswarte, F.E.I., Clark, J.H., Wilson, A.J. *et al.* (2007) Toward an integrated straw-based biorefinery. *Biofuels, Bioproducts and Biorefining*, **1** (4), 245–254.

Diao, Y., Walawender, W., and Fan, L. (2002) Activated carbons prepared from phosphoric acid activation of grain sorghum. *Bioresource Technology*, **81**, 45–52.

Dorado, M.P., Lin, S.K.C., Koutinas, A. *et al.* (2009) Cereal-based biorefinery development: utilisation of wheat milling by-products for the production of succinic acid. *Journal of Biotechnology*, **143** (1), 51–59.

Dunford, N.T. (1995) Use of supercritical carbon dioxide for edible oil processing, in *Department of Agricultural, Food and Nutritional Science*, University of Alberta, Edmonton, AB, Canada.

Dunford, N.T. (2004) Effects of oil and oilseed processing techniques on bioactive compounds, in *Nutritionally Enhanced Oil and Oilseed Processing* (eds N.T. Dunford and H.B. Dunford), AOCS Press, Champaign, IL, pp. 25–37.

Dunford, N.T. (2009) Wheat germ oil, in *Gourmet and Health-Promoting Specialty Oils* (eds R.A. Moreau and A. Kamal-Eldin), AOCS Press, Urbana, IL, pp. 359–376.

Dunford, N.T. and Dunford, H.B. (2004) *Nutritionally Enhanced Oil and Oilseed Processing*, AOCS Press, Champaign, IL.

Dunford, N.T. and Edwards, J. (2010) Nutritional bioactive components of wheat straw as affected by genotype and environment. *Bioresource Technology*, **101** (1), 422–425.

Dunford, N.T., King, J.W., and inventors (January 13 2004) Supercritical Fluid Fractionation Process for Phytosterol Ester Enrichment in Vegetable Oils., U.S. patent 6,677,469 B1.

Dunford, N.T. and King, W.J. (2000) Phytosterol enrichment of rice bran oil by a supercritical carbon dioxide fractionation technique. *Journal of Food Science*, **65**, 1395–1399.

Dunford, N.T. and Martinez, J. (2003) Nutritional components of supercritical carbon dioxide extracted wheat germ oil. 6th International Symposium on Supercritical Fluids. Versailles, France, pp. 273–278.

Dunford, Nurhan T. and Su, Aihua. 2010. Effect of Canola Oil Quality on Biodiesel Conversion Efficiency and Properties. *Trans ASABE*, **53** (3), 993–997.

Dunford, N.T., Teel, J.A., and King, J.W. (2002) A continuous countercurrent supercritical fluid deacidification process for phytosterol ester fortification in rice bran oil. *Food Research Intenational*, **36** (2), 175–181.

Dunford, N.T. and Zhang, M. (2003) Pressurized solvent extraction of wheat germ oil. *Food Research International*, **36**, 905–909.

Eisenmenger, M., Dunford, N.T., Eller, F. *et al.* (2006) Pilot scale supercritical carbon dioxide extraction and fractionation of wheat germ oil. *Journal of the American Chemical Society*, **10** (2), 863–868.

Elliot, D.C. (2004) *Chemicals from Biomass*, Elsevier Inc., San Diego.

Ellis, P.R., Cochrane, P.M., Dale, M.F.B. *et al.* (1998) Starch production and industrial use. *Journal of the Science of Food and Agriculture*, **77**, 289–311.

FAO (2010) Food and Agricultural Organization (FAO) Faostat Data Base. FAO.

FDA (2003a) Federal Register - 68 Fr 41433 July 11, 2003: Food Labeling; Trans Fatty Acids in Nutrition Labeling; Consumer Research to Consider Nutrient Content and Health Claims and Possible Footnote or Disclosure Statements; Final Rule and Proposed Rule. US Food and Drug Administration, Department of Health and Human Services. pp. 41433–41506.

FDA (2003b) Qualified Health Claims: Letter of Enforcement Discretion-Nuts and Coronory Heart Disease. US Food and Drug Administration, http://www.cfsan.fda.gov/~dms/qhcnuts2.html, accessed 6 November 2010.

FDA (2005) FDA Allows Barley Products to Claim Reduction in Risk of Coronary Heart Disease. http://www.fda.gov/NewsEvents/Newsroom/PressAnnouncements/2005/ucm108543.htm. Access date: October 3, 2010.

Goncalves, L.M.D., Xavier, A.M.R.D., Almeida, J.S., and Carrondo, M.J.T. (1991) Concomitant substrate and product inhibition kinetics in lactic acid production. *Enzyme and Microbial Technology*, **13**, 314–319.

Goodrum, J.W. and Geller, D.P. (2004) Influence of fatty acid methyl esters from hydroxylated vegetable oils on diesel fuel lubricity. *Bioresource Technology*, **96**, 851–855.

Graboski, M.S. and McCormick, R.L. (1998) Combustion of fat and vegetable oil derived fuels in diesel engines. *Progress in Energy and Combustion Science*, **24**, 125–164.

Griffey, C., Brooks, W., Kurantz, M. *et al.* (2009) Grain composition of virginia winter barley and implications for use in feed, food, and biofuels production. *Journal of Cereal Science*, **51** (1), 41–49.

Guilbert, S., Gontard, N., Morel, M.H. *et al.* (2002) Formulation and properties of wheat gluten films and coatings, in *Protein-Based Films and Coatings* (ed. A. Gennadios), CRC Press, Boca Raton, FL, pp. 69–122.

Gupta, G. (2007) *Practical Guide for Vegetable Oil Processing*, AOCS Press, Urbana, IL.

Gutfinger, T. and Letan, A. (1974) Quantitative changes in some unsaponifiable components of soya bean oil due to refining. *Journal of Agricultural and Food Chemistry*, **25**, 1143–1147.

Hammond, E.G., Johnson, L.A., Su, C. *et al.* (2005) Soybean oil, in *Bailey's Industrial Oil and Fat Products* (ed. F. Shahidi), John Wiley & Sons, Inc., Nwe Jersey, USA.

Haumann, B.F. (1998) Peanuts finds niche in healthy diet. *Inform*, **9** (8), 746–752.

Higgs, J. (2002) The beneficial role of peanuts in the diet- an update and rethink! peanuts and their role in CHD. *Nutrition & Food Science*, **32**, 214–218.

Ingledew, W.M., Jones, A.M., Bhatty, R.S., and Rossnagel, B.G. (1995) Fuel alcohol production from hull-less barley. *Cereal Chemistry*, **72** (2), 147–150.

Irmak, S., Dunford, N.T., and Milligan, J. (2005) Policosanol contents of beeswax, sugar cane and wheat extracts. *Food Chemistry*, **95**, 312–318.

Jenkins, T. (2008) Toward a biobased economy: examples from the UK. *Biofuels, Bioproducts and Biorefining.*, **2**, 133–143.

Jonnala, R.S., Dunford, N.T., and Dashiell, K.E. (2005) New high oleic peanut cultivars grown in the southwestern U.S. *Journal of the American Oil Chemists' Society*, **85**, 125–128.

Karinen, R.S. and Krause, A.O.I. (2006) New biocomponents from glycerol. *Applied Catalysis A: General*, **306**, 128–133.

Khan, L.M. and Hanna, M.A. (1983) Expression of oil from oilseeds-a review. *Journal of Agricultural Engineering Research*, **28** (6), 495–503.

Koutinas, A.A., Wang, R., Campbell, G.M., and Webb, C. (2008) A whole crop biorefinery system: a closed system for the manufacture of non-food products from cereals, in *Biorefineries-Industrial Processes and Products* (eds B. Kamm, P.R. Gruber, M. Kamm, and R.P.) Wiley-VCH, Weinheim, Germany, pp. 165–191.

Koutinas, A.A., Wang, R., and Webb, C. (2004) Evaluation of wheat as generic feedstock for chemical production. *Industrial Crops and Products*, **20**, 75–88.

Koutinas, A.A., Xu, Y., Wang, R., and Webb, C. (2007) Polyhydroxybutyrate production from a novel feedstock derived from a wheat-based biorefinery. *Enzyme and Microbial Technology*, **40** (5), 1035–1044.

Kris-Etherton, P.M., Zhao, G., Binkoski, A.E. *et al.* (2001) The effects of nuts on coronary heart disease risk. *Nutrition Review*, **59**, 103–111.

Lang, X., Macdonald, D.G., and Hill, G.A. (2001) Recycle bioreactor for bioethanol production from wheat starch II. Fermentation and economics. *Energy Sources*, **23**, 427–436.

Lappas, A.A., Bezergianni, S., and Vasalos, I.A. (2009) Production of biofuels via co-processing in conventional refining processes. *Catalysis Today*, **145** (1–2), 55–62.

Liu, K. (1997) *Soybeans: Chemisty, Technology, and Utilization*, Chapman & Hall., New York, NY.

MacMasters, M.M., Hinton, J.J.C., and Bradbury, D. (1971) Microscopic structure and composition of the wheat kernel, in *Wheat Chemistry and Technology*, 2nd edn (ed. Y. Pomeranz), Am. Assoc. Cereal Chem., St. Paul, MN, pp. 51–114.

Maher, K.D. and Bressler, D.C. (2007) Pyrolysis of triglyceride materials for the production of renewable fuels and chemicals. *Bioresource Technology*, **98**, 2351–2368.

Mäkinen, K. and Nuutila, A.M. (2004) Barley seed as a production host for industrially important proteins. *AgBiotechNet*, **6** (119), 1N–8N.

McCartney, D.H., Block, H.C., Dubeski, P.L., and Ohama, A.J. (2006) Review: the composition and availability of straw and chaff from small grain cereals for beef cattle in western Canada. *Canadian Journal of Animal Science*, **86**, 443–455.

Mckean, W.T. and Jacobs, R.S. (1997) Wheat Straw as a Paper Fiber Source. Prepared for Recycling Technology Assistance Partnership (ReTAP) by The Clean Washington Center and Domtar Inc., http://www.cwc.org/paper/pa971rpt.pdf. Accessed on August 1, 2010.

Mellan, I. (1977) *Acids. Industrial Solvents Handbook*, 2nd edn, Noyes Data Corporation, N.J., pp. 435–448.

Mendes, M.F., Pessoa, F.L.P., and Uller, A.M.C. (2002) An economic evaluation based on an experimental study of the vitamin E concentration present in deodorizer distillate of soybean oil using supercritical CO_2. *The Journal of Supercritical Fluids*, **23** (3), 257–265.

Milne, T.A., Evans, R.J., and Nagle, N. (1990) Catalytic conversion of microalgae and vegetable oils to premium gasoline, with shape-selective zeolites. *Biomass*, **21** (3), 219–232.

Monk, R.L., Miller, F.R., and McBee, G.G. (1984) Sorghum improvement for energy production. *Biomass*, **6** (1–2), 145–153.

Mrema, G.C. and McNulty, P.B. (1985) Mathematical model of mechanical oil expression from oilseeds. *Journal of Agricultural Engineering Research*, **31**, 361–370.

Munck, L. (1995) New milling technologies and products: whole plant utilization by milling and separation of the botanical and chemical components, in *Sorghum and Millets: Chemistry and Technology* (ed. D.A.V. Dendy), American Association of Cereal Chemists, St. Paul, MN, USA, pp. 223–281.

Murphy D.J. (1992) Modifying oilseed crops for non-edible products. *Trends in Biotechnology*, **10**, 84–87.

Nagao, T., Kobayashi, T., Hirota, Y. *et al.* (2005) Improvement of a process for purification of tocopherols and sterols from soybean oil deodorizer distillate. *Journal of Molecular Catalysis B: Enzymatic*, **37** (1–6), 56–62.

Newman, R.K. and Newman, C.W. (2008) *Barley for Food and Health: Science, Technology and Products*, John Wiley & Sons, Inc.

O'Kennedy, M.M., Grootboom, A., and Shewry, P.R. (2006) Harnessing sorghum and millet biotechnology for food and health. *Journal of Cereal Science*, **44** (3), 224–235.

Oscarsson, M., Andersson, R., Salomonsson, A.C., and Åman, P. (1996) Chemical composition of barley samples focusing on dietary fibre components. *Journal of Cereal Science*, **24** (2), 161–170.

Paster, M., Pellegrino, J.L., and Carole, T.M. (2003) *Industrial Bioproducts: Today and Tomorrow*, Prepared by Energetics, Incorporated for The U.S., Department of Energy, Office of Energy Efficiency and Renewable Energy, Office of the Biomass Program, Washington, D.C.

Posner, E.S. and Hibbs, A.N. (2005) *Wheat Flour Milling*, 2nd edn, American Association of Cereal Chemists, Inc, St. Paul.

Pryde, E.H. and Rothfis, J.A. (1989) Industrial and nonfood uses of vegetable oils, in *Oil Crops of the World* (eds G. Robbelen, R.K. Downey, and A. Ashri), McGraw-Hill Publishing Company, NY, p. 553.

Przybylski, R., Mag, T.K., Eskin, N.A.M., and McDonald, B.E. (2005) Canola oil, in *Bailey's Industrial Oil and Fat Products* (ed. F. Shahidi), John Wiley & Sons, Inc., NJ, USA, pp. 61–121.

Putt, E.D. (1997) Early history of sunflower, in *Sunflower Technology and Production* (ed. A.A. Scheiter), ASA, CSSA and ASSA, Madison, WI, pp. 1–19.

Saha, B.C., Iten, L.B., Cotta, M.A., and Wu, Y.V. (2005) Dilute acid pretreatment, enzymatic saccharification and fermentation of wheat straw to ethanol. *Process Biochemistry*, **40**, 3693–3700.

Schwab, A.W., Bagby, M.O., and Freedman, B. (1987) Preparation and properties of diesel fuels from vegetable oils. *Fuel*, **66** (10), 1372–1378.

Septiano, W., Dunford, N.T., Wilkins, W., and Edwards, J. (2010) Ethanol production from winter hulless barley. *Biological Engineering*, **2** (4), 211–219.

Sharma, R.K. and Bakhshi, N.N. (1993) Catalytic conversion of fast pyrolysis oil to hydrocarbon fuels over HZSM-5 in a dual reactor system. *Biomass and Bioenergy*, **5** (6), 445–455.

Singh, V., Moreau, R.A., and Hicks, K.B. (2003) Yield and phytosterol composition of oil extracted from grain sorghum and its wet-milled fractions. *Cereal Chemistry*, **80**, 126–129.

Sohn, M., Himmelsbach, D.S., Barton, F.E. *et al.* (2007) Near-infrared analysis of ground barley for use as a feedstock for fuel ethanol production. *Applied Spectroscopy*, **61**, 1178–1183.

Song, Y. and Jane, J. (2000) Characterization of barley starches of waxy, normal, and high amylose varieties. *Carbohydrate Polymers*, **41** (4), 365–377.

Sosulski, F.W. (1983) Rapeseed protein for food use, in *Developments in Food Proteins* (ed. B.J.F. Hudson), Applied Science Publishers, London and New York, pp. 109–132.

Stumborg, M., Wong, A., and Hogan, E. (1996) Hydroprocessed vegetable oils for diesel fuel improvement. *Bioresource Technology*, **56** (1), 13–18.

Szczodrak, J. (1988) The enzymatic hydrolysis and fermentation of pretreated wheat straw to ethanol. *Biotechnology and Bioengineering*, **32**, 771–776.

Taylor, J.R.N. and Dewar, J. (2001) Developments in sorghum food technologies, in *Advances in Food and Nutrition Research*, Academic Press, pp. 217–264.

Taylor, J.R.N., Schober, T.J., and Bean, S.R. (2006) Novel food and non-food uses for sorghum and millets. *Journal of Cereal Science*, **44** (3), 252–271.

Tyson, K.S. (2005) DOE Analysis of fuels and coproducts from lipids. *Fuel Processing Technology*, **86** (10), 1127–1136.

USDA (2011) Oilseeds: World Markets and Trade. United States Department of Agriculture, Foreign Agricultural Service.

Van Nieuwenhuyzen, W. (1976) Lecithin production and properties. *Journal of the American Oil Chemists' Society*, **53**, 425–427.

Webb, C. and Wang, R. (1997) Development of a generic fermentation feedstock from whole wheat flour, in *Cereals: Novel Uses and Processes* (eds G.M. Campbell, C., Webb and S.L. McKee), Plenum Press, New York, p. 205.

Weller, C.L., Gennadios, A., Saraiva, R.L., and Cuppett, S.L. (1998) Grain sorghum wax as an edible coating for gelatine-based candies. *Journal of Food Quality*, **21**, 117–128.

Wendel, A. (2000) Lecithin: the first 150 years. Part I: from discovery to early commercialization. *Inform*, **11** (8), 885–890, 892.

Werpy, T. and Petersen, G. (2004) Top Value Added Chemicals from Biomass. Results of screening for potential candidates from sugars and synthesis gas. Produced by staff at the Pacific Northwest National Laboratory (PNNL) and the National Renewable Energy Laboratory (NREL) http://www1.eere.energy.gov/biomass/pdfs/35523.pdf.

Xie, M. (2009) Aqueous enzymatic extraction of wheat germ oil, *Biosystems and Agricultural Engineering*, Oklahoma State University, Stillwater, OK, p. 83.

Xie, M. and Dunford, N.T. (2010) Optimization of aqueous enzymatic oil extraction form wheat germ. American Oil Chemists Society Annual Meeting and Expo. Phoenix, AR: AOCS.

Yang, P. and Seib, P.A. (1995) Low-input wet-milling of grain sorghum for readily accessible starch and animal feed. *Cereal Chemistry*, **72** (5), 498–503.

Yoshie-Stark, Y., Wada, Y., and Wasche, A. (2008) Chemical composition, functional properties, and bioactivities of rapeseed protein isolates. *Food Chemistry*, **107** (1), 32–39.

Yu, L. (2008) *Wheat Antioxidants*, Wiley-Interscience, Hoboken, NJ.

Zhang, Y., Gao, F., Zhang, S.-P. *et al.* (2011) Simultaneous production of 1,3-dihydroxyacetone and xylitol from glycerol and xylose using a nanoparticle-supported multi-enzyme system with in situ cofactor regeneration. *Bioresource Technology*, **102** (2), 1837–1843.

6

Bioactive Soy Co-Products

Arvind Kannan, Srinivas Rayaprolu and Navam Hettiarachchy
Department of Food Science, University of Arkansas, Fayetteville, Arkansas, USA

6.1 Introduction

By the beginning of the twenty-first century fossil fuels have become an essential part of providing efficient energy on the planet, but their reserves in the world are declining as the demands are increasing rapidly. Developed countries like the United States and the European Union recognized the need for exploring alternative processes and methods to those that use fossil fuels for energy. These processes can be developed based on sustainable, efficient and renewable resources that could produce energy for the future generations. Conventional and renewable energy sources have been sought, but their exploration have been limited due to various factors, including location, high infrastructure investments and cost efficiency. Meanwhile, scientists have developed biorefineries that use agricultural or plant-based biomass to produce fuels like ethanol by utilizing fermentation as a biomass conversion process. Biorefineries are facilities that integrate biomass conversion processes to produce fuel, power and chemicals from biomass (Demirbas, 2009). The importance of biorefineries has grown rapidly in recent years due to the increased financial burden on the economy of procuring fossil fuels. Biorefineries, apart from producing biofuels, also produce large amounts of waste which is thought to be uneconomical and is sold as feed at a cheaper price. This chapter discusses the various co-products from biorefineries and assesses the advantages of utilizing these economically important co-products in various food applications. It demonstrates the feasibility of producing certain nutritional and medicinal bioactive products that add value to the output of biorefineries.

Biorefinery Co-Products: Phytochemicals, Primary Metabolites and Value-Added Biomass Processing, First Edition.
Edited by Chantal Bergeron, Danielle Julie Carrier and Shri Ramaswamy.
© 2012 John Wiley & Sons, Ltd. Published 2012 by John Wiley & Sons, Ltd.

6.1.1   Industrial Agricultural Biomass

Advancements in fields including genetics, biotechnology, process chemistry and engineering have lead to new technologies for converting renewable biomass to valuable fuels and other products in a biorefinery. Nature provides biomass energy sources that include agricultural products, animal waste, fuel wood, charcoal and other biologically derived fuel sources. Biomass from agriculture currently accounts for about 14% of world energy consumption, especially in many developed and developing countries (Kaygusuz and Türker, 2002). Corn, soy, sugar cane and sugar beet are some of the crops that are utilized in biorefineries. Biomass upgrading processes include fractionation, liquefaction, pyrolysis, hydrolysis, fermentation and gasification to produce fuels from organic matter. More than 1 billion dry tons of harvested biomass could be fed to biorefineries to replace 30% of the annual fuel consumption in the US (Perlack *et al.*, 2005). Many of the bio-based industry products currently used are results of direct physical or chemical treatment and processing of agricultural biomass, for example cellulose, starch, oil, protein, lignin and terpenes. Lignocellulosic feedstock is involved in the production of three major chemical fractions: hemicellulose/polyoses, which are predominantly sugar polymers of pentose sugars; cellulose and lignin, a phenolic polymer. Starch can undergo reactions including plasticization (a mixed-polymerization reaction), chemical modifications, including etherification into carboxymethyl starch; esterification into fatty acid esters; reductive amination into ethylenediamine, hydrogenative splitting into sorbitol and glycols, and biotechnological conversion into poly-3-hydroxybutyric acid (Kamm, Gruber, and Kamm, 2007; Stocker, 2008). (Figure 6.1) Mono-, sesqui-, and diterpenes have also been documented as products released from biomass (Rupar and Sanati, 2005). Biotechnological research and advancement has created processes and methods to produce ethanol, butanol, acetone, lactic acid and amino acids, including glutamic acid, lysine and tryptophan in biorefineries (Kamm and Kamm, 2004).

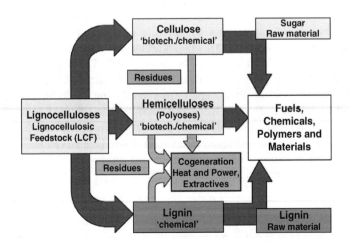

Figure 6.1   *Lignocellulosic feedstock biorefining. (Adapted from Kamm et al., 2006). Adapted from Kamm B. et al., 2006 © Helmholtz Centre Potsdam GFZ German Research Centre for Geosciences (2006).*

6.1.2 Processing of Co- and Byproducts

The agriculture biomass that is utilized in the biorefineries to produce fuels leaves behind large quantities of co- and byproducts. The term "chemurgy" (Hale, 1934) is used to denote utilization of such renewable agricultural resources in the industry. Several technologies, including bioengineering, polymer chemistry, food science and agriculture were developed by integrating different fields to meet the tasks of converting the biomass into value-added products and co-products.

6.1.3 Value Addition and Sustainability

Value addition and sustainability are the primary objectives of biorefinery systems. While the primary product from the refinery is utilized for energy production, its co- and byproducts are being researched for value-added uses. Sustainable economic growth requires safe and sustainable resources for industrial production, a long-term and confident investment and finance system, ecological safety, and sustainable life and work perspectives for the public. Fossil resources are not regarded as sustainable, however, and their availability is more than questionable in the long-term. Because of the increasing price of fossil resources, the feasibility of their utilization is declining. It is essential to establish solutions which reduce the rapid consumption of fossil resources, which are not renewable (petroleum, natural gas, coal, minerals). An approach is the stepwise conversion of large parts of the global economy to a sustainable bio-based economy with bioenergy, biofuels and bio-based products. Alternative raw materials, including wind, sun, biomass and nuclear sources, are used for establishing energy production systems to cater to the demands of the industry. The principle behind these establishments is based on conversion of sustainable material like plants by applying industrial biotechnology that helps in fuel generation (Kamm, Gruber, and Kamm, 2007).

6.2 Co-Products Obtained from Industrial Biorefineries

6.2.1 Cereal- and Legume-Based Industrial Co-Products

The industrial co-products of biorefineries are primarily used as feed for livestock, although recent advances in research and technologies to extract ingredients allow their use in food products. Some of the agricultural biomass that is processed in the biorefineries and the co-products are discussed in the following sections.

6.2.2 Legume Co-Products – Soy

Soybeans are popular legume crops used in biorefineries to manufacture biodiesel from the soybean oil. The production of biodiesel from soybeans occurs in two stages: the first treatment involves the removal of the oil from soybeans, and then the soybean oil is converted into biodiesel. The first stage is often called crushing, and the most common method employed to convert the oil into biodiesel is a process known as transesterification. Oil extraction and transesterification processes result in the release of two important co-products, soybean meal and crude glycerine, respectively (Pradhan *et al.*, 2009).

Oil is extracted from soybeans by crushing using mechanical extruders, or more commonly by using hexane. Soybeans are initially cleaned, heated and dried (to 10% moisture content) and passed through mechanical rollers which crack the beans into pieces. The soybean hulls removed by aspiration may be further ground or may be blended with the soybean meal that is obtained at the end of the extraction and sold as animal feed. The dehulled beans are further conditioned by heating, ground into flakes, and fed to the oil extraction unit where the oil from the beans is separated by desolving in hexane. The oil and hexane mixture is treated with steam to remove the hexane, which is recycled for additional processing. Final heating and cooling of the oil is done with hot air and cooling water. The crude soybean oil is degummed and may be deodorized, bleached and neutralized. The oil-depleted, dried soybean flakes are ground to a uniform size to make soybean meal. Further, the hexane is used in a series of chemical reactions through separations and distillations to yield biodiesel (Figure 6.2).

While the lipid content in the soybean meal/flour is utilized for biodiesel processing (Alcantara *et al.*, 2000; Vasudevan and Briggs, 2008), the biomass that remains is rich in carbohydrates, proteins and minerals like potassium, calcium, magnesium, phosphorus and sulfur. These co-products can be utilized to prepare proteins and peptides that have potential uses in food products as nutritional additives. Proteins including leguminins (11S globulin fraction) and vicilins (7S globulin) extracted from soy beans were studied for properties including gelation, emulsification, solubility and hydrophobicity for enhanced functionality during preparation that could be advantageous when added as ingredients to food products (Boatright and Hettiarachchy, 1995; Wu, Hettiarachchy, and Qi, 1998). The protein isolates were extracted under alkaline conditions and iso-electrically precipitated (pH 4.5). Enzymatic (papain, pancreatin and trypsin) modifications were conducted under optimized conditions and the degree of hydrolysis was measured (Kim Lee, Peter, and Rhee, 1990), while chemical modification was conducted using alkali (pH 10). The modified proteins were then tested for surface hydrophobicity based on the 1-anilino-8-naphthalene sulfonate (ANS) binding method (Hayakawa and Nakai, 1985); nitrogen solubility (Bera and Mukherjee, 1989); emulsifying activity and emulsifying stability by the turbidimetric method of Pearce and Kinsella (1978) and foaming properties (Kato *et al.*, 1983). Significantly improved physico-chemical properties including foaming, emulsifying, water-holding and solubilizing activities that can add better functionality when these proteins are used in food systems. Enzyme-modified protein performed better in comparison to alkali-treated. Results showed decreased solubility of the protein with addition of lipid, especially phospholipids, while enzymatic modifications to the protein enhanced the solubility and emulsifying properties, and decreased surface hydrophobicity in comparison to un-treated protein. These enzymatic and chemical methods of modification to soy protein have been developed for use in both the food and pharmaceutical industries (Hettiarachchy, Kalapathy, and Myers, 1995; Were, Hettiarachchy, and Kalapathy, 1997; Qi, Hettiarachchy, and Kalapathy, 1997; Wu, Hettiarachchy, and Qi, 1998). Soy protein was also studied for its unique adhesion properties, which were utilized in preparing edible films as a protective hurdle against disease-causing microorganisms (Ko *et al.*, 2001). Antimicrobials incorporated into edible films were developed with soy protein and were used in preventing contamination of agricultural and poultry products (Xie *et al.*, 2002; Eswaranandam, Hettiarachchy, and Johnson, 2004; Sivarooban, Hettiarachchy, and Johnson, 2006; Hettiarachchy and Eswaranadam, 2007).

The health-promoting effects of soybean consumption have been more recently linked to the biological activities of a specific group of phenolic compounds known as isoflavonoids (McCue and Shetty, 2004). Genistein, daidzein and glycitein are the three major isoflavonoids found in soybean and soy products, the properties of which have been extensively studied

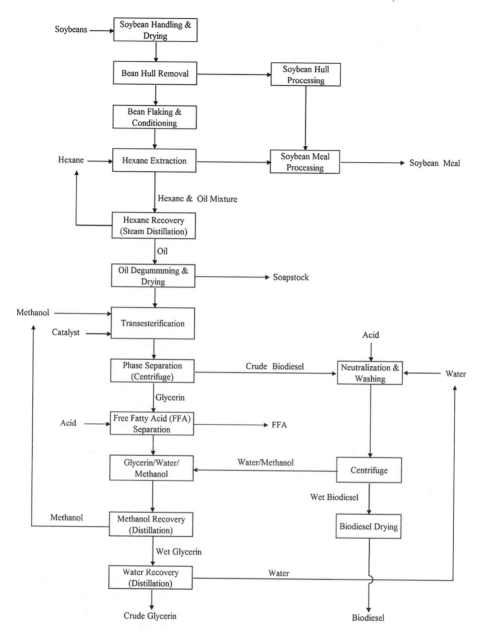

Figure 6.2   *Soybean crushing process and biodiesel conversion. Adapted from Pradhan et al., 2009.*

(Park and Surh, 2004; Khan, Afaq, and Mukhtar, 2008). De Lumen (2005) studied and characterized lunasin, a peptide from soybean for anticancer effects. Similar di- and oligopeptides include the kidney membrane hydrolysate and plasma protease digests from fermented soy (natto), Asp-Leu-Pro and Asp-Gly peptides derived from alkaline hydrolysate of soy protein, which showed ACE inhibitory activity (Gibbs *et al.*, 2004; Wu and Ding, 2002).

Such and several protein co-products can be derived from legume crop plants where abundant waste/byproducts are being produced. These co-products can be utilized for value addition by evaluating for functional and bioactive properties.

6.3 Technologies Used to Extract Co-Products

6.3.1 Extractive Distillation

Extractive distillation is the process of distillation in the presence of a miscible, high-boiling, relatively non-volatile component and a solvent, that does not react with the other components in the mixture. The method is used for mixtures having a low relative volatility, nearing unity. Such mixtures cannot be separated by simple distillation, because the volatility of the two components in the mixture is almost the same, causing them to evaporate at nearly the same temperature at a similar rate, making normal distillation impractical.

Distillation is the most widely used separation technique in the chemical-processing industries. Not all liquid mixtures can be separated by ordinary fractional distillation, however. When the components of a system to be separated have relative volatilities of close to 1.00 (i.e., a close boiling mixture), separation becomes difficult and expensive because a large number of trays and a high reflux ratio are necessary. Both the equipment and utility costs will increase markedly and separation by ordinary distillation can become uneconomical. If the mixture forms azeotropes (or constant boiling mixtures), defined as mixtures of two or more liquids in a specific ratio where simple distillation cannot change composition (Petrucci and Herring, 2007), a different problem arises – the azeotropic composition can limit the separation. The method of extractive distillation uses a separation solvent which is generally non-volatile, has a high boiling point and is miscible with the mixture, but does not form an azeotropic mixture. The solvent interacts differently with the components of the mixture, thereby causing their relative volatilities to change. This enables the new three-part mixture to be separated by normal distillation. The original component with the greatest volatility separates out as the top product. The bottom product consists of a mixture of the solvent and the other component, which can then be easily separated because the solvent does not form an azeotrope with it. The bottom product can be separated by any of the methods available.

6.3.2 Adsorption

Adsorption is the adhesion of atoms, ions, biomolecules or molecules of gas, liquid or dissolved solids as an accumulation on the surface of an adsorbent material, which can be achieved using electrostatic forces, a technique that is commonly used in chemical engineering (Ferrari *et al.*, 2010). This process creates a film of molecules or atoms on the surface of the adsorbent. The most common industrial adsorbents are activated carbon, silica gel and alumina, because they present enormous surface areas per unit weight (Babel and Kurniawan, 2003). Similar to surface tension, adsorption is a consequence of surface energy, wherein all the ionic, covalent or metallic bonding requirements of the constituent atoms of the material are filled by other atoms. However, atoms on the surface of the adsorbent are not wholly surrounded by other adsorbent atoms and therefore can attract adsorbates. Weak van der Waals forces and covalent bonding respectively cause physisorption or chemisorption onto the adsorbents or electrostatic attraction may be the cause (Ferrari *et al.*, 2010). Adsorption techniques like solid-phase microextraction have been used to analyze flavors in sensory science (Yang and Peppard, 1994).

Protein adsorption is a fundamental process having applications in the field of biomaterials. Several biomaterial surfaces are formed due to interaction with biological material such as blood or serum, or proteins. Therefore, it is perceived that the living cells interact with the adsorbed protein layer rather than with the biomaterial. This protein layer acts as a mediator between biomaterials and cells, eventually transcribing the physical and chemical properties of the biomaterial into "biological language" (Wilson *et al.*, 2005). An example of such a phenomenon exists with cell membrane receptors that bind to proteins' bioactive sites, and these receptor–protein binding events are transduced through the cell membrane. This in turn stimulates specific intracellular signal-transduction processes that then impart functionality in the form of cell adhesion, shape, growth and differentiation. Protein adsorption is a process influenced by many surface properties, such as surface wettability, surface chemical composition (Sivaraman, Fears, and Latour, 2009) and surface nanometer-scale morphology (Scopelliti *et al.*, 2010).

6.3.3 Membrane Separation

Membrane separation processes for extracting ingredients for food applications involve ultrafiltration and reverse osmosis methods. Solutions containing different molecular weight solutes are separated using ultrafiltration by pressure activation. Separation of solute molecules of varying sizes from solution is the method of reverse osmosis (Sivasankar, 2005).

6.3.4 Supercritical and Subcritical Fluid Extractions

The definition of a supercritical fluid is the state of a substance or solvent where temperature and pressure exceed the critical point values. These critical points denote high density with high solvency and high compressibility, with large variability of solvency due to small changes in temperature and pressure (Mukhopadhyay, 2009). These advantageous properties are useful in extraction and processing methods using supercritical fluids. Supercritical fluid extraction (SCFE) using carbon dioxide (CO_2) as a solvent has emerged as a highly popular technology today over the conventional techniques for extraction of natural products, due to rapid, contamination-free, tailor-made extracts having superior quality and shelf-life, and the high potency of active ingredients.

Water and ethanol have shown efficient properties as solvents during extraction of biochemical molecules at high pressure, above the boiling point and below the critical temperature (Srinivas *et al.*, 2009). These are called subcritical fluids and have been successfully used as solvents in the extraction of several food ingredients, phytochemicals and nutraceuticals (Garcia-Marino *et al.*, 2006; Howard and Pandjaitan, 2008; Mannila and Wai, 2003). Although both super- and subcritical fluid extraction technologies are very effective methods, supercritical fluid extraction is one of the most used in various fields, including food technology, natural products, byproduct recovery and pharmaceuticals (Herrero *et al.*, 2010). More on SCFE is found in this book in the chapter devoted to sugar-cane processing.

6.4 Bioactivities and Nutritional Value in Biorefinery Co-Products

Biorefinery co-products are of high value since they are good sources of compounds, like proteins and polyphenols, which are not utilized in the industrial process. The ingredients in the co-products can be utilized efficiently to prepare products that yield benefits like functionality in food products and health promotion, as well as being economical in production.

We have researched on the preparation (using enzymes and chemicals), separation, purification, characterization and identification of proteins from rice endosperm and rice bran (Hettiarachchy, Griffin, and Gnanasambandam, 1996; Wang *et al.*, 1999; Anderson, Hettiarachchy, and Ju, 2000; Tang, 2002; Tang, Hettiarachchy, and Shelhammer, 2002; Tang *et al.*, 2003a, 2003b; Paraman *et al.*, 2006; Beck *et al.*, 2006); however, these proteins lacked desirable functional properties and modification-enhanced emulsification properties (Paraman, Hettiarachchy, and Christian, 2007a; Paraman *et al.*, 2007b). Wang and Gonzalez de Mejia (2005) reviewed bioactive peptides from soybeans that are released during gastrointestinal digestion or food processing of proteins. These peptides may be of various molecular sizes, but the small peptides were found to have hormone-like activities. Even though most studied bioactive peptides are derived from milk and dairy products (Gill *et al.*, 2000), soybean protein and derived peptides also have angiotensin converting enzyme (ACE) inhibitory activity and have potential activity in preventing other chronic diseases. Other bioactives from soy include genistein, an isoflavone, which has been tested in animal studies against bladder cancer (Singh *et al.*, 2006). Research also proved that soy protein and isoflavones may enhance endothelial function and alleviate events leading to lesion and thrombus formation (Potter, 2004). Endoprotease-digested fermented soy hydrolysates were found to have ACE inhibitory, anti-thrombotic, surface tension and antioxidant properties (Gibbs *et al.*, 2004; Farzamirad and Aluko, 2008).

6.4.1 Anti-Disease Properties

Natural Management and Preventing Risk of Chronic Diseases

Soybeans are grown for commercial edible oil. During this process soybean meal is generated as an underutilized co-product after oil extraction from seeds. Adding value to this co-product and making it functional has potential for commercialization. Proteins and peptides (protein fragments) obtained from the soy meal may have health beneficial effects, including anti-cancer, anti-hypertension and anti-Alzheimer's properties. Bioreactor fermentation of soy meal can also release nutrients, including peptides for evaluating bioactive properties and they may also have potential as a prebiotic to grow probiotic bacteria. The value-added soy meal will find applications in new food product formulations with enhanced value. We anticipate that the soy meal can be utilized to sustain agriculture and crop production because it can no longer be considered and treated as a waste co-product, but can be used in several food, ingredient and nutraceutical areas due to its benefits. It would be potentially profitable to soybean growers and biorefineries, and also as a natural alternative to synthetic drugs used for treating chronic illnesses due to its health-beneficial components.

Some commercial bio-based products, including protein co-products, have been successfully used to promote health and reduce the risk of some complicated diseases. Peptides and proteins have been used to control mild hypertension and hyperlipidaemia, which are risk factors for atherosclerosis and diabetes. Some of these products have been approved for presentation of health claim on labels by governments. Examples of companies that are involved in these bioactives are: American Oriental Bioengineering Company – anti-liver-cancer soybean peptides – and Pharming (Dutch Biotech Company). Such companies specialize in developing innovative products for the treatment of genetic disorders, speciality products, and herbal and vitamin supplements. The technologies include innovative platforms for the production of protein therapeutics, technology and processes for the purification and formulation of these products, as well as technology in the field of DNA repair (e.g. DNAge).

In addition, *in vitro* and animal experiments have implied that progression of cancer could be suppressed by food-derived peptides. However, there are few data on the control of other degenerative diseases of the joint, skin, bone, liver, digestive bowel, neuron, and so on, by food-derived peptides and proteins. Recently, some animal experiments and preclinical trials have suggested the potential of food-derived proteins and peptides for improvement of these complicated diseases.

We have investigated polyphenolic extracts and their constituents derived from selected plant materials (Cai, Hettiarachchy, and Jalaluddin, 2004) for antioxidant, antimicrobial and antimutagenic activities. Protein hydrolysates and peptide(s) have been derived from heat-stabilized de-fatted rice bran as anticancer agents against multiple human cancer cell lines (Kannan *et al.*, 2008; Kannan, Hettiarachchy, and Narayan, 2009; Kannan *et al.*, 2010). An ongoing research project is evaluating protein hydrolysates and peptides derived from soybean meal co-products for multiple bioactivities. Peptides derived by enzymatic hydrolysis of meals from local grown soybean lines are being tested for activity against human colon, liver, lung, breast and prostate cancer cell lines using the MTS cell proliferation inhibition assay. These peptides will be representative of biorefinery co- and byproducts of soybean biodiesel preparation.

Several industrial processes may generate byproducts rich in proteins that when hydro-lyzed can yield peptides of biological and nutritional importance. Bioactive peptides from soybean with cancer-preventive effects have been reported. Azuma *et al.* (2000) and Kanamoto *et al.* (2001) demonstrated that a high-molecular-weight fraction (HMF) of proteinase-treated soybean protein isolate suppressed colon and liver tumorgenesis in experimental animals. Several whey protein hydrolysates obtained from peptic, tryptic and proteinase-k digests have proven hypotensive when fed to spontaneously hypertensive rats (SHRs) (Costa *et al.*, 2005). Anorectic peptides from soybean (e.g., Leu-Pro-Tyr-Pro-Arg and Pro-Gly-Pro peptides) have been shown to exert anti-obesity activity through decreasing food intake and body weight (Wang and Gonzalez de Mejia, 2005). These authors also found that soybean protein hydrolysates decreased serum or hepatic triacylglycerol levels and body weight in rats. A synthetic peptide [epsilon]-polylysine has been demonstrated to have an anti-obesity function in mice by inhibiting intestinal absorption of dietary fat (Tsujita *et al.*, 2006).

6.4.2 Food Products

Incorporation of biorefinery co-products into food products is essential to promote the health benefits of their bioactivities. In addition there is abundant scope for biotechnological products to be developed through industrial biotechnology processes that utilize raw materials and convert them to value-added bioactive components. These processes involve biocatalysis, microbial fermentation or cell culture to produce application-orientated functional co-products that can promote value to foods.

6.4.3 Alternative Medicine

Alternate medical therapies are common in urban communities and are mainly based on the economic condition of the general population and used to improve general health. Preventative and curative medicine is becoming increasingly expensive, with a high insurance burden on individuals. The National Centre for Complementary and Alternative Medicine (NCCAM) of the National Institute of Health defines and regulates complementary, integrative or alternative medicine which is a non-standard method that supplements standard medicinal practices in treating diseases. Some of the common alternative therapies are acupuncture, chiropractic and herbal medicines. Diseases with high mortality like cancers have also been treated with alternative medicine (Cassileth and Deng, 2004; Ernst and Cassileth, 1998), even though medical professionals have been discussing the controversies of these methods. Alternative medicine has also been studied in other diseases that affect organs like the liver and have shown positive effects (Seeff *et al.*, 2001). Quantification of use depends on the types of complementary and alternative medicine (CAM) used and the time frame asked. Although much of the use does not appear to be maladaptive, a small percentage of individuals have enthusiastically adapted CAM in ways that would not be endorsed by most allopathic physicians.

6.5   Modern Technologies for Efficient Delivery – Nanoencapsulation

Nanotechnology is the manipulation of matter at dimensions of approximately 0.1 to 100 nanometers, which is a delicate but effective process. This unique phenomenon enables novel applications for delivering certain molecules at the required site of action without losing their effectiveness. Nanotechnology is a combination of nanoscale science, engineering and technology, and involves certain processes like imaging, measuring, modelling and manipulating matter at nano length scales (Quintanilla-Carvajal *et al.*, 2010). Recent rapid advances in nanotechnology and microencapsulation have opened new opportunities in medicine for the diagnosis and therapy of various diseases. Research has proved that bioactive peptides and their synthetic analogs have become prominent class of moieties that are used in therapy. Some medicines are generally ineffective when administered orally and thus nanoencapsulation has become an important tool for delivering these drugs, while preventing breakdown of the effective molecule in the drug (Reis *et al.*, 2006). Biorefineries utilize the carbohydrates in the biomass, leaving the protein as co-product. Food scientists have shown that protein can be an effective material for nanoencapsulation and protect the vital drug or ingredient from disintegration (Zimet and Livney, 2009).

Nanoencapsulation is basically used to protect the core material and to release it when it is required. Applications for this include targeted drug-delivery systems, timed-release drug delivery, cosmetics for branded perfumed clothing, food additives and food enhancements without altering texture and taste, and enhancing the shelf life and stability of products like vitamins.

6.5.1   Issues – Stability, Bioavailability and Toxicity

It will be of paramount importance for the derived co-products to be highly amenable to food matrices with increased stability and bioavailability. These issues generally present as limitations to the large-scale production of co-products owing to their sometimes unstable nature. Although certain modifications can improve their stability, bioavailability and toxicity concerns may exist.

Several investigations to examine stability and bioavailability are being undertaken before application of such co-products in food matrices or other appropriate systems. For example, encapsulants like proteins that are usually used to coat the co-products before delivery into food are regularly tested for physical and chemical degradation patterns. With large biorefinery plants, the amount of co-products generated will be huge, requiring the equivalent use of modifiable agents necessary to confer stability. In such cases, there is potential for losing bioavailability. Biorefinery procedures should thus focus on minimizing such usage of extensive modifiers, and include steps that can improve stability of the co-products as they are formed within the larger process.

Several companies are manufacturing encapsulants for various uses as refining agents for various products. Such commercial production of encapsulants has also stemmed a green revolution in product applications, whereby use of encapsulants provides long-term sustainability and shelf life to the product. Application of cutting-edge food-preservation technologies such as nanoencapsulation and microencapsulation are revolutionizing the food encapsulation industry. This is because such technologies are driven towards not only preservation, but also towards masking ingredients, aroma, flavor and taste. The global food encapsulation market is projected to reach about \$39 billion by the year 2015, according to reports from Global Industry Analysts, Inc.

6.6 Conclusion and Future Prospects

This chapter provides an overview of important co- and byproducts of raw materials used in biorefineries. It also deals with the description of extraction and separation methods from the biomass produced by the biorefinery and utilization of the ingredients in it for their nutritional and health-promoting activities. Driven by several forces, including climate change, the need for producing more value-added co-products from biorefinery plants has increased rapidly in recent years. Technologies and processes for efficient, green and sustainable production of biomass and co-products are being investigated and implemented with significant aid from governments. Such processes and technologies have also gained momentum in the food industry, where natural food resources are being modified to increase value and for use as functional food ingredients in several food formulations. It will be exciting to broaden research on such value-added products as alternatives for the treatment and management of chronic illnesses.

Utilizing state-of-the-art extraction methods and testing for activity or functionality in food products can lead to the production of value-added co-products from biorefineries that can be economical to the industry. Extraction, purification and possible further chemical modification of biomass components will result in the liberation of value-added co-products. Examples include carrageenan from kelp, biodiesel from vegetable oils and resin acids from the extracts of trees. In these processes, much of the original chemical structure and value of the biomass is retained in the final product. These functional or bioactive co-products can be commercialized, bringing more value per dollar for the biorefineries and the agriculturists. There is a growing demand for commercialization of value-added biorefinery co-products in the United States and the European Union with a high market potential.

Other commercial productions include treatment of raw materials (plants) to produce sugars from plant polysaccharides mainly for use in microbial fermentations from which the products of fermentation can be recovered. Examples include fuel ethanol from corn by yeast fermentation, poly-lactic acid from *Lactobacillus* fermentations, and other co-products obtained by patented fermentation approaches as that of DuPont.

The market potential for utilizing and harnessing raw materials for value-added products has seen a rapid increase in the last decade. Due to several driving forces, including climate change, governments around the world are now providing huge budgets to stimulate research and development, and the commercial introduction of bio-based fuels, chemicals and materials. Additionally, in the last three to five years, this area has attracted massive investment from well-recognized venture capitalists, major corporations like DuPont, Rohm and Haas, and Cargill, and oil companies such as Petro-Canada, Shell, BP and Chevron. The food and nutraceutical industry has also shown growth in novel product formulations, promoting value-added ingredients derived from co-products obtained from large biorefinery processes.

References

Anderson, A., Hettiarachchy, N.S., and Ju, Z.Y. (2001) Physicochemical properties of pronase- treated rice glutelin. *Journal of the American Oil Chemists Society*, **78** (1), 1–6.

Alcantara, R., Amores, J., Canoira, L. *et al.* (2000) Catalytic production of biodiesel from soy-bean oil, used frying oil and tallow. *Biomass and Bioenergy*, **18** (6), 515–527.

Azuma, B., Machida, K., Saeki, Y. *et al.* (2000) Preventive effect of soybean resistant proteins against experimental tumorigenesis in rat colon. *Journal of Nutritional Science and Vitaminology*, **46**, 23–29.

Babel, S. and Kurniawan, T.A. (2003) Low-cost adsorbents for heavy metals uptake from contaminated water: a review. *Journal of Hazardous Materials*, **97** (1–3), 219–243.

Beck, M.I., Hettiarachchy, N.S., Leuenberger, B.H. *et al.* (2006) Protective Hydrocolloid for Active ingredients. Filed in the United States Patent and Trademark office on December 8, 2005 Case #: 25125 US/PO 60/748,192.

Bera, M.B. and Mukherjee, R.K. (1989) Solubility, emulsifying, and foaming properties of rice bran protein concentrates. *Journal of Food Science*, **54**, 142–145.

Boatright, W.L. and Hettiarachchy, N.S. (1995) Spray-Dried Soy Protein Isolate Solubility, Gelling Characteristics, and Extractable Protein as Affected by Antioxidants. *Journal of Food Science*, **60** (4), 806–809.

Cai, R., Hettiarachchy, N.S., and Jalaluddin, M. (2003) High-performance liquid chromatography determination of phenolic constituents in 17 varieties of cowpeas. *Journal of Agricultural and Food Chemistry*, **51**, 1623–1627.

Cassileth, B.R. and Deng, G. (2004) Complementary and alternative therapies for cancer. *The Oncologist*, **9** (1), 80–89.

Costa, E.L., Almeida, A.R., Netto, F.M., and Gontijo, J.A. (2005) Effect of intraperitoneally administered hydrolyzed whey protein on blood pressure and renal sodium handling in awake spontaneously hypertensive rats. *Brazilian Journal of Medical and Biological Research*, **38**, 1817–1824.

De Lumen, B.O. (2005) Lunasin: A cancer-preventive soy peptide. *Nutrition Reviews*, **63**, 16–21.

Demirbas, M.F. (November 2009) Biorefineries for biofuel upgrading: A critical review. *Applied Energy*, **86** (Supplement 1), S151–S161.

Ernst, E. and Cassileth, B.R. (1998) The prevalence of complementary/alternative medicine in cancer: a systematic review. *Cancer*, **83** (4), 777–782.

Eswaranandam, S., Hettiarachchy, N.S., and Johnson, M.G. (2004) Antimicrobial activity of citric, lactic, malic, or tartaric acids and nisin-incorporated soy protein film against listeria monocytogenes, Escherichia coli O157: H7, and Salmonella gaminara. *Journal of Food Science*, **69** (3), FMS79–FMS84.

Farzamirad, V. and Aluko, R.E. (2008) Angiotensin-converting enzyme inhibition and free-radical scavenging properties of cationic peptides derived from soybean protein hydrolysates. *International Journal of Food Sciences and Nutrition*, **59**, 428–437.

Ferrari, L., Kaufmann, J., Winnefeld, F., and Plank, J. (2010) Interaction of cement model systems with superplasticizers investigated by atomic force microscopy, zeta potential, and adsorption measurements. *Journal of Colloid and Interface Science*, **347** (1), 15–24.

Garcia-Marino, M., Rivas-Gonzalo, J.C., Ibanez, E., and Garcia-Moreno, C. (2006) Recovery of catechins and proanthocyanidins from winery byproducts using subcritical water extraction. *Analytica Chimica Acta*, **563**, 44–50.

Gibbs, B.F., Zougmanb, A., Massea, R., and Mulligan, C. (2004) Production and characterization of bioactive peptides from soy hydrolysate and soy-fermented food. *Food Research International*, **37** (2), 123–131.

Gill, H.S., Doull, F., Rutherfurd, K.J., and Cross, M.L. (2000) Immunoregulatory peptides in bovine milk. *The British Journal of Nutrition*, **84**, S111–S117.

Hale, W.J. (1934) *The Farm Chemurgic: Farmward the Star of Destiny Lights Our Way*, University of California: The Stratford company., p. 201.

Hayakawa, S. and Nakai, S. (1985) Relationships of hydro hobicity and net charge to the solubility of milk and soy proteins. *Journal of Food Science*, **50**, 486–491.

Herrero, M., Mendiolaa, J.A., Cifuentesa, A., and Ibanez, E. (2010) Supercritical fluid extraction: Recent advances and applications. *Journal of Chromatography A*, **1217** (2010), 2495–2511.

Hettiarachchy, N.S., Griffin, V.K., and Gnanasambandam, R. (1996) Preparation and functional properties of a protein isolate from defatted wheat germ. *Cereal Chemistry*, **73** (3), 363–367.

Hettiarachchy, N.S., Kalapathy, U., and Myers, D.J. (1995) Alkali-modified soy protein with improved adhesive and hydrophobic properties. *Journal of Oil Chemical Society*, **72** (12), 1461–1464.

Hettiarachchy, N.S. and Eswaranadam, S. (2007) U.S. and International Patent filed by the University of Arkansas "Organic acids incorporated edible antimicrobial films". Patent No.: US 7160580.

Howard, L. and Pandjaitan, N. (2008) Pressurized liquid extraction of flavonoids from spinach. *Journal of Food Science*, **73** (3), C151–C157.

Kamm, B. and Kamm, M. (2004) Principles of biorefineries. Mini review. *Applied Microbiology and Biotechnology*, **64**, 137–145.

Kamm, B., Gruber, P.R., and Kamm, M. (2007) *Biorefineries-Industrial Processes and Products*, Wiley-VCH Verlag GmbH & Co. doi: 10.1002/14356007.l04 l01.

Kamm, B., Schneider, B.U., Hüttl, R.F. *et al.* (2006) Lignocellulosic feedstock biorefinery –combination of technologies of agroforestry and a biobased substance and energy economy. *Forum der Forschung*, **19**, 53–62.

Kanamoto, R., Azuma, B., Miyamoto, Y. *et al.* (2001) Soybean resistant proteins interrupt an enterohepatic circulation of bile acids and suppress liver tumorigenesis induced by azoxymethane and dietary deoxycholate in rats. *Bioscience, Biotechnology, and Biochemistry*, **65**, 999–1002.

Kannan, A., Hettiarachchy, N., Johnson, M.G., and Nannapaneni, R. (2008) Human colon and liver cancer cell proliferation inhibition by peptide hydrolysates derived from heat-stabilized defatted rice bran. *Journal of Agricultural and Food Chemistry*, **56** (24), 11643–11647.

Kannan, A., Hettiarachchy, N., and Narayan, S. (2009) Colon and breast anti-cancer effects of peptide hydrolysates derived from rice bran. *The Open Bioactive Compounds Journal.*, **2**, 17–20.

Kannan, A., Hettiarachchy, N., Lay, J.O., and Liyanage, R. (2010) Human cancer cell proliferation inhibition by a pentapeptide isolated and characterized from rice bran. *Journal of Peptides*, **31** (9), 1629–1634.

Kato, A., Takahashi, A., Matsudomi, N., and Kobayashi, K. (1983) Determination of foaming properties of proteins by conductivity measurement. *Journal of Food Science*, **48**, 62–65.

Kaygusuz, K. and Türker, M.F. (2002) Biomass energy potential in Turkey. Technical note. *Renewable Energy*, **26** (4), 661–678.

Khan, N., Afaq, F., and Mukhtar, H. (2008) Cancer chemoprevention through dietary antioxidants: progress and promise. *Antioxidant & Redox Signaling*, **10**, 475–510.

Kim Lee, S.Y., Peter, S.W., and Rhee, K. (1990) Functional properties of proteolytic enzyme modified soy protein isolate. *Journal of Agricultural and Food Chemistry*, **38**, 651–656.

Ko, S., James, M.E., Hettiarachchy, N.S., and Johnson, M.G. (2001) Physical and chemical properties of edible films containing nisin and their action against listeria monocytogenes. *Journal of Food Science*, **66** (7), 1006–1011.

Mannila, M. and Wai, C.M. (2003) Pressurized water extraction of naphtodianthrones in St. John's wort (Hypercium PerforatumL.). *Green Chemistry*, **5**, 387–391.

McCue, P. and Shetty, K. (2004) Health benefits of soy isoflavonoids and strategies for enhancement: a review. *Critical Reviews in Food Science and Nutrition*, **44**, 361–367.

Mukhopadhyay, M. (2009) Extraction and processing with supercritical fluids. *Journal of Chemical Technology and Biotechnology*, **84**, 6–12.

Paraman, I., Hettiarachchy, N.S., Schaefer, C., and Beck, M.I. (2006) Physicochemical properties of rice endosperm proteins extracted by chemical and enzymatic methods. *Cereal Chemistry*, **83** (6), 663–667.

Paraman, I., Hettiarachchy, N.S., and Christian, S. (2007a) Glycosylation and deamidation of rice endosperm protein for improved emulsifying properties. *Cereal Chemistry*, **84** (6), 593–599.

Paraman, I., Hettiarachchy, N.S., Schaefer, C., and Beck, M.I. (2007b) Hydrophobicity, solubility, and emulsifying properties of enzyme-modified rice endosperm protein. *Cereal Chemistry*, **84** (4), 343–349.

Park, O.J. and Surh, Y.-H. (2004) Chemopreventive potential of epigallocatechin gallate and genistein: evidence from epidemiological and laboratory studies. *Toxicology Letters*, **150**, 43–56.

Pearce, K.N. and Kinsella, J.E. (1978) Emulsifying properties of proteins: Evaluation of a turbidimetric technique. *Journal of Agricultural and Food Chemistry*, **26**, 716–723.

Perlack, R.D., Wright, L.L., Turhollow, A.F., and Graham, R.L. (2005) Biomass as Feedstock for a Bioenergy and Bioproducts Industry: The Technical Feasibility of a Billion-Ton Annual Supply. DOE/GO-102005-2135, Oak Ridge National Laboratory, Oak Ridge, Tennessee. Accessed 12 September 2007 http://feedstockreview.ornl.gov/pdf/billion_ton_vision.pdf.

Petrucci, H. and Herring, M. (2007) *General Chemistry: Principles & Modern Applications*, 9th edn, Pearson Education, Inc., Upper Saddle River, NJ.

Pihlanto-Leppala, A. (2001) Bioactive peptides derived from bovine whey proteins: opioid and ace-inhibitory peptides. *Trends in Food Science & Technology*, **11**, 347–356.

Potter, S.M. (2004) Soy Protein and Cardiovascular Disease: The Impact of Bioactive Components in Soy. *Nutrition Reviews*, **56** (8), 231–235.

Pradhan, A., Shrestha, D.S., McAloon, A. *et al.* (2009) Energy Life-Cycle Assessment of Soybean Biodiesel. USDA-ERS Report Number 845. Accessed from: Web at http://www.usda.gov/oce/global_change/index.htm.

Rupar, K. and Sanati, M. (2005) The release of terpenes during storage of biomass. *Biomass and Bioenergy*, **28** (1), 29–34.

Qi, M., Hettiarachchy, N.S., and Kalapathy, U. (1997) Solubility and emulsifying properties of soy protein isolates modified by pancreatin. *Journal of Food Science*, **62** (6), 1110–1115.

Quintanilla-Carvajal, M.X., Camacho-Díaz, B.H., Meraz-Torres, L.S. *et al.* (2010) Nanoencapsulation: A new trend in food engineering processing. *Food Engineering Reviews*, **2**, 39–50.

Reis, C.P., Neufeld, R.J., Ribeiro, A.J., and Veiga, F. (2006) Nanoencapsulation II. Biomedical applications and current status of peptide and protein nanoparticulate delivery systems. *Nanomedicine: Nanotechnology, Biology, and Medicine*, **2**, 53–65.

Scopelliti, P.E., Borgonovo, A., Indrieri, M. *et al.* (2010) The effect of surface nanometre-scale morphology on protein adsorption. *PLoS ONE*, **5** (7), e11862. doi: 10.1371/journal.pone.0011862.

Seeff, L.B., Lindsay, K.L., Bacon, B.R. *et al.* (2001) Complementary and alternative medicine in chronic liver disease. *Hepatology*, **34** (3), 595–603.

Singh, A.V., Franke, A.A., Blackburn, L.G., and Zhou, J.R. (2006) Soy phytochemicals prevent orthotopic growth and metastasis of bladder cancer in mice by alterations of cancer cell proliferation and apoptosis and tumor angiogenesis. *Cancer Research*, **66**, 1851–1858.

Sivaraman, B., Fears, K.P., and Latour, R.A. (2009) Investigation of the effects of surface chemistry and solution concentration on the conformation of adsorbed proteins using an improved circular dichroism method. *Langmuir*, **25** (5), 3050–3056.

Sivarooban, T., Hettiarachchy, N.S., and Johnson, M.G. (2006) Inhibition of listeria monocytogenes by nisin combined with grape seed extract or green tea extract in soy protein film coated on turkey frankfurters. *Journal of Food Science*, **71** (2), M39–M44.

Sivasankar, B. (2005) *Bioseparations – Principles and Techniques*, Prentice-Hall of India, New Delhi.

Srinivas, K., King, J.W., Monrad, J.K. *et al.* (2009) Optimization of subcritical fluid extraction of bioactive compounds using hansen solubility parameters. *Journal of Food Science*, **74** (6), E342–E354.

Stocker, M. (2008) Biofuels and biomass-to-liquid fuels in the biorefinery: catalytic conversion of lignocel-lulosic biomass using porous materials. *Angewandte Chemie International Edition*, **47** (48), 9200–9211.

Tang, S. (2002) OT Optimization of protein extraction system and protein functionalities for heat- stabilized defatted rice bran. Ph.D Thesis, p. University of Arkansas, Fayetteville. Arkansas.

Tang, S., Hettiarachchy, N.S., and Shelhammer, T.H. (2002) Protein extraction from heat-stabilized defatted rice bran. 1. Physical processing and enzyme treatments. *Journal of Agricultural and Food Chemistry*, **50**, 7444–7448.

Tang, S., Hettiarachchy, N.S., Eswaranandam, S., and Crandall, P. (2003a) Protein extraction from heat stabilized defatted rice bran. II. The role of amylase, celluclast, and viscoyme. *Journal of Food Science*, **68** (2), 471–475.

Tang, S., Hettiarachchy, N.S., Horax, R., and Eswaranandam, S. (2003b) Physical properties and functionality of rice bran protein hydrolysates prepared from heat-stabilized defatted rice bran with the aid of enzymes. *Journal of Food Science*, **68** (1), 152–157.

Tsujita, T., Takaichi, H., Takaku, T. *et al.* (2006) Antiobesity action of epsilon-polylysine, a potent inhibitor of pancreatic lipase. *Journal of Lipid Research*, **47**, 1852–1858.

Vasudevan, P.T. and Briggs, M. (2008) Biodiesel production—current state of the art and challenges. *The Journal of Industrial Microbiology and Biotechnology*, **35** (5), 421–430.

Wang, M., Hettiarachchy, N.S., Qi, M. *et al.* (1999) Preparation and functional properties of rice bran protein isolate (from non-heat treated defatted rice bran). *Journal of Agricultural and Food Chemistry*, **47** (2), 411–416.

Wang, W. and Gonzalez de Mejia, E. (2005) A new frontier in soy bioactive peptides that may prevent age-related chronic diseases. Comprehensive reviews in food science and food safety. *Institute of Food Technologists*, **4**, 63–78.

Were, L., Hettiarachchy, N.S., and Kalapathy, U. (1997) Modified soy proteins with improved foaming and water hydration properties. *Journal of Food Science*, **62** (4), 821–824.

Wilson, C.J., Clegg, R.E., Leavesley, D.I., and Pearcy, M.J. (2005) Mediation of biomaterial-cell interactions by adsorbed proteins: A Review. *Tissue Engineering*, **11** (1), 1–18.

Wu, J. and Ding, X. (2002) Characterization of inhibition and stability of soy-protein-derived angiotensin I-converting enzyme inhibitory peptides. *Food Research International*, **35** (4), 367–375.

Wu, W.U., Hettiarachchy, N.S., and Qi, M. (1998) Hydrophobicity, solubility, and emulsifying properties of soy protein peptides prepared by papain modification and ultrafiltration. *Journal of the American Oil Chemists Society*, **75** (7), 845–850.

Xie, L., Hettiarachchy, N.S., Ju, Z.Y. *et al.* (2002) Edible film coating to minimize eggshell breakage and reduce post-wash bacterial contamination measured by dye penetration in eggs. *Journal of Food Science*, **67** (1), 280–284.

Yang, X., and Peppard, T. (1994) Solid-phase microextraction for flavor analysis. *Journal of Agricultural and Food Chemistry*, **42**, 1925–1930.

Zimet, P. and Livney, Y.D. (2009) Beta-lactoglobulin and its nanocomplexes with pectin as vehicles for ω-3 polyunsaturated fatty acids. *Food Hydrocolloids*, **23** (4), 1120–1126.

7

Production of Valuable Compounds by Supercritical Technology Using Residues from Sugarcane Processing

Juliana M. Prado and M. Angela A. Meireles
*LASEFI/DEA/FEA/UNICAMP, University of Campinas (UNICAMP),
Campinas, SP, Brazil*

7.1 Introduction

Brazil presents an enviable program of producing and using ethanol for energy generation. Today, 44% of the Brazilian energy matrix is renewable, and 13.5% is derived from sugarcane (Soccol *et al.*, 2010). The Brazilian sugarcane system of agroenergy is considered the most efficient in the world (Goldemberg, 2007).

Brazil is the major producer of sugarcane in the world, so that sugar and ethanol production is a very important sector of the Brazilian economy. As an example, the 2008–2009 sugarcane harvest reached 659 million tons, from which 27.5 billion litres of ethanol and 31 million tons of sugar were obtained. It should be noted that 60% of the sugar production was exported to other countries (UNICA, 2011). Because of the size of the Brazilian sugar and ethanol industry, any process that is aimed at using by-products will have a positive impact on the country's economy (Shintaku, 2006). The processing of sugarcane residues is favoured

Biorefinery Co-Products: Phytochemicals, Primary Metabolites and Value-Added Biomass Processing, First Edition.
Edited by Chantal Bergeron, Danielle Julie Carrier and Shri Ramaswamy.
© 2012 John Wiley & Sons, Ltd. Published 2012 by John Wiley & Sons, Ltd.

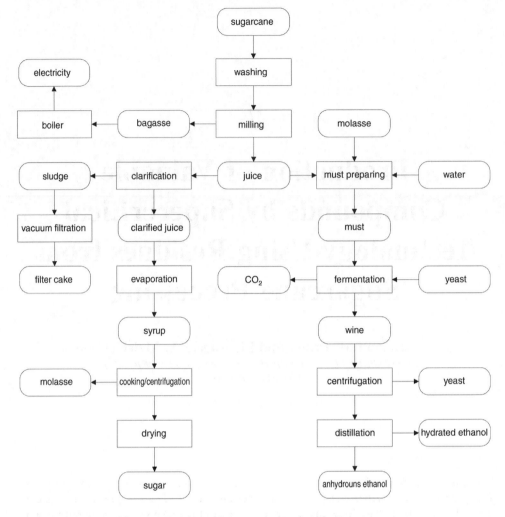

Figure 7.1   *Scheme of sugar and ethanol production process.*

because this can be integrated into a pre existing unit, resulting in lower investment and transportation costs (Soccol *et al.*, 2010).

Figure 7.1 presents a simplified scheme of the sugar and ethanol production process. Several residue streams are generated from this industry, and adding value to them improves the economic outlook of the whole process. An important stream, that represents one third of the residue output, is the generation of bagasse, which is produced after the milling step (Soccol *et al.*, 2010). Bagasse is usually burnt to generate electricity, but it could be used instead for generating non-pollutant energy, through its conversion to fermentable hexoses for the production of ethanol. Another residue, the filter cake, is currently used as a fertilizer, but some valuable phytochemicals present in it could be recovered before the residue is ultimately disposed of.

The global impetus devoted to searching for ecologically based technologies has promoted the development of industrial products and applications that are environmentally benign. Some good examples of environmentally benign technologies are those based on supercritical fluids, which use solvents that are harmless to the environment and to human health. Using green technologies such as those based on supercritical fluids is ideal to add value to residue streams such as those present in the sugar to ethanol industry. In this chapter we will present the processing of two residues using supercritical technologies: (1) extraction of long-chain fatty alcohols from the filter cake; and (2) hydrolysis of bagasse for the production of fermentable hexoses.

7.2 Supercritical Fluid Extraction of Filter Cake

7.2.1 Supercritical Fluid Extraction

Supercritical fluid extraction (SFE) is considered to be an emergent technology; it has proven to be technically and economically feasible, and presents several advantages when compared to traditional extraction methods, especially because of the green label and the selectivity of the process.

SFE is a physico-chemical separation process, based on the contact of a fixed bed of raw material with a solvent that is in its supercritical state, and this results in the removal of a solute or mixture of solutes from the solid phase. SFE consists mainly of two steps: (1) extraction and (2) separation of the extract from the solvent. Figure 7.2 presents a simplified scheme of the SFE process.

The solvent (stored in R2 and pressurized by B1) continuously flows (C1) through the bed of particles (E). When deemed necessary the addition of co-solvent (stored in R3 and pumped by B2), it is mixed to the supercritical solvent (MI) before the inlet of the extractor (E). The solutes are extracted by a combination of convection and diffusion principles. After the extraction step (C2), there is the separation step, where the pressure is reduced, leading to a decrease in the solvent power of the fluid, and the precipitation of the solute in the separator (S), where its quantity can be measured (C3). After the separation step the solvent can be recirculated (C4) in the system. In this case, there is a buffer tank (R2) which feeds the extractor (C5), whilst another reservoir (R1) stores the solvent for refilling the line (C7). In industrial plants, the extraction process may be conducted at multiple stages, and depressurization may be fractionated in multiple separators, resulting in extracts with different compositions.

Most applications of SFE use carbon dioxide (CO_2) as the solvent (Díaz-Reinoso *et al.*, 2006). CO_2 is a generally recognized as safe (GRAS) solvent, and does not leave residues either in the extract or in the exhausted raw material, which is particularly important in the food and pharmaceutical industries. CO_2 is a powerful solvent for a wide range of compounds, relatively inert, inexpensive, non-flammable, recyclable, and available at high purity. Due to its moderate critical pressure (7.4 MPa), industrial plants using CO_2 require lower investment costs when compared to other supercritical solvents, such as water and methanol. Because of its low critical temperature (304.2 K), CO_2 can be used to process thermosensitive compounds.

7.2.2 Extraction of Long-Chain Fatty Alcohols from Filter Cake

Filter cake is the residue generated in the sugarcane juice clarification step of sugar production (Figure 7.1). After clarification, the mud obtained is filtered for recovery of the residual sugar;

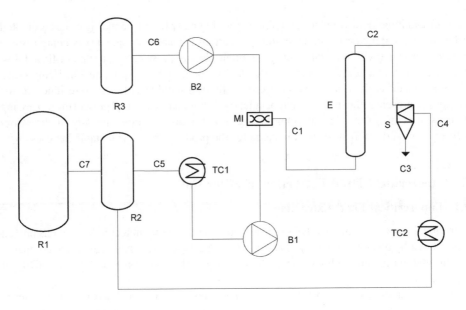

B1 – CO_2 pump
B2 – co-solvent pump
C1 – solvent inlet stream (CO_2 + co-solvent) in the extractor
C2 – solvent + extract outlet stream from the extractor
C3 – extract (+ co-solvent) outlet stream
C4 – CO_2 recycling stream
C5 – CO_2 feeding stream
C6 – co-solvent feeding stream

C7 – CO_2 refilling stream
E – extractor
MI – CO_2 + co-solvent mixer
R1 – CO_2 reservoir
R2 – CO_2 buffer tank
R3 – co-solvent reservoir
S – separator
TC1 – heat exchanger for keeping CO_2 in liquid state
TC2 – heat exchanger for CO_2 recycling

Figure 7.2   *SFE process scheme.*

the residue is the filter cake, which is generally used as a fertilizer (Shintaku, 2006). However, a lipophilic material, containing long-chain fatty alcohols and phytosterols, can be recovered from this residue. These compounds can be used as nutraceuticals or pharmaceuticals. Their recovery by SFE is technically feasible (Shintaku and Meireles, 2008). The product obtained by SFE can be compared to commercially available policosanol products (Shintaku, 2006), which are usually extracted and purified with toxic organic solvents, such as hexane and dichloromethane (Laguna Granja *et al.*, 1994; Vieira, 2003; de Lucas *et al.*, 2007; Rozário, 2008).

Policosanol is a mixture of alcohols with chain lengths varying from 24 to 34 carbon atoms. Octacosanol, containing 28 carbon atoms, is the main constituent (Rozário, 2008) and can be used as a low density lipoprotein (LDL) control, as a protectant against atherosclerosis and hepatic injury progression, and as antiplatelet, anti-ischaemic and antithrombotic agent, with good tolerance by the human body (Pons *et al.*, 1997; Más *et al.*, 1999; Gouni-Berthold and Berthold, 2002; Hargrove, Greenspan, and Hartle, 2004; Noa and Mas, 2005; Castaño

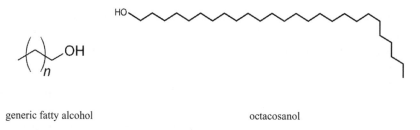

generic fatty alcohol   octacosanol

Figure 7.3   *Generic fatty alcohol and octacosanol structures.*

et al., 2006; Lowther and Sbordone, 2007; Ohta *et al.*, 2008; Rozário, 2008). Because octacosanol presents antiangiogenic effects (Thippeswamy, Sheela, and Salimath, 2008), its use was suggested as an alternative to aspirin (Taylor, Rapport, and Lockwood, 2003). Figure 7.3 presents the generic structure of fatty alcohols and octacosanol.

LASEFI researchers have recovered policosanol from sugarcane filter cake by SFE, determining global yields, kinetic parameters, scale-up and fractionation (Shintaku, 2006; Shintaku and Meireles, 2008; Prado, Prado, and Meireles, 2011). Shintaku (2006) reported SFE global yields (mass of extract per mass of feed raw material) with pressures varying from 10 MPa to 35 MPa, temperatures between 303 K and 333 K, and a fixed solvent to feed ratio (S/F) of 100. Maximum global yields ranged from 3 to 6% depending on the origin of the raw material, indicating that the initial feedstock affected the resulting extraction yields. For the filter cake extraction by a low-pressure solvent extraction (LPSE) process, global yields were 2 to 4% after extraction and purification (Vieira, 2003). Thus, SFE presented similar yields compared to those obtained by LPSE, and with no need for a purification step.

Shintaku (2006) determined that a temperature of 333 K and a pressure of 25 MPa resulted in the highest yield. The mass transfer rate increased with both temperature and pressure. Therefore, when extraction kinetics were considered together with global yield data, the conditions of 333 K and 35 MPa were determined as optimal.

Similar extraction conditions were used by Prado, Prado, and Meireles (2011) to study the scale-up of the filter cake extraction process. Prado, Prado, and Meireles (2011) kept the S/F constant from analytical to pilot scale, and determined that the extraction curves were of a similar shape and that the yields were 15% higher in a 15-fold scale-up assay.

One of the advantages of the SFE process, when compared to classical extraction methods, is the possibility of obtaining fractions that are enriched with target compounds without adding unit operations, by simple manipulation of separation conditions, also known as fractional separation. Prado, Prado, and Meireles (2011) studied fractional separation of filter cake SFE. For extraction at a temperature of 333 K and at a pressure of 35 MPa, the extract was fractionated in three separators in series (S_1, S_2, and S_3) operating at 323 K and 10 MPa (S_1), 303 K and 7 MPa (S_2) and 313 K and 3 MPa (S_3). S_1, S_2 and S_3 yields were 67, 18, and 15% of total extraction yield, respectively.

With respect to the chemical composition of the extracts, Shintaku (2006) determined that the major components of SFE extracts were eicosanol, docosanol, tetracosanol, hexacosanol, heptacosanol, octacosanol and triacontanol fatty alcohols, and β-sitosterol and stigmasterol phytosterols. Their proportions varied with operational conditions. Shintaku (2006) reported that octacosanol, the compound that is documented to display hypocholesterolemic effects, represented 2 to 4% of the SFE extract composition, which is similar to that of commercial policosanol. In LPSE extracts, octacosanol contents of up to 12% were found after the purification step, also conducted with organic solvents (Vieira, 2003).

Prado, Prado, and Meireles (2011) determined that octacosanol was more concentrated in the extract (up to 7%) at the beginning of the process, and up to 90 minutes thereafter, implying that longer extraction times do not improve the quality of the final product. The LPSE process took 8 h, and further purification adds more than 2 h to the process (Vieira, 2003), which is almost seven times slower than the SFE process. Additionally, when fractional separation was employed, the fraction obtained in S_1 was more concentrated in octacosanol, by up to 8.3% of the total extract (Prado, Prado, and Meireles, 2011). In addition to obtaining directly an octacosanol-enriched extract without using toxic solvents and with shorter extraction times, the SFE extract is a white powder (Prado, Prado, and Meireles, 2011) as compared to the LPSE extract which is a green to brown colour (Vieira, 2003). The dark colour may present a drawback for commercial applications because, from a sensorial point of view, acceptance of a green to brown extract may prove problematic. SFE yield and extract quality was determined to be comparable to that of the LPSE product and commercially available preparations. Moreover, the SFE-produced extracts present several advantages over traditional extraction techniques: fewer unit operations; no use of toxic solvents; shorter extraction time; product readily available for application just after processing; and the fact that exhausted raw material can still be used as fertilizer, since it is dry and contains no traces of toxic residues. With the technical feasibility demonstrated, the next step is to evaluate the economical potential of the SFE process.

7.3   Process Simulation for Estimating Manufacturing Cost of Extracts

7.3.1   Process Simulation

The development of an industrial process requires the identification of the necessary equipment, process parameters, and final products. In today's market, flexibility and low cost are important to consider while developing a new process (Prado, 2009). To transfer any technology from the academic environment to industry, tools that allow rapid and accurate estimation of technical and economical feasibility are critical. SFE technology is very particular in terms of transposing extraction parameters from matrix to matrix because each raw material must be individually studied. Specific pressures, temperatures, and solvent to feed ratios must be determined for each feedstock and targeted extracted compound. Therefore, simple tools that need the least amount of experimental information possible should be used to optimize the process.

Process simulators are tools that are often used in many engineering fields, especially when evaluating scenarios and optimizing integrated processes. They can be used when selecting projects based on economical evaluation or other critical demands of the process. There are many advantages to using simulators for developing processes: they may be used to decrease the time and cost of developing the process; to compare multiple processes with the same calculation basis; to process and analyze iteratively a large amount of information; and to study the interactions amongst the inlet and outlet streams of several integrated unit operations (Rouf et al., 2001; Prado, 2009).

SFE process simulation may be done by applying mathematical models (Alonso et al., 2002; Bravi et al., 2002; Diaz and Brignole, 2009), by using computational methods, like neural networks (Fullana, Trabelsi, and Recasens, 2000), or by using commercial simulation packages (Matilha, Cardozo-Filho, and Wolff, 2001; Leal *et al.*, 2006; Leal, Takeuchi, and Meireles, 2007; Takeuchi *et al.*, 2008).

7.3.2 Manufacturing Cost

After three decades of development, no commercial SFE plant is located in Latin America. The high cost of an SFE plant when compared to that of a classical low-pressure steam distillation or LPSE plant discourages this type of investment (Meireles, 2008). However, SFE capital costs have been decreasing due to competition amongst manufacturers (Perrut, 2007). Moreover, when a large quantity of raw material is processed, like in the coffee and tea decaffeination industries, SFE operational costs are below US$ 3.00/kg raw material (Brunner, 2005). An important economical drawback with respect to the use of traditional extraction methods is the high operational costs that are associated with steam generation and solvent evaporation (del Valle and Aguilera, 1999). Therefore, to make possible the construction of an industrial SFE plant in Brazil, economical evaluations are important, especially when competing technologies that present lower investment costs are available.

LASEFI researchers have developed simple methodologies for estimating the cost of manufacturing (COM) of extracts, so that SFE can be considered alongside other technologies when choosing the most suitable commercial extraction method. Rosa and Meireles (2005) presented a methodology for estimating the COM of SFE extracts. Despite high investment costs, the COM of SFE extracts is extremely competitive with the COM of extracts obtained by traditional extraction techniques when all costs involved in the process are taken into account (Rosa and Meireles, 2005; Pereira and Meireles, 2007a, 2007b; Pereira, Rosa, and Meireles, 2007; Leal, 2008; Navarro-Díaz *et al.*, 2009; Prado, Leal, and Meireles, 2009a; Prado *et al.*, 2010).

Later, other economic analyses were conducted using the commercial simulator SuperPro Designer, for both high-pressure and low-pressure extraction methods. The simulator COM estimation is more accurate, and also shows the economic feasibility of SFE in Brazil for several raw materials (Prado *et al.*, 2009b; Takeuchi *et al.*, 2009; Veggi, 2009; Veggi *et al.*, 2009; Albuquerque and Meireles, 2010; Leal *et al.*, 2010; Santos, Veggi, and Meireles, 2010; Veggi and Meireles, 2010).

SFE process optimization depends, amongst others, on the yield, time, amount of solvent used, and chemical composition of the product. Interrupting the extraction before the bed is completely exhausted of solute may be economically advantageous due to shorter processing times (Rosa and Meireles, 2005). Economical evaluation allows for determining the balance between yield and cost, considering the product's quality.

7.3.3 Manufacturing Cost Estimation of Sugarcane Wax

In this section we present the COM estimation for SFE of sugarcane filter cake using the commercial simulator SuperPro Designer v6.0. The methodology developed by Prado *et al.* (2009b) was adapted. An SFE unit was built with the tools available in the simulator databanks (Figure 7.4). The plant design was based on the pilot equipment used by Prado, Prado, and Meireles (2011) for determining experimental data, which was used for COM estimation (Table 7.1). Yield obtained in S_3 was discarded as a product stream, since in this

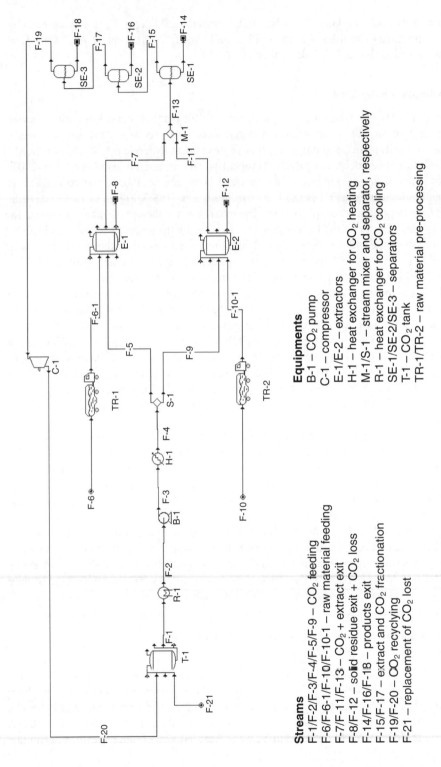

Equipments

B-1 – CO$_2$ pump
C-1 – compressor
E-1/E-2 – extractors
H-1 – heat exchanger for CO$_2$ heating
M-1/S-1 – stream mixer and separator, respectively
R-1 – heat exchanger for CO$_2$ cooling
SE-1/SE-2/SE-3 – separators
T-1 – CO$_2$ tank
TR-1/TR-2 – raw material pre-processing

Streams

F-1/F-2/F-3/F-4/F-5/F-9 – CO$_2$ feeding
F-6/F-6-1/F-10/F-10-1 – raw material feeding
F-7/F-11/F-13 – CO$_2$ + extract exit
F-8/F-12 – solid residue exit + CO$_2$ loss
F-14/F-16/F-18 – products exit
F-15/F-17 – extract and CO$_2$ fractionation
F-19/F-20 – CO$_2$ recycling
F-21 – replacement of CO$_2$ lost

Figure 7.4   *Scheme of SFE process built in SuperPro Designer simulator, used for economic evaluation.*

Table 7.1 *Experimental data used for COM estimation, temperature = 333 K, pressure = 35 MPa, feed mass = 1339 g (w.b.), CO_2 flow rate = 1.84×10^{-3} kg/s (Prado, Prado, and Meireles, 2011). Reprinted with permission from J.M. Prado et al., 2011 © Elsevier (2011).*

Time (min)	S/F (d.b.)	Yield accumulated (%, d.b.)			Total yield (%, d.b.)
		S_1	S_2	S_3	
30	2.44	0.90	0.11	0.03	1.05
60	4.82	1.37	0.24	0.11	1.73
120	9.63	1.75	0.48	0.26	2.49
180	14.42	1.93	0.53	0.42	2.88

separator mostly water was recovered, and the chemical composition of this fraction was not determined.

The economic data used in the simulation are presented in Table 7.2. An economic evaluation was performed considering the fact that the SFE unit would be installed in Brazil. Three extraction units of different scales were evaluated, all with the same design (Figure 7.4); extractor volumes ranging from 5 to 500 L were assessed (Table 7.2). The criterion for scale-up from 5 L to 50 L to 500 L consisted in keeping the S/F constant, as recommended by Prado, Prado, and Meireles (2011).

The COM is influenced by a series of factors that can be divided into the following categories: direct costs, fixed costs, and general expenses. The direct costs take into account factors that depend directly on the production rate, such as the raw material, the utilities, and the operation costs. The fixed costs do not directly depend on the production rates, and are considered even when the operation is interrupted; some examples are taxes, insurance, and

Table 7.2 *Economic parameters used for COM estimation.*

Industrial units	
2 extractors of 5 L	US$ 100,000.00[a]
2 extractors of 50 L	US$ 300,000.00[a]
2 extractors of 500 L	US$ 1,150,000.00[a]
Depreciation rate	10%/year
Labour	US$ 4.00/h[b,c]
2 extractors of 5 L	1 operator[a]
2 extractors of 50 L	2 operators[a]
2 extractors of 500 L	3 operators[a]
Raw materials	
Filter cake	0
Pre-processing	US$ 40.00/ton[c]
CO_2 (2% loss)	US$ 0.15/kg[b,c]
Utilities	
Electricity	US$ 0.092/kWh[a]
Cooling water	US$ 0.19/ton[a]
Steam	US$ 4.20/ton[a]

[a] Prado (2009).
[b] Market value in 2010.
[c] Prices in Brazilian currency, exchange rate was US$ 1.00 = R$ 1.75 (June/2010).

depreciation. General expenses are those deemed necessary to keep the business operating, and include administration and marketing costs, and selling, as well as research and development expenses (Turton *et al.*, 2003).

For the calculations, the plant was assumed to operate 24 h with three shifts per day, for 330 days, with 30 days per year reserved for maintenance, totalling 7920 h operation per year (Rosa and Meireles, 2005). The number of operators needed per shift varied according to the capacity of the plant, as shown in Table 7.2. Labour and peripheral charges not directly associated with production were estimated by the simulator.

Raw material cost was directly related to the plant material and the CO_2 lost during the process. CO_2 losses are mainly due to the depressurization of the extractor at the end of each batch (Perrut, 2007). The cost of the filter cake was disregarded, based on the fact that the SFE plant would most likely be located within the confines of a sugar and alcohol processing plant (Shintaku, 2006). Pre-processing costs were represented by the drying and the comminution of the raw material. Utility costs were encountered for operation of the heat exchangers and the electricity used in the process. The utilities needs for each of the operations were estimated by the simulator's energy balance. The cost of waste treatment was neglected, since the residue stream of the SFE process consists of the exhausted filter cake, which can be used as fertilizer. The CO_2 lost during depressurization is in small quantities and requires no special processing (Brunner, 2005). COM estimations need to include transportation costs. Final COM estimations for the SFE production of extract from the filter cake are presented in Table 7.3.

One piece of technical information that can be obtained from cost analysis is optimum batch time. To evaluate the optimized cycle time, yield, chemical composition, and cost data should be considered. The extract quality has been confirmed up to 120 min (Prado, Prado, and Meireles, 2011). By economic evaluation, 60 minutes is the batch time presenting the lowest

Table 7.3 *Economic evaluation of filter cake processing by SFE.*

Time (min)	Productivity (kg/year)				Operational cost (US$/year)	COM (US$/kg extract)	CRM (%)	COL (%)	FCI (%)	CQC (%)	CUT (%)
	S_1	S_2	S_3	Total							
5 L											
30	74	9	2	85	120 000.00	1445.78	0.46	60.56	29.17	9.08	0.73
60	81	15	7	103	120 000.00	1250.00	0.37	60.58	29.19	9.09	0.78
120	66	18	10	94	120 000.00	1428.57	0.28	60.61	29.19	9.09	0.82
180	53	14	12	79	120 000.00	1791.04	0.25	60.62	29.20	9.09	0.84
50 L											
30	741	91	25	857	287 000.00	344.95	1.93	50.75	36.66	7.61	3.06
60	806	145	67	1018	287 000.00	301.79	1.55	50.83	36.74	7.62	3.26
120	655	180	97	932	286 000.00	342.51	1.21	50.92	36.79	7.64	3.44
180	530	144	116	790	286 000.00	424.33	1.05	50.96	36.82	7.64	3.52
500 L											
30	7413	906	247	8566	798 000.00	95.92	6.93	27.39	50.57	4.11	11.00
60	8057	1452	665	10 174	793 000.00	83.39	5.62	27.57	50.91	4.13	11.77
120	6552	1797	974	9323	788 000.00	94.38	4.40	27.74	51.21	4.16	12.49
180	5301	1443	1163	7907	786 000.00	116.55	3.84	27.82	51.35	4.17	12.82

COM: cost of manufacturing; CRM: cost of raw material; COL: cost of labour; FCI: fixed cost of investment; CQC: cost of quality control; CUT: cost of utilities.

COM. Therefore, it can be concluded that for filter cake SFE at 333 K and 35 MPa, 60 minutes and an S/F of 4.8 present the best relationship between cost, yield, and product quality.

As for economic parameters, Shintaku (2006) estimated COM for the SFE from the filter cake by assuming the use of an industrial unit with three extractors of 7740 L each, which would be necessary to process all the filter cake generated by one sugar and alcohol processing plant. For this plant size, the COM was estimated to be US$ 13.63/kg for Class 1 estimation, and US$ 229.29/kg for Class 5 estimation. According to the Association for the Advancement of Cost Engineering International, cost estimations can be divided in five classes (1 to 5). Class 5 estimation is based on the lowest definition level of the project, while Class 1 estimation is closer to the full definition of the project, that is, high maturity. Therefore, Shintaku (2006) demonstrated the tendency for a substantial reduction in COM with increasing project detail. The COM estimation presented in Table 7.3 is for Class 3. For 5 L extractors, the minimum COM was of US$ 1250.00/kg, and it decreased to US$ 84.00/kg for 500 L. Therefore, as the operation scale increased, the COM substantially decreased.

In the market place, the most usual commercialized form of policosanol is as tablets containing 10–20 mg of active ingredient; the mass of a commercial tablet is 320 mg, resulting in 3–6% policosanol per tablet. The range in price is US$ 310 to US$ 1090/kg of tablet containing 3% policosanol. The COM estimated for SFE of filter cake using the SuperPro Designer simulator was US$ 84/kg extract, to which packaging and transportation costs need to be added. Considering the fact that policosanol concentrations in SFE extract can be up to 8.3% (Prado, Prado, and Meireles, 2011), that the COM estimated is below the selling price range of the product, and that as the definition level of the project increases, COM usually decreases, SFE of filter cake for recovery of policosanol in Brazil is economically feasible.

7.4 Hydrolysis of Bagasse with Sub/Supercritical Fluids

Ethanol feedstock can be classified into three types: (1) sucrose-containing feedstock, such as sugar beet, sweet sorghum, and sugarcane; (2) starchy materials, such as wheat, corn, and barley; and (3) lignocellulosic biomass, such as wood, straw, and grasses. The use of food crops such as wheat, corn, sugar beet, and sugarcane for ethanol production could cause conflict with food production (Soccol *et al.*, 2010). Moreover, due to environmental and availability problems, it is necessary to substitute fossil sources of energy by renewable and non-pollutant forms. Using indigenous and abundant raw materials for supplying energy may avoid the hegemony of certain countries in the energy sector (Mazaheri *et al.*, 2010). Therefore, second-generation energy production is based on using lignocellulosic biomass as a source for producing bioethanol.

Biomass as raw material for producing second-generation biofuels via supercritical technology has been studied, but there is still no commercial plant, since the process needs to be optimized in order to become feasible. Associating raw material availability with a clean technology is an encouraging scenario for launching Brazil's production of second-generation ethanol.

7.4.1 Biomass Conversion

For the production of fermentable sugars, which are then used for producing bioethanol (Zhao *et al.*, 2009a), raw materials rich in polysaccharides are used. The reaction of these biopolymers with water at high temperature, that is, hydrolysis, leads to their conversion into oligomers and/or monomers (Rogalinski *et al.*, 2008).

As covered previously, lignocellullosic material contains mostly cellulose (35–50%), hemicellulose (20–35%), and lignin (10–25%) (Schacht, Zetzl, and Brunner, 2008). Cellulose is more resistant to hydrolysis than hemicellulose, and the higher the lignin content, the harder it is to degrade the cellulosic material, due to the lignocellulose structure and to cellulose crystallinity (Mok and Antal, 1992; Zhao *et al.*, 2009a). Therefore, the differences in composition associated to the nature of the raw materials influence the hydrolysis process.

7.4.2 Polysaccharide Hydrolysis

The products generated by polysaccharide hydrolysis are pentose and hexose oligomers and monomers. One example of cellulose hydrolysis is presented in Figure 7.5. In addition to oligosaccharides and monosaccharides, degradation products, such as acetic acid, furfural, and hydroxymethyl-furfural are formed, and these are toxic to the microorganisms that ferment the monomers to produce bioethanol (Miyafuji *et al.*, 2005; Nakata, Miyafuji, and Saka, 2006; Schacht, Zetzl, and Brunner, 2008). Thus, the process should be optimized so that monomer and oligomer yields are maximized, while degradation product formation is minimized. This optimization is not easily achieved, because the sugar degradation rates can be higher than the conversion of lignocellulosic material to oligomers and monomers (Sasaki *et al.*, 1998, 2000; Schacht, Zetzl, and Brunner, 2008).

Several hydrolysis methods are presented in the literature. The leading technologies include using acid, basic, and enzymatic catalyzers, or steam explosion (Van Walsum and Shi, 2004; Teymouri *et al.*, 2005; Schacht, Zetzl, and Brunner, 2008; Zhao *et al.*, 2009b). The main problems associated with the use of acid hydrolysis are corrosion, the use of toxic solvents, and the necessity for neutralizing the medium after the reaction, generating solid residues (Van Walsum and Shi, 2004; Schacht, Zetzl, and Brunner, 2008). Enzymatic hydrolysis is a slow process (Miyazawa and Funazukuri, 2006; Schacht, Zetzl, and Brunner, 2008). Steam explosion leads to a limited decomposition of the lignocellulosic structure and the formation of high amounts of degradation products (Schacht, Zetzl, and Brunner, 2008).

The objective of a system processing lignocellulosic biomass by supercritical technology is to convert material rich in dry matter without the addition of toxic products and with a strong focus on energetic efficiency. Sub/supercritical water hydrolysis, also known as hydrothermal treatment or hydrothermolysis, has been tested and presents encouraging results, because this is a rapid and selective process, tuneable according to process temperature, pressure, and time to various feedstocks. Another advantage of hydrothermolysis is that the feedstock material may be used at high moisture contents because the reaction environment is water.

7.4.3 Hydrothermolysis

Water at high temperature, from 473 to 643 K, and pressure, between 4 MPa and 22 MPa, presents unique properties because ionic products, the dielectric constant and the density can be modified, enabling the deconstruction of the plant cell wall (Rogalinski *et al.*, 2008, Zhao *et al.*, 2009b). At these pressure and temperature conditions, water can act as an acid catalyzer, accelerating the hydrolysis process. Conducting the hydrothermolysis process at temperatures and pressures above 647 K and 22 MPa, respectively, results in a supercritical water process.

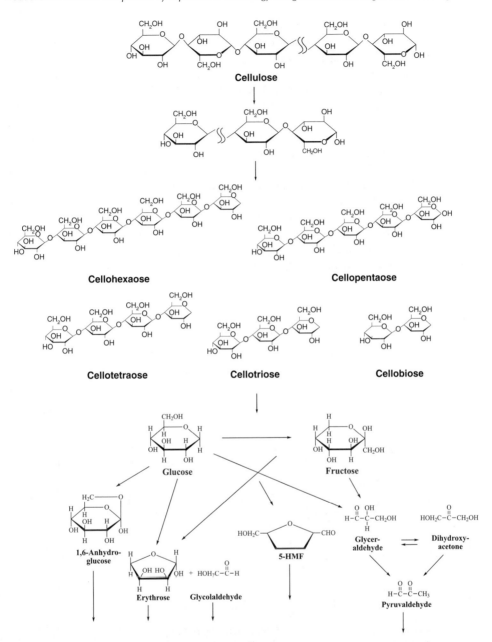

Figure 7.5 *The main reaction pathway of cellulose hydrolysis (reprinted from* The Journal of Supercritical Fluids, ***13***, *M. Sasaki, B. Kabyemala, R. Malaluan, S. Hirose, N. Takeda, T. Adschiri and K. Arai, Cellulose hydrolysis in subcritical and supercritical water, Copyright (1998), with permission from Elsevier). Reprinted with permission from M. Sasaki et al., 1998 © Elsevier (1998).*

For plant cell wall deconstruction, there is no clear advantage as to whether it is better to use subcritical water, supercritical water, or a combination of both conditions (Ehara and Saka, 2005; Zhao *et al.*, 2009a, 2009b).

It is possible to add CO_2 to the sub/supercritical aqueous medium in order to increase acidity and consequently accelerate the hydrolysis process. The CO_2 reacts with water forming carbonic acid, which acts as a catalyzer (Rogalinski *et al.*, 2008); this is less corrosive when compared to conventional acid hydrolysis (Schacht, Zetzl, and Brunner, 2008). After depressurization, the CO_2 is easily separated and recirculated into the process without generating solid residues (Miyazawa and Funazukuri, 2005; Schacht, Zetzl, and Brunner, 2008). CO_2 addition to sub- and supercritical water results in accelerated hydrolysis, increased monomer yields, decreased product molecular weight distribution, and reduced formation of degradation products (Van Walsum and Shi, 2004; Miyazawa and Funazukuri, 2005; Rogalinski *et al.*, 2008; Schacht, Zetzl, and Brunner, 2008).

Figure 7.6 presents a simplified scheme of a continuous hydrothermolysis set-up. Water (stored in R3) and the feed solution (stored in R4) are pre-heated (in TC3 and TC4, respectively), and then mixed (MI2) right before they enter the reactor (RE), through

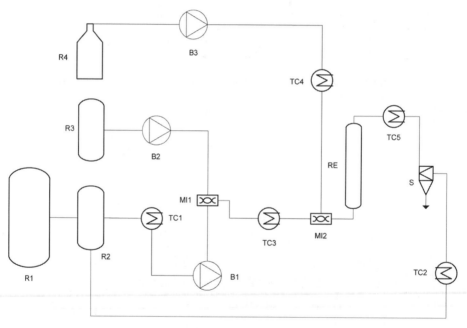

B1 – CO_2 pump	R3 – co-solvent reservoir
B2 – water pump	RE – reactor
B3 – feed solution pump	S – separator
MI1 – CO_2 + water mixer	TC1 – heat exchanger for keeping CO_2 in liquid state
MI2 – solvent + feed solution mixer	TC2 – heat exchanger for CO_2 recycling
R1 – CO_2 reservoir	TC3 – heat exchanger for solvent heating
R2 – CO_2 buffer tank	TC4 – heat exchanger for feed heating
	TC5 – heat exchanger for cooling product

Figure 7.6   *Scheme of continuous sub/supercritical hydrolysis equipment.*

which they flow continuously. When CO_2 is added (stored in R2), it is directly mixed with the water (MI1) before the solvent stream meets the feed stream (MI2). After flowing through the reactor, the solution temperature is reduced (TC5), which stops the reaction, and then the pressure is reduced, which allows recovery of the product in a separator (S). The CO_2 can be recirculated into the system. In this case, there is a buffer tank (R2) which feeds the reactor, while another reservoir (R1) stores CO_2 for refilling the line.

In an hydrothermolysis process, the most important process parameters that affect yield and product quality are: temperature, pressure, time, and solid:solvent proportion (Zhao *et al.*, 2009a, 2009b). However, temperature and time interactions represent the main parameters to be optimized, since their combination determines the maximum polymer with minimum degradation products (Van Walsum and Shi, 2004; Schacht, Zetzl, and Brunner, 2008; Petersen, Larsen, and Thomsen, 2009).

7.4.4 Hydrothermolysis of Sugarcane Bagasse

Most hydrothermolysis studies have been conducted on relatively simple materials such as pure cellulose or starch. Hydrothermolysis application to more complex raw materials poses some difficulties because the hydrolysis rates and yields depend on the initial cell wall composition and structure (Oomori *et al.*, 2004). Hydrothermolysis studies using industrial residues, such as ginger bagasse (Moreschi, Petenate, and Meireles, 2004), corn stalk and stover (Van Walsum and Shi, 2004; Zhao *et al.*, 2009b), sugarcane bagasse (Moreschi, 2004), guar gum (Miyazawa and Funazukuri, 2006), rice bran (Schacht, Zetzl, and Brunner, 2008), organic domestic waste (Schacht, Zetzl, and Brunner, 2008), and wheat straw (Petersen, Larsen, and Thomsen, 2009; Zhao *et al.*, 2009b) have been reported. Hydrothermolysis process parameters have to be optimized for each raw material, since different lignin contents influence the optimum hydrolysis conditions (Mok and Antal, 1992; Garrote, Domínguez, and Parajó, 1999; Zhao *et al.*, 2009b), and therefore, each raw material represents a technological challenge (Soccol *et al.*, 2010).

Sugarcane bagasse is composed of 19–24% lignin, 27–32% hemicelllose, 32–44% cellulose, and 4.5–9.0% ash (Soccol *et al.*, 2010). The cellulosic material can be converted to fermentable sugars by means of supercritical technology.

Mok and Antal (1992) solubilized approximately 40% of sugarcane bagasse with water at 473 to 503 K and 34.5 MPa for 0 to 15 min, recovering almost 100% of the total hemicellulose, 40% of the lignin, and 7% of the cellulose. Allen *et al.* (1996) fractionated sugarcane bagasse with compressed water for 45 to 240 s at 463 to 503 K and 5 MPa, resulting in over 50% solubilization of the biomass, including 100% of the hemicellulose, 60% of the acid-insoluble lignin, and 10% of the cellulose; small amounts of the furfural degradation product (1%) were obtained at a process temperature of 493 K.

Sasaki, Adschiri, and Arai (2003) fractionated sugarcane bagasse by gradually increasing subcritical water temperatures from 473 K to 553 K at 15 MPa, and obtained different fractions rich in hydrolysates of: (1) lignin and hemicellulose (galactose, arabinose, and xylose) at temperatures of 473 K to 503K; and (2) cellulose (glucose and cellobiose) at temperatures of 503 to 553 K. These preliminary studies show that recovery of fermentable sugars from sugarcane bagasse is technically feasible. However, further optimization of operational conditions and economic evaluation of the process are still needed so that it can be scaled-up to an industrial level.

7.5   Conclusions

The sugar and ethanol industry represents an important sector of the Brazilian economy, and therefore adding value to its residues by means of green technology presents an excellent opportunity to add revenue to this important industry. LASEFI research on the SFE recovery of policosanol from the filter cake, as well as its economic feasibility were presented in this chapter. The reuse of bagasse for production of fermentable sugars via hydrolysis with sub- and supercritical water was also presented. This process has shown encouraging results, but is still under development so that it can be proven technically and economically viable.

Acknowledgements

J. M. Prado thanks FAPESP for a post-doctoral fellowship (2010/08684-8).

References

Albuquerque, C.L.C. and Meireles, M.A.M. (2010) Estimate of the cost of manufacturing (COM) of natural colorants obtained by supercritical fluid extraction. II Iberoamerican Conference on Supercritical Fluids, Natal, Brazil, SC 113.

Allen, S.G., Kam, L.C., Zemann, A.J., and Antal, M.J. Jr. (1996) Fractionation of sugar cane with hot, compressed, liquid water. *Industrial & Engineering Chemistry Research*, **35**, 2709–2715.

Alonso, E., Cantero, F.J., García, J., and Cocero, M.J. (2002) Scale-up process of supercritical extraction with adsorption of solute onto active carbon. Application to soil remediation. *The Journal of Supercritical Fluids*, **24**, 123–135.

Bravi, M., Bubbico, R., Manna, F., and Verdone, N. (2002) Process optimization in sunflower oil extraction by supercritical CO_2. *Chemical Engineering Science*, **57**, 2753–2764.

Brunner, G. (2005) Supercritical fluids: technology and application to food processing. *Journal of Food Engineering*, **67**, 21–33.

Castaño, G., Arruzazabala, M.L., Fernández, L. *et al.* (2006) Effects of combination treatment with policosanol and omega-3 fatty acids on platelet aggregation: a randomized, double-blind clinical study. *Current Therapeutic Research*, **67**, 174–192.

de Lucas, A., García, A., Alvarez, A., and Gracia, I. (2007) Supercritical extraction of long-chain *n*-alcohols from sugar cane crude wax. *The Journal of Supercritical Fluids*, **41**, 267–271.

del Valle, J.M. and Aguilera, J.M. (1999) High pressure CO_2 extraction. Fundamentals and applications in the food industry. *Food Science and Technology International*, **5**, 1–24.

Diaz, M.S. and Brignole, E.A. (2009) Modeling and optimization of supercritical fluid processes. *The Journal of Supercritical Fluids*, **47**, 611–618.

Díaz-Reinoso, B., Moure, A., Domínguez, H., and Parajó, J.C. (2006) Supercritical CO_2 extraction and purification of compounds with antioxidant activity. *Journal of Agricultural and Food Chemistry*, **54**, 2441–2469.

Ehara, K. and Saka, S. (2005) Decomposition behavior of cellulose in supercritical water, subcritical water, and their combined treatments. *Journal of Wood Science*, **51**, 148–153.

Fullana, M., Trabelsi, F., and Recasens, F. (2000) Use of neural net computing for statistical and kinetic modelling and simulation of supercritical fuid extractors. *Chemical Engineering Science*, **55**, 79–95.

Garrote, G., Domínguez, H., and Parajó, J.C. (1999) Hydorthermal processing of lignocellulosic materials. *Holz Als Roh-Und Werkstoff*, **57**, 191–202.

Goldemberg, J. (2007) Ethanol for a sustainable energy future. *Science*, **315**, 808–810.

Gouni-Berthold, I. and Berthold, H.K. (2002) Policosanol: clinical pharmacology and therapeutic significance of a new lipid-lowering agent. *American Heart Journal*, **143**, 356–365.

Hargrove, J.L., Greenspan, P., and Hartle, D.K. (2004) Nutritional significance and metabolism of very long chain fatty alcohols and acids from dietary waxes. *Experimental Biology and Medicine*, **229**, 215–226.

Laguna Granja, A., Magraner Hernandez, J., Carbajal Quintana, D. *et al.* (1994) Mixture of higher primary aliphatic alcohols, its obtention from sugar cane wax and its pharmaceutical uses, Patent US 5663156.

Leal, P.F. (2008) Comparative study of cost of manufacturing and funcional properties of volatile oils obtained by supercritical extraction and steam distillation, Universidade Estadual de Campinas, PhD Thesis, Available from http://www.fea.unicamp.br/alimentarium/ver_documento.php?did=727 (in Portuguese).

Leal, P.F., Takeuchi, T.M., Orestes, T. *et al.* (2006) Simulation of flash tank operation conditions to improve anethole obtaining from two different sources by supercritical technology. VII Iberoamerican Conference on Phase Equilibria and Fluid Properties for Process Design, Morélia, Mexico, pp. 866–878.

Leal, P.F., Takeuchi, T.M., and Meireles, M.A.A. (2007) Comparison of two anethole sources (Foeniculum vulgare and Croton zethntneri Pax Et Hoffm) in terms of supercritical extraction: economical evaluation. In Iberoamerican Conference on Supercritical Fluids, Iguassu Falls, Brazil, SC-028.

Leal, P.F., Kfouri, M.B., Alexandre, F.C. *et al.* (2010) Brazilian Ginseng extraction via LPSE and SFE: global yields, extraction kinetics, chemical composition and antioxidant activity. *The Journal of Supercritical Fluids*, **54**, 38–45.

Lowther, S. and Sbordone, J. (2007) Food product containing policosanols, Patent CA 2544227.

Más, R., Rivas, P., Izquierdo, J.E. *et al.* (1999) Pharmacoepidemiologic study of policosanol. *Current Therapeutic Research*, **60**, 458–467.

Matilha, A., Cardozo-Filho, L., and Wolff, F. (2001) Simulação do processo de desterpenação supercrítica do óleo essencial de laranja. *Acta Scientiarum - Technology*, **23**, 1433–1437 (in Portuguese).

Mazaheri, H., Lee, K.T., Bhatia, S., and Mohamed, A.R. (2010) Subcritical water liquefaction of oil palm fruit press fiber for the production of bio-oil: effect of catalysts. *Bioresource Technology*, **101**, 745–751.

Meireles, M.A.A. (2008) Extraction of bioactive compounds from Latin American plants, in *Supercritical Fluid Extraction of Nutraceuticals and Bioactive Compounds* (ed. J. Martinez), CRC Press – Taylor & Francis Group, Boca Raton.

Miyafuji, H., Nakata, T., Ehara, K., and Saka, S. (2005) Fermentability of water-soluble portion to ethanol obtained by supercritical water treatment of lignocellulosics. *Applied Biochemistry and Biotechnology*, **121–124**, 963–972.

Miyazawa, T. and Funazukuri, T. (2005) Polysaccharide hydrolysis accelerated by adding carbon dioxide under hydrothermal conditions. *Biotechnology Progress*, **21**, 1782–1785.

Miyazawa, T. and Funazukuri, T. (2006) Noncatalytic hydrolysis of guar gum under hydrothermal conditions. *Carbohydrate Research*, **341**, 870–877.

Mok, W.S.-L. and Antal, M.J. Jr. (1992) Uncatalyzed solvolysis of whole biomass hemicellulose by hot compressed liquid water. *Industrial & Engineering Chemistry Research*, **31**, 1157–1161.

Moreschi, S.R.M. (2004) Hydrolysis by using subcritical water and CO_2, of the cellulose and starch present in the bagasse of the supercritical extraction of the ginger (*Zingiber Officinale* Roscoe): oligossacharides production, Universidade Estadual de Campinas, PhD Thesis, Available from http://www.fea.unicamp.br/alimentarium/ver_documento.php?did=86 (in Portuguese).

Moreschi, S.R.M., Petenate, A.J., and Meireles, M.A.A. (2004) Hydrolysis of ginger bagasse starch in subcritical water and carbon dioxide. *Journal of Agricultural and Food Chemistry*, **52**, 1753–1758.

Nakata, T., Miyafuji, H., and Saka, S. (2006) Bioethanol from cellulose with supercritical water treatment followed by enzymatic hydrolysis. *Applied Biochemistry and Biotechnology*, **129–132**, 476–485.

Navarro-Díaz, H.J., Carvalho, G.H., Leal, P.F. *et al.* (2009) Obtaining of extracts from *Origanum vulgare* and *Cordia verbenacea* via supercritical technology and steam distillation: process and economical study. 9th International Symposium on Supercritical Fluids, Arcachon, France, P18.

Noa, M. and Mas, R. (2005) Protective effect of policosanol on atherosclerotic plaque on aortas in monkeys. *Archives of Medical Research*, **36**, 441–447.

Ohta, Y., Ohashi, K., Matsura, T. *et al.* (2008) Octacosanol attenuates disrupted hepatic reactive oxygen species metabolism associated with acute liver injury progression in rats intoxicated with carbon tetrachloride. *Journal of Clinical Biochemistry and Nutrition*, **42**, 118–125.

Oomori, T., Khajavi, S.H., Kimura, Y. *et al.* (2004) Hydrolysis of disaccharides containing glucose residue in subcritical water. *Biochemical Engineering Journal*, **18**, 143–147.

Pereira, C.G. and Meireles, M.A.A. (2007a) Economic analysis of rosemary, fennel and anise essential oils obtained by supercritical fluid extraction. *Flavour and Fragrance Journal*, **22**, 407–413.

Pereira, C.G. and Meireles, M.A.A. (2007b) Evaluation of global yield, composition, antioxidant activity and cost of manufacturing of extracts from lemon verbena (*Aloysia triphylla* [L'Hérit.] Britton) and mango (*Mangifera indica* L.) leaves. *Journal of Food Process Engineering*, **30**, 150–173.

Pereira, C.G., Rosa, P.V.T., and Meireles, M.A.A. (2007) Extraction and isolation of indole alkaloids from *Tabernaemontana catharinensis* A.DC: Technical and economical analysis. *The Journal of Supercritical Fluids*, **40**, 232–238.

Perrut, M. (2007) Industrial applications of supercritical fluids: development status and scale-up issues, in *I Iberoamerican Conference on Supercritical Fluids*, Iguassu Falls, Brazil.

Petersen, M.Ø., Larsen, J., and Thomsen, M.H. (2009) Optimization of hydrothermal pretreatment of wheat straw for production of bioethanol at low eater consumption without addition of chemicals. *Biomass & Bioenergy*, **33**, 834–840.

Pons, P., Illnait, J., Más, R. *et al.* (1997) A comparative study of policosanol versus probucol in patients with hypercholesterolemia. *Current Therapeutic Research*, **58**, 26–35.

Prado, I.M. (2009) Use of simulator in scale up and economic viability study of supercritical fluid extraction from natural sources, Universidade Estadual de Campinas, Master's Thesis, Available from http://www.fea.unicamp.br/alimentarium/ver_documento.php?did=727 (in Portuguese).

Prado, J.M., Leal, P.F., and Meireles, M.A.A. (2009a) Comparison of manufacturing cost of thyme extract obtained by supercritical fluid extraction and steam distillation. 9th International Symposium on Supercritical Fluids, Arcachon, France, P19.

Prado, I.M., Albuquerque, C.L.C., Cavalcanti, R.N., and Meireles, M.A.A. (2009b) Use of commercial process simulator to estimate the cost of manufacturing (COM) of carotenoids obtained via supercritical technology from palm and buriti trees. 9th International Symposium on Supercritical Fluids, Arcachon, France, P20.

Prado, J.M., Assis, A.R., Maróstica-Júnior, M.R., and Meireles, M.A.A. (2010) Manufacturing cost of supercritical-extracted oils and carotenoids from Amazonian plants. *Journal of Food Process Engineering*, **33**, 348–369.

Prado, J.M., Prado, G.H.C., and Meireles, M.A.A. (2011) Scale-up study of supercritical fluid extraction process. *The Journal of Supercritical Fluids*, **56**, 231–237.

Rogalinski, T., Liu, K., Albrecht, T., and Brunner, G. (2008) Hydrolysis kinetics of biopolymers in subcritical water. *The Journal of Supercritical Fluids*, **46**, 335–341.

Rosa, P.T.V. and Meireles, M.A.A. (2005) Rapid estimation of the manufacturing cost of extracts obtained by supercritical fluid extraction. *Journal of Food Engineering*, **67**, 235–240.

Rouf, S.A., Douglas, P.L., Moo-Young, M., and Scharer, J.M. (2001) Computer simulation for large scale bioprocess design. *Biochemical Engineering Journal*, **8**, 229–234

Rozário, C.H.R. (2008) Processo simplificado de obtenção de policosanol a partir de cera de cana-de-açúcar, Patent BR PI0702137-2 (in Portuguese).

Santos, D.T., Veggi, P.C., and Meireles, M.A.A. (2010) Extraction of antioxidant compounds from jabuticaba (*Myrciaria cauliflora*) skins: yield, composition and economical evaluation. *Journal of Food Engineering*, **101**, 23–31.

Sasaki, M., Kabyemala, B., Malaluan, R. *et al.* (1998) Cellulose hydrolysis in subcritical and supercritical water. *The Journal of Supercritical Fluids*, **13**, 261–268.

Sasaki, M.S., Fang, Z., Fukushima, Y. *et al.* (2000) Dissolution and hydrolysis of cellulose in subcritical and supercritical water. *Industrial & Engineering Chemistry Research*, **39**, 2883–2890.

Sasaki, M., Adschiri, T., and Arai, K. (2003) Fractionation of sugarcane bagasse by hydrothermal treatment. *Bioresource Technology*, **86**, 301–304.

Schacht, C., Zetzl, C., and Brunner, G. (2008) From plant materials to ethanol by means of supercritical fluid technology. *The Journal of Supercritical Fluids*, **46**, 299–321.

Shintaku, A. and Meireles, M.A.A. (2008) Processo de extração de componentes ativos da cera contida na torta de filtro resultante do processamento da cana de açúcar empregando processo de extração supercrítica, Patent BR PI0701341-8 (in Portuguese).

Shintaku, A. (2006) Obtaining extract from the residue of the sugar cane and alcohol industry using supercritical CO$_2$: process parameter and estimative of the cost of manufacturing, Universidade Estadual de Campinas, Master's Thesis, Available from http://www.fea.unicamp.br/alimentarium/ver_documento.php?did=575 (in Portuguese).

Soccol, C.R., Vandenberghe, L.P.S., Medeiros, A.B.P. *et al.* (2010) Bioethanol from lignocelluloses: Status and perspectives in Brazil. *Bioresource Technology*, **101**, 4820–4825.

Takeuchi, T.M., Leal, P.F., Favareto, R. *et al.* (2008) Study of the phase equilibrium formed inside the flash tank used at the separation step of a supercritical fluid extraction unit. *The Journal of Supercritical Fluids*, **43**, 447–459.

Takeuchi, T.M., Pereira, C.G., Braga, M.E.M. *et al.* (2009) Low-pressure solvent extraction (solid-liquid extraction, microwave-assisted, and ultrasound assisted) from condimentary plants, in *Extracting Bioactive Compounds for Food Products: Theory and Application* (ed. M.A.A. Meireles), CRC Press – Taylor & Francis Group, Boca Raton.

Taylor, J.C., Rapport, L., and Lockwood, G.B. (2003) Octacosanol in human health. *Nutrition*, **19**, 192–195.

Teymouri, F., Laureano-Perez, L., Alizadeh, H., and Dale, B. (2005) Optimization of the ammonia fiber explosion (AFEX) treatment parameters for enzymatic hydrolysis of corn stover. *Bioresource Technology*, **96**, 2014–2018.

Thippeswamy, G., Sheela, M.L., and Salimath, B. (2008) Octacosanol isolated from *Tinospora cordifolia* downregulates VEGF gene expression by inhibiting nuclear translocation of NF-κB and its DNA binding activity. *European Journal of Pharmacology*, **588**, 141–150.

Turton, R., Bailie, R.C., Whiting, W.B., and Shaeiwitz, J.A. (2003) *Analysis, Synthesis and Design of Chemical Processes*, Prentice Hall, New Jersey.

UNICA - União da Agroindústria Canavieira de São Paulo (2010) Available from www.unica.com.br, Accessed in September.

Van Walsum, G.P., and Shi, H. (2004) Carbonic acid enhancement of hydrolysis in aqueous pretreatment of corn stover. *Bioresource Technology*, **93**, 217–226.

Veggi, P.C. (2009) Obtaining vegetable extracts by different extraction methods: experimental study and process simulation, Universidade Estadual de Campinas, Master's Thesis, Available from http://www.fea.unicamp.br/alimentarium/ver_documento.php?did=737 (in Portuguese).

Veggi, P.C. and Meireles, M.A.A. (2010) Economical evaluation of chamomile (*Chamomilla recutita* [L.] Rauschert) extract obtained via supercritical technology and low pressure solvent extraction. II Iberoamerican Conference on Supercritical Fluids, Natal, Brazil, SC 05.

Veggi, P.C., Prado, I.M., Vaz, N. *et al.* (2009) Manufacturing cost of extracts from jackfruit (*Astocarpus heterophyllus*) leaves obtained via supercritical technology and solvent extraction. 9th International Symposium on Supercritical Fluids, Arcachon, France, P21.

Vieira, T.M.F.S. (2003) Obtenção de cera de cana de açúcar a partir de subproduto da indústria sucro-alcooleira: extração, purificação e caracterização, Universidade Estadual de Campinas, PhD Thesis, Available from http://www.fea.unicamp.br/alimentarium/ver_documento.php?did=207 (in Portuguese).

Zhao, Y., Lu, W.-J., Wang, H.-T., and Li, D. (2009a) Combined supercritical and subcritical process for cellulose hydrolysis to fermentable hexoses. *Environmental Science & Technology*, **43**, 1565–1570.

Zhao, Y., Lu, W.-J., Wang, H.-T., and Yang, J.-L. (2009b) Fermentable hexoses production from corn stalks and wheat straw with combined supercritical and subcritical hydrothermal technology. *Bioresource Technology*, **100**, 5884–5889.

8

Potential Value-Added Co-products from Citrus Fruit Processing[*]

John A. Manthey

USDA-ARS, US Horticultural Research Laboratory, Fort Pierce, Florida, USA

8.1 Introduction

Recent efforts to convert agricultural waste streams to ethanol have included work with citrus peel, due to its high carbohydrate content. These studies have characterized and optimized enzymatic hydrolyses to liquefy the main citrus peel carbohydrate fractions and to maximize monomeric sugar production for conversion to ethanol by fermentation for fuel production (Grohmann and Baldwin, 1992; Grohmann, Cameron, and Buslig, 1995; Wilkins *et al.*, 2005; Wilkins *et al.*, 2007a). Yet, due to their high vitamin and phytonutrient contents, citrus fruits have also been highly regarded for their health-promoting properties, and because of these positive attributes and potentially important influences on the human diet, the chemical compositions of these fruit crops have been studied for decades. Perhaps one can ultimately trace a major portion of these studies back to the early work of Szent Györgyi and co-workers, who first reported the capillary sparing properties of vitamin C and of bioflavonoids typical of those in citrus (Rusznyák and Szent-Györgyi, 1936; Armentano *et al.*, 1936; Bentsáth,

---

[*] Mention of a trademark or proprietary product is for identification only and does not imply a guarantee or warranty of the product by the US Department of Agriculture. The US Department of Agriculture prohibits discrimination in all its programs and activities on the basis of race, color, national origin, gender, religion, age, disability, political beliefs, sexual orientation, and marital or family status.

The contribution of John A. Manthey has been written in the course of his official duties as a US government employee and is classified as a US government work, which is in the public domain in the United States of America.

Biorefinery Co-Products: Phytochemicals, Primary Metabolites and Value-Added Biomass Processing, First Edition. Edited by Chantal Bergeron, Danielle Julie Carrier and Shri Ramaswamy.

Rusznyák, and Szent-Györgyi, 1936). For the citrus industry, studies continue to this day on the identification, taste properties, and purported health benefits of many such phytonutrients. Of particular relevance is the study of the chemical compositions of citrus peels – from which most processing byproducts are derived. Most notable in peels are the polysaccharides, carotenoids, oils, and numerous phytonutrients, mainly the phenolics and limonoids. Hence, total peel processing schemes now include both fuel ethanol production and possibly other bio-based materials, along with integrated speciality compound recoveries as value-added materials.

8.2 Fruit Processing and Byproduct Streams

The largest source of citrus peel is from orange juice production. Current worldwide production of orange juice still centers in Florida and Brazil, although juice productions in China, India, Mexico, and elsewhere are showing sharp annual increases. Smaller amounts of grapefruit and lemon processing also occur. As a close approximation, half of the citrus crop allocated to juice production constitutes peel byproduct (Braddock, 1999), and in Florida, the production of orange and grapefruit peel waste during the 2008/2009 growing season approximated 3.14 million metric tons annually (Citrus Summary, 2008–2009). The major portion of this citrus waste is converted to dried peel pellets, typically sold as cattle feed, and because of its low value, its production has traditionally been considered mainly a means of waste peel disposal. Mass balances of the main byproduct streams during commercial juice processing have been calculated for orange, grapefruit, and lemon. Based on a model of grapefruit, 1000 boxes of fruit (38 555 kg) yields 1458 kg of concentrated peel extract (molasses) (1215 l) at 28% moisture, and 2815 kg of dried citrus peel pellets (8.0% moisture content) (Hendrickson and Kesterson, 1965). Similar numbers are obtained for Valencia oranges (Kesterson and Braddock, 1976).

From the inception of commercial citrus juice processing, there has been the realization of a need to develop additional uses for the waste portion of this crop, both for disposal purposes and for commercial gain (USDA, 1962). Extensive descriptions and reviews of citrus processing byproduct streams have been published (Sinclair, 1972; Kesterson and Braddock, 1976; Licandro and Odio, 2002; Braddock, 1999), and extensive work has been done to develop alternate, value-added products from citrus peel waste, where numerous schemes have been proposed. One such scheme is reproduced from Tucker and Long (1968) in Figure 8.1, and although this scheme was proposed over 43 years ago, the main byproduct streams depicted, that is, citrus molasses, cold-pressed peel oil, D-limonene, bioflavonoids, and peel polysaccharides, still remain the main targets of research. As in most byproduct schemes, the dried peels and membranes are proposed sources of peel oils, polysaccharides (currently speciality pectins), and phytochemicals, particularly the flavanone glycosides and the polymethoxylated flavones. Molasses is also a source of phytochemicals (i.e., flavonoids and limonoids – for various nutraceutical applications) and carbohydrates, which are used as feedstocks for bland syrups and ethanol production (Grohmann *et al.*, 1999). Numerous other potential products have been investigated, although few are in production. To better achieve commercialization of these waste streams, current research is aimed at developing speciality products of high retail value, a number of which are discussed in the following sections. The carbohydrate and phytochemical compositions of the main fruit-processing waste streams, from which these speciality products will be derived, are also discussed in the following sections.

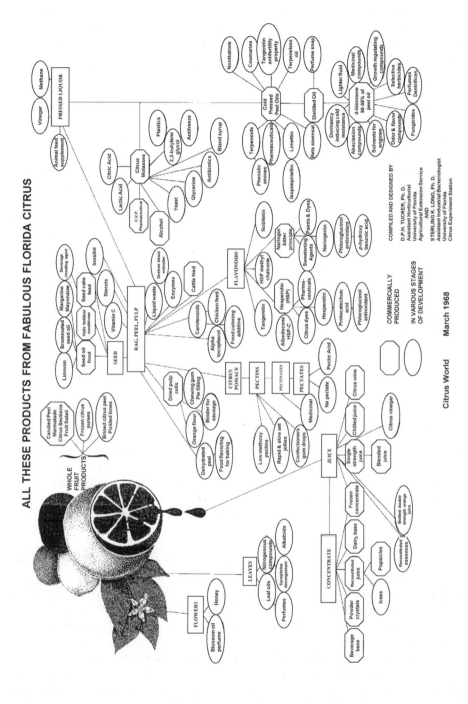

Figure 8.1 *By-product conversion scheme proposed by Tucker and Long (1968). Reproduced from Tucker, D.P.H. and Long, S.K. 1968. More by-products research needed. Citrus World, 4(10): 13–17.*

8.2.1   Polysaccharide Compositions of Dried Peel Pellets and Peel Molasses

The high contents of the soluble and insoluble carbohydrates in dried peel pellets and peel molasses make these materials attractive potential feedstocks for conversion into value-added products. A detailed compositional study of peel juice and molasses was reported by Grohmann *et al.* (1999). Soluble sugars, that is, glucose, fructose, and sucrose account for 60–70% of the dry solids of molasses. Dried peel pellets – composed of peel and membrane fragments – reflect the typical polysaccharide composition of plant cell walls, including pectin, hemicellulose, and cellulose, but unlike most other plant tissues, dried peel pellets have very low levels of lignin. The pectin content in citrus peel has been summarized by Sinclair (1984), and is typically between 2.4 to 5.0% wet weight. As a rule of thumb, the carbohydrate composition of dried peel pellets contains 15–25% pectin, 30–40% simple sugars (mainly glucose, fructose, and sucrose), 8–10% cellulose, and 5–7% hemicellulose (Grohmann, Cameron, and Buslig, 1995). Other constituents include 8–10% phenolics, lignin, and other secondary metabolites, 5–10% ash (Chapman *et al.*, 1972), and 5% protein (Kirk and Davis, 1954). Detailed analyses of the carbohydrate fractions of dried Valencia and Navel orange peel give compositions of 42.7 and 44.8% alcohol-insoluble solids, respectively (Sinclair and Crandall, 1953), consisting of cell wall polysaccharides such as celluloses, lignin, pectin, and hemicelluloses. On average only between 53 and 70% of the alcohol-insoluble solids were recovered as carbohydrates in the peel of various citrus varieties. The alcohol-soluble solids, containing mono- and disaccharides, with smaller amounts of essential oils, waxes, amino acids, and other undetermined substances, comprised 57 and 55% of dried Valencia and Navel orange peel, respectively. For Valencia orange the total sugar content in glucose equivalents of the alcohol-soluble solids was 55.4% dry weight, and 72.3% for Navel orange peel. Compositional analyses similar to Valencia and Navel oranges were also characterized for grapefruit and lemon (Sinclair and Crandall, 1949a, 1949b).

8.2.2   Phytochemical Compositions of Dried Peel Pellets and Peel Molasses

The processing of waste peel generated during commercial juice production involves the co-production of a concentrated aqueous peel extract, termed molasses, which contains high concentrations of the phytochemicals in wet and dried peel (Manthey and Grohmann, 1996, 2001; Hasegawa *et al.*, 1996). Because of this, the phytochemical compositions of dried peel pellets and molasses are jointly discussed. The primary groups of phytochemicals in citrus peel byproducts include several classes of phenolic compounds, including flavonoids, hydroxy-cinnamates, and in some *Citrus* species, coumarins and furanocoumarins (psoralens). Amongst the flavonoids are flavanone and flavone glycosides, and diverse sets of poly-methoxylated flavones (Figure 8.2). Each of the commercial *Citrus* species contains distinctive collections of flavanone and flavone glycosides (Albach and Redman, 1969; Kanes *et al.*, 1993). Of particular importance are the differences between grapefruit, which contains the bitter flavonoid neohesperidosides, and oranges and lemons which contain non-bitter flavonoid rutinosides (Horowitz and Gentili, 1977) (Figure 8.3). An important distinction between lemons, and oranges and grapefruits is the predominance of eriocitrin and other flavonoids with B-ring ortho-dihydroxy structures in lemons, which are more powerful antioxidants than the predominant flavonoids in oranges (hesperidin and narirutin) and grapefruit (naringin and neohesperidin). Concentrations of the flavanone glycosides in citrus peels and byproducts have been widely reported (Kanes *et al.*, 1993; Manthey and

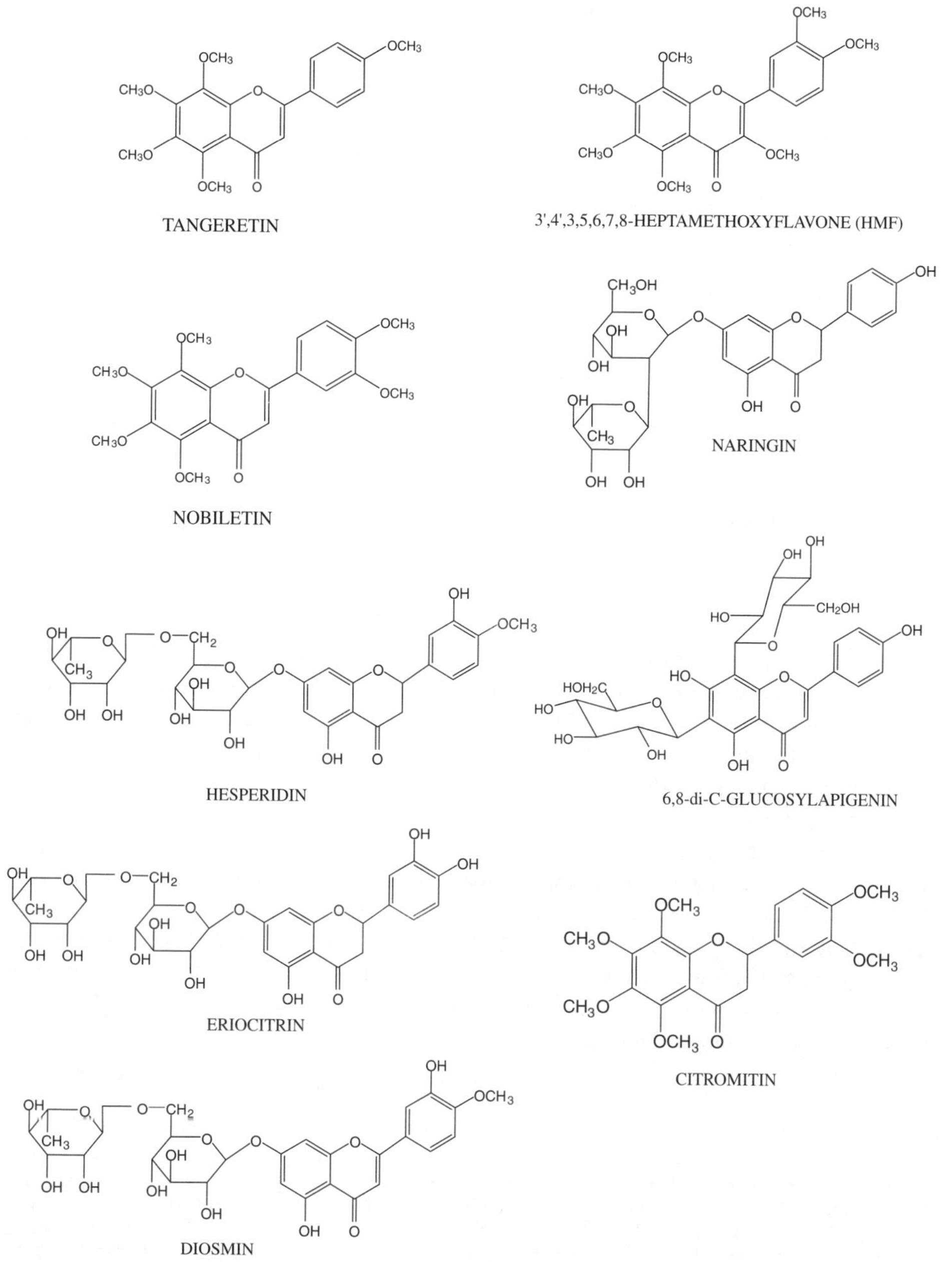

TANGERETIN

3',4',3,5,6,7,8-HEPTAMETHOXYFLAVONE (HMF)

NOBILETIN

NARINGIN

HESPERIDIN

6,8-di-C-GLUCOSYLAPIGENIN

ERIOCITRIN

CITROMITIN

DIOSMIN

Figure 8.2   *Common flavonoids in citrus by-products.*

Grohmann, 1996; Sinclair, 1972), and citrus peels have long been recognized as promising commercial sources of these compounds (Poore, 1934). Hesperidin occurs at 2–3% dry weight in orange peel, with narirutin-4′-glucoside, narirutin, and isosakuranetin rutinoside occurring at 835, 1998, and 1858 ppm, respectively, in dried Valencia orange peel (Manthey and

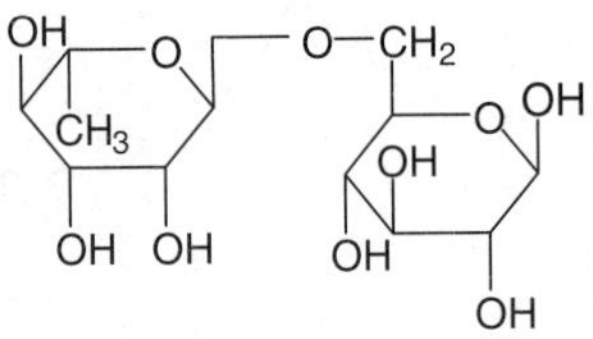

RUTINOSE (6-O-alpha-L-rhamnopyranosyl-beta-D-glucopyranose)

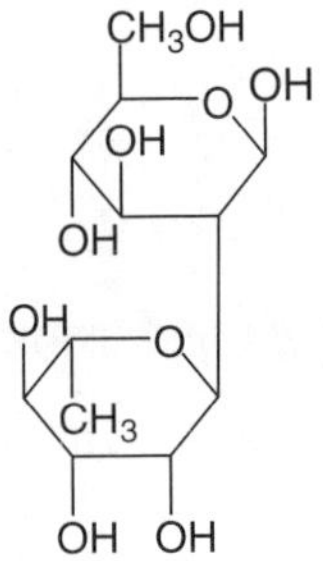

NEOHESPERIDOSE (2-O-alpha-L-rhamnopyranosyl-beta-D-glucopyranose)

Figure 8.3   *Structures of nonbitter rutinose and bitter neohesperidose.*

Grohmann, 1996). Similar values were observed in Hamlin, Temple, and Ambersweet orange peels. The tissue distributions of the main flavonoid glycosides in Hamlin sweet orange fruit are compared in Table 8.1, and the flavonoid composition of representative early/mid season dried peel pellets is shown in Table 8.2, where the total content of the major flavonoid glycosides averaged 1.8% dry weight.

Methods of commercial-scale recoveries of the flavanone glycosides have been extensively studied, where most procedures involve aqueous extraction in the presence of calcium hydroxide, then recrystalization after adjustment of pH to ~4.0 (Higby, 1947; Baier, 1948; Hendrickson and Kesterson, 1954, 1956). Alternatively, hot water extraction is a process that allows subsequent pectin extraction as an additional product from citrus peel (Hendrickson and Kesterson, 1964; Crandall and Kesterson, 1976; El-Nawawi, 1995). Total flavanone glycoside yields from such wet peel extractions give compositions of approximately 5% flavonoids on a dry weight basis.

Table 8.1   *Average amounts (mg) of major flavonoid glycosides in different tissues of Hamlin sweet orange fruit (n = 8).*

Tissue (ave. g dwt)	6,8-diC-gluAP[a]	Narirutin-4'-glu[b]	Narirutin	Hesperidin	Isosakuranetin rutinoside
Flavedo	4.4	7.6	14.3	302.4	10.5
Albedo	1.4	17.0	42.1	266.6	22.2
Membrane	0.9	5.6	17.8	45.6	11.8
Juice serum	6.2	13.8	13.5	13.9	1.6
Total	12.9	44.0	87.7	628.5	46.1

[a] 6,8-di-C-glucosyl apigenin.
[b] narirutin-4'-glucoside.

Table 8.2 *Flavonoid composition of early/mid season dried peel pellets. Values are reported in ppm and are averages of triplicate samples prepared from a single collection of pellets.*

Compound	Value (ppm)
6,8-di-C-glucosyl apigenin	332 ± 2
Narirutin-4′-glucoside	962 ± 24
Hesperidin-4′-glucoside	861 ± 17
Narirutin	1448 ± 23
Hesperidin	$13\ 991 \pm 126$
Isosakuranetin rutinoside	803 ± 16
Sinensetin	351 ± 5
Quercetagetin hexamethylether	83 ± 2
Nobiletin	407 ± 8
Tetramethylscutellarein	220 ± 19
Heptamethoxyflavone	290 ± 30
Tangeretin	73 ± 6

Unlike the flavanone and flavone glycosides, the polymethoxylated flavones (PMFs) in citrus byproducts occur as non-conjugated aglycones. These compounds occur in particularly high concentrations in tangerine and orange peels, and to a lesser extent in grapefruit. The PMFs are extensively methoxylated, and are typically less polar than the glycosidated flavonoids in citrus. Often the PMFs exhibit higher biological activities in animal cells (see below) than the other flavonoid glycosides, and currently, these compounds are primary targets in value-added materials for health applications. Levels of the main PMFs in the outer pigmented portion of the peel (flavedo) of several orange varieties are shown in Table 8.3. Total levels of the PMFs in dried peel pellets are similar to the levels in the original dried peel. This occurs in spite of the fact that high levels of these compounds also occur in peel oil recovered during juice processing (see below). Levels of the PMFs in 20° Brix orange peel molasses were shown to occur between 200 to 300 ppm (Manthey and Grohmann, 2001), but none were detected in molasses prepared from lemon, and only trace levels were detected in grapefruit

Table 8.3 *Concentrations (ppm) of the major polymethoxylated flavones in dried sweet orange (Rhode Red, Dulce Riberia, Hamlin) and grapefruit (Ruby Red) peel flavedo.*

[a]Compound	Rhode Red	Dulce Riberia	Hamlin	Ruby Red
SIN	903 ± 91	742 ± 31	453 ± 33	trace
QHME	355 ± 71	179 ± 7	213 ± 45	trace
NOB	1124 ± 202	1010 ± 202	856 ± 217	300 ± 44
HMF	707 ± 165	515 ± 110	399 ± 115	213 ± 25
TMS	505 ± 85	216 ± 80	251 ± 86	trace
TAN	233 ± 45	410 ± 160	178 ± 22	129 ± 18

[a] These compounds were absent in the inner, white albedo portion of the peel, as well as in the segment membranes and juice vesicles.

Abbreviations: (SIN) sinensetin (5,6,7,3′,4′-pentamethoxyflavone), (QHME) quercetagetin hexamethyl ether (3,5,6,7,3′,4′-hexamethoxyflavone), (NOB) nobiletin (5,6,7,8,3′,4′-hexamethoxyflavone), (HMF) 3,5,6,7,8,3′,4′-heptamethoxyflavone, (TMS) tetra-methylscutellariein (5,6,7,4′-tetramethoxyflavone), and (TAN) tangeretin (5,6,7,8,4′-pentamethoxyflavone).

Table 8.4 *Concentrations of hydrolyzable hydroxycinnamic acids in 20° Brix ultrafiltered molasses from six commercial citrus cultivars after alkaline hydrolysis (analyses done in triplicate) (Reproduced from J. Agric. Food Chem. 2001, **49**, 3268–3273).*

Cultivar	Total concentration (μg mL^{-1})		
	Ferulic	p-Coumaric	Sinapic
Valencia orange (late-season)	702 ± 15	284 ± 20	156 ± 20
Early/mid season orange[a]	527 ± 11	168 ± 9	71 ± 2
Dancy tangerine	1146 ± 34	270 ± 9	55 ± 2
Sunburst tangerine	1266 ± 19	197 ± 5	trace
Marsh grapefruit	236 ± 7	37 ± 1	64 ± 3
Lemon	94 ± 1	337 ± 2	63 ± 4

[a] Mainly Hamlin sweet orange.

molasses. Very low levels were detected in molasses prepared from Sunburst and Clementine tangerines, whilst, in contrast, Dancy tangerine 20° Brix molasses contained 711 and 90 ppm nobiletin and tangeretin, respectively.

Also prevalent in citrus peel-derived byproducts are the hydroxycinnamates – compounds containing conjugated forms of hydroxycinnamic acids, that is, *p*-coumaric, ferulic, caffeic, and sinapic acids. Numerous hydroxycinnamates are detected in molasses (Manthey and Grohmann, 2001; Peleg *et al.*, 1991) and a small number of these compounds have been identified (Risch and Herrmann, 1988b). These include galactaric acid esters of *p*-coumaric and ferulic acids (Risch *et al.*, 1987) as well as 2′-(*E*)-*O*-*p*-coumaroyl-, 2′-(*E*)-*O*-feruloyl, and 2,4-(*E,E*)-*O*-diferuloyl aldaric acid esters (Risch, Herrmann, and Wray, 1988a). The majority of the hydroxycinnamates in molasses occur as highly polar compounds, although, less polar hydroxycinnamates also appear. Enriched hydroxycinnamate fractions are recoverable by selective binding to anion exchange resins, and almost all of the molasses hydroxycinnamates, with the main exception of feruloylputrescine, hydrolyze under alkaline conditions to yield almost exclusively free ferulic, *p*-coumaric, and sinapic acids. Hence, analyses of hydroxycinnamates in molasses are made by quantitations of these hydrolyzed free acids. As shown in Table 8.4, the levels of ferulic, *p*-coumaric, and sinapic acids in alkaline-hydrolyzed 20° Brix Valencia molasses were 702 ± 15, 284 ± 20, and $156 \pm 20 \,\mu$g mL^{-1}, respectively (Manthey and Grohmann, 2001). Molasses prepared from two varieties of tangerine contained higher levels of hydroxycinnamates than the levels in Valencia and early-season orange molasses. Far lower levels of hydroxycinnamates were observed in lemon and grapefruit molasses. Unlike the other citrus varieties, where ferulic acid was the predominant hydroxycinnamic acid following alkaline hydrolysis, the main hydroxycinnamic acid in alkaline-hydrolyzed lemon molasses was *p*-coumaric acid.

Distributions of the hydrolyzable hydroxycinnamates were measured in the different portions of three varieties of sweet orange fruit (Table 8.5). Far higher concentrations of ferulic acid (2451–3103 ppm dwt) and *p*-coumaric acid (391–1427 ppm dwt) occurred in the flavedo than in the albedo, membranes, residue, and juice serum. Only trace levels of hydrolyzable *p*-coumaric acid is detected in the albedo (unpigmented portion of the peel). In contrast, the levels of hydrolyzed sinapic acid (132–499 ppm dwt) appear more evenly distributed between the flavedo, albedo, and segment membranes. These compounds significantly contribute to the antioxidants in peel-derived materials (Manthey, 2004).

Table 8.5 Hydrolyzable hydroxycinnamic acid content (ppm) of the fruit tissues of samples from three orange varieties (analyses done in triplicate) (Reproduced from J. Agric Food Chem. 2001, **49**, 3268–3273).

Variety	Ferulic	p-Coumaric	Sinapic
		Flavedo	
Rhode Red Valencia	3103 ± 39	1360 ± 20	270 ± 40
Dulce Riberia	2451 ± 89	1427 ± 158	trace
Hamlin	2480 ± 65	392 ± 18	499 ± 14
		Albedo	
Rhode Red Valencia	402 ± 9	trace	400 ± 16
Dulce Riberia	309 ± 7	trace	254 ± 15
Hamlin	394 ± 8	trace	295 ± 20
		Membrane	
Rhode Red Valencia	353 ± 2	trace	132 ± 10
Dulce Riberia	299 ± 16	trace	159 ± 3
Hamlin	295 ± 23	84 ± 12	207 ± 17
		Residue	
Rhode Red Valencia	218 ± 8	trace	trace
Dulce Riberia	242 ± 17	trace	trace
Hamlin	220 ± 23	trace	trace
		Juice Serum	
Rhode Red Valencia	61 ± 1	7 ± 1	7 ± 1
Dulce Riberia	49 ± 1	9 ± 1	9 ± 1
Hamlin	45 ± 2	9 ± 1	7 ± 1

An interesting fact regarding the total phenolic composition of citrus peel byproducts is that much of the total contents of these compounds occur in as yet poorly defined fractions of minor constituents. An analysis of orange-peel phenolic compounds fractionated by size-exclusion/ adsorption chromatography revealed broad collections of compounds, consisting of minor-occurring, yet uncharacterized flavone glycosides, complex hydroxycinnamate conjugates, and many other miscellaneous compounds, which contributed significantly to the total phenolic content and antioxidant levels of dried peel (Manthey, 2004). Conjugates of limocitrin, limocitrol, and chrysoeriol, as well as of methoxylated flavanones were amongst some of the flavonoid glycosides occurring in trace levels (Sawabe and Matsubara, 1999; Iwase et al., 2001; Manthey and Buslig, 2003; Manthey, 2004). Similar observations were made with the fractionation of citrus tissue extracts by centrifugal partition chromatography (Manthey, 2010).

In addition to the phytochemicals in bulk peel waste, the peel oils recovered from juice processing are also sources of a number of classes of potential value-added compounds. Methods of oil extraction and recovery during citrus juice processing have been recently reviewed by Di Giacomo and Di Giacomo (2002). Compositions of the volatile and non-volatile constituents of citrus oils have also been reviewed (Shaw, 1977; Dugo et al., 1994, 2002; Dugo and McHale, 2002). Yields of oil obtained under commercial juice processing conditions depend on a number of factors, particularly relating to peel attributes (Di Giacomo and Di Giacomo, 2002). Yields of cold-pressed peel oil per ton of fruit can vary from 0.9–3.2 kg for lemon, 0.5–3.6 kg for orange, and 0.5–0.9 kg for grapefruit and tangerine oils (USDA, 1962). Total oil yields per specific citrus cultivar are reported by Braddock (1999), although current commercial procedures typically yield only 60–75% of the total oil present in the peel, which agrees with the earlier reported values above (USDA, 1962). These oils may be concentrated by vacuum distillation to reduce their D-limonene contents from approximately 50 to 90%, yielding folded oils that are mainly used by the food and beverage flavoring industries. Pressed peels and co-produced waste streams can also be steam distilled to yield further quantities of oil, the majority of this oil yielding D-limonene or citrus stripper oil. Worldwide production of D-limonene is estimated to be between 50 000 and 75 000 metric tons, mostly from Florida and Brazil (Braddock, 1999).

Qualitatively, a minimum of 109 compounds have been identified as constituents of cold-pressed orange peel oil (Shaw, 1977), where these constituents occur mainly as mono- and sesquiterpene hydrocarbons, and their oxygenated alcohols, aldehydes, esters, hydrocarbons, ethers, ketones, and so on (Dugo *et al.*, 2002). Although the volatile constituents of citrus oils comprise the vast majority of these materials, there is also particular interest in the non-volatile constituents of citrus oils, including the PMFs, coumarins, furanocoumarins, and phytosterols. The PMFs and phytosterols are major constituents of the non-volatiles of orange and tangerine oils, with only very low levels of coumarins and furanocoumarins (Dugo and McHale, 2002). In orange oil, the major PMFs include sinensetin (5,6,7,3′,4′-pentamethoxyflavone), querce-tagetin hexamethyl ether (3,5,6,7,3′,4′-hexamethoxyflavone), nobiletin (5,6,7,8,3′,4′-hexam-ethoxyflavone), tetra-methylscutellarein (5,6,7,4′-tetramethoxyflavone), 3,5,6,7,8,3′,4′-heptamethoxyflavone, and tangeretin (5,6,7,8,4′-pentamethoxyflavone) (Horowitz and Gentili, 1977; Dugo *et al.*, 1994). Numerous other minor PMF constituents have been reported (Li, Lo, and Ho, 2006a). In tangerine, the primary PMFs include nobiletin and tangeretin, with other trace compounds. Coumarins and furanocoumarins of cold-pressed lemon oil have been extensively reviewed (Dugo and McHale, 2002), and have been shown to contain numerous mixed methoxy/hydroxy, geranyloxy, and isopentenyl derivatives. Particularly prevalent in lemon and grapefruit oils are 5- and 8-substituted linear furanocoumarins (Stanley and Vannier, 1957; Stanley and Jurd, 1971; Tatum and Berry, 1979). In bergamot oil, only bergapten, citropten, bergamottin, and bergaptol are observed (Benincasa *et al.*, 1990; Dugo and McHale, 2002).

Other health-promoting compounds recoverable from citrus byproducts with potential commercial value include the tetracyclic triterpenoid limonoids. Limonin (Figure 8.4) is the

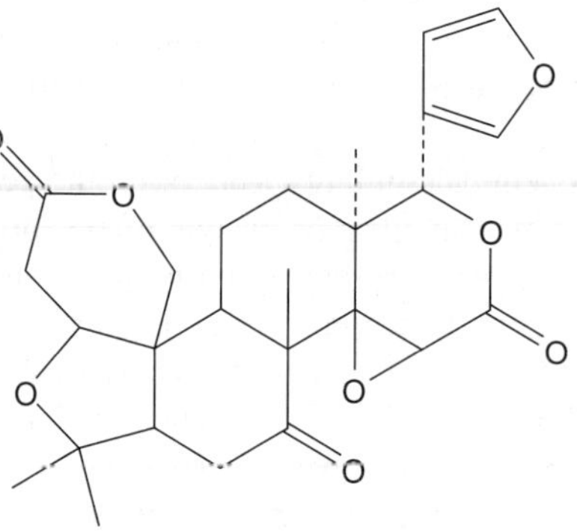

LIMONIN

Figure 8.4   *Structure of limonin.*

most abundant limonoid in citrus, and occurs at the highest percentages in the seeds (0.5 to 1% dwt) (Fong *et al.*, 1993), but at lower concentrations in the waste peel. Other limonoid aglycones prevalent in seeds are nomilin, obacunone, and nomilinic acid, although many other minor compounds also occur. In contrast, the peel is highest in concentrations of the limonoid glucosides, reaching up to 40–50 mg limonin glucoside per fruit for Valencia oranges (Fong *et al.*, 1992). Total limonoid glucoside concentrations in orange peel molasses have been reported as high as 7960 ppm (Schoch, Manners, and Hasegawa, 2001), but actual values are highly dependent on peel source. Hasegawa *et al.* (1996) reported a limonoid glucoside content of approximately $2\,g\,L^{-1}$ from 40° Brix orange peel molasses. Methods have been successfully developed for commercial recoveries of the limonoid glucosides by supercritical fluid extraction (Yu *et al.*, 2006) and through the use of ion-exchange and styrene/divinyl-benzene resins (Schoch, Manners, and Hasegawa, 2002). The latter method was shown to yield approximately 5 g of a water-soluble powder containing over 60% by weight limonoid glucosides per liter of molasses (40–50° Brix). Limonoids have been found to exhibit insect antifeedant and growth-regulating activities (Ruberto *et al.*, 2002; Roy and Saraf, 2006). These compounds have also attracted attention for their potential prevention of certain cancers and other chronic diseases in humans (Lam and Hasegawa, 1989; Miller *et al.*, 1994, 2004; Poulose, Harris, and Patil, 2005; Poulose, Harris, and Patil, 2006; Tanaka *et al.*, 2001). Yet, it is to be noted that recent studies have also explored the influence of citrus limonoids as P-glycoprotein substrates, which may raise the possibility of food/drug interactions in humans (Lien, Wang, and Lien, 2002; El-Readi *et al.*, 2010).

8.3 Polysaccharides as Value-Added Products

8.3.1 Dietary Fiber

As described in the preceding sections, the peel residue byproduct streams generated during orange juice processing are pectin-rich sources of dietary fiber. A study by Grigelmo-Miguel and Martin-Belloso (1998) showed that the total content of dietary fiber of orange residues occurred at 35.4–36.9% weight – of which approximately 16% (peel dry weight) occurred as pectin, 17% as cellulose and hemicelluloses, and 3% lignin. Dietary fiber derived from orange peel processing waste exhibited excellent water (~7–10 g water per g fiber) and oil (~1 g oil per g fiber) absorption. Fiber-rich fractions of orange peel exhibit higher water and oil-holding capacities, higher cation exchange properties than those of cellulose, and also exhibit increased *in vitro* glucose absorption and retardation of glucose diffusion (Chau, Huang, and Lee, 2003; Chau and Huang, 2003). Applications of such dietary fibers include clouding agents in beverages, thickener and gelling agents, and a low-calorie bulk ingredient, for both humans and livestock (Fernandez-Lopez *et al.*, 2004; Hon, Oluremi, and Anugwa, 2009). An increasing number of reports provide evidence of the potential health benefits of incorporation of orange pulp in animal diets. Potential health benefits of extruded orange peel fiber were further suggested by the observed decreased *in vitro* starch hydrolysis in the presence of extruded orange peel fiber and high glucose sequestration by the fiber (Chau, Huang, and Lee, 2003; Céspedes, Martinez-Bustos, and Chang, 2010). These potential hypoglycaemia effects of orange peel fiber suggest that dietary fibers produced from orange processing waste may be useful in beneficially controlling glucose levels in the blood. Elsewhere, micronized orange peel fiber lowered the concentrations of serum triglycerides by 15.6–17.8% and serum cholesterol by 15.7 to 17.0% in hamsters fed diets supplemented with cholesterol (2.0 g kg^{-1} of diet) and micronized citrus insoluble fiber (59 g kg^{-1} of diet) (Wu, Wu, and Chau, 2009).

8.3.2   Peel Hydrolysis and Ethanol Production

In addition to dietary uses of citrus waste peel, attention has also been drawn to the potential conversion of bulk orange peel waste into ethanol. To date, though, conversion of orange peel to ethanol is incomplete, and many of the other potential co-products remain to be exploited. Much of the research on peel-to-ethanol conversion has dealt with acid pretreatments, linonene removal, and selections of optimal enzymes for the hydrolysis of different peel components. Dilute sulfuric acid pretreatments (0.06 and 0.5%) at elevated temperatures (100–140° C) solubilized a large portion of the total carbohydrates in orange peel, including mainly soluble sugars and sugars derived from hemicelluloses, whilst the pectin components remained resistant to hydrolysis. However, these pretreatments significantly increased the rates of subsequent cellulolytic and pectinolytic hydrolyses of orange peel (Grohmann, Cameron, and Buslig, 1995). Extensive studies have also been done to characterize and optimize enzymatic hydrolysis to liquefy citrus peel carbohydrate fractions and maximize monomeric sugar production for conversion to ethanol by fermentation (Grohmann and Baldwin, 1992; Grohmann, Cameron, and Buslig, 1995; Wilkins *et al.*, 2005; Wilkins *et al.*, 2007a). The subsequent peel fermentation, however, is inhibited by D-limonene, the major component of orange oil, which occurs at roughly 1% of peel wet weight. Hence, pretreatments to optimize D-limonene removal prior to fermentation have been pursued (Widmer, Zhou, and Grohmann, 2010). Steam explosion/steam stripping has been demonstrated to sterilize and significantly disrupt peel tissues, making them more accessible to enzymes for hydrolysis, and to remove D-limonene to levels suitable for simultaneous saccharification and fermentation by added exogenous enzymes and active dry yeast (*Saccaromyces cerevisiae*) (Wilkins, Widmer, and Grohmann, 2007b). Sterilization during steam treatment is important as the microflora present in the peel waste will use available sugars if not treated, thus lowering the ethanol yield. Ethanol yields obtained after optimal pretreatment of citrus peel waste, followed by simultaneous saccharification and fermentation ranged from 83–89% of the theoretical yields calculated from the amount of fermentable sugars released after hydrolysis of the pretreated waste with excess enzyme and no fermentation. With the use of fresh orange processing waste, one can obtain 32–39 g ethanol per liter of fermented hydrolysate. These advances in the simultaneous fermentation and saccharification will ultimately yield processes applicable to commercial-scale ethanol production from citrus waste peels.

8.3.3   Speciality Pectins

Beyond dietary fiber and ethanol production, a third focus of new product development from the citrus peel polysaccharides lies with pectin, with an estimated worldwide market value of at least 400 million Euros (Savary *et al.*, 2003). As a structurally complex polysaccharide, pectin is composed of at least five different sugar moieties, with typically 80–90% of its dry weight comprised of galacturonic acid. Most of the galacturonic acid in pectin occurs in homogalacturonan regions, which are unbranched linear polymers of galacturonic acid residues. Many of the galacturonic acid residues in these regions are methyl esterified at the C6 position. Although the basic properties of pectin have long been known, recent advances have been made in understanding the complex fine structures of pectin. Recent advances have also been made in techniques allowing specific modifications to be made to these highly complex fine structures of pectin, thus allowing for the development of novel, high value applications for these pectin-derived materials (Sörensen, Pedersen, and Willats, 2009). Elements of pectin structures that confer functionality include molecular weight, side-chain branching, degree of methylation (DM), degree of amidation, presence of demethylated blocks, and demethylated block average size and number per pectin molecule (May, 1990; Luzio, 2003; Willats *et al.*, 2001; Willats, Knox, and Mikkelsen, 2006). Each of these structural attributes can be modified to create materials filling broad scopes of functionalities, which can then be applied to a wide array of high-value food- and pharmaceutical-based applications. Spatial distributions of methylation and of demethylated homogalacturonan blocks can be predictably created or modified within pectin by uses of specific plant, bacterial, and fungal pectin methy-lesterases and through classical chemical means (Cameron *et al.*, 2008 and references therein). In addition to the above-mentioned structural attributes, special distributions of methylation – or demethylation – critically influence water-holding capacity (Willats *et al.*, 2001), as well as other pectin functionalities (Cameron *et al.*, 2008; Luzio and Cameron, 2008). Expanding the capabilities to create specific fine structures in pectin will dramatically increase the scope of applications of speciality pectins, thus creating new, high-value uses for this major citrus peel component.

8.4   Phytonutrients as Value-Added Products

Perhaps the most notable amongst the potential value-added materials in citrus processing waste are the many diverse non-volatile secondary natural products, including primarily flavonoids, hydroxycinnamates, and other related compounds. Amongst the flavonoids are the glycosides of the flavanones and flavones, and the non-glycosidated polymethoxylated flavone (Figure 8.2). Numerous studies into the potential health benefits of these two broad categories of compounds have been done, and along with the hydroxycinnamates, each will be discussed separately.

8.4.1   Flavonoid Glycosides

In Vitro Antioxidant Studies of Hesperidin/Naringin

The primary biological attributes of the flavonoid glycosides in citrus peels are linked to their antioxidant, anti-inflammatory, and hypolipidemic effects, and extensive work has been done to commercialize these peel components. These components, typically labelled "citrus bioflavonoids" widely occur as components in vitamins, herbal supplements, teas, and many other food products. Extensive studies have been made of the *in vitro* antioxidant actions of these compounds and of citrus peel extracts (Freeman, Eggett, and Parker, 2010; Yi *et al.*, 2008; Wilmsen, Spada, and Salvador, 2005; Yu *et al.*, 2005), particularly of hesperidin and naringin. The antioxidant actions of the citrus flavanones are often linked to their inhibition of enzymes (e.g., lipoxygenase, cyclooxygenase, phosphodiesterase, and others) involved in inflammation (Manthey, Guthrie, and Grohmann, 2001; Rathee *et al.*, 2009). Hesperidin has been shown to prevent *tert*-butyl-hydroperoxide-induced oxidative stress in human hepatocytes in a dose-dependent manner. Western blot data suggest an induction by hesperidin of hepatocyte cytoprotectivehaem oxygenase-1 expression, thus leading to protection against *in vitro* oxidative-stress-related injury (Chen *et al.*, 2010a). The flavanone aglycone, hesperetin, effectively scavenges peroxynitrite and exhibited cytoprotection from cell damage by peroxynitrite *in vitro* (Kim *et al.*, 2004). Cytoprotection of osteoblasts by hesperetin occurred against diabetes-induced oxidative damage, suggesting that citrus flava-nones may be useful as a supplement for minimizing oxidative injury in diabetes-related bone disease (Choi and Kim, 2008).

In Vivo *Antioxidant Studies of Hesperidin/Naringin*

Dietary uptake of the citrus peel flavonoids, hesperidin and naringin, has been shown to beneficially mediate the expression of inflammation markers in a variety of chronic human diseases. Hesperidin administered ($100 \, \text{mg kg}^{-1}$, p.o) to type-2 diabetic rats for 28 days resulted in restored glutathione levels, superoxide dismutase and catalase activity levels and reduced lipid peroxidation and nitrite, in heart tissue after isoproterenol-induced myocardial infarction, in comparison to non-hesperidin-treated animals (Jagdish, Mehul, and Nehal, 2010). Admin-istration of hesperidin ($25 \, \text{mg kg}^{-1}$ body weight) to Swiss albino mice significantly reduced much of the oxidative stress tissue damage in benzo(a)pyrene-induced lung cancer (Kamaraj *et al.*, 2009). Similar activation of key antioxidant defence enzymes was previously detected following administration of hesperetin to 7,12-dimethylbenz(a)anthracene-dosed mice (Choi, 2008). Neuroprotective effects of hesperetin in the brains of mice administered hesperetin at 10 and $50 \, \text{mg kg}^{-1}$ body weight were found to correlate with significant inhibition of oxidative stress. Hesperetin significantly activated the catalase, total superoxide dismutase, and gluta-thione peroxidase and reductase activities (Choi and Ahn, 2008). Indicators of decreased oxidative stress following uptake of the citrus flavanone glycosides have also been described in a number of other oxidative stress models (Rapavi *et al.*, 2007; Tirkey *et al.*, 2005; Miyake *et al.*, 1998). Increases in blood plasma antioxidant activity following hesperidin and naringin uptake have also been reported (Gorenstein *et al.*, 2007). The identit(ies) of the active chemical species, most probably metabolites of hesperidin, that are responsible for these protective effects of orally administered hesperidin remain to be determined.

8.4.2 Polymethoxylated Flavones (PMFs)

Unlike hesperidin, naringin, and many of the other flavonoid glycosides that occur largely in the albedo and segment membranes of citrus peels, the PMFs consist of a group of non-glycosylated flavones which occur in the oil glands of the pigmented flavedo portion of peels. Due to their contrasting chemical structures and increased lipophilicity relative to the more polar glycosides, the PMFs are typically treated and studied as a separate class of compounds. Similar to the glycosides, PMFs exhibit a number of beneficial biological actions in animal cells, and similarly occur as valuable co-products of citrus fruit processing. These biological actions mainly focus on the regulation of lipid profiles, glucose homoeostasis, inhibition of inflammation, and antiproliferative effects towards human cancer cells.

Hypolipidemic Effects and Modulation of Lipid Profiles

The hypolipidemic effects of the PMFs were initially characterized by their inhibition of apoprotein B secretion by human liver HepG2 cells (Kurowska and Manthey, 2002).

Apoprotein B is the structural protein of the low-density lipoprotein (LDL). The IC_{50} values of tangeretin (2.5 ppm) and nobiletin (4.9 ppm) were ten times lower than for the flavanone aglycones, hesperetin (43 ppm) and naringenin (48 ppm). Subsequent feeding trials in hypercholesterolemic hamsters showed dramatic drops in serum total cholesterol (-25%), VLDL + LDL (-39%), and triglycerides (-48%), but no change in high density lipoprotein (Kurowska and Manthey, 2004) when administered diets containing 1.0% tangeretin. In this study, a diet containing a 1.0% mixture of PMFs, in the proportions occurring in orange peel, produced similar results. The modes of action of the PMFs associated with these dramatic decreases in total cholesterol, LDL, VLDL, and triglycerides were subsequently characterized in a series of studies, where multiple sites of action were identified. Inhibition of apoprotein B production/secretion by HepG2 liver cells involved blockage of the incorporation of radiolabeled precursors into cellular cholesterol, cholesterol esters, and triglycerides. Enzyme control points regulating diacylglycerol acyltransferase, activation of the peroxisome proliferator-activated receptor (PPAR), and the microsomal triglyceride transfer protein were also influenced (Kurowska *et al.*, 2004). Nobiletin was uniquely able to inhibit *in vitro* macrophage foam cell production mediated by the cell surface SR-A receptor, thus further preventing atherosclerotic plaque formation (Whitman *et al.*, 2005). Nobiletin also stimulates the transcription of the hepatic LDL receptor (LDLR) gene at levels far lower than required for the flavanones, hesperetin and naringin (Morin *et al.*, 2008), and these cardioprotective properties of the citrus peel PMFs have been validated in human clinical trials (Roza, Xian-Liu, and Guthrie, 2007; Oben *et al.*, 2008).

Anti-Inflammatory Actions

The anti-inflammatory actions are another application of the PMFs in human health, and were initially documented in an early study by Freedman and Merritt (1963). An orally administered PMF-containing extract of orange peel inhibited the inflammation response in the Arthus reaction in guinea pigs – an immune-driven response. The PMF fraction exhibited a higher inhibitory activity than cortisone, the positive control. Analysis of the anti-inflammatory activities of the fractionated extract showed that two PMFs, one of which was nobiletin, possessed significantly higher inhibitory activities than either hesperidin or naringin.

Biochemical mechanisms of the anti-inflammatory actions of the PMFs were subsequently investigated *in vitro* (for reviews, Manthey, Guthrie, and Grohmann, 2001; Benevente-Garcia and Castillo, 2008). Early studies reported potent cAMP phosphodiesterase inhibition by a number of methoxylated flavones, including PMFs in citrus (Petkov, Nokolov, and Uzunoy, 1981; Nikaido *et al.*, 1988). Glycosidation of these flavones dramatically blocked their inhibitory actions. The involvement of phosphodiesterase inhibition was further demonstrated *in vitro* where citrus PMFs dramatically inhibited induction of tumor necrosis factor-α (TNF-α) expression in bacterial lipopolysaccharide-activated human monocytes (Manthey *et al.*, 1999). TNF-α is a key protein signalling molecule responsible for cell activation in the inflammation cascade. Also monitored was the induction of other proinflammatory cytokine proteins, including macrophage inflammatory protein-1α (MIP-1α), and interleukins-1β, 6, 8 and 10. The PMFs inhibited MIP-1α and IL-10, but were inactive as inhibitors of IL-1β, -6 and -8. The most active compounds, nobiletin, sinensetin, 3,5,6,7,8,3',4'-heptamethoxyflavone (HMF), and 5-desmethylnobiletin had IC_{50} values between 5 and 10 µM. Very little cytotoxicity was observed with these compounds at these low concentrations. In contrast, hydroxylated flavones were less active, but more cytotoxic, and glycosylation destroyed the inhibitory activity. This suggests that glucuronidated metabolites of citrus PMFs may also be inactive as phosphodiesterase inhibitors.

In this study, the active PMFs were shown to strongly inhibit human phosphodiesterase-4 (PDF-4), and to induce a substantial elevation of cAMP levels in human monocytes, similar to the action of the known PDF-4 inhibitor3-isobutyl-1-methylxanthine. PDF-4 activity is centrally involved in early stages of inflammation, and the above inhibition of the expression of the proinflammatory cytokines likely results from the inhibition by this enzyme. The inhibition of cytokine production by the PMFs was subsequently studied *in vivo*. Injections of bacterial lipopolysaccharides into mice produce dramatic rises in blood serum levels of TNF-α. Prior intraperitoneal injections of 3,5,6,7,8,3′,4′-heptamethoxyflavone (HMF), dissolved in vegetable oil at 33 and 100 mg kg^{-1} body weight significantly inhibited the TNF-α production induced by the bacterial lipopolysaccharides by 30 and 45%, respectively (Manthey and Bendele, 2008). In comparison, prior treatment with the positive control, dexamethasone, produced a 66% inhibition of the TNF-α production. In sharp contrast, three daily doses of HMF (suspended in vegetable oil) at 50, 100, and 200 mg kg^{-1} body weight, with the final dose given one hour prior to the LPS challenge, produced no inhibition of TNF-α production. Significantly, HMF administration by intraperitoneal injection as a suspension produced only a non-significant 11% inhibition, which is in sharp contrast to the inhibition observed by the previously mentioned treatments with HMF *dissolved* in vegetable oil. A similar dichotomy of results for oral and intraperitoneal injection administrations occurred in carrageenan-induced paw oedema inflammation assays in mice (Manthey and Bendele, 2008).

Results obtained in this study indicated that efficacy strongly relied upon the dissolved state of HMF in vegetable oil, and the apparent absence of activity following oral administration due to poor absorption of these compounds. HPLC-ESI-MS analyses showed that HMF levels reached 14.2 ± 4.7 and $8.6 \pm 1.6 \,\mu\text{g} \, \text{g}^{-1}$ serum following the 100 and 33 mg kg^{-1} body weight intraperitoneal injections of HMF dissolved in vegetable oil (Manthey and Bendele, 2008). Yet, in sharp contrast, the HMF levels in the mice administered HMF (100 mg kg^{-1}) by intraperitoneal injection as a suspension in vegetable oil reached only $0.035 \pm 0.024 \,\mu\text{g} \, \text{g}^{-1}$ serum. This sharply contrasting bioavailability of HMF dissolved in vegetable oil or as a suspension is in line with the sharply contrasting effects of these doses on the TNF-α in the mice following the LPS challenges.

The lack of inhibition by HMF administered orally in these mouse inflammation studies appears to contradict the previously mentioned findings of Freedman and Merritt (1963). Yet, the Freedman and Merritt study involved an entirely different inflammation model, test compounds, and animal species. Hence, it will be important to repeat with guinea pig the Arthus reaction used by Freedman and Merritt to verify the positive effects of orally administered nobiletin. Differences in sites of action in the Arthus reaction, as well as differences in PMF bioavailabilities, may critically influence the biological actions of these compounds in different inflammation settings. And in fact, preliminary trials to repeat the Arthus reaction in guinea pigs, using HMF (substituted for nobiletin), as described by Freedman and Merritt (1963), have shown weak positive inhibition of markers expressed in this inflammation model (Manthey and Cesar, 2008). Further work to validate these findings will also help reconcile them with the results obtained in a recent human clinical trial, where a formulation of *Phellodendron* extract and citrus PMFs, orally administered, proved effective against inflammation associated with osteoarthritis (Oben *et al.*, 2009). The inflammation biomarker C-reactive protein was significantly reduced in normal and overweight treatment groups. US patent (Nos. 6184246 and 6987125)-protected products of this formulation are currently commercialized. The efficacy of orally administered PMFs has also been validated in a study of PMF influence on insulin resistance in hamsters (Judy *et al.*, 2010), and cholesterol lowering in hamsters (Kurowska and Manthey, 2004) and humans (Roza, Xian-Liu, and Guthrie, 2007).

Successful development of products derived from PMFs depends on their efficacy and safety (Delaney *et al.*, 2001; VanHoecke *et al.*, 2005). As expected, and as indicated in previous inflammation studies, the pharmacokinetics and bioavailabilities of the PMFs appear to impact levels of responses observed with these compounds. Yet, little has been done to improve formulations that increase oral PMF uptake. However, it was observed recently that nobiletin dissolved in vegetable oil was significantly more bioavailable than nobiletin suspended in vegetable oil (Manthey *et al.*, 2011). Moreover, the pharmacokinetics of nobiletin was notably different using the two formulations. Results also suggested that there were significant differences in the bioavailabilities of different PMFs, and if validated, this may influence future compositions of PMF products. Knowledge of the pharmacokinetics of the metabolites of the orally administered PMFs may also play roles in future product development. A substantial amount of work has been done to characterize metabolites of nobiletin (Yasuda *et al.*, 2003; Li *et al.*, 2006b), tangeretin (Nielsen *et al.*, 2000; Manthey *et al.*, 2011), and HMF (Manthey and Bendele, 2008). Biological studies have now shown that some of the metabolite aglycones express activity levels similar to the parent compounds (Eguchi *et al.*, 2007; Li *et al.*, 2007; Lai *et al.*, 2008; Al Rahim *et al.*, 2009; Xiao *et al.*, 2009), and this may significantly contribute to the total activity expressed by specific doses of PMFs.

Recent studies have also pointed to exciting new applications, and thus, potentially new markets for the citrus PMFs, including applications pertaining to bone resorption in osteoporosis and rheumatoid arthritis (Murakami *et al.*, 2007), and to Alzheimer's disease. In particular, nobiletin has been shown to improve memory deficits, to decrease the β-amyloid burden and plaques, and to influence a number of key biological enzyme and protein markers of Alzheimer's disease (Onozuka *et al.*, 2008; Yamakuni, Nakajima, and Ohizumi, 2010). Mechanisms relating to the prevention of bone loss and inhibition of collagen-induced arthritis are likely to parallel those of inflammation, and thus may enhance the potential marketing of the PMFs in additional inflammation settings.

8.4.3 Hydroxycinnamates

The hydroxycinnamates and related phenolic acids represent classes of largely untapped, unexplored resources of potentially health-beneficial compounds recoverable from citrus fruit processing waste. The biological actions of hydroxycinnamates in citrus peel waste may likely parallel those found in other foods and agricultural crops. The metabolism and bioavailablity of such dietary hydroxycinnamates have been extensively reviewed (Williamson *et al.*, 2000; Kern *et al.*, 2003; Crozier, Jaganath, and Clifford, 2009; Zhao and Moghadasian, 2010). An *in vitro* model suggests that the human small intestinal epithelium plays a role in the metabolism and bioavailability of these phenolic compounds (Kern *et al.*, 2003). The biological actions of dietary hydroxycinnamates and of their non-conjugated hydroxycinnamic acids focus on their antioxidant, hypolipidemic and anti-inflammatory effects in animal cells (Shahidi and Chandrasekara, 2010). Recent studies show that metabolites of these dietary compounds exert influences on cell signalling cascades critical to cellular biology of inflammation, lipid regulation, cell growth, and apoptosis (Crozier, Jaganath, and Clifford, 2009; Cheng *et al.*, 2008; Clifford, 2004). Increased bioavailablity of phenolic acids in humans led to decreased expression of proinflammatory cytokines in *ex vivo* bacterial lipopolysaccharide-stimulated blood (Mateo Anson *et al.*, 2011). Ferulic acid showed potent dose-dependent inhibition of gamma radiation induction of the adhesion molecules, intercellular cell adhesion molecule-1 and vascular cell adhesion molecule-1, in human umbilical vein endothelial cells (Ma *et al.*, 2010). Evidence was shown of these inhibitory effects of ferulic acid in blocking c-Jun N-terminal kinase. Ferulic acid has also been recently shown to potentially exert a chondroprotective effect in the treatment of osteoarthritis. Inflammation in cartilage stimulates the expression of the inflammatory cytokines interleukin-1β and TNF-α (Hedbom and Häuselmann, 2002). Chondrocytes pretreated with 40 µM ferulic acid exhibited inhibition of the above hydrogen peroxide-induced cytokines and metalloproteinase gene expression at the mRNA level (Chen *et al.*, 2010b). This suggests that ferulic acid and possibly other structurally related compounds might be useful against osteoarthritis. Evidence of the usefulness of long-term administration of ferulic acid and related compounds against Alzheimer's disease has also been shown (Yan *et al.*, 2001; Sultana *et al.*, 2005).

The intimate association between cellular oxidation and atherosclerosis is also modulated by hydroxycinnamates. Oxidized low-density lipoprotein is a trigger for vascular injury to endothelial cells lining vascular tissue. Sodium ferulate significantly reduces oxidized low-density lipoprotein induction of 32 genes encoding chemokines, inflammatory factors, growth factors and nuclear receptors, thus, promoting endothelial function under atherosclerotic conditions (Zhang *et al.*, 2009). In a feeding trial with apolipoprotein-E-deficient mice, administration of ferulic acid led to significantly lower plasma concentrations of total cholesterol, and apolipoprotein B compared to control groups (Kwon *et al.*, 2010). Numbers of mice exhibiting aortic fatty plaques were also much lower in the ferulic acid fed mice. Also observed were significantly higher activity levels of antioxidant enzymes (superoxide dismutase, catalase, glutathione reductase and peroxidase, and paraoxonase) in the hepatocytes and erythrocytes of the ferulic acid-fed mice. The structurally related *p*-coumaric acid also produced reductions in total cholesterol levels in rats fed cholesterol-rich diets (Lee *et al.*, 2003). Administration of *p*-coumaric further increased plasma concentrations of high-density lipoprotein, but decreased the hepatic cholesterol and triglyceride levels. Administration of this compound also significantly lowered hepatic oxidative stress. These findings suggest that the hydroxycinnamates, following metabolism to the aglycone chemical species, may exert similar antiatherogenic activities in animals, and thus may complement those already described for the main flavanones and PMFs in citrus processing waste.

8.5   Fermentation and Production of Enhanced Byproducts

Finally, whilst the contents and biological actions of these major classes of materials and compounds in citrus processing waste have been widely studied, the production of new materials may also be achieved in the future as results of peel waste fermentation. As mentioned previously, fermentation is already a key component of peel-to-ethanol conversions, and these steps may also be used to generate novel products. Microbial metabolism of flavonoids typically yields ring-fission products and these bioconversions have been reviewed by Barz and Hösel (1975). A characterization of flavonoid catabolism by *Pseudomonas putida* has been reported (Pillai and Swarup, 2002), and new metabolites with intact flavonoid ring structures have also been characterized in fermented citrus peels. Fermentation of pumelo (*C. grandis*) peel produced significant increases in the antioxidant activity and contents of minor flavonoids in the fermented peel compared to the untreated peel (Hyon *et al.*, 2009; Jang *et al.*, 2010). Increased antioxidant activities, as well as unique, new compounds, including 8-hydroxyhesperetin and 8-hydroxynaringenin, were produced in citrus peels fermented with three species of *Aspergillus* (Miyake *et al.*, 2004a). Lemon peel fermented by *Aspergillus saitoi* similarly contained higher antioxidant activity content, as well as exhibiting significant suppressive effects against primary oxidative stress markers in rats following exhaustive exercise (Miyake *et al.*, 2004b). These findings illustrate the potential in employing microbial fermentation to produce modified citrus peel components with enhanced, targeted biological applications.

8.6   Conclusion

The citrus peel waste generated from juice processing contains a wealth of components that can be conceivably recovered as value-added products and also serve as a feedstock for energy production. The use of citrus peel waste as a source of high-value compounds has been in place for decades, but up to the present, little attempt has been made by the industry to integrate recoveries to profit from these potential product profiles. Research is in progress to develop new speciality, high-value products, and to explore rationales for the expanded use of peel components, whilst integrating this extraction step into a biorefinery scenario. The body of evidence firmly points to potential health benefits of compounds in citrus processing waste, and conclusive clinical trials to validate these actions will continue at an expanded pace. Speciality pectin production and fermentation of peel-to-ethanol production provide a blanket approach to using the whole peel carbohydrate component, and these steps with integrated recoveries of flavonoids, limonoids, hydroxycinnamates, and other constituents, hold promise for future production of high-value product profiles from this agricultural source.

References

Albach, R.F. and Redman, G.H. (1969) Composition and inheritance of flavanones in citrus fruit. *Phytochemistry*, **8**, 127–143.

Al Rahim, M., Nakajima, A., Saigusa, D. *et al.* (2009) 4'-demethylnobiletin, a bioactive metabolite of nobiletin enhancing PKA/ERK/CREB signaling, rescues learning impairment associated with NMDA receptor antagonism via stimulation of the ERK cascade. *Biochemistry*, **48**, 7713–7721.

Armentano, L., Bentsáth, A., Berés, T. *et al.* (1936) Uber den Einfluss von Substanzen der Flavongruppe auf die Permeabilität der kapillaren; Vitamin P. *Deutsche Medizinische Wochenschrift*, **62**, 1325–1328.

Barz, W. and Hösel, W. (1975) Metabolism of flavonoids, in *The Flavonoids. Part 2* (eds J.B. Harborne, T.J. Mabry, and H. Mabry), Academic Press, Inc., New York.

Baier, W.E. (May 25 1948) Process for recovery of hesperidin. U.S. Pat. 2442110.

Benevente-Garcia, O. and Castillo, J. (2008) Update on uses and properties of citrus flavonoids: New findings in anticancer, cardiovascular, and anti-inflammatory activity. *Journal of Agricultural and Food Chemistry*, **56**, 6185–6205.

Benincasa, M., Buiarelli, F., Cartoni, G.P., and Coccioli, F. (1990) Analysis of lemon and bergamot essential oils by HPLC with microbore columns. *Chromatographia*, **30**, 271–276.

Bentsáth, A., Rusznyák, S., and Szent-Györgyi, A. (1936) Vitamin nature of flavones. *Nature*, **138**, 798.

Braddock, R.J. (1999) *Handbook of Citrus By-Products and Processing Technology*, Wiley-Interscience Publication., New York.

Cameron, R.G., Luzio, G.A., Goodner, K., and Williams, M.A.K. (2008) Demethylation of a model homogalacturonan with a salt independent pectin methylesterase from citrus: I. Effect of pH on demethylated block size, block number and enzyme mode of action. *Carbohydrate Polymers*, **71**, 287–299.

Céspedes, M.A.L., Martinez-Bustos, F., and Chang, Y.K. (2010) The effect of extruded orange pulp on enzymatic hydrolysis of starch and glucose retention index. *Food and Bioprocess Technology*, **3**, 684–692.

Chapman, H.L. Jr., Ammerman, C.B., Baker, F.S. Jr. *et al.* (1972) Citrus feeds for beef cattle. *Florida Agricultural Experimental Station*, Gainesville. Bull., No. 751.

Chau, C.F. and Huang, Y.L. (2003) Comparison of the chemical composition and physicochemical properties of different fibers prepared from the peel of *Citrus sinensis* L Cv. Liucheng. *Journal of Agricultural and Food Chemistry*, **51**, 2615–2618.

Chau, C.F., Huang, Y.L., and Lee, M.H. (2003) *In vitro* hypoglyceric effects of different insoluble fiber-rich fractions prepared from the peel of *Citrus sinensis* L cv. Liucheng. *Journal of Agricultural and Food Chemistry*, **51**, 6623–6626.

Chen, M., Gu, H., Ye, Y. *et al.* (2010a) Protective effects of hesperidin against oxidative stress of *tert*-butyl hydroperoxide in human hepatocytes. *Food and Chemical Toxicology*, **48**, 2980–2987.

Chen, P.M., Yang, S.H., Chou, C.H. *et al.* (2010b) The condroprotective effects of ferulic acid on hydrogen peroxide-stimulated chondrocytes: inhibition of hydrogen peroxide-induced proinflammatory cytokines and metalloproteinase gene expression at the mRNA level. *Inflammation Research*, **59**, 587–595.

Cheng, C.Y., Su, S.Y., Tang, N.Y. *et al.* (2008) Ferulic acid provides neuroprotection against oxidative stress-related apoptosis after cerebral ischemia/reperfusion injury by inhibiting ICAM-1 mRNA expression in rats. *Brain Research*, **1209**, 136–150.

Choi, E.J. (2008) Antioxidative effects of hesperetin against 7,12-dimethylbenz(a)anthracene-induced oxidative stress in mice. *Life Sciences*, **82**, 1059–1064.

Choi, E.J. and Ahn, W.S. (2008) Neuroprotective effects of chronic hesperetin administration in mice. *Archives of Pharmacal Research*, **31**, 1457–1462.

Choi, E.M. and Kim, Y.H. (2008) Hesperetin attenuates the highly reducing sugar-triggered inhibition of osteoblast differentiation. *Cell Biology and Toxicology*, **24**, 225–231.

Citrus Summary (2008–09) Maitland, FL: Florida Dept. of Agriculture and Consumer Services and USDA National Agriculture Statistics Service.

Clifford, M.N. (2004) Diet-derived phenols in plasma and tissues and their implications for health. *Planta Medica*, **70**, 1103–1114.

Crandall, P.G. and Kesterson, J.W. (1976) Recovery of naringin and pectin from grapefruit albedo. *Proceedings of the Florida State Horticultural Society*, **89**, 189–191.

Crozier, A., Jaganath, I.B., and Clifford, M.N. (2009) Dietary phenolics: Chemistry, bioavailablity and effects on health. *Natural Product Reports*, **26**, 1001–1043.

Delaney, B., Phillips, K., Buswell, D. *et al.* (2001) Immunotoxicity of a standardized citrus polymethoxylated flavone extract. *Food and Chemical Toxicology*, **39**, 1087–1094.

Di Giacomo, A. and Di Giacomo, G. (2002) Essential oil production, in *The Genus Citrus. Medicinal and Aromatic Plants – Industrial profiles* (eds G. Dugo and A. Di Giacomo), Taylor and Francis, Boca Raton.

Dugo, G., Cotroneo, A., Verzera, A., and Bonaccorsi, I. (2002) Composition of the volatile fraction of cold-pressed citrus peel oils, in *The Genus Citrus. Medicinal and Aromatic Plants – Industrial profiles* (eds G. Dugo and A. Di Giacomo), Taylor and Francis, Boca Raton.

Dugo, P. and McHale, D. (2002) The oxygen heterocyclic compounds of citrus essential oils, in *The Genus Citrus. Medicinal and Aromatic Plants – Industrial profiles* (eds G. Dugo and A. Di Giacomo), Taylor and Francis, Boca Raton.

Dugo, P., Mondello, L., Cogliandro, E. *et al.* (1994) On the genuiness of citrus essential oils. Part 46. Polymethoxylated flavones of the non-volatile residue of Italian sweet orange and mandarin essential oils. *Flavour and Fragrance*, **9**, 105–111.

El-Nawawi, S.A. (1995) Extraction of citrus glucosides. *Carbohydrate Polymers*, **27**, 1–4.

El-Readi, M.Z., Hamdan, D., Farrag, N. *et al.* (2010) Inhibition of P-glycoprotein activity by limonin and other secondary metabolites from *Citrus* species in human colon and leukemia cell lines. *European Journal of Pharmacology*, **626**, 139–145.

Eguchi, A., Marakami, A., Li, S. *et al.* (2007) Suppressive effects of demethylated metabolites of nobiletin on phorbol ester-induced expression of scavenger receptor genes in THP-1 human monocytic cells. *Biofactors*, **31**, 107–116.

Fernandez-Lopez, J., Fernandez-Gines, J.M., and Aleson-Carbonell, L., Sendra, E., Sayas-Barbera, E., Perez-Alvarez, J.A. (2004) Application of functional citrus by-products to meat products. *Trends in Food Science & Technology*, **15** 176–185.

Freeman, B.L., Eggett, D.L., and Parker, T.L. (2010) Synergistic and antagonistic interactions of phenolic compounds found in navel oranges. *Journal of Food Science*, **75**, C570–C576.

Freedman, L. and Merritt, A.J. (1963) Citrus flavonoid complex: chemical fractionation and biological activity. *Science*, **139**, 344–345.

Fong, C.H., Hasegawa, S., Coggins, C.W. Jr. *et al.* (1992) Contents of limonoids and limonin 17-*β*-glucopyranoside in fruit tissue of Valencia orange during fruit growth and maturation. *Journal of Agricultural and Food Chemistry*, **40**, 1178–1181.

Fong, C.H., Hasegawa, S., Miyake, M. *et al.* (1993) Limonoids and their glucosides in Valencia orange seeds during fruit growth and development. *Journal of Agricultural and Food Chemistry*, **41**, 112–115.

Grigelmo-Miguel, N. and Martin-Belloso, O. (1998) Characterization of dietary fiber from orange juice extraction. *Food Research International*, **31**, 355–361.

Gorenstein, S., Leontowicz, H., Leontowicz, M. *et al.* (2007) Effect of hesperidin and naringin on the plasma lipid profile and plasma antioxidant activity in rats fed a cholesterol-containing diet. *Journal of the Science of Food and Agriculture*, **87**, 1257–1262.

Grohmann, K. and Baldwin, E.A. (1992) Hydrolysis of orange peel with pectinase and cellulose enzymes. *Biotechnology Letters*, **14**, 1169–1174.

Grohmann, K., Cameron, R.G., and Buslig, B.S. (1995) Fractionation and pretreatment of orange peel by dilute acid hydrolysis. *Bioresource Technology*, **54**, 129–141.

Grohmann, K., Manthey, J.A., Cameron, R.G., and Buslig, B.S. (1999) Purification of citrus peel juice and molasses. *Journal of Agricultural and Food Chemistry*, **47**, 4859–4867.

Hasegawa, S., Fong, C.H., Miyake, M., and Keithly, J.H. (1996) Limonoid glucosides in orange molasses. *Journal of Food Science*, **61**, 560–561.

Hedbom, E. and Häuselmann, H.J. (2002) Molecular aspects of pathogenesis in osteoarthritis: the role of inflammation. *Cellular and Molecular Life Sciences*, **59**, 45–53.

Hendrickson, R. and Kesterson, J.W. (1954) Recovery of citrus glucosides. *Proceedings of the Florida State Horticultural Society*, **67**, 199–201.

Hendrickson, R. and Kesterson, J.W. (1956) Purification of naringin. *Proceedings of the Florida State Horticultural Society*, **69**, 149–152.

Hendrickson, R. and Kesterson, J.W. (1964) Hesperidin in Florida oranges. *University of Florida Agricultural Experiment Station Technical Bulletin*, **684**.

Hendrickson, R. and Kesterson, J.W. (1965) Byproducts of Florida Citrus. *University of Florida Agricultural Experiment Station Technical Bulletin*, **698**.

Higby, R.H. (May 27 1947) Method for recovery of flavanone glucosides. U.S. Pat. 2421061.

Hon, F.M., Oluremi, O.I.A., and Anugwa, F.I.O. (2009) The effect of dried sweet orange (*Citrus sinensis*) fruit pulp meal on the growth performance of rabbits. *The Pakistan Journal of Nutrition*, **8**, 1150–1155.

Horowitz, R.M. and Gentili, B. (1977) Flavonoid constituents in citrus, in *Citrus Science and Technology*, vol. 1 (eds S. Nagy, P.E. Shaw, and M.K. Veldhuis), Avi Publishing Co., Westport, CN.

Hyon, J.S., Kang, S.M., Han, S.W. *et al.* (2009) Flavonoid component changes and antioxidant activities of fermented *Citrus grandis* Osbeck peel. *Journal of the Korean Society of Food Science and Nutrition*, **38**, 1310–1316.

Iwase, Y.M., Takahashi, M., Takemura, Y. *et al.* (2001) Isolation and identification of two new flavanones and chalcone from *Citrus kinokuni*. *Chemical & Pharmaceutical Bulletin*, **49**, 1356–1358.

Jagdish, K., Mehul, S., and Nehal, S. (2010) Effect of hesperidin on serum glucose, hba1c and oxidative stress in myocardial tissue in experimentally induced myocardial infarction in diabetic rats. *Pharmacognosy Journal*, **2**, 185–189.

Jang, H.D., Chang, K.S., Chang, T.C., and Hsu, C.L. (2010) Antioxidant potentials of buntan pumelo (*Citrus grandis* Osbeck) and its ethanolic and acetified fermentation products. *Food Chemistry*, **118**, 554–558.

Judy, W., Stogsdill, W., Judy, D. *et al.* (2010) Efficacy of diabetinol on glycemic control in insulin resistant hamsters and subjects with impaired fasting glucose – a pilot study. *Journal of Functional Foods*, **2**, 171–178.

Kamaraj, S., Ramakrishnan, G., Anandakumar, P. *et al.* (2009) Antioxidant and anticancer efficacy of hesperidin in benzo(a)pyrene induced lung carcinogenesis in mice. *Investigational New Drugs*, **27**, 214–222.

Kanes, K., Tisserat, B., Berhow, M., and Vandercook, C. (1993) Phenolic composition of various tissues of Rutaceae species. *Phytochemistry*, **32**, 967–974.

Kim, J.Y., Jung, K.J., Choi, J.S., and Chung, H.Y. (2004) Hesperetin: A potent antioxidant against peroxynitrite. *Free Radical Research*, **38**, 761–769.

Kirk, W.G. and Davis, G.K. (1954) Citrus pellets for beef cattle. *University of Florida Agricultural Experiment Station Technical Bulletin*, **538**, 1–16.

Kern, S.M., Bennett, R.N., Needs, P.W. *et al.* (2003) Characterization of metabolites of hydroxycinnamates in the *in vitro* model of human small intestinal epithelium Caco-2 cells. *Journal of Agricultural and Food Chemistry*, **51**, 7884–7891.

Kesterson, J.W. and Braddock, R.J. (1976) By-products and specialty products of Florida citrus, in *Bulletin 784. Institute of Food and Agricultural Sciences*, U. of Florida, Gainesville, FL.

Kurowska, E.M. and Manthey, J.A. (2002) Regulation of lipoprotein metabolism in HepG2 cells by citrus flavonoids. *Advances in Experimental Medicine and Biology*, **505**, 173–179.

Kurowska, E.M. and Manthey, J.A. (2004) Hypolipidemic effects and absorption of citrus polymethoxylated flavones in hamsters with diet-induced hypercholesterolemia. *Journal of Agricultural and Food Chemistry*, **52**, 2879–2886.

Kurowska, E.M., Manthey, J.A., Casachi, A., and Theriaubt, A.G. (2004) Modulation of HepG2 cell net apolipotrotein B secretion by the citrus polymethoxylated flavone, tangeretin. *Lipids*, **39**, 143–151.

Kwon, E.Y., Do, G.M., Cho, Y.Y. *et al.* (2010) Anti-atherogenic property of ferulic acid in apolipoprotein E-deficient mice fed Western diet: comparison with clofibrate. *Food and Chemical Toxicology*, **48**, 2298–2303.

Lai, C.S., Li, S., Chai, C.Y. *et al.* (2008) Anti-inflammatory and antitumor promotional effects of a novel urinary metabolite, 3′,4′-didemethylnobiletin, derived from nobiletin. *Carcinogenesis*, **29**, 2415–2424.

Lam, L.K.T. and Hasegawa, S. (1989) Inhibition of benzo[a]pyrene-induced forestomach neoplasia in mice by citrus limonoids. *Nutrition and Cancer*, **12**, 43–47.

Lee, J.S., Bok, S.H., Park, Y.B. *et al.* (2003) 4-Hydroxycinnamate lowers plasma and hepatic lipids without changing antioxidant enzyme activities. *Annals of Nutrition and Metabolism*, **47**, 144–151.

Li, S., Sang, S., Pan, M.H. *et al.* (2007) Anti-inflammatory property of the urinary metabolites of nobiletin in mouse. *Bioorganic & Medicinal Chemistry Letters*, **17**, 5177–5181.

Li, S., Lo, C.Y., and Ho, C.T. (2006a) Hydroxylated polymethoxyflavones and methylated flavonoids in sweet orange (*Citrus sinensis*) peel. *Journal of Agricultural and Food Chemistry*, **54**, 4176–4185.

Li, S., Wang, Z., Sang, S. *et al.* (2006b) Identification of nobiletin metabolites in mouse urine. *Molecular Nutrition & Food Research*, **50**, 291–299.

Licandro, G. and Odio, C.E. (2002) Citrus by-products, in *The Genus Citrus. Medicinal and Aromatic Plants – Industrial profiles* (eds G. Dugo and A. Di Giacomo), Taylor and Francis, Boca Raton.

Lien, E.J., Wang, T., and Lien, L.L. (2002) Phytochemical and SAR analyses of limonoids in citrus and Chinese herbs: Their benefits and risks of drug interactions. *Chinese Pharmaceutical Journal*, **54**, 77–85.

Luzio, G.A. (2003) Effect of block desterification on the psuedoplastic properties of pectin. *Proceedings of the Florida State Horticultural Society*, **116**, 425–429.

Luzio, G.A. and Cameron, R.G. (2008) Demethylation of a model homogalacturonan with the salt-independent pectin methylesterase from citrus: Part II. Structure-function analysis. *Carbohydrate Polymers*, **71**, 300–309.

Ma, Z.C., Hong, Q., Wang, Y.G. *et al.* (2010) Ferulic acid attenuates adhesion molecule expression in gamma-radiated human umbilical vascular endothelial cells. *Biological & Pharmaceutical Bulletin*, **33**, 752–758.

Manthey, J.A. (2004) Fractionation of orange peel phenols in ultrafiltered molasses and mass balance studies of their antioxidant levels. *Journal of Agricultural and Food Chemistry*, **52**, 7586–7592.

Manthey, J.A. (2010) Fractionation of secondary metabolites of orange (*Citrus sinensis* L.) Leaves by fast centrifugal partition chromatography. *Proceedings of the Florida State Horticultural Society*, **23**, 248–251.

Manthey, J.A. and Bendele, P. (2008) Anti-inflammatory activity of an orange peel polymethoxylated flavone, 3′,4′,3,5,6,7,8-heptamethoxyflavone, in the rat carrageenan/paw edema and mouse lipopolysaccharide-challenge assays. *Journal of Agricultural and Food Chemistry*, **56**, 9399–9403.

Manthey, J.A. and Buslig, B.S. (2003) HPLC-MS analylsis of methoxylated flavanones in orange oil residue. *Proceedings of the Florida State Horticultural Society*, **116**, 410–413.

Manthey, J.A. and Cesar, T.C. (2008) A citrus polymethoxylated flavone, 3′,4′,3,5,6,7,8-heptamethoxylflavone, exhibits activity in the Arthus reaction. Abstract of the 236th American Chemical Society meeting, AGFD 136.

Manthey, J.A., Cesar, T.C., Jackson, E., and Mertens-Talcott, S. (2011) Pharmacokinetic study of nobiletin and tangeretin in rat serum by high-performance liquid chromatography-electrospray ionization-mass spectrometry. *Journal of Agricultural and Food Chemistry*, **59**, 145–151.

Manthey, J.A. and Grohmann, K. (1996) Concentrations of hesperidin and other orange peel flavonoids in citrus processing byproducts. *Journal of Agricultural and Food Chemistry*, **44**, 811–814.

Manthey, J.A. and Grohmann, K. (2001) Phenols in citrus peel byproducts. Concentrations of hydroxycinnamates and polymethoxylated flavones in citrus molasses. *Journal of Agricultural and Food Chemistry*, **49**, 3268–3273.

Manthey, J.A., Grohmann, K., Montanari, A. *et al.* (1999) Polymethoxylated flavones derived from citrus suppress tumor necrosis factor-α expression by human monocytes. *Journal of Natural Products*, **62**, 441–444.

Manthey, J.A., Guthrie, N., and Grohmann, K. (2001) Biological properties of citrus flavonoids pertaining to cancer and inflammation. *Current Medicinal Chemistry*, **8**, 135–153.

Mateo Anson, N., Aura, A.M., Selinheimo, E. *et al.* (2011) Bioprocessing of wheat bran in whole wheat bread increases the bioavailability of phenolic acids in men and exerts anti-inflammatory effects ex vivo. *Journal of Nutrition*, **141**, 137–143.

May, C.D. (1990) Industrial pectins: Sources, production and applications. *Carbohydrate Polymers*, **12**, 79–99.

Miller, E.G., Gonzales-Sanders, A.P., Couvillon, A.M. *et al.* (1994) Citrus limonoids as inhibitors of oral carcinogenesis. *Food Technology*, **48**, 4.

Miller, E.G., Porter, J.L., Binnie, W.H. *et al.* (2004) Further studies on the anticancer activity of citrus limonoids. *Journal of Agricultural and Food Chemistry*, **52**, 4908–4912.

Miyake, Y., Fukumoto, S., Sakaida, K., and Osawa, T. (2004a) Antioxidative activity of the citrus peels fermented by three species of genus *Aspergillus*. *Nippon Shokuhin Kagaku Kaishi*, **51**, 181–184.

Miyake, Y., Minato, K.I., Fukumoto, S. *et al.* (2004b) Radical-scavenging activity *in vitro* of lemon peel fermented with *Aspergillus saitoi* and its suppressive effect against exercise-induced oxidative damage in rat liver. *Food Science and Technology Research*, **10**, 70–74.

Miyake, Y., Yamamoto, K., Tsujihara, N., and Osawa, T. (1998) Protective effects of lemon flavonoids on oxidative stress in diabetic rats. *Lipids*, **33**, 689–695.

Morin, B., LaNita, A.N., Zalasky, K.M. *et al.* (2008) The citrus flavonoids hesperetin and nobiletin differentially regulate low density lipoprotein receptor gene transcription in HepG2 liver cells. *Journal of Nutrition*, **138**, 1274–1281.

Murakami, A., Song, M., Katsumata, S.I. *et al.* (2007) Citrus nobiletin suppresses bone loss in ovariectomized ddY mice and collagen-induced osteoclastogenesis regulation. *BioFactors*, **30**, 179–192.

Nikaido, T., Ohmoto, T., Sankawa, U. *et al.* (1988) Inhibition of adenosine 3′,5′-cyclic monophosphate phosphodiesterase by flavonoids. *Chemical & Pharmaceutical Bulletin*, **36**, 654–661.

Nielsen, S.E., Breinholt, V., Cornett, C., and Dragsted, L.O. (2000) Biotransformation of the citrus flavone tangeretin in rats: Identification of metabolites with intact flavone nucleus. *Food and Chemical Toxicology*, **38**, 739–746.

Oben, J., Enonchong, E., Kothari, S. *et al.* (2009) *Phellodendron* and *Citrus* extracts benefit joint health in osteoarthritis patients: a pilot, double-blind, placebo-controlled study. *Nutrition Journal*, **8**, 38.

Oben, J., Enonchong, E., Kothari, S. *et al.* (2008) *Phellodendron* and *Citrus* extracts benefit cardiovascular health in osteoarthritis patients: a pilot, double-blind, placebo-controlled pilot study. *Nutrition Journal*, **7**, 16.

Onozuka, H., Nakajima, A., Matsuzaki, K. *et al.* (2008) Nobiletin, a citrus flavonoids, improves memory impairment and AB pathology in a transgenic mouse model of Alzheimer's disease. *Journal of Pharmacology and Experimental Therapeutics*, **326**, 739–744.

Peleg, H., Naim, M., Rouseff, R.L., and Zehavi, U. (1991) Distribution of bound and free phenolic acids in oranges and grapefruit. *Journal of the Science of Food and Agriculture*, **57**, 417–426.

Petkov, E., Nokolov, N., and Uzunoy, P. (1981) Inhibitory effect of some flavonoids and flavonoid mixtures on cyclic AMP phosphodiesterase activity of rat heart. *Planta Medica*, **43**, 183–186.

Pillai, B.V.S. and Swarup, S. (2002) Elucidation of the flavonoids catabolism pathway in *Pseudomonas putida* PML2 by comparative metabolic profiling. *Applied and Environmental Microbiology*, **68**, 143–151.

Poulose, S.M., Harris, E.D., and Patil, B.S. (2005) Citrus limonoids induce apoptosis in human neuroblastoma cells and have radical scavenging activity. *Journal of Nutrition*, **135**, 870–877.

Poulose, S.M., Harris, E.D., and Patil, B.S. (2006) Antiproliferative effects of citrus limonoids against human neuroblastoma and colonic adenocarcinoma cells. *Nutrition and Cancer*, **56**, 103–112.

Poore, H.D. (1934) Recovery of naringin and pectin from grapefruit residue. *Industrial & Engineering Chemistry*, **26**, 637–639.

Rapavi, E., Kocsis, I., Fehér, E. *et al.* (2007) The effect of citrus flavonoids on the redox state of alimentary-induced fatty livers in rats. *Natural Products Research*, **21**, 274–281.

Rathee, P., Chaudhary, H., Rathee, S. *et al.* (2009) Mechanism of action of flavonoids as anti-inflammatory agents: A review. *Inflammation and Allergy – Drug Targets*, **8**, 229–235.

Risch, B., Herrmann, K., Wray, V., and Grotjahn, L. (1987) 2′-(E)-O-p-coumaroylgalactaric acid and 2′-(E)-O-feruloylgalactaric acid in citrus. *Phytochemistry*, **26**, 509–510.

Risch, B., Herrmann, K., and Wray, V. (1988a) (E)-O-p-coumaroyl-(E)-O-feruloyl-derivatives of glucaric acid in citrus. *Phytochemistry*, **27**, 3327–3329.

Risch, B. and Herrmann, K. (1988b) Hydroxycinnamic acid derivatives in citrus fruits. *Zeitschrift für Lebensmittel-Untersuchung und -Forschung*, **186**, 530–534.

Roza, J.M., Xian-Liu, Z., and Guthrie, N. (2007) Effect of citrus flavonoids and tocotrienols on serum cholesterol levels in hypercholesterolemic subjects. *Alternative Therapies in Health and Medicine.*, **13**, 44–48.

Roy, A. and Saraf, S. (2006) Limonoids: Overview of significant bioactive triterpenes distributed in plants kingdom. *Biological & Pharmaceutical Bulletin*, **29**, 191–201.

Ruberto, G., Renda, A., Tringali, C. *et al.* (2002) Citrus limonoids and their semisynthetic derivatives as antifeedant agents against Spodoptera frugiperda larvae. A structure-activity relationship study. *Journal of Agricultural and Food Chemistry*, **50**, 6766–6774.

Rusznyák, S.P. and Szent-Györgyi, A. (1936) Vitamin P: Flavonols as vitamins. *Nature*, **138**, 27.

Savary, B.J., Hotchkiss, A.T., Fishman, M.L. *et al.* (2003) Development of a Valencia orange pectin methylesterase for generating novel pectin products, in *Advances in Pectin and Pectinase Research* (eds F. Voragen, H. Schols, and R. Visser), Kluwer Academic Publishers., The Netherlands.

Sawabe, A. and Matsubara, Y. (1999) Bioactive glycosides in citrus fruit peels. *Studies for Plant Sciences*, **6**, 261–274.

Schoch, T.K., Manners, G.D., and Hasegawa, S. (2001) Analysis of limonoid glucosides from *Citrus* by electrospray ionization liquid chromatography-mass spectrometry. *Journal of Agricultural and Food Chemistry*, **49**, 1102–1108.

Schoch, T.K., Manners, G.D., and Hasegawa, S. (2002) Recovery of limonoid glucosides from citrus molasses. *Journal of Food Science*, **67**, 3159–3163.

Shahidi, F. and Chandrasekara, A. (2010) Hydroxycinnamates and their *in vitro* and *in vivo* antioxidant activities. *Phytochemistry Review*, **9**, 147–170.

Shaw, P.E. (1977) Essential Oils, in *Citrus Science and Technology*, vol. 1 (eds S. Nagy, P.E. Shaw, and M.K. Veldhuis), Avi Publishing Co., Westport, CN.

Sinclair, W.B. (1972) *The Grapefruit. Its Composition, Physiology, and Products*, University of California, Division of Agricultural Sciences, Los Angeles, CA.

Sinclair, W.B. (1984) *The Biochemistry and Physiology of the Lemon and other Citrus Fruits*, University of California, Division of Agricultural Sciences, Oakland, CA.

Sinclair, W.B. and Crandall, P.R. (1949a) Carbohydrate fractions of grapefruit peel. *Botanical Gazette*, **111**, 153–165.

Sinclair, W.B. and Crandall, P.R. (1949b) Carbohydrate fractions of lemon peel. *Plant Physiology*, **24**, 681–705.

Sinclair, W.B. and Crandall, P.R. (1953) Polyuronide fraction and soluble and insoluble carbohydrates of orange peel. *Botanical Gazette*, **115**, 162–173.

Sörensen, I., Pedersen, H.L., and Willats, W.G.T. (2009) An array of possibilities for pectin. *Carbohydrate Research*, **344**, 1872–1878.

Stanley, W.L. and Jurd, L. (1971) Citrus coumarins. *Journal of Agricultural and Food Chemistry*, **19**, 1106–1110.

Stanley, W.L. and Vannier, S.H. (1957) Chemical composition of lemon oil. I. Isolation of a series of substituted coumarins. *Journal of the American Chemical Society*, **79**, 3488–3491.

Sultana, R., Ravagna, A., Mohmmad-Abdul, H. *et al.* (2005) Ferulic acid ethyl ester protects neurons against amyloid beta-peptide(1-42)-induced oxidative stress and neurotoxicity: relationship to antioxidant activity. *Journal of Neurochemistry*, **92**, 749–758.

Tanaka, T., Maeda, M., Kohno, H. *et al.* (2001) Inhibition of azoxymethane-induced colon carcinogenesis in male F344 rats by the citrus limonoids obacunone and limonin. *Carcinogenesis*, **22**, 193–198.

Tatum, J.H. and Berry, R.E. (1979) Coumarins and psoralens in grapefruit peel oil. *Phytochemistry*, **18**, 500–502.

Tirkey, N., Pilkhwal, S., Kuhad, A., and Chopra, K. (2005) Hesperidin, a citrus bioflavonoid, decreases the oxidative stress produced by carbon tetrachloride in rat liver and kidney. *BMC Pharmacology*, **5**, 2.

Tucker, D.P.H. and Long, S.K. (1968) More by-products research needed. *Citrus World*, **4**, 13–17.

US Department of Agriculture (1962) Chemistry and Technology of Citrus, Citrus Products, and Byproducts. Agricultural Handbook No.98. Washington, D.C.

VanHoecke, B.W., Delporte, F., Van Braeckel, E. *et al.* (2005) A safety study of oral tangeretin and xanthohumol administration to laboratory mice. *In Vivo*, **19**, 103–108.

Whitman, S.C., Kurowska, E.M., Manthey, J.A., and Daugherty, A. (2005) Nobiletin, a citrus flavonoid isolated from tangerines, selectively inhibits class A scavenger receptor-mediated metabolism of acetylated LDL by mouse macrophages. *Atherosclerosis*, **178**, 25–32.

Widmer, W., Zhou, W., and Grohmann, K. (2010) Pretreatment effects on orange processing waste for making ethanol by simultaneous saccharification and fermentation. *Bioresource Technology*, **101**, 5242–5249.

Wilkins, M.R., Widmer, W.W., Cameron, R.G., and Grohmann, K. (2005) Effect of seasonal variation on enzymatic hydrolysis of Valencia orange peel. *Proceedings of the Florida State Horticultural Society*, **119**, 419–422.

Wilkins, M.R., Widmer, W.W., Grohmann, K., and Cameron, R.G. (2007a) Hydrolysis of grapefruit peel waste with cellulose and pectinase enzymes. *Bioresource Technology*, **98**, 1596–1601.

Wilkins, M.R., Widmer, W.W., and Grohmann, K. (2007b) Simultaneous saccharification and fermentation of citrus peel waste by *Saccharomyces cerevisiae* to produce ethanol. *Process Biochemistry*, **42**, 1614–1619.

Willats, W.G., Knox, J.P., and Mikkelsen, J.D. (2006) Pectin: New insights into an old polymer are starting to gel. *Trends in Food Science & Technology*, **17**, 97–104.

Willats, W.G., Orfila, C., Limberg, G. *et al.* (2001) Modulation of the degree and pattern of methyl-esterification of pectic homogalacturonan in plant cell walls. Implications for pectin methyl esterase action, matrix properties, and cell adhesion. *The Journal of Biological Chemistry*, **276**, 19404–19413.

Williamson, G., Day, A.J., Plumb, G.W., and Couteau, D. (2000) Human metabolic pathways of dietary flavonoids and cinnamates. *Biochemical Society Transactions*, **28**, 16–22.

Wilmsen, P.K., Spada, D.S., and Salvador, M. (2005) Antioxidant activity of the flavonoid hesperidin in chemical and biological systems. *Journal of Agricultural and Food Chemistry*, **53**, 4757–4761.

Wu, S.C., Wu, S.H., and Chau, C.F. (2009) Improvement of the hypocholesterolemic activites of two common fruit fibers by micronization processing. *Journal of Agricultural and Food Chemistry*, **57**, 5610–5614.

Xiao, H., Yang, C.S., Li, S. *et al.* (2009) Monodemethylated polymethoxyflavones from sweet orange (*Citrus sinensis*) peel inhibit growth of human lung cancer cells by apoptosis. *Molecular Nutrition & Food Research*, **53**, 389–406.

Yasuda, T., Yoshimura, Y., Yabuki, H. *et al.* (2003) Urinary metabolites of nobiletin orally administered to rats. *Chemical & Pharmaceutical Bulletin*, **51**, 1426–1428.

Yamakuni, T., Nakajima, A., and Ohizumi, Y. (2010) Preventive action of nobiletin, a constituent of Aurantii Nobilis Pericarpium with anti-dementia activity, against amyloid-beta peptide-induced neurotoxicity expression and memory impairment. *Yakuga. Zasshi*, **130**, 517–520.

Yan, J.J., Cho, J.Y., Kim, H.S. *et al.* (2001) Protection against beta-amyloid peptide toxicity *in vivo* with long-term administration of ferulic acid. *British Journal of Pharmacology*, **133**, 89–96.

Yi, Z., Yu, Y., Liang, Y., and Zeng, B. (2008) *In vitro* antioxidant and antimicrobial activities of the extract of Pericarpium Citri Reticulatae of a new *Citrus* cultivar in its main flavonoids. *Food Science and Technology*, **41**, 597–603.

Yu, J., Dandekar, D.V., Toledo, R.T. *et al.* (2006) Supercritical fluid extraction of limonoid glucosides from grapefruit molasses. *Journal of Agricultural and Food Chemistry*, **54**, 6041–6045.

Yu, J., Wang, L., Walzem, R.L. *et al.* (2005) Antioxidant activity of citrus limonoids, flavonoids, and coumarins. *Journal of Agricultural and Food Chemistry*, **53**, 2009–2014.

Zhang, D., Bi, Z., Li, Y. *et al.* (2009) Sodium ferulate modified gene expression profile of oxidized low-density lipoprotein-stimulated human umbilical vein endothelial cells. *Journal of Cardiovascular Pharmacology and Therapeutics*, **14**, 302–313.

Zhao, Z. and Moghadasian, M.H. (2010) Bioavailability of hydroxycinnamates: A brief review of in vitro and in vivo studies. *Phytochemistry Reviews*, **9**, 133–145.

9

Recovery of Leaf Protein for Animal Feed and High-Value Uses

Bryan D. Bals, Bruce E. Dale and Venkatesh Balan

Biomass Conversion Research Laboratory, Department of Chemical Engineering and Materials Science, Michigan State University, Lansing, Michigan, USA

9.1 Introduction

Feeding a population that is steadily increasing in size and wealth is one of the primary challenges facing the world in the twenty-first century [1]. Not only is the world's population continuing to grow, but increasing wealth exacerbates this problem, as wealthy populations tend to demand more meat in their diets [2]. Animal feed requirements (in terms of energy and protein) are nearly an order of magnitude greater in the United States than actual human requirements in order to account for the daily maintenance of the animals [3]. Thus, it is clear that the challenge of food security is one of providing enough animal feed, rather than direct human consumption.

There is a strong belief that developing a bioenergy economy will compete with food and feed resources [4,5]. While wastes, forest residues, and agricultural residues can provide some energy, most studies suggest that they will only supply a small portion of the world's total energy needs [6,7]. If a large-scale bioenergy future is envisioned, dedicated energy crops will be required, which may compete with food and feed production. If managed improperly, this could lead to less feed available for animals, clearly an undesired outcome. Instead, the agricultural sector must be re-created to support both increased feed needs as well as biofuels.

While animal nutrition is complex and requires meeting several nutritional requirements, the two dominant requirements are caloric intake and protein [8]. Ruminant animals (which can digest cellulosic materials) such as cattle obtain much of their protein from green leafy forages, primarily alfalfa or cool season grasses. Non-ruminants such as swine and poultry cannot digest fibre, and thus must obtain protein from other sources. Soybeans are the primary

Biorefinery Co-Products: Phytochemicals, Primary Metabolites and Value-Added Biomass Processing, First Edition.
Edited by Chantal Bergeron, Danielle Julie Carrier and Shri Ramaswamy.

Table 9.1   *Potential sources of leaf proteins*

Species	Location	Yield (Mg ha^{-1})	Protein (kg Mg^{-1})	Protein Yield (Mg ha^{-1})	Reference
Soybean	U.S. 5 yr average	2.48	386	0.96	USDA NASS
Alfalfa	U.S. 5 yr average	7.52	120–200	0.90–1.50	USDA NASS
Switchgrass	Quebec	8.4[a]	112[a]	0.94	[16]
Mixed species	Pennsylvania	9.1–10.3	209–238	1.90–2.45	[95]
Triticale[b]	Iowa	7.6	70	0.53	[18]
Red Clover[b]		7.33	213	1.56	[96]

[a] Only the first cutting of switchgrass is considered for protein and biomass yield.
[b] Potential double crops.

protein supplement in the United States, as soybeans fix nitrogen in the soil to provide a natural fertilizer for corn, are high in protein (~40% of the total dry weight), and are complementary to corn grain in terms of its amino acid content [9]. Typically, soybean oil is extracted from the seeds and the remaining material crushed to produce a 48% protein soybean meal.

Leaf protein concentrate (LPC) is protein separated from the fibre of leafy materials, and has the potential to be a high-value replacement for soy. Leaf protein is predominantly RuBisCO protein, a large enzyme that converts CO_2 into plant biomass and is the most common protein in the world [10]. This protein is high in lysine, the most limiting amino acid in non-ruminant diets. Other amino acids tend to be complementary to corn, as amino acids that are high in corn are low in leaf proteins and vice versa [9]. This is important given that corn is the primary energy source for non-ruminants, and thus corn protein is an essential part of the diet.

Despite their widespread use, soybeans may not be the most land-efficient method of producing protein. While the final product is high in protein, the total amount of soybean produced per ha is low, as seen in Table 9.1. Alfalfa, in contrast, can produce large quantities of biomass per ha, although the concentration of protein within the biomass is much lower. More protein is produced per ha with alfalfa compared to soybean, and more non-protein biomass is also produced, which can be used for biofuel production. Alfalfa, like soy, is a legume, and thus can fix nitrogen within the soil and at a greater rate than soybeans [11]. Besides alfalfa, grasses can also be used as a potential source of protein, although total protein productivity is generally low. An alternative is to mix grasses and legumes for increased protein and biomass production. Native prairies or mixed species tend to have improved environmental performances compared to monocultures [12], and thus may also be a viable source of protein. Several studies have also suggested switchgrass, which has gained prominence as a potential bioenergy crop, as a source of protein due to its high yields [3,13]. However, in this situation the switchgrass must be harvested at multiple points throughout the growing season to capture the protein in earlier harvests [14–16]. Studies on the amount of biomass available during an early harvest vary, but it is expected to be only 60% of the yield expected from a single late autumn harvest [16].

One final alternative for leaf protein is the possibility of using double crops [17,18]. These crops are grown in the autumn after harvesting corn or soy, and grow quickly over spring prior to planting corn. Thus, they do not require any new land, and so do not compete with food crops. Double cropping is widely practiced, although generally for soil conservation rather than animal feed. Double crops can be either legumes (such as clover or vetch), which generally have low yields and high protein content, or grasses (such as winter wheat or triticale), which generally have high yields and low protein content.

9.2   Methods of Separating Protein

9.2.1   Mechanical Pressing

The most common method of separating protein from cell wall components is mechanical pressing, in which fresh material is pulped and the protein rich juice is pressed out of the fibre. This approach has been studied since at least the 1940s, when concerns over food supply led to alternate attempts to find protein for human use [19]. Large-scale experiments in England in the 1950s provided information on the type of equipment that could be used in a commercial facility. In the 1970s, a demonstration facility was developed in Colorado in conjunction with the United States Department of Agriculture (USDA), using a technology called the Pro-Xan process [20]. Work in New Zealand at a pilot facility improved the pulping process, leading to greater overall yields. To our knowledge, Desialis, a French company, is the only commercial producer of LPCs operating today.

There are three primary steps during this process, as shown in the generalized process flow diagram in Figure 9.1. The first step involves pulping to break open the cells, allowing the cytoplasmic proteins to be released. The cells and any excess water are thoroughly mixed, allowing the proteins to disburse from inside the cells and be solubilized into a protein-rich juice. This juice is then separated from the fibrous mat via mechanical pressing, which generally reduces the moisture content of the fibre to approximately 70% [21]. This fibre can then be dried or further processed in a biorefinery. The protein in the juice can be concentrated through a variety of methods. The most common method is heat coagulation, in which the juice is heated to precipitate the protein, which is then separated from the de-proteinated juice (whey) through filtration or centrifugation. The whey can be added to the fibre, partially recycled through the process, or further processed in a biorefinery. The remaining protein is then dried and ready for commercial use.

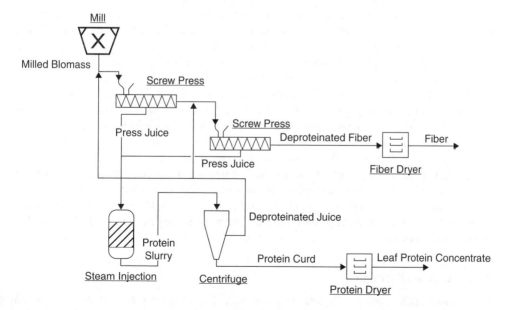

Figure 9.1   *Representative process flow diagram for the mechanical pressing and concentration of leaf proteins.*

The first step, pulping, is a key factor in the eventual yield of LPC, as it determines the concentration of protein in the juice. Pulping effectiveness is often measured as the degree of cell disruption (DCD), or the protein nitrogen in the juice divided by the total protein nitrogen [22]. Additional water can be added at this point to improve mixing and eventual pressing at the expense of increasing downstream separation costs. Likewise, ammonia or sodium hydroxide can be used to increase protein solubility, although this may also increase the solubility of undesirable material such as phenolics [23].

Several mills have been considered for pulping the biomass. Sugarcane rollers were initially considered, but discarded due to low capacity and high power requirements [24]. Extruders, which force material through a small orifice, require less energy input. Nelson *et al.* [25] created design calculations for desired throughput and die size for extruders, allowing for rapid initial design specifications. Alternatively, the Pro-Xan process used a hammer mill [21], which uses free swinging blades to beat the biomass. Carroud *et al.* [22] tested multiple hammer mills and configurations, and found the DCD ranged from 48–56% while energy consumption ranged from 4.5–8.2 kWh per wet Mg. MacDonald *et al.* refined this approach by adding a disk mill after the first pressing to fully disrupt the biomass [26]. Energy use was approximately 100 kWh per dry MT, but cell disruption was nearly complete, producing overall protein juice yields in excess of 70% [27].

After pulping, the mixture is dewatered using a press. Initially, the moisture content is between 75–85% of the total weight, which is reduced to no lower than 60% after pressing. This step can be combined with the pulping process, although it is difficult to do on a commercial scale [28]. In the laboratory, this process is often done by squeezing the juice out of cheesecloth or other similar material, although the final moisture content remains high in these situations [29]. Rollers and conveyers were eliminated as options due to uneven beds and short residence times [19]. Belt presses can be used on both the lab and commercial scales [30,31], but improper charging is also an issue [32]. Instead, screw presses have been used in the Pro-Xan facility [20], as well as in New Zealand [26], and have been proposed for a facility in Germany [33]. In tests with alfalfa, a twin screw press consistently extracted more protein and other nutrients (xanthophylls and carotene) than a single screw press, rollers, or a V-press [34].

The primary design considerations of a press are the energy usage, the mass throughput, and the final moisture content of the remaining fibrous residue (the press cake). The press cake remains fairly wet (~70%), and the relative amount of moisture that is removed is the key component in determining yields in this step [35]. Because of this, designs often include multiple screw presses [20] and adding water or whey back to the cake between presses [35]. Another concern is the thickness of the fibrous mat, as chloroplastic proteins can be filtered by the fibre if the mat is too thick [30]. A summary of typical protein solubility yields is shown in Table 9.2. Unfortunately, much of the protein is unrecoverable in this format, either due to incomplete cell disruption or incomplete solid/liquid separation. These yields do not include protein concentration, and thus are much higher than final LPC yields. Although protein in the press cake has been substantially removed, this method can only achieve up to 70% soluble protein in the best cases.

9.2.2 Aqueous Extraction

While mechanical pressing is well characterized, it requires fresh, wet biomass as an input [20], which is not available year-round in most climates. An alternative process that

Table 9.2 *Typical protein extraction results from mechanical pressing*

Species	Moisture Content	Number of Presses	Protein	Protein Extracted	Reference
Ryegrass	86%	1	N/A	45.5%	[31]
Alfalfa	78%	1	20.4%	59.1%	[97]
Alfalfa	78%	2	20.0%	38.8%	[20]
Alfalfa	85%	1	20.9%	47.0%	[21]
Red Clover	84%	2	19.9%	45.8%	[98]
S. japonica	68%	1	13.3%	24.6%	[99]
Alfalfa	81%	3	22.6%	69.1%	[27]

can use feedstock that has been dried and stored (and thus applicable to year-round processing) is to extract the protein with a solvent. After the protein is solubilized, the solids and liquids are separated and the protein can be concentrated in a variety of methods, similar to the mechanical pressing approach. No attempts have been made to date to scale up this aqueous extraction method.

Multiple studies on optimizing extraction conditions have been published, focusing on key conditions such as pH, residence time, temperature, and liquid to solid ratio. Proteins tend to be more soluble as the pH increases, although little additional benefit is seen in increasing the pH above 10 [13,36]. Sodium hydroxide is the most commonly used alkali [29,36,37], but ammonia [13] and calcium hydroxide [38,39] have also been tested. Extraction is generally performed at slightly elevated temperatures, although room temperature [40] extractions have also been performed. Fernandez *et al.* [37] also extracted *Atriplex* leaves at 5 °C, although no reason for the cold temperature was given. Most studies, however, show a slight improvement in extraction efficiency as temperatures increase from 30–70 °C [36,41]. The residence time is generally brief, ranging from 15 minutes [29] to five hours [36], but often in the 30–60 minute range [38,41]. Finally, greater liquid to solid ratios tend to improve protein extraction, with most studies focusing on 10–20:1 ratios [13,29,39]. However, Shen *et al.* [36] found increased protein extracted up to a 40:1 ratio, while, conversely, Fernandez *et al.* [42] saw improved protein yields as the liquid:solid ratio decreased to 5:1.

Yields for protein extraction from leaves can vary significantly based on the type of material being extracted and the conditions present, as seen in Table 9.3. In general, extraction yields range from 30–55% of the total protein in the plant, similar to that of mechanical pressing. Urribarri *et al.*, for example, were able to obtain 53% yield of protein from dwarf elephant grass at high temperatures, but were only able to obtain 29% from cassava leaves using the same technique [39,41]. Likewise, Nanda *et al.* performed extraction on multiple species, with yields ranging from 35% for Egyptian bean leaf and 57% for alfalfa [29]. For some crops, multiple harvests were also tested, and a general trend of decreasing extraction yields as the age of the plant increases was observed. This is likely due to increased density of cell wall components, limiting mass transfer of protein out of the cells and into solution.

9.2.3 Leaf/Stem Separation

While the two previous methods can provide for a concentrated protein product, they both have considerable capital and operating expense. A third method, leaf/stem separation, is inexpensive, but does not fully concentrate the protein. In alfalfa, most of the protein and important nutrients, such as xanthophylls, are in the leaves, while most of the fibre is in the stems [43].

Table 9.3   *Typical protein extraction results from aqueous extraction*

	Agent	pH	L:S[a]	Temp[b]	Protein	Extracted	Reference
Switchgrass	Ammonia	10	10:1	50 °C	7.3%	43.4%	[13]
Bermudagrass	Ca(OH)$_2$	11.5	20:1	50 °C	10.0%	33.9%	[38]
Atriplex	NaOH	10	5:1	5 °C	26.9%	41.2%	[42]
Hyacinth (early)	NaOH	11	22:1	22 °C	16.8%	53.6%	[29]
Hyacinth (late)	NaOH	11	22:1	22 °C	13.5%	35.7%	[29]
Alfalfa	NaOH	11	22;1	22 °C	16.0%	57.4%	[29]
Tea	NaOH	0.1M	40:1	40 °C	25%	56.4%	[36]
Elephantgrass	Ca(OH)$_2$	12.6	10:1	90 °C	12.3%	52.7%	[39]
Cassava	Ca(OH)$_2$	10	10:1	90 °C	18.6%	20.0%	[41]

[a] L:S: Liquid to solid ratio.
[b] Temp: Temperature of the extraction.

Initially, leaf/stem separation was designed to allow more control over ruminant diets, as the leaves and stems could be recombined in different proportions to fine-tune the amount of protein and fibre in diets [44]. However, it is also possible to use the leaf portion as a protein supplement for non-ruminants, while the stems are used as a biofuel feedstock [45]. The leaf fraction will have two to three times as much protein as the stem, as well as roughly 50% less fibre [46,47]. However, the protein concentration in these leaf fractions was only 27–28%, considerably less than 48% soybean meal.

Mechanical sieving is one method of separating leaves and stems, but cannot be used for wet material due to clogging pores. Unfortunately, drying can be inconsistent and lead to over-drying and damage to the nutritive value of the leaves [48]. Alternatively, air separation can be used, due to different drag coefficients [43]. This can be performed in an air tunnel [43] or, alternatively, in a modified drum dryer, thus combining it with the drying process [48,49]. By separating the leaves and the stems in the dryer, better control over their drying rates can also be had, thereby eliminating the concern of overdrying the leaves. A final option is a separate harvest of leaves and stems [46]. In this case, the leaves are stripped from the plant on the field, and the stems can be left to grow new leaves or to be harvested separately.

9.2.4   Post-Fermentation Recovery

In the corn ethanol industry, dry distillers' grains and solubles (DDGS) are a valuable co-product, accounting for up to 25% of the total revenue of a refinery [50,51]. Being the residue remaining after yeast fermentation of corn starch, it is composed of nearly 30% protein, thus giving it value as a protein supplement. A similar approach – using the fermentation residue as an animal feed – has been proposed for alfalfa as well [45]. By converting the residual fibre as ethanol, the protein content can increase to approximately 30% of the mass, and the production of yeast can also provide additional nutrients. However, the presence of indigestible lignin (up to 18% of the initial weight of alfalfa) [52] and the potential use of genetically modified organisms for ethanol production [53] may prevent this concept from commercial production.

An alternative is to capture the protein in soluble form after enzymatic hydrolysis and fermentation of the cell wall components. This method can be combined with one of the previous approaches for maximum effect. Bajracharya and Mudgett [54] performed solid-state fermentation on the fibrous residue of alfalfa after undergoing mechanical pressing. Cellulase-producing fungus was used to break down the cell walls of the residue, and an additional pressing released 36% of the remaining protein. Bals *et al.* [13] considered the amount of protein from switchgrass released into solution after cellulose hydrolysis. A total of 60% of the protein was solubilized, but virtually complete solubilization was observed if an aqueous extraction was performed prior to hydrolysis.

9.3 Protein Concentration

In the most common methods of LPC production – mechanical pressing and aqueous extraction – the initial operation is to separate the protein from the fibre by solubilizing the protein followed by solid/liquid separation. While the juice may be fed directly to cattle [55], the protein is most often concentrated and dried prior to consumption for storage purposes. This concentration step may have the added advantage of further purifying the protein, eliminating other soluble compounds that may not have as much nutritional value. However, the protein may not all be separated from the juice, decreasing the final yield. Multiple techniques are possible, including heat coagulation, acid precipitation, filtration, and drying. A generalized summary of these methods is shown in Table 9.4.

9.3.1 Steam Injection

One method of concentration that has been demonstrated at pilot scales is heat coagulation. The juice is heated in order to denature the proteins, opening up hydrophobic sites and causing the proteins to coagulate and fall out of solution. This can be achieved through direct steam injection, causing a rapid temperature rise to decrease total residence time. The protein curd produced can then be separated from the deproteinated juice ("whey") by centrifugation or filtration, and is then dried in a spray or fluidized bed dryer.

Besides being used as a method to concentrate the protein, this approach can also further fractionate the protein into low- and high-quality products. Chloroplastic (green) protein

Table 9.4 *Summary of leaf protein purification methods*

Method	Subsequent Steps	Major Advantages	Major Disadvantages
Heat coagulation	• Filtration/Centrifugation	• Fractionation of green and white proteins possible	• High energy cost
Acid precipitation	• Drying • Filtration/Centrifugation • Drying	• High protein recovery • Low capital and energy cost	• Potential heat damage • Poor selectivity • Potentially low recovery • Inability to recycle solvent
Salt precipitation	• Filtration/Centrifugation • Drying	• Low temperature • Fractionation of green and white proteins possible	• Poor selectivity • Difficulty in recycling solvent
Ultrafiltration	• Diafiltration • Drying	• Low capital cost • High protein quality • High protein concentration	• Potential for fouling from lignin • Must replace filters regularly • Potentially low recovery
Spray Drying	• None	• Retains non-protein nitrogen	• Capital and energy intensive • Retains salts/other undesirable compounds

coagulates at ~60 °C, and is not as high a quality as the cytoplasmic (white) protein for animal or human feeding purposes. White protein coagulates at ~90 °C, and thus a two-stage heat coagulation step can add further value to the process. This approach was attempted at the Pro-Xan facility, and achieved yields of 15% of the total crude protein as green protein and 9% as white. The white protein product was 89% protein, evidence of its high value as a purified protein product [56]. Since some white protein is also precipitated with the green protein and is a function of the juice pH as well [57], care must be taken to maximize the yield of pure white protein in this approach.

9.3.2 Acid Precipitation

Another precipitation technique is to change the protein solubility properties of the solvent. This can be performed through either "salting out" the protein or, more commonly, adding acid. Minimal protein solubility occurs at pH 3.5–4.5 [58], and so adjusting the pH to this range will allow most proteins to fall out of solution without any heat damage or loss of solubility [59]. Miller *et al.* [60] tested acid precipitation of alfalfa juice, and found protein recovery to be slightly less than heat coagulation and ultrafiltration. Furthermore, acid-precipitated LPC had lower protein and higher ash contents than those from other concentration methods. An alternative method used an anionic polyelectrolyte to precipitate the chloroplastic protein prior to using acid for the cytoplasmic protein [61], although heat coagulation can also be used to precipitate the green protein [62]. While these approaches are effective at removing protein at low temperature, the large amount of non-protein material precipitated, as well as the inability to recycle the solvent (particularly with alkaline extraction) may hamper its potential in a commercial facility.

9.3.3 Ultrafiltration

While precipitation techniques have been extensively tested, they alter the properties of the protein isolate [59]. An alternative is to concentrate the protein via ultrafiltration, which selectively retains protein and other large molecules, while allowing water and smaller solubles through. Ultrafiltration is commonly used in the food industry [63], and can conceivably perform protein concentration faster and with lower energy consumption than heat coagulation [59]. Following filtration, the concentrated protein can be dried and stored, or can undergo diafiltration to remove any remaining undesirable solubles [64]. One concern with ultrafiltration is fouling by large polyphenolic compounds such as soluble lignin [65,66], which can reduce its effectiveness.

Studies tend to show that the concentration of protein in the final product for ultrafiltration alone is comparable to heat coagulation, but when combined with diafiltration the results can be greatly improved. Fernandez *et al.* [64], for example, obtained a true protein concentration of 71% from alkaline-extracted *Atriplex* leaves when diafiltration was used, while Knuckles *et al.* [67] obtained a 67% alfalfa protein concentrate in a pilot-scale system. In contrast, Koschuh *et al.* [68] and Ostrowski [69] obtained filtered concentrates that were similar to heat-coagulated samples (between 40–60% of total dry matter). Despite the value of the final product, however, total protein recovery can be fairly low. One study [65] recovered 30–50% of the protein in switchgrass extracts, while Lamsal *et al.* [70] recovered 40–50% in alfalfa juice. If costs can remain low, this technique may also be used in conjunction with precipitation. Gao *et al.* recovered 24% of the protein from pea protein whey after acid precipitation, significantly increasing the total protein recovered [71].

9.3.4 Spray Drying

A unique approach is to simply spray dry the entire juice, collecting all of the soluble biomass as protein concentrate. This approach was used by Hartman *et al.* [72] on both alfalfa and pea vines. The dried material included all soluble components, resulting in a protein concentration ranging from 18% for the pea vine to 35% in alfalfa. A subsequent ethanol extraction selectively removed chlorophyll and other non-protein components, increasing the protein concentration to 26 and 43%, respectively. While this approach can increase protein recovery and produce an acceptable concentration of protein (at least for alfalfa), no statement on the economic impact of drying a large quantity of juice was given. Alternatively, spray drying can be used in conjunction with ultrafiltration [73], although this eliminates the advantage of including all protein and amino acids in the LPC.

9.4 Uses for Leaf Protein

9.4.1 Leaf Protein as Animal Feed

The primary use of leaf protein is as a protein supplement for animals. LPCs would likely replace soybean meal, and thus would need to have similar properties as soybean meal if replaced on a 1:1 weight basis. Of primary importance is the digestibility of the protein, as well as the amino acid balance, particularly lysine and methionine. In addition, the presence of other components, whether harmful (such as polyphenolic compounds) or beneficial (such as pigmentation) is also a concern. While the efficacy of protein supplementation is often measured by the growth rate of the livestock, other characteristics, such as the quality of the carcass and protein efficiency ratio (growth per unit protein fed) can also be considered.

In general, leaf protein was an effective protein substitute in multiple trials, as seen in Table 9.5. Leaf protein replaced either soybean meal or fishmeal in the diets of rats, chicken, fish, and swine. Under most situations, LPCs could not completely replace the initial protein meal in diets without hampering growth. Effective replacement of 25–60% of the protein meal (whether soy or fishmeal) was common in these reports. The difference is often due to poor lysine and methionine utilization; when supplementation of these amino acids is included, the efficacy of LPCs increases substantially [74].

One key factor that was discussed and investigated in multiple studies is the impact of heat damage, particularly during protein coagulation and drying. Cheeke *et al.*, for example, determined that freeze-dried alfalfa LPC had similar protein digestibility as soybean meal, but commercially dried LPC using a conventional oven was significantly poorer [75]. This was due not only to poor digestibility, but also lower lysine availability and decreased palatability, leading to lower uptake. Lu *et al.* considered the temperature of protein coagulation, and determined that LPC coagulated at 60 °C was more digestible than protein at 80 °C [76]. However, this may be due in part to an increase in the relative amount of non-protein nitrogen vs. true protein in the 80 °C sample compared to 60 °C. When a later study increased the coagulation temperature to 100 °C, however, no further reduction in digestibility occurred [77]. Multiple studies have also concluded that heat coagulation decreases the digestibility of protein.

Table 9.5   *Review of animal feeding trial results using leaf protein*

Leaf protein	Animal	Recommended replacement	Notes	Reference
Glyricidia	Broiler Chicks	25% of fishmeal		[100]
Alfalfa	Swine	100% of soy meal	Drying using heat decreased the value of LPCs	[75]
P. stratiotes	Rats	3% of wheat flour	Improved protein efficiency and weight gain	[101]
Cassava	Broiler Chicks	40% of fishmeal	Poor palatability compared to fishmeal	[102]
Alfalfa	Sheep	N/A	Spray drying improves protein retention in ruminants	[77]
Alfalfa	Rats	70% of soy meal	Requires AA supplements to completely replace soy meal	[74]
Alfalfa	Tilapia fish	35% of fishmeal	No deleterious effects of pigmentation was observed	[103]
Clover	Broiler Chicks	50% of soy meal	Similar growth as on soybean diets	[104]

Hanczakowski *et al.* determined that protein precipitated by heat had a lower digestibility than that precipitated by acid, aluminium sulfate, or Prestol (a flocculant) [78]. However, the biological value of heat-coagulated LPC was similar to the other methods of precipitation. Lu *et al.* found that LPC produced by fermentation was highly digestible, albeit lower in protein and energy content than traditional coagulation methods [77]. Likewise, Pandey *et al.* also observed high protein digestibility from LPC produced by fermentation, but their method increased the amount of protein recovered to nearly the same level as heat coagulation [79].

9.4.2   Leaf Protein for Human Consumption

As stated previously, leaf protein can be broadly categorized as low-quality green protein and high-quality white protein. LPC employed in human trials has modest digestibility that spans the range 65–90%, while the white fraction is much more digestible [80]. The balance of amino acids in the white fraction was preferable to the green fraction as well as soybean protein extract, but was not equivalent to egg protein, particularly for cysteine and methionine [81]. Amino acid profiles can change during processing, whether due to Maillard reactions with reducing sugars or amino acids degrading at high temperature (particularly lysine, cysteine, and glycine) [82]. Human feed trials of LPC also showed some deleterious effects because of the presence of anti-nutritional compounds like phytates, saponins, tannins and a number of phytoestrogens [81].

9.4.3   Leaf Protein for Enzyme Production

Expressing proteins in plants has been developed for 20 years, and it is now possible to express virtually any protein of interest in large quantities and extract it directly from plant tissue. Of particular interest is the production of cellulosic enzymes due to the high cost of industrial commercial biomass degrading enzymes (produced in microbes), and thus co-producing carbohydrates and enzymes in a single organism such as switchgrass could substantially reduce bioenergy costs [83]. In the past two decades several articles have been published reporting expression of glycosyl hydrolases in model plants or crops, such as alfalfa, *Arabidopsis*, maize, rice, tobacco, wheat, and barley for applications in converting biomass to biofuels. Some of them include endoglucanase (E1, E2, Cel 6A, Cel 6B, Egl, CelA, Cel 6G); cellobiohydrolase (E3); exoglucanase (CBHI); xylanase (XylA, B, Z, II); glucanase and amylases. A comprehensive review has been published recently [84] on this topic. The expression levels of the glycosyl hydrolysases were relatively low (2–3%) and several strategies are currently employed to improve the expression level. However, the extreme conditions present during biomass pretreatment can deactivate the enzymes, negating their effectiveness [85]. Protein extraction prior to pretreatment can preserve the enzymes, and the extract can be added back to the biomass as the hydrolysate medium [86].

9.4.4   Leaf Protein for Bio-Based Chemicals

Proteases can be used as an alternative or as a supplement to alkaline extraction to solubilize protein as peptides, a process that has been studied on distillers' grains [87,88]. This peptide extraction is influenced by both protease concentration and the extraction temperature. Acid hydrolysis is well known to further convert extracted peptides and proteins to individual amino acids. Because the amino acids are small and highly soluble, concentration in the form of precipitation or ultrafiltration is not possible. Instead, the amino acids can be used as building blocks for producing bio-based chemicals [89], either using a biological or a catalytic route. Bio-based chemicals not only reduce fossil energy use by replacing petroleum as the feedstock, but also require less energy to be produced [89,90]. However, chromatographic separation technologies are still under development to economically obtain the pure amino acids needed to make this approach feasible. The potential for this market is vast; Scott *et al.* estimate that 100 million Mg of proteins/amino acid as raw materials in the chemical industry could equate to the replacement of about 2000–3000 PJ of fossil resources [89]. While no effort on this approach has been applied to leaf proteins, it remains a possibility.

Some of the extracted proteins (zeins from corn, gluten from wheat, and glycinin from soybean) are used for making higher-value plastics or for making thin films [91,92]. Since protein-based plastics are biodegradable, they offer significant environmental benefits, while decomposing their materials after use. They also have unique advantages in medical applications, as there is less risk of disease transmission than from other sources. The processing of film coatings or other protein-based materials has three main steps: (i) breaking of intermolecular bonds (non-covalent and covalent, if necessary) by using chemical or physical processes, (ii) arranging and orienting mobile polymer chains in the desired conformation; and (iii) allowing the formation of new intermolecular bonds and interactions stabilizing the three-dimensional network [91]. Several processing conditions (heat treatment with pressure and shear) are followed to change the gluten protein functionality.

9.5   Integration with Biofuel Production

9.5.1   Advantages of Biofuel Integration

As stated previously, the value of the deproteinated fibre is low relative to the protein, and the cost of drying the fibre is high. Thus, the clearest approach to biofuel integration is to use the deproteinated fibre as a source of biofuel via either the thermochemical or biochemical routes [26,93]. If performed at the same location as protein extraction, there is no need to dry the fibre prior to biochemical pretreatment, and thus this method is generally the preferred approach. Any feedstock can be used for bioenergy, but grasses tend to be easier to convert to fuels while producing less protein than legumes [52]. Conventional dedicated energy crops such as switchgrass would require a change in farming practices to allow an early-season, high-protein harvest [16]. The protein extraction portion of the refinery would thus only be in operation part of the year, when high-protein biomass is abundant, while biofuel production could be performed year-round with deproteinated fibre and late-harvest grasses.

The concept of using deproteinated fibre as a biofuel source gained interest in the 1980s. Buchanan et al. [94] proposed a complete fractionation of herbaceous biomass, speculating that $320\,kg\,Mg^{-1}$ protein meal and $100\,kg\,Mg^{-1}$ ethanol were obtainable, along with other speciality products. Dale [45] looked specifically at protein, ethanol, and a catchall residue as the only products from alfalfa with multiple processing possibilities, obtaining $80\,kg\,Mg^{-1}$ of protein and $150\,kg\,Mg^{-1}$ ethanol. The idea was also proposed by Sinclair [26], who measured the composition of deproteinated alfalfa. The cellulose content was 39%, significantly higher than the initial alfalfa. Since high cellulosic content is a desired trait in biofuel feedstocks, there are synergistic effects to combined protein and biofuel production. Likewise, because multiple feedstocks can be used at the refinery, the seasonality of protein extraction becomes less important. By co-locating ethanol and protein production, some capital expenditures (such as the site itself) can be ameliorated over a year rather than the short growing season.

Furthermore, because alkaline pretreatments act to disrupt cell walls and solubilize lignin without heavily degrading protein, it can act to improve protein solubility, particularly with respect to alkaline extraction. As shown by Urribari et al., with both dwarf elephant grass [39] and cassava leaves [41], white protein extraction using calcium hydroxide increased after the biomass was pretreated with ammonia at 80–90 °C. However, the increase in extraction yield was less for cassava leaves than for the elephant grass, and it was speculated that this was due to higher concentration of lignin. Ammonia pretreatment followed by protein extraction was also attempted by de la Rosa et al. [38], and again showed that pretreatment improved protein extraction yields. In contrast, Bals et al. [13] showed no improvement in extraction yields for switchgrass. High-temperature ammonia pretreatments partially solubilize the hemicellulose, and so may reduce sugar yields during later fibre hydrolysis [13]. It is possible, however, to use the whey after protein concentration as the hydrolysate media, thus preserving much of the hemicellulose fraction and increasing sugar yields.

9.5.2 Analysis of Integration Economics

Although leaf protein extraction has been considered for years, it has rarely been produced commercially. This is likely due to an economic disadvantage to soybean meal, which has historically been priced at $200–300\,Mg^{-1}$ for 48% protein soy meal. An economic analysis was developed for the Pro-Xan demonstration facility that showed promise for alfalfa protein extraction [20], yet commercial development did not materialize. This analysis predicted an annual return on investment of 29–45% for a 40 wet ton h^{-1} facility operating for 130 days per year. This analysis may have been based on optimistic selling prices, with $92 ton^{-1} expected for deproteinated fibre and $430 ton^{-1} for the protein concentrate in 1976 dollars. The high selling price of LPC was due to the added value of xanthophylls; the authors estimated the selling price would be $210 ton^{-1} without factoring in the value of xanthophylls. Length of season was also a major factor in the economics, reflecting the uncertainty in protein production year round. Dale *et al.* [45] also speculated that the price for the fibre may be too high, suggesting a market for deproteinated fibre pellets (such as biofuel production) was important in preventing commercial LPC production.

The concept of combining leaf protein extraction with biofuel production was first considered in detail in a New Zealand study [27]. The biorefinery was assumed to be operating year round using alfalfa as a feedstock, and used mechanical pressing with a disk mill for improved protein release to obtain the LPC. The deproteinated fibre was then converted into ethanol, with several pretreatments considered (ammonia fibre expansion or AFEX, sodium hydroxide, steam, and no pretreatment). The process was deemed to be uneconomical (with ethanol selling process above $10 L^{-1}$) due to the assumption of exceptionally high costs for enzymes. The protein extraction portion itself was considered to be revenue neutral, with the LPC competing with high-value protein supplements and selling for upwards of $600 Mg^{-1}$.

The most in-depth economic analysis of protein extraction integrated with biofuel production was developed at Dartmouth University [3]. In this model, an ammonia extraction is performed on switchgrass immediately before and after AFEX pretreatment to remove 84% of the total protein. The extraction was assumed to be a multi-stage batch process with solvent recycling and a pneumatic press to increase the solvent yield after the initial extraction. After the second extraction, the remaining ammonia in the wet fibre is evaporated and concentrated, and the fibre is subsequently converted to ethanol. The protein is filtered to 30% of the total weight and then spray dried. The authors expressed concern regarding heat damage to the protein after pretreatment, and suggested that further improvements could lead to extraction of all the protein prior to pretreatment.

This study was part of an overall project to evaluate possible technologies within a biorefinery [93]. In this study, a detailed material and energy analysis of the technologies was conducted to provide an economic assessment of each approach. Integrating protein extraction with biofuel production, as opposed to fuel and energy production only, decreased the energy output of the refinery due primarily to increased energy requirements for the feed beyond its pure energy use. Furthermore, the high protein content in the switchgrass decreased carbo-hydrate and lignin content, decreasing the amount of bioenergy that could be produced. However, the economic benefit to the refinery was substantial; adding protein extraction to the refinery allowed the refinery to be profitable at a lower ethanol selling price ($0.17 L^{-1}$) compared to four other bioenergy configurations (ranging from $0.18-0.22 L^{-1}$). Even when historically low protein selling prices were considered, the profitability to the refinery was comparable to the bioenergy-only configurations.

The previous two economic analyses were explicitly compared with each other by Bals *et al.* [65]. In this analysis, it was assumed that the protein extraction facility is co-located with the biofuel production centre, and that the protein extraction would not significantly impact the economics of biofuels. Furthermore, the deproteinated juice was used as the hydrolysate medium for fibre hydrolysis, allowing 6 g water to exit the protein process per g fibre. For the alkaline extraction process, the ammonia was stripped from the extract and assumed to be re-concentrated in the pretreatment process. Furthermore, the hydrolysate was filtered to recover any additional proteins released after fibre hydrolysate. In this analysis, mechanical pressing was found to be superior to aqueous extraction under the baseline assumptions used (with an overall profit of $26 Mg^{-1}$ biomass vs. $14 Mg^{-1}$ biomass for aqueous extraction).

However, if the yield of extraction and filtration improved, then the profit of aqueous extraction increased to $37 Mg^{-1}, consistent with the Dartmouth examination.

9.6   Conclusions

Leaf protein extraction has been studied for several decades, but has generally been deemed uncompetitive with traditional protein crops. With renewed interest in bio-based fuels and chemicals, leaf protein may become a viable industry due to its positive impact on economics and land-use issues. Leaf protein, once separated from the fibre, can be used as animal feed or upgraded to human nutritional supplements or precursors to plastics or other chemicals. In addition, the techniques of leaf protein extraction can be applied to genetically modified crops that produce enzymes or other value-added products. These approaches are generally well suited to integration with biofuel production, and could provide a valuable co-product for a green biorefinery.

References

1. Godfray, H.C.J., Beddington, J.R., Crute, I.R. *et al.* (2010) Food security: The challenge of feeding 9 billion people. *Science*, **327**, 812–818.
2. McMichael, A.J., Powles, J.W., Butler, C.D., and Uauy, R. (2007) Food, livestock production, energy, climate change, and health. *Lancet*, **370**, 1253–1263.
3. Dale, B.E., Allen, M.S., Laser, M., and Lynd, L.R. (2009) Protein feeds coproduction in biomass conversion to fuels and chemicals. *Biofuels, Bioproducts and Biorefining*, **3**, 219–230.
4. Rathmann, R., Szklo, A., and Schaeffer, R. (2010) Land use competition for production of food and liquid biofuels: An analysis of the arguments in the current debate. *Renewable Energy*, **35**, 14–22.
5. Tenenbaum, D.J. (2008) Food vs. Fuel: Diversion of crops could cause more hunger. *Environmental Health Perspectives*, **116**, A254–A257.
6. Kim, S. and Dale, B.E. (2004) Global potential bioethanol production from wasted crops and crop residues. *Biomass Bioenergy*, **26**, 361–375.
7. Kadam, K.L. and McMillan, J.D. (2003) Availability of corn stover as a sustainable feedstock for bioethanol production. *Bioresource Technology*, **99**, 17–25.
8. Ensminger, M.E. and Olentine, C.G. (1978) *Feeds and Nutrition*, Ensminger Publishing, Clovis.
9. National Research Council (2001) *Nutrient requirements of dairy cattle: Seventh revised edition, 2001*, National Academy Press, Washington.
10. Kromus, S., Kamm, B., Kamm, M. *et al.* (2006) The green biorefinery concept – fundamentals and potential, in *Biorefineries-Industrial Processes and Products: Status Quo and Future Directions* (ed. B. Kamm), Wiley VCH, Weinheim.
11. Yang, J.Y., Drury, C.F., Yang, X.M. *et al.* (2010) Estimating biological N2 fixation in canadian agricultural land using legume yields. *Agriculture, Ecosystems & Environment*, **137**, 192–201.
12. Groom, M.J., Gray, E.M., and Townsend, P.A. (2008) Biofuels and biodiversity: Principles for creating better policies for biofuel production. *Conservation Biology*, **22**, 602–609.
13. Bals, B., Teachworth, L., Dale, B., and Balan, V. (2007) Extraction of proteins from switchgrass using aqueous ammonia within an integrated biorefinery. *Applied Biochemistry and Biotechnology*, **143**, 187–198.
14. Monti, A., Bezzi, G., Pritoni, G., and Venturi, G. (2008) Long-term productivity of lowland and upland switchgrass cytotypes as affected by cutting frequency. *Bioresource Technology*, **99**, 7425–7432.
15. Fike, J.H., Parrish, D.J., Wolf, D.D. *et al.* (2006) Switchgrass production for the upper southeastern USA: Influence of cultivar and cutting frequency on biomass yields. *Biomass Bioenergy*, **30**, 207–213.

16. Madakadze, I.C., Stewart, K.A., Peterson, P.R. *et al.* (1999) Cutting frequency and nitrogen fertilization effects on yield and nitrogen concentration of switchgrass in a short season area. *Crop Science*, **39**, 552–557.

17. Snapp, S.S., Swinton, S.M., Labarta, R. *et al.* (2005) Evaluating cover crops for benefits, costs and performance within cropping system niches. *Agronomy Journal*, **97**, 322–332.

18. Heggenstaller, A.H., Anex, R.P., Liebman, M. *et al.* (2008) Productivity and nutrient dynamics in bioenergy double-cropping systems. *Agronomy Journal*, **100**, 1740–1748.

19. Pirie, N.W. (1959) The large-scale separation of fluids from fibrous pulps. *Journal of Biochemical and Microbiological Technology and Engineering*, **1**, 13–25.

20. Enochian, R.V., Kohler, G.O., Edwards, R.H. *et al.* (1980) Producing Pro-Xan (leaf protein concentrate) from alfalfa: Economics of an emerging technology. Agricultural Economic Report No. 445, US Dept. of Agricul.

21. Fiorentini, R. and Galoppini, C. (1981) Pilot plant production of an edible alfalfa protein concentrate. *Journal of Food Science*, **46**, 1514–1520.

22. Carroad, P.A., Anaya-Serrano, H., Edwards, R.H., and Kohler, G.O. (1981) Optimization of cell disruption for alfalfa leaf protein concentration (pro-xan) production. *Journal of Food Science*, **46**, 383–386.

23. Fafunso, M. and Byers, M. (1977) Effect of pre-press treatments of vegetation on the quality of the extracted leaf protein. *Journal of the Science of Food and Agriculture*, **28**, 375–380.

24. Telek, L. (1983) *Plants: The Potentials For Extracting Protein, and other Useful Chemicals*, Office of Technology Assessment OTA BP F23, Washington DC.

25. Nelson, F.W., Barrington, G.P., and Bruhn, H.D. (1981) Throughput variables for rotary extrusion macerators. *Transactions of American Society of Agricultural Engineers*, **24**, 1146–1148.

26. Sinclair, S. (2009) Protein extraction from pasture: The plant fractionation bioprocess and adaptability to farming systems, Ministry of Agriculture and Forestry, New Zealand, SFF No. C08/001.

27. McDonald, R.M., Ritchie, J.M., Donnely, P.E. *et al.* (1981) Economics of the leaf-protein extraction process under New Zealand conditions.

28. Morrison, J.E. and Pirie, N.W. (1961) The large-scale production of protein from leaf extracts. *Journal of the Science of Food and Agriculture*, **12**, 1–5.

29. Nanda, C.L., Ternouth, J.H., and Kondos, A.C. (1975) An improved technique for plant protein extraction. *Journal of the Science of Food and Agriculture*, **26**, 1917–1924.

30. Davys, M.N.G. and Pirie, N.W. (1965) A belt press for separating juices from fibrous pulps. *Journal of Agricultural Engineering*, **10**, 142–145.

31. Davys, M.N.G., Pirie, N.W., and Street, G. (1969) A laboratory-scale press for extracting juice from leaf pulp. *Biotechnology and Bioengineering*, **11**, 529–538.

32. Pirie, N.W. (1987) An economical unit for pressing juice from fibrous pulps. *Journal of Agricultural Engineering*, **38**, 217–222.

33. Kamm, B., Hille, C., Schönicke, P., and Dautzenberg, G. (2010) Green biorefinery demonstration plant in havelland (germany), *Biofuels. Bioproducts and Biorefining*, **4**, 253–262.

34. Knuckles, B.E., Bickoff, E.M., and Kohler, G.O. (1972) Pro-xan process: Methods for increasing protein recovery from alfalfa. *Journal of Agricultural and Food Chemistry*, **20**, 1055–1057.

35. Edwards, R.H., De Fremery, D., and Kohler, G.O. (1978) Use of recycled dilute alfalfa solubles to increase the yield of leaf protein concentrate from alfalfa. *Journal of Agricultural and Food Chemistry*, **26**, 738–741.

36. Shen, L., Wang, X., Wang, Z. *et al.* (2008) Studies on tea protein extraction using alkaline and enzyme methods. *Food Chemistry*, **107**, 929–938.

37. Fernandez, S.S., Menendez, C.J., Mucciarelli, S., and Padilla, A.P. (2007) Saltbush Atriplex lampa leaf protein concentrate by ultrafiltration for use in balanced animal feed formulations. *Journal of the Science of Food and Agriculture*, **87**, 1850–1857.

38. De La Rosa, L., Reshamwala, S., Latimer, V. *et al.* (1994) Integrated production of ethanol fuel and protein from coastal bermudagrass. *Applied Biochemistry and Biotechnology*, **45–46**, 483–497.

39. Urribarri, L., Ferrer, A., and Colina, A. (2005) Leaf protein from ammonia-treated dwarf elephant grass (Pennisetum purpureum schum cv. Mott). *Applied Biochemistry and Biotechnology*, **121–124**, 721–730.

40. Lu, P.S. and Kinsella, J.E. (1972) Extractability and properties of protein from alfalfa leaf meal. *Journal of Food Science*, **37**, 94–99.

41. Urribarrí, L., Chacón, D., González, O., and Ferrer, A. (2009) Protein extraction and enzymatic hydrolysis of ammonia-treated cassava leaves (Manihot esculenta crantz). *Applied Biochemistry and Biotechnology*, **153**, 94–102.

42. Fernandez, S.S., Padilla, A.P., and Mucciarelli, S. (1999) Protein extraction from Atriplex lampa leaves: Potential use as forage for animals used for human diets. *Plant Foods for Human Nutrition*, **54**, 251–259.

43. Bilanski, W.K., Graham, W.D., Mowat, D.N., and Mkomwa, S.S. (1989) Separation of alfalfa silage into stem and leaf fractions in a horizontal airstream. *Transactions of American Society of Agricultural Engineers*, **32**, 1684–1690.

44. Adapa, P.K., Schoenau, G.J., and Arinze, E.A. (2005) Fractionation of alfalfa into leaves and stems using a three pass rotary drum dryer. *Biosystems Engineering*, **91**, 455–463.

45. Dale, B.E. (1983) Biomass refining: Protein and ethanol from alfalfa. *Industrial and Engineering Chemistry Product Research and Development*, **22**, 466–472.

46. Shinners, K.J., Herzmann, M.E., Binversie, B.N., and Digman, M.F. (2007) Harvest fractionation of alfalfa. *Transactions of American Society of Agricultural Engineers*, **50**, 713–718.

47. Bourquin, L.D. and Fahey, G.C. Jr. (1994) Ruminal digestion and glycosyl linkage patterns of cell wall components from leaf and stem fractions of alfalfa, orchardgrass, and wheat straw. *Journal of Animal Science*, **72**, 1362–1374.

48. Adapa, P.K., Schoenau, G.J., Tabil, L., and Sokhansanj, S. (2004) Fractional drying of alfalfa leaves and stems: Review and discussion, in *Dehydration of Products of Biological Origin* (ed. A.S. Mujumdar), Science Publishers, Oxford.

49. Gan-Mor, S., Wiseblum, A., and Regev, R. (1986) Separation of leaves from stems with a perforated rotating drum under suction. *Journal of Agricultural Engineering*, **34**, 275–284.

50. Ladisch, M. and Dale, B. (2008) Distillers grains: On the pathway to cellulose conversion. *Bioresource Technology*, **99**, 5155–5156.

51. Perkis, D., Tyner, W., and Dale, R. (2008) Economic analysis of a modified dry grind ethanol process with recycle of pretreated and enzymatically hydrolyzed distillers' grains *Bioresour. Technology*, **99**, 5243–5249.

52. Dien, B.S., Jung, H.-J.G., Vogel, K.P. *et al.* (2006) Chemical composition and response to dilute-acid pretreatment and enzymatic saccharification of alfalfa, reed canarygrass, and switchgrass. *Biomass Bioenergy*, **30**, 880–891.

53. Dien, B.S., Cotta, M.A., and Jeffries, T.W. (2003) Bacteria engineered for fuel ethanol production: Current status. *Applied Microbiology and Biotechnology*, **63**, 258–266.

54. Bajracharya, R. and Mudgett, R.E. (1979) Solid-substrate fermentation of alfalfa for enhanced protein recovery. *Biotechnology and Bioengineering*, **21**, 551–560.

55. Barber, R.S., Braude, R., Mitchell, K.G. *et al.* (1980) Value of freshly produced lucerne juice as a source of supplemental protein for the growing pig. *Animal Feed Science and Technology*, **5**, 215–220.

56. Edwards, R.H., Miller, R.E., De Fremery, D. *et al.* (1975) Pilot plant production of an edible white fraction leaf protein concentrate from alfalfa. *Journal of Agricultural and Food Chemistry*, **23**, 620–626.

57. De Fremery, D., Miller, R.E., Edwards, R.H. *et al.* (1973) Centrifugal separation of white and green protein fractions from alfalfa juice following controlled heating. *Journal of Agricultural and Food Chemistry*, **21**, 886–889.

58. Wang, J.C. and Kinsella, J.E. (1976) Functional properties of novel proteins: Alfalfa leaf protein. *Journal of Food Science*, **41**, 286–292.

59. Lillford, P.J. (1983) Extraction processes and their effect on protein functionality. *Plant Foods for Human Nutrition*, **32**, 401–409.

60. Miller, R.E., De Fremery, D., Bickoff, E.M., and Kohler, G.O. (1975) Soluble protein concentrate from alfalfa by low-temperature acid precipitation. *Journal of Agricultural and Food Chemistry*, **23**, 1177–1179.

61. Fiorentini, R. and Galoppini, C. (1983) The proteins from leaves. *Plant Foods for Human Nutrition*, **32**, 335–350.

62. Hernandez, A., Martinez, C., and Alzueta, C. (1989) Effects of alfalfa leaf juice and chloroplast-free juice ph values and freezing upon the recovery of white protein concentrate. *Journal of Agricultural and Food Chemistry*, **37**, 28–31.

63. Jönsson, A.S. and Trägårdh, G. (1990) Ultrafiltration applications. *Desalination*, **77**, 135–179.

64. Fernández, S.S., Menéndez, C.J., Mucciarelli, S., and Pérez Padilla, A. (2003) Concentration and desalination of zampa (Atriplex lampa) extract by membrane technology. *Desalination*, **159**, 153–160.

65. Bals, B. and Dale, B.E. (2011) Economic comparison of multiple techniques for recovering leaf protein in biomass processing. *Biotechnology and Bioengineering*, **108**, 530–537.

66. Susanto, H., Feng, Y., and Ulbricht, M. (2009) Fouling behavior of aqueous solutions of polyphenolic compounds during ultrafiltration. *Journal of Food Engineering*, **91**, 333–340.

67. Knuckles, B.E., Edwards, R.H., Miller, R.E., and Kohler, G.O. (1980) Pilot scale ultrafiltration of clarified alfalfa juice. *Journal of Food Science*, **45**, 730–734.

68. Koschuh, W., Povoden, G., Thang, V. *et al.* (2004) Production of leaf protein concentrate from ryegrass (lolium perenne x multiflorum) and alfalfa (medicago sauva subsp. Sativa). Comparison between heat coagulation/centrifugation and ultrafiltration. *Desalination*, **163**, 253–259.

69. Ostrowski, H.T. (1978) Membrane filtration for isolation, fractionation and purification of food-and feed-grade proteins from pasture herbage. *Journal of Food Processing and Preservation*, **3**, 59–84.

70. Lamsal, B.P. and Koegel, R.G. (2005) Evaluation of a dynamic ultrafiltration device in concentrating soluble alfalfa leaf proteins. *Transactions of American Society of Agricultural Engineers*, **48**, 691–701.

71. Gao, L., Nguyen, K.D., and Utioh, A.C. (2001) Pilot scale recovery of proteins from a pea whey discharge by ultrafiltration. *Lebensmittel-Wissenschaft & Technologie*, **34**, 149–158.

72. Hartman, G.H., Akeson, W.R., and Stahmann, M.A. (1967) Leaf protein concentrate prepared by spray-drying. *Journal of Agricultural and Food Chemistry*, **15**, 74–79.

73. Knuckles, B.E. and Kohler, G.O. (1982) Functional properties of edible protein concentrates from alfalfa. *Journal of Agricultural and Food Chemistry*, **30**, 748–752.

74. Myer, R.O. and Cheeke, P.R. (1975) Utilization of alfalfa meal and alfalfa protein concentrate by rats. *Journal of Animal Science*, **40**, 500–508.

75. Cheeke, P.R., Kinzell, J.H., De Fremery, D., and Kohler, G.O. (1977) Freeze-dried and commercially-prepared alfalfa protein concentrate evaluation with rats and swine. *Journal of Animal Science*, **44**, 772–777.

76. Lu, C.D., Jorgensen, N.A., Straub, R.J., and Koegel, R.G. (1981) Quality of alfalfa protein concentrate with changes in processing conditions during coagulation. *Journal of Dairy Science*, **64**, 1561–1570.

77. Lu, C.D., Jorgensen, N.A., Pope, A.L., and Straub, R.J. (1982) Digestion and nutrient flow in the gastrointestinal tract of sheep fed alfalfa protein concentrate prepared by various methods. *Journal of Animal Science*, **55**, 690–699.

78. Hanczakowski, P. and Skraba, B. (1984) The effect of different precipitating agents on quality of leaf protein concentrate from lucerne. *Animal Feed Science and Technology*, **12**, 11–17.

79. Pandey, V.N. and Srivastava, A.K. (1991) Yield and quality of leaf protein concentrates from monochoria hastata (l.) solms. *Aquat Botany*, **40**, 295–299.

80. Hernandez, T., Martinez, C., Hernandez, A., and Urbano, G. (1997) Protein quality of alfalfa protein concentrates obtained by freezing. *Journal of Agricultural and Food Chemistry*, **45**, 797–802.

81. Chiesa, S. and Gnansounou, E. (2011) Protein extraction from biomass in a bioethanol refinery - possible dietary applications: Use as animal feed and potential extension to human consumption. *Bioresource Technology*, **102**, 427–436.

82. Walker, A.F. (1979) Determination of protein and reactive lysine in lead-protein concentrates by dye-binding. *The British Journal of Nutrition*, **42**, 445–454.
83. Sticklen, M. (2006) Plant genetic engineering to improve biomass characteristics for biofuels. *Current Opinion in Biotechnology*, **17**, 315–319.
84. Taylor Ii, L.E., Dai, Z., Decker, S.R. *et al.* (2008) Heterologous expression of glycosyl hydrolases in planta: A new departure for biofuels. *Trends in Biotechnology*, **26**, 413–424.
85. Teymouri, F., Alizadeh, H., Laureano-Pérez, L. *et al.* (2004) Effects of ammonia fiber explosion treatment on activity of endoglucanase from Acidothermus cellulolyticus in transgenic plant. *Applied Biochemistry and Biotechnology*, **116**, 1183–1191.
86. Oraby, H., Venkatesh, B., Dale, B. *et al.* (2007) Enhanced conversion of plant biomass into glucose using transgenic rice-produced endoglucanase for cellulosic ethanol. *Transgenic Research*, **16**, 739–749.
87. Cookman, D.J. and Glatz, C.E. (2009) Extraction of protein from distiller's grain. *Bioresource Technology*, **100**, 2012–2017.
88. Bals, B., Brehmer, B., Dale, B., and Sanders, J. (2011) Protease digestion from wheat stillage within a dry grind ethanol facility. *Biotechnology Progress*. doi: 10.1002/btpr.521.
89. Scott, E., Peter, F., and Sanders, J. (2007) Biomass in the manufacture of industrial products—the use of proteins and amino acids. *Applied Microbiology and Biotechnology*, **75**, 751–762.
90. Brehmer, B. and Sanders, J. (2009) Implementing an energetic life cycle analysis to prove the benefits of lignocellulosic feedstocks with protein separation for the chemical industry from the existing bioethanol industry. *Biotechnology and Bioengineering*, **102**, 767–777.
91. B. Lagrain, B. Goderis, K. Brijs, and J.A. Delcour, (2010) Molecular basis of processing wheat gluten toward biobased materials. *Biomacromolecules*, **11**, 533–541.
92. Mooney, B.P. (2009) The second green revolution? Production of plant-based biodegradable plastics. *The Biochemical Journal*, **418**, 219–232.
93. Laser, M., Jin, H., Jayawardhana, K. *et al.* (2009) Projected mature technology scenarios for conversion of cellulosic biomass to ethanol with coproduction, thermochemical fuels, power, and/or animal feed protein. *Biofuels, Bioproducts and Biorefining*, **3**, 231–246.
94. Buchanan, R.A., Otey, F.H., and Hamerstrand, G.E. (1980) Multi-use botanochemical crops, an economic analysis and feasibility study. *Industrial and Engineering Chemistry Product Research and Development*, **19**, 489–496.
95. Deak, A., Hall, M.H., Sanderson, M.A., and Archibald, D.D. (2007) Production and nutritive value of grazed simple and complex forage mixtures. *Agronomy Journal*, **99**, 814–821.
96. Smith, D. (1965) Forage production of red clover and alfalfa under differential cutting1. *Agronomy Journal*, **57**, 463–465.
97. Edwards, R.H., de Fremery, D., Mackey, B.E., and Kohler, G.O. (1978) Factors affecting juice extraction and yield of leaf protein concentrated from ground alfalfa. *Transactions of American Society of Agricultural Engineers*, **21**, 55–62.
98. Byers, M. and Sturrock, J.W. (1965) The yields of leaf protein extracted by large-scale processing of various crops. *Journal of the Science of Food and Agriculture*, **16**, 341–355.
99. González, G., Alzueta, C., Barro, C., and Salvador, A. (1988) Yield and composition of protein concentrate, press cake, green juice and solubles concentrate from wet fractionation of sophora japonica, I. Foliage. *Animal Feed Science and Technology*, **20**, 177–188.
100. Agbede, J.O. and Aletor, V.A. (2003) Evaluation of fish meal replaced with leaf protein concentrate from glyricidia in diets for broiler chicks: Effect on performance, muscle growth, haematology and serum metabolites. *International Journal of Poultry Science*, **2**, 242–250.
101. Dewanji, A. and Matai, S. (1996) Nutritional evaluation of leaf protein extracted from three aquatic plants. *Journal of Agricultural and Food Chemistry*, **44**, 2162–2166.
102. Fasuyi, A.O. and Aletor, V.A. (2005) Protein replacement value of cassava (manihot esculenta, crantz) leaf protein concentrate (clpc) in broiler starter: Effect on performance, muscle growth, haematology and serum metabolites. *International Journal of Poultry Science*, **4**, 339–349.

103. Olvera-Novoa, M.A., Campos, S.G., Sabido, M.G., and Martínez Palacios, C.A. (1990) The use of alfalfa leaf protein concentrates as a protein source in diets for tilapia (Oreochromis mossambicus). *Aquaculture (Amsterdam, Netherlands)*, **90**, 291–302.

104. Szymczyk, B., Gwiazda, S., and Hanczakowski, P. (1996) The nutritive value for rats and chicks of unextracted and defatted leaf protein concentrates from red clover and italian ryegrass. *Animal Feed Science and Technology*, **63**, 297–303.

10

Phytochemicals from Algae

Liam Brennan[1], Anika Mostaert[2], Cormac Murphy[3] and Philip Owende[1,4]

[1]*Charles Parsons Energy Research Programme, Bioresources Research Centre, School of Agriculture, Food Science and Veterinary Medicine, University College Dublin, Belfield, Dublin, Ireland*
[2]*School of Biology and Environmental Science, University College Dublin, Belfield, Dublin, Ireland*
[3]*School of Biomolecular and Biomedical Science, University College Dublin, Belfield, Dublin, Ireland*
[4]*School of Informatics and Engineering, Institute of Technology Blanchardstown, Dublin, Ireland*

10.1 Introduction

Algae are a large, heterogeneous group of organisms belonging to diverse evolutionary lineages. The algae are traditionally represented by both eukaryotic and prokaryotic forms, with great variation in size, ranging from microscopic single cells (5 to 100 μm in diameter) to complex, multicellular organisms, such as large seaweeds up to 70 metres in length.

They are generally defined as simple, aquatic, photosynthetic, oxygen-producing protists (eukaryotes) and cyanobacteria (prokaryotes). There are exceptional cases occurring in relatively dry, terrestrial habitats, and non-photosynthetic species, that is, heterotrophic, and therefore requiring an external source of organic compounds and nutrients as an energy source (Lee, 1980). For most algae, however, photosynthesis is a key component of their survival, whereby they convert CO_2 and H_2O, with the aid of solar radiation, into sugar ($C_6H_{12}O_6$) and O_2, which is then used in respiration to produce the energy required to support growth (Falkowski and Raven, 1997; Zilinskas Braun and Zilinskas Braun, 1974).

The prokaryotic cyanobacteria lack characteristics of the eukaryotic algal forms (e.g., membrane-bound organelles), but unique amongst prokaryotes, cyanobacteria produce chlorophyll *a* (together with accessory pigments) enabling photosynthesis (Graham, Graham, and Wilcox, 2009). The eukaryotic algae are taxonomically distinguished mainly by their pigment

Biorefinery Co-Products: Phytochemicals, Primary Metabolites and Value-Added Biomass Processing, First Edition.
Edited by Chantal Bergeron, Danielle Julie Carrier and Shri Ramaswamy.
© 2012 John Wiley & Sons, Ltd. Published 2012 by John Wiley & Sons, Ltd.

composition, life cycle, morphology, storage products and basic cellular structure (Khan *et al.*, 2009). They include microscopic phytoplanktonic protists such as the Dinophyta (dinoflagellates) and Bacillariophyta (diatoms) (Graham, Graham, and Wilcox, 2009), as well as macroscopic protists such as the marine seaweeds. Seaweeds are non-vascular, multicellular marine macrophytes that inhabit coastal regions, commonly within rocky, intertidal habitats (Rorrer and Cheney, 2004). They are classified into three distinct groups, encompassing about 10 000 species: Heterokontophyta (Brown algae), Rhodophyta (Red algae) and Chlorophyta (Green algae) (McHugh, 2002).

Ecologically, algae occupy the base of food chains in aquatic ecosystems, and they contribute a major component of the Earth's atmospheric oxygen and organic carbon (Graham, Graham, and Wilcox, 2009). While the diminishing fossil fuel reserves originated largely from this organic carbon source, much research is now focused on using algae as a reliable, alternative source of renewable energy to replace fossil fuels. Algae are also economically important as sources of products such as food, fertilisers and hydrocolloids, as well as, amongst many others, phytochemicals with nutraceutical and pharmaceutical benefits. The algae-based phytochemical industry is developing rapidly, the driving force being the modern market trend towards natural, sustainable products.

10.1.1   Phytochemical Recovery from Biofuel-Destined Algal Biomass

It is arguable that the potential application of microalgae in multiple product recovery far exceeds that of macroalgae since available evidence indicates that macroalgae are applicable to a limited range of commercially viable compounds, for example, phycocolloids (see Section 10.2.6), whereas microalgae produce a wide range of commercial compounds (highlighted in Section 10.2). Therefore, although salient aspects of macroalgae biomass resources are discussed, this chapter primarily focuses on microalgae.

With appropriate technological development and deployment, microalgae-derived liquid biofuel is considered to be the only viable competitor to petroleum-based fuels due to several intrinsic advantages compared to biofuels from terrestrial bio-energy crops, namely: (1) the higher growth rates of algae (about 37 tonnes ha^{-1} per annum have been recorded), primarily due to higher photosynthetic efficiencies compared with terrestrial plants; (2) higher lipid productivity ($\approx$75% dry weight for some algae species), with higher proportions of triacylglycerol, a precursor to efficient biodiesel processing; (3) microalgae production absorbs CO_2 (production utilises about 1.83 kg of CO_2 per kg of dry algal biomass yield), therefore having potential for CO_2 recycling and air quality improvement; (4) capability of growing in wastewater, which offers the duel potential for integrating the treatment of organic effluent (phycoremediation) with biofuel production and (5) inherent potential for processing of valuable co-products that could enhance the economics of the production systems (Brennan and Owende, 2010). The residual algal biomass fraction after oil fraction extraction ($\approx$50% dry weight), contains valuable proteins with possible uses for livestock, poultry and fish feed additives and "phytochemicals" (Spolaore *et al.*, 2006). The oil fraction also contains components with nutraceutical and pharmaceuticals value (e.g., docosahexaenoic acid, ω-3 and ω-6 oils). Table 10.1 summarises the nutritional contents of biomass of some important algal species, compared to a range of conventional foodstuffs.

Phytochemicals refer to organic components from natural sources with potential nutritional, pharmaceutical and industrial applications, such as production of biodegradable

Table 10.1   *Gross chemical composition of different algae species compared to other food sources.*

Algae Species/Food Product	Constituents (% dry weight)		
	Lipids	Proteins	Carbohydrates
Algae:			
Anabaena cylindrica	4–7	43–56	25–30
Botryococcus braunii	25–75	—	—
Chlamydomonas reinhardtii	21	48	17
Chlorella emersonii	25–63	—	—
Chlorella protothecoides	14.6–57.8	—	—
Chlorella pyrenoidosa	2	57	26
Chlorella vulgaris	11–22	51–58	12–17
Dunaliella bioculata	8	49	4
Dunaliella primolecta	23.1	—	—
Dunaliella salina	6–25	57	32
Dunaliella tertiolecta	15–71	—	—
Euglena gracilis	14–20	39–61	14–18
Haematococcus pluvialis	25	—	—
Isochrysis galbana	7–40	—	—
Nannochloropsis oculata	22.7–31	—	—
Phaeodactylum tricornutum	18–57	—	—
Porphyridium cruentum	9–18.8	28–39	40–57
Prymnesium parvum	22–38	28–45	25–33
Scenedesmus dimorphus	16–40	8–18	21–52
Scenedesmus obliquus	11–55	50–56	10–17
Skeletonema costatum	13.5–51.3	—	—
Spirogyra sp.	11–21	6–20	33–64
Spirulina maxima	6–9	60–71	13–16
Spirulina platensis	4–16.6	46–63	8–14
Tetraselmis maculata	3	52	15
Foodstuffs:			
Baker's yeast	1	39	38
Rice	2	8	77
Egg	41	47	4
Milk	28	26	38
Meat muscle	34	43	1
Soya products	20	37	30

— data not available at the time of compiling chapter.
The figures provided were adopted from multiple sources including: Becker (1994), Balat and Balat (2010), Griffiths and Harrison (2009), Mata, Martins, and Caetano (2010).

plastic, detergents, cleaners and non-toxic polymers. It is widely accepted that concurrent extraction of valuable phytochemicals is imperative for economic and sustainable large-scale production of biofuel-directed microalgae (Lardon *et al.*, 2009; U.S. DOE, 2010; Wijffels and Barbosa, 2010; Wijffels, Barbosa, and Eppink, 2010). The range of valuable nutritional products includes polyunsaturated fatty acids, anti-oxidants, foods and feed, fertilizer, and other speciality applications. The chemical composition of an alga species is dependent on, and affected by, a wide range of factors culture conditions (e.g., temperature, pH, CO_2, dissolved oxygen (DO), illumination, nutrients, etc.), and on the growth phase at harvest (Brennan and Owende, 2010). These environmental factors can be modulated to induce the desired chemical composition in algal biomass (Becker, 1994). Phytochemical recovery will therefore be crucial to the development of sustainable and economically feasible exploitation of algae-derived biofuels, particularly for bulk production in an integrated product chain.

10.1.2 Algae Biomass Utilisation

Algae have been utilised by humans for over 2500 years where a wide range of species have been used for their food, fodder, fertiliser and medicinal value (Barsanti and Gualtieri, 2006). The phytochemical potential of algae was recognised as early as 500 BC in China, where the indigenous population consumed macroalgae for food (Potvin and Zhang, 2010). The Aztec civilisation (region presently Mexico) utilised algae for their nutritional qualities around 1300 AD. Wild blooms of the cyanobacteria *Spirulina* were harvested to make the dry cake, *tecuitlatl*, which was sold in markets and commonly consumed with maize and other cereals or with a tomato- and chilli-pepper-based sauces (Farrar, 1966). The use of naturally blooming *Spirulina* as a staple food in Africa around Lake Chad is believed to date back to the ninth century (Abdulqader, Barsanti, and Tredici, 2000). Algae are now firmly recognised as a vital and strategic resource for the biotechnology sector, with significant potential applications and therefore directed-production for bioenergy, food, pharmaceuticals and nutraceuticals (Demmig-Adams and Adams, 2002; Wijffels, Barbosa, and Eppink, 2010).

In 2006, the worldwide aquaculture macroalgae production was estimated at approximately 15.1 million tonnes (worth about €5.4 billion) (FAO, 2008). The expansion of the macroalgae aquaculture industry has increased steadily since 1970, at an annual growth rate of 8% (FAO, 2008), with regions of significant production being mainly in Asia (Table 10.2). There are three dominant macroalgae species, namely: the Japanese kelp (*Laminaria japonica* – 4.9 million tonnes), Wakame (*Undaria pinnatifida* – 2.4 million tonnes) and Nori (*Porphyra tenera* – 1.5 million tonnes). The main applications of macroalgae production include use as sea vegetables, and the extraction of phycocolloids and phycosupplements (Table 10.3).

Over the last three decades, there has been growing interest in diversified microalgae applications in biofuel, functional foods and the pharmaceutical industries (Plaza *et al.*, 2009; U.S. DOE, 2010). Commercial large-scale production of microalgae started in the early 1960s in Japan with the culturing of *Chlorella* as a food additive, and later production expanded in countries such as USA, India, Israel and Australia (Borowitzka, 1999; Pulz and Scheinbenbogan, 1998; Spolaore *et al.*, 2006). By 2004, industrial-scale microalgae production had grown to 7000 tonnes per annum dry matter (Pulz and Gross, 2004), from which a range of

Table 10.2 *Worldwide production of macroalgae by aquaculture for 2006 (FAO, 2008).*

Country	Production (billion tonnes)	Percent of World production	Value (Million € for 2006)
China	10.9	72.1	3900
Philippines	1.50	10	—
Indonesia	0.91	6	—
Republic of Korea	0.77	5.1	—
Japan	0.49	3.2	825
Rest of the World	0.53	3.6	—

Table 10.3   *Recorded commercial applications of macroalgae in 2004 (adapted from Barringgton, Chopin, and Robinson (2009)).*

Application/Utility	Production (billion tonnes)		Value (Million €)
	Raw material (wet)	Final products	
Sea-vegetables:			
Kombu (*Laminaria*)	4.520	1.080	2000
Nori (*Porphyra*)	1.400	0.142	1000
Wakame (*Undaria*)	2.520	0.166	760
Other	0.150	0.033	134
Phycocolloids:			
Carrageenans	0.528	0.033	223
Alginates	0.600	0.030	159
Agars	0.127	0.008	102
Phycosupplements:			
Soil additives	1.100	0.220	22
Agrichemicals	0.020	0.002	7
Animal feeds	0.100	0.020	7
Other	0.003	>0.001	2

products are derived. The major commercial activity has been focused on two main products, namely β-carotene from *Dunaliella salina* and astaxanthin from *Haematococcus pluvialis* (Jin and Melis, 2003). Other applications include: (1) food as an ingredient in health food products and as a protein/vitamin supplement in feeds for poultry, cattle, pigs and aquaculture animals; (2) therapeutics as antibiotic sources and for the regulation of cholesterol synthesis; (3) pigments such as phycobilins for use as food colour, in diagnostics, cosmetics and analytical reagents; (4) phycocolloids as gums, viscocifiers and ion exchangers; (5) sources of fatty acids, lipids, waxes, sterols, hydrocarbons, amino acids, enzymes, vitamin C and E and (6) long-chain hydrocarbons and esterified lipids as combustible oils (Becker, 1994).

10.2   Commercial Applications of Algal Phytochemicals

Table 10.4 shows a range of the current commercial applications of microalgae-derived compounds. The uses in human health food supplements are currently limited to three species, namely *Spirulina*, *Dunaliella* and *Chlorella*, due to the strict food safety regulations in most regions, market demand and specific preparation requirements (Pulz and Gross, 2004). As a food supplement they are generally marketed in tablet or powder form. However, there has been concern that some cyanobacteria contain the neurotoxin β-N-methylamino-L-alanine (BMAA) which has been linked to Alzheimer's disease, Lou Gehrig's disease and amyptrophic lateral sclerosis-Parkinsonism dementia complex (Cox *et al.*, 2005).

Specific microalgae species are also suitable for animal and aquaculture feed supplements. Their noted benefits include improved immune response and improved fertility in animals (Pulz and Gross, 2004). The enhancement of iodine content in meat products can be achieved through feeding animals with a diet supplemented with algae (He, Hollwich, and Rambeck, 2002). Benefits as feed in aquaculture (molluscs, shrimp and fish) include enhancement of immune systems, inducement of essential biological activities

Table 10.4 Algae-derived commercially valuable products.

Extractible Compound	Microalgae source	Commercial Product	Potential Application(s)
Carotenoids	Dunaliella salina, Haematococcus pluvialis	β-Carotene	Anti-oxidant activity, cancer prevention, food colorants, cosmetic and body-care products, precursor of vitamin A
	Haematococcus pluvialis	Astaxanthin	Anti-oxidant, immunomodulation and cancer prevention
	Haematococcus pluvialis, Chlorella vulgaris	Cantaxanthin	Anti-oxidant, immunomodulation and cancer prevention
	Chlorella pyrenoidosa, Haematococcus pluvialis	Lutein	Anti-oxidant activity
	Chlorella ellipsoidea	Violaxanthin	Anti-oxidant activity
Fatty Acids	Phaeodactylum tricornutum, Monodus subterraneus, Phorphyridium cruentum	EPA	Reduce risk of heart disease, nutritional supplements, aquaculture feed
	Haematococcus pluvialis, Dunaliella salina, Spirulina platensis	Oleic acid	Anti-oxidant activity
	Dunaliella salina, Spirulina platensis	γ-Linolenic acid	Anti-oxidant activity, infant formula for full-term infants, nutritional supplements
	Dunaliella salina	Palmitic acid	Anti-microbial activity
	Spirulina platensis	Palmitoleic acid	Reduce risk of heart disease
	Spirulina platensis, Crypthecodinium sp., Schizochytrium sp.	DHA	Reduce risk of heart disease, Infant formula for full-term/preterm infants, nutritional supplements, aquaculture feed
	Porphyridium sp.	Arachidonic acid	Infant formula for full-term/preterm infants, nutritional supplements
Proteins	Spirulina platensis, Porphyridium spp.	Phycobiliproteins	Immunomodulation activity, anti-cancer activity, hepatoprotective, anti-inflammatory, anti-oxidant properties, food colouring and fluorescent labels
		Recombinant proteins	
		Heavy isotope labelled metabolites	Anti-viral, anti-bacterial, anti-inflammatory and anti-coagulation
Phycocolloids	Chlorella pyrenoidosa, Porphyridium spp.	Sulfated polysaccharide	Anti-viral, anti-tumour, anti-hyperlipidemia and anti-coagulant
	Chlorella vulgaris	Insoluble fibre	Reduce total and LDL cholesterol
	Gelidium amansii	Agar	
	Laminaria hyperborean, Ascophyllum nodosum,	Alginates	
	Chondrus crispus	Carrageenans	
Feed	Spirulina platensis, Chlorella sp., Chlamydomonas sp.	Whole-cell dietary supplements	
	Tetraselmis sp., Nannochloropsis sp., Isochrysis sp., Nitzschia sp.	Whole-cell aquaculture feed	
Phycosupplements	Porphyra tenera	Nori (sea vegetable)	
	Ascophyllum nodsum	Fertiliser	
		Biochar	

The figures provided were adapted from multiple sources, including: Plaza et al. (2009), US DOE (2010), Spolaore et al. (2006), Hejazi and Wijffels (2004), Lorenz and Cysewski (2000), Ratledge (2004), Lyons (2000) and Rosenberg et al. (2008).

(Muller-Feuga, 2000) and improvement of quality of culturing medium ("green-water" technique) (Chuntapa, Powtongsook, and Menasveta, 2003).

10.2.1   Proteins

Protein is considered to be the major organic constituent of algal biomass with compositions ranging between 6 and 77% dry weight (Lavens and Sorgeloos, 1996). As such, mass culture of microalgae has previously been suggested as a potential alternative protein source for human consumption (Becker, 2004; Spoehr and Milner, 1949). Algae are capable of synthesising all essential amino acids required by humans (Becker, 2004). Table 10.5 shows the amino acid profiles of some species, which compare favourably with conventional food items, with the exception of minor disparities amongst sulfur-containing amino acids, such as methionine.

10.2.2   Lipids (i.e., Polyunsaturated Fatty Acids)

Algae contain a wide range of lipids and fatty acids (Table 10.6), which are components of the cell membrane, and function as metabolites and as energy storage components. The lipid content generally varies between 1–40%, but can be as high as 85% of dry weight. Microalgae are the primary source of the most chemically important algal biomass derivative, the polyunsaturated fatty acids (PUFAs), which are recognised for many health benefits. PUFAs are passed into the human food chain through consumption of marine foods (Mansour *et al.*, 2005) and are essential for human growth, development and physiology (Hu *et al.*, 2008; Wen and Chen, 2001). They provide health benefits against a range of medical conditions, including coronary heart disease (Rodolfi *et al.*, 2003), inflammatory diseases, such as rheumatoid arthritis, and hypertension (Sidhu, 2003). Since animals and higher plants lack the necessary enzymes to synthesize PUFAs, microalgae remain the primary source (Rubio-Rodríguez *et al.*, 2010), with the commercially important PUFAs being eicosapentaenoic acid, arachidonic acid and docosahexaenoic acid. Species that synthesise these PUFAs include: *Phaeodactylum* and *Nannochloropsis* (eicosapentaenoic acid), *Porphyridium* (arachidonic acid) and *Crypthecodinium* (docosahexaenoic acid).

Eicosapentaenoic Acid (EPA)

EPA (20:5 ω-3 shown in Figure 10.1) is a PUFA with the last double bond located at the third carbon atom from the methyl terminal and the configuration of all the double bonds is *cis* (Wen and Chen, 2003). EPA is a precursor of a group of eicosanoids that are crucial in regulating developmental and physiology in humans (Wen and Chen, 2003). It has been reported for potential use in the treatment of brain disorders (Fenton, Hibbeln, and Knable, 2000; Peet, 2004), and for certain cancer conditions (Tisdale, 1999). However, a possible side effect related to the thinning of artery walls has been reported (Ward and Singh, 2005).

Main algal sources of EPA include the photoautotrophic marine diatom *Phaeodactylum tricornutum* and the green unicellular microalga *Nannochloropsis* sp. *P. tricornutum.* PUFA contents range from 30 to 45% of the total cell dry weight, of which EPA accounts for 20 to

Table 10.5 Amino acid profile of different algae compared to some conventional protein rich foodstuffs (adapted from Becker (1994) and Lourenço et al. (2002)).

	Amino acid (g per 100 g protein)															
	Isoleucine	Leucine	Valine	Lysine	Phenylalanine	Tyrosine	Methionine	Threonine	Alanine	Arginine	Aspartic acid	Glutamic acid	Glycine	Histidine	Proline	Serine
Algae:																
Chlorella vulgaris	3.2	9.5	7.0	6.4	5.5	2.8	1.3	5.3	9.4	6.9	9.3	13.7	6.3	2.0	5.0	5.8
Dunaliella bardawil	4.2	11.0	5.8	7.0	5.8	3.7	2.3	5.4	7.3	7.3	10.4	12.7	5.5	1.8	3.3	4.6
Spirulira platensis	6.7	9.8	7.1	4.8	5.3	5.8	2.5	6.2	9.5	7.3	11.8	10.3	5.7	2.2	4.2	5.1
Ulva faciata	3.9	7.6	5.7	5.1	5.1	3.3	0.9	5.1	8.5	5.6	13.0	12.6	6.5	2.4	4.6	5.8
Codium spongiosum	4.4	8.4	6.6	6.8	5.4	2.3	0.8	5.4	8.2	4.0	12.0	14.1	6.1	2.3	4.6	5.3
Sargassum vulgare	4.8	8.5	5.8	5.4	5.3	2.2	2.2	4.8	7.2	4.3	10.9	17.6	5.7	2.1	4.6	5.1
Padina gymnospora	4.7	8.8	5.7	5.7	5.6	2.5	1.0	5.4	7.2	5.3	13.1	13.4	6.3	2.5	4.6	5.4
Dictyota menstrualis	4.7	8.7	5.7	5.4	5.4	2.6	1.3	5.3	6.9	5.4	13.8	12.7	6.1	2.2	4.8	6.1
Chnoospora minima	4.2	8.1	5.9	5.3	5.1	2.1	2.2	5.4	8.1	4.2	12.2	14.8	6.2	2.2	4.5	6.2
Porphyra acanthophora	4.4	8.6	6.8	6.7	5.0	2.5	1.2	6.2	9.4	5.1	13.3	13.7	7.5	3.2	4.9	5.7
Laurencia flagellifera	4.6	7.7	6.0	10.2	4.7	3.7	0.5	5.4	6.8	4.3	13.0	15.3	5.6	1.5	4.2	5.1
Gracilaria domingensis	4.1	8.8	5.6	5.7	5.7	2.3	0.7	6.1	8.1	4.7	12.2	12.6	6.6	2.9	5.1	5.3
Foodstuff:																
Egg	6.5	8.8	7.2	5.3	5.8	4.2	3.2	5.0	—	6.2	11.0	12.6	4.2	2.4	4.2	6.9
Soya bean	5.3	7.7	5.3	6.4	5.0	3.7	1.3	4.0	5.0	7.4	1.3	19.0	4.5	2.6	5.3	5.8

Table 10.6 Fatty acid composition of different algae (adapted from Thompson (1996), Becker (1994), Reitan et al. (1997) and Jiang, Chen, and Liang (1999)).

Fatty Acid	Weight Proportion (% of total fatty acids)											
	Spirulina platensis	Crypthecodinium cohnii	Scenedesmus obliquus	Chlorella vulgaris	Dunaliella bardawil	Chlamydomonas reinhardtii	Botryococcus braunii	Codium fragile	Nannochloropsis sp.	Dunaliella parva	Tetraselmis suecica	Phaeodactylum tricornutum
Lauric acid (12:0)	0.04	1.6	0.3	—	—	—	—	—	—	—	—	—
Myristic acid (14:0)	0.7	13	0.6	0.9	—	—	—	—	—	—	—	7.5
Pentadecanoic acid (15:0)	—	—	—	1.6	—	—	—	—	—	—	—	0.6
Palmitic acid (16:0)	45.5	19.8	16.0	20.4	41.7	20.0	18.0	28.0	40.0	19.0	24.0	17.4
Palmitoleic acid (16:1)	9.6	—	8.0	5.8	7.3	4.0	2.0	2.0	28.0	2.0	2.0	18.4
Hexadecatetraenic acid (16:4)	—	—	26.0	—	3.7	22.0	—	—	—	11.0	8.0	0.9
Heptadecanoic acid (17:0)	0.3	—	—	15.3	2.9	—	—	—	—	—	—	—
Stearic acid (18:0)	1.3	12.8	0.3	15.3	2.9	—	1.0	1.0	—	2.0	—	0.9
Oleic acid (18:1)	3.8	0.8	8.0	6.6	8.8	7.0	27.0	11.0	3.0	6.0	16.0	3.4
Linoleic acid (18:2)	14.5	—	6.0	1.5	15.1	6.0	12.0	6.0	1.0	11.0	14.0	1.6
α-Linoleic acid (18:3)	0.3	—	28.0	—	20.5	30.0	18.0	27.0	—	44.0	7.0	0.7
Eicosanotrienoic acid (20:3)	0.4	—	—	20.8	—	—	—	—	—	—	—	—
Arachidonic acid (20:4)	—	—	—	—	—	—	—	3.0	3.0	—	2.0	0.5
Eicosapentaenoic acid (20:5)	—	—	—	—	—	—	3.0	2.0	20.0	—	5.0	26.4
Docosahexaenoic acid (22:6)	—	51	—	—	—	—	—	—	—	—	—	—

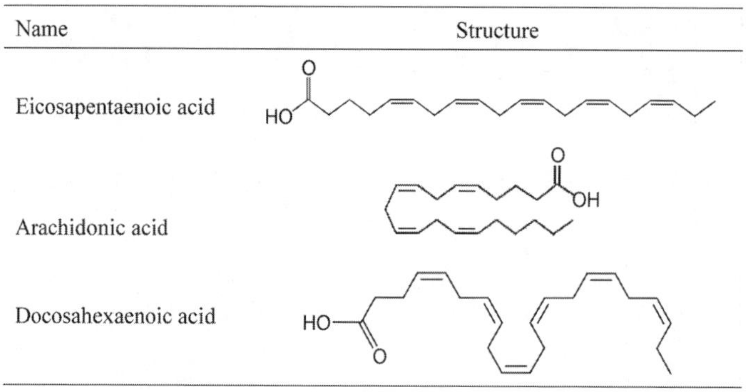

Name	Structure
Eicosapentaenoic acid	
Arachidonic acid	
Docosahexaenoic acid	

Figure 10.1   *Structure of major PUFAs from algae.*

40% of the total fatty acids (Fajardo *et al.*, 2007; Molina Grima *et al.*, 1996). *Nannochloropsis* sp. is also known to produce high amounts of EPA, and is commonly cultivated in fish hatcheries as feed for rotifers and to create a "green-water effect" in fish larvae tanks to improve the survival rate of fish larvae.

Arachidonic Acid (AA)

AA (20:4 ω-6 shown in Figure 10.1) is the principal ω-6 fatty acid in the brain and has a major role as a structural lipid (Ward and Singh, 2005). The red microalga *Porphyridium cruentum* has been shown to produce significant amounts of several major PUFAs, dominated by AA (Plaza *et al.*, 2009). *P. cruentum* is characterised by spherical unicells, 5–13 μm in diameter and can occur as a mass of cells aggregated within a mucilaginous matrix consisting of water-soluble sulfated polysaccharides (Graham, Graham, and Wilcox, 2009).

Docosahexaenoic Acid (DHA)

DHA (22:6 ω-3 shown in Figure 10.1) is a major structural component of the brain and eye retina of humans, and is recognised to support good cardiovascular health (Ward and Singh, 2005). The marine dinoflagellate *Crypthecodinium cohnii* is a primary producer of DHA that is incorporated into the food chain (Harrington *et al.*, 1970; Ward and Singh, 2005).

10.2.3   Vitamins

Microalgae are considered to be valuable sources of several major vitamins and therefore the biomass is of high nutritional value (Becker, 1994). As with all algal phytochemicals, the synthesis and accumulation of vitamins is dependent on the environmental factors, while processing methods, including biomass harvesting and drying, can have a considerable effect on the viability of the vitamin content. For example, the temperature-unstable vitamins B_1, B_2, C and nicotinic acid, may denature during the drying process, thus destroying their nutritional and economic values. Table 10.7 compares the vitamin content of some important algae species to some conventional food based high-vitamin sources. Algae have higher values for

Table 10.7 *Vitamin contents of different algae species compared to some conventional vitamin rich food sources (Data sources: Becker (1994) and Brown et al. (1999)).*

	Vitamins (μg mg^{-1} of dry weight biomass)										
	A	B$_1$	B$_2$	B$_6$	B$_{12}$	C	E	Nicotinate	Biotin	Folic acid	Pantothenic acid
Algae:											
Spirulina platensis	840.0	44.0	37.0	3.0	7.0	80	120.0	—	0.3	0.4	13.0
Chlorella pyrenoidosa	480.0	10.0	36.0	23.0	—	—	—	240.0	0.15	—	20.0
Scenedesmus quadricauda	554.0	11.5	27.0	—	1.1	396.0	—	108.0	—	—	46.0
Tetraselmis sp.	2.2	109.0	26.0	5.8	1.9	3000	70.0	—	1.3	—	—
Pavlova pinguuis	<0.25	36.0	50.0	8.4	1.7	1300	140.0	—	1.9	—	—
Stichococcus sp.	<0.25	29.0	25.0	17.0	1.9	2500	160.0	—	1.3	—	—
Nannochloropsis sp.	<0.25	70.0	25.0	3.6	1.7	2500	290.0	—	1.1	—	—
Foodstuff:											
Liver	60.0	3.0	29.0	7.0	0.65	310.0	10.0	136.0	1.0	2.9	73.0
Spinach	130.0	0.9	1.8	1.8	—	470.0	—	5.5	0.007	0.7	2.8
Baker's yeast	—	7.1	16.5	21.0	—	—	112.0	4.0	5.0	53.0	—

vitamins A, B$_1$ and B$_2$, and overall they can be considered an excellent source of vitamins for humans.

10.2.4 Carotenoids

Carotenoids are red, yellow or orange lipid-soluble compounds that are synthesised by photosynthetic organisms like higher plants and algae, as well by non-photosynthetic organisms such as animals, fungi and many bacteria (Del Campo, García-González, and Guerrero, 2007). Their chemical structure is derived from C$_{40}$ isoprenoids, which are composed of eight isoprene units, and over 700 carotenoids have been detected in natural sources (Naik *et al.*, 2003). Carotenoids can be divided into two major groups: (1) carotenes, which are pigments composed of oxygen-free hydrocarbons and (2) xanthophylls, the oxygenated derivatives (Becker, 2004). A typical concentration of carotenoids in algae is about 0.1–0.2% of dry weight, but some species can accumulate up to 14% (Becker, 2004). Carotenoids play multiple roles in photosynthetic organisms, contributing to light harvesting, and maintaining structure and function of photosynthetic complexes during instances of photoinhibition (Demmig-Adams and Adams, 2002). They also quench singlet oxygen ($^1O_2$) and free radicals, which can damage metabolising tissues (Ye, Jiang, and Wu, 2008).

There is increasing interest in commercial applications of natural carotenoids such as those from microalgae due to health concerns surrounding synthetic alternatives. For example, high-dose synthetic β-carotene supplementation appears to increase the risk of lung cancer amongst smokers (Tanvetyanon and Bepler, 2008). Potential utilisation of natural carotenoids includes use in pharmaceutical, nutritional and cosmetic industries. For example, their bioactive properties are essential for the prevention and treatment of many types of diseases, for example, cardiovascular disease (Shaish *et al.*, 2006; Ye, Jiang, and Wu, 2008), cataract (Suresh Kumar *et al.*, 2003; Ye, Jiang, and Wu, 2008) and risk reduction associated with certain cancers (Mayne *et al.*, 1994; Michaud *et al.*, 2000; van Poppel, 1993; Ye, Jiang, and

Wu, 2008). The most common algae-derived carotenoids are β-carotene from *Dunaliella salina* (García-González *et al.*, 2005), and astaxanthin from *Haematococcus pluvialis* (Huntley and Redalje, 2007).

β-Carotene

β-carotene is a natural carotenoid pigment that has important applications in food, pharmaceutical and cosmetic industries, with an estimated market value of €195 million (Del Campo, García-González, and Guerrero, 2007). β-carotene is commercially used as a food additive, food colourant, anti-oxidant and also as a colourant in numerous cosmetic and body-care products (Mojaat *et al.*, 2008). β-carotene has numerous health benefits, as it is an effective precursor of vitamin A, which is necessary for the proper function of vision and of epithelial tissues (Hosseini Tafreshi and Shariati, 2009). In addition, it has been shown to inhibit or prevent various types of tumours in humans, including skin cancer, epidermoid cancers and cancers of the gastrointestinal tract (Hosseini Tafreshi and Shariati, 2009).

The biflagellate microalga *Dunaliella* has been shown to produce and accumulate large amounts of β-carotene (up to 14% of dry weight) within cellular oil globules (Jin and Melis, 2003). *Dunaliella* is a unicellular Chlorophyta alga characterised by an ovoid cell 2–8 μm wide and 5–15 μm in length (Ben-Amotz, 1999). *Dunaliella* cells lack a rigid polysaccharide cell wall, instead, they are enclosed by a thin elastic membrane which permits rapid cell volume changes. *Dunaliella* can survive in a wide range of marine habitats such as oceans, brine lakes, salt marshes and is capable of thriving in salt concentrations from as low as 0.1 M to > 4 M NaCl (Ben-Amotz, 2004). Its unusual halotolerance is attributed, in part, to the synthesis of large amounts of glycerol to balance osmolarity, but *Dunaliella* also produces high levels of β-carotene in response to high light intensities (Çelekli and Dönmez, 2006). Species that have been shown to overproduce β-carotene include *D. tertiolecta* (Fazeli *et al.*, 2006), *D. salina* (Mojaat *et al.*, 2008), and *D. bardawil* (Mogedas *et al.*, 2009).

β-carotene from *Dunaliella* is composed mainly of all-*trans* (Figure 10.2) and *9-cis* stereoisomers which accumulate within oily globules in the interthylakoid spaces of the chloroplast periphery (Ben-Amotz, 2004). The accumulation of β-carotene and ratio of stereoisomers produced depends greatly on algal division time; any stress condition that slows the rate of cell division, will in turn increase β-carotene production in *Dunaliella*; variable stress conditions, such as high light intensity (Çelekli and Dönmez, 2006; Hejazi and Wijffels, 2003; Mogedas *et al.*, 2009), high salt concentration (Fazeli *et al.*, 2006), nutrient deficiency (Çelekli and Dönmez, 2006; Phadwal and Singh, 2003), pH (Çelekli and Dönmez, 2006) and extreme temperatures (García, Freile-Pelegrín, and Robledo, 2007).

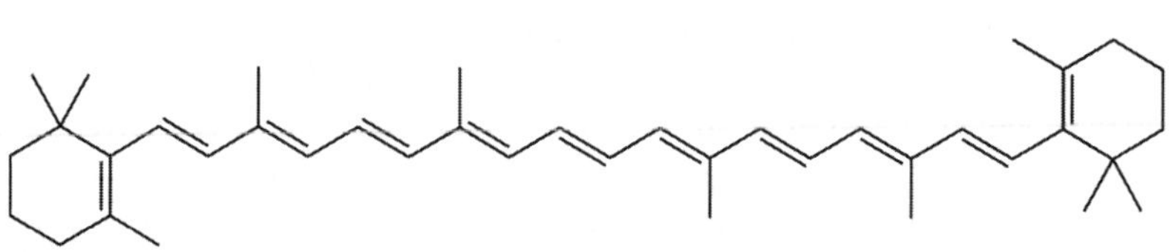

Figure 10.2   *Structure of all-trans β-carotene stereoisomer.*

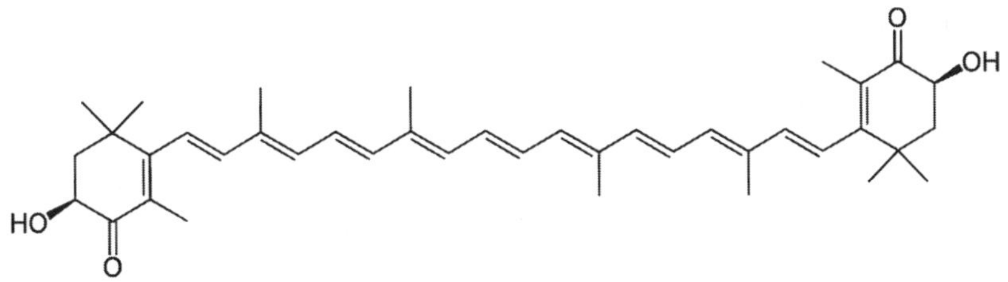

Figure 10.3   Structure of astaxanthin.

Astaxanthin

Astaxanthin (3, 3'-dihydroxy-β-carotene-4, 4'-dione shown in Figure 10.3) is a natural carotenoid pigment that gives marine invertebrates (crabs, lobsters and shrimps) and some fish (salmon and trout) a distinctive orange-red colour (Johnson and An, 1991). It possesses several important bioactive characteristics, including anti-oxidant, immune response enhancement and anti-cancer properties (Guerin, Huntley, and Olaizola, 2003; Kobayashi, Kakizono, and Nagai, 1991; Waldenstedt *et al.*, 2003).

Haematococcus pluvialis, a green unicellular microalga, has received considerable commercial attention due to its natural production of astaxanthin (Lorenz and Cysewski, 2000). Under optimal growth conditions, its cells are green and ellipsoid, two flagella provide motility and cells divide rapidly (Jaime *et al.*, 2010). Under stressed conditions (e.g., high light, nutrient deficiency or elevated salt concentration), the vegetative cells synthesise astaxanthin and at the same time undergo morphological changes resulting in the formation of large, non-motile, red palmelloid cells (Hagen, Siegmund, and Braune, 2002). Astaxanthin pigments are generally accumulated in lipid globules outside the chloroplast and their main functions include protecting the cell from photoinhibition by reducing the amount of light available to the light-harvesting complex (Lee and Zhang, 1999). Astaxanthin exists in *H. pluvialis* in mono- and di-ester forms, which constitute up to 98% of the total carotenoids profile, and reach up to 4% of the total cellular dry weight (Boussiba *et al.*, 1999).

10.2.5 Phycobiliproteins

The phycobiliproteins are water-soluble, proteinaceous, photosynthetic accessory pigments (photosynthetic pigments that capture light energy and channel it to chlorophyll *a*). They are an important component of the assemblage phycobilisome protein complexes (Becker, 2004). Phycobiliproteins absorb light over the visible spectrum and transfer the excitation energy into the reaction centres in the photosynthetic membrane (Bermejo *et al.*, 2002). They are found in the algal classes Rhodophyta and Cryptophyta, and are classified according to their absorption properties: phycoerythrins (PE, λ_{max} 540–570 nm), phycocyanins (PC, λ_{max} 610–620 nm) and allophycocyanins (APC, λ_{max} 650–655 nm) (Viskari and Colyer, 2003).

Their primary commercial utilisation has been as natural dyes and more recently for their health and pharmaceutical applications (Spolaore *et al.*, 2006). Phycobiliproteins are used in cosmetics (e.g., lipsticks and eyeliners), and as a food colourants (chewing gums, jellies, dairy products, etc.) (Bermejo *et al.*, 2002). More recent applications include fluorescent labels for cell organelles in highly sensitive fluorescent techniques such as flow cytometry (Viskari and

Colyer, 2003), and therapeutics relating to its immunomodulating and anti-cancer activities (Bermejo *et al.*, 2002).

The phototrophic cyanobacterium *Spirulina (Arthospira) platensis* is a major source of phycobiliproteins, notably C-phycocyanin (C-PC) and allophycocyanin, with reported production of more that 20% of dry weight (Becker, 2004). Another algal source of phycobiliproteins is the red unicellular microalgae *Porphyridium* spp. (Bermejo *et al.*, 2002). *Porphyridium* is characterised by spherical cells 5–13 μm in diameter, which can occur aggregated within a mucilaginous matrix consisting of water-soluble sulfated polysaccharides (Graham, Graham, and Wilcox, 2009). *Porphyridium cruentum* has been shown to contain four types of phycobiliproteins, namely: allophycocyanin, R-phycoerythrin (R-PC), b-phycoerythrin (b-PC), and B-phycoerythrin (B-PC) (Bermejo, Talavera, and Alvarez-Pez, 2001).

10.2.6 Phycocolloids

Phycocolloids make up the main structural component of many algal cell walls. Phycocolloids from algae have a wide range of industrial applications such as the utilisation as thickening and gelling agents in the food and pharmaceutical industries, as a gel substrate in biological culture media, as sizing agents in the textile industry, and most extensively as agar, carrageenan and alginate in the food and cosmetic industries (Cardozo *et al.*, 2007).

Agar

Agar is a complex mixture of polysaccharide galactans containing mostly $\alpha(1–4)$-3,6-anhydro-L-galactose and $\beta(1–3)$-D-galactose residues (Cardozo *et al.*, 2007). It is composed of two main fractions: agarose, a high strength gel, and agaropectin, a sulfated polysaccharide of low gel strength (Marinho-Soriano and Bourret, 2003; Yaphe, 1984). Agar is derived from species of red macroalgae (Rhodophyta), most commonly from *Gracilaria* and *Gelidium* (McHugh, 1991). The quality of agar extracted depends on a number of factors, including growth conditions (Bird, 1988; Villanueva *et al.*, 1999), physiological factors (Lignell and Pedersén, 1989; Marinho-Soriano *et al.*, 1999) and methods of extraction (Durairatnam, 1987). Gel-quality-dependent applications include (Cardozo *et al.*, 2007): (1) low-quality agars, used as gelling agents in food products (frozen foods, bakery icings, dessert gels and fruit juices) and for industrial applications, such as paper sizing/coating, adhesives, castings and textile printing; (2) medium-quality agars, predominantly used as a gel substrate in biological culture media and as a bulking agent for pharmaceuticals and (3) high-quality agar, used for electrophoresis, immunodiffusion and gel chromatography in molecular biology (Cardozo *et al.*, 2007).

Carrageenan

The carrageenans are a family of linear galactans, composed of alternating $\alpha(1–4)$-3,6-anhydro-D-galactose and $\beta(1–3)$-D-galactose residues (van de Velde, Pereira, and Rollema, 2004). They are obtained from red macroalgae (Rhodophyta), such as *Kappaphycus alvarezii* (Cardozo *et al.*, 2007). Natural carrageenans consist of different sulfated polysaccharide mixtures, the three most commercially important being the λ, κ and ι-carrageenans (van de Velde *et al.*, 2005). Applications are mainly in the food industry where it is used as an emulsifier/stabiliser due to its thickening and suspension-forming properties (Cardozo *et al.*, 2007). Other applications are based on the potential anti-tumour, anti-viral, anti-coagulant and immunomodulation activities exhibited by carrageenans (Schaeffer and Krylov, 2000; Zhou *et al.*, 2004; Zhou *et al.*, 2005).

Alginate

Alginic acid, or alginate, is derived from the brown seaweed (Heterokontophyta). It is composed of linear polysaccharides containing 1,4-linked β-D-mannuronic and α-L-guluronic acid residues, arranged in a non-regular, blockwise order along the chain (Andrade *et al.*, 2004). The main application is in sizing cotton yarn in the textile industry. Other applications stem from its capacity to be used as a gelling agent, to chelate metal ions and to form highly viscous solutions (Cardozo *et al.*, 2007).

10.2.7   Phycosupplements

Phycosupplements refer to algal-based substances that supply plant nutrients or amend soil fertility as a means of increasing crop production. For centuries, macroalgae biomass has been used as fertilizer, that is, to enhance soil structure and mineral composition (Riley, 2002; U.S. DOE, 2010, Kumar *et al.*, 2010). Another potential application is in biochar production. The incorporation of biochar into soil enhances the soil structure, increases fertility and productivity, and neutralises acid (Fowles, 2007). It is also considered to be a viable long-term carbon sink (Lehmann, 2007), despite pending uncertainties about the life-cycle emissions of the biochar production process (Reijnders, 2009; Reijnders and Huijbregts, 2008). Algal biochars display high pH, ash, nitrogen and inorganic nutrients such as P, K, Ca and Mg, while being comparatively low in carbon, surface area and cation exchange capacity (Bird *et al.*, 2011). The capability to neutralise acidic soils provides a direct application in crop productivity enhancement (Bird *et al.*, 2011).

10.3   Production Techniques for Algal Phytochemicals

10.3.1   Microalgae Biomass Production

The commercial viability of algal phytochemical production is dependent on the economics of artificial algal biomass production methods. Under natural growth conditions, phototrophic algae absorb sunlight, and assimilate carbon dioxide from the air, and nutrients from the aquatic habitats. Therefore, as far as possible, artificial production should attempt to replicate and enhance the optimum natural growth conditions. Commercial algae production under natural growth conditions has the advantage of sunlight availability as a free natural resource (Janssen *et al.*, 2003). However, production may be limited due to diurnal cycles and seasonal variations, thereby limiting the production to areas with high solar radiation. On the other hand, artificial lighting allows for continuous production, but at significantly higher energy input (Brennan and Owende, 2010). Algae can assimilate CO_2 from three different sources, namely: CO_2 from the atmosphere; CO_2 in discharge gases from heavy industry and CO_2 from soluble carbonates (Wang *et al.*, 2008). Under natural growth conditions, they assimilate CO_2 from the air which contains 360 ppmv CO_2, although most species can thrive in substantially higher levels of CO_2, typically up to 150 000 ppmv (Bilanovic *et al.*, 2009; Chiu *et al.*, 2009). Inorganic nutrients required for commercial algae production include a nitrogen source and phosphorus for all species, and silicon for diatoms (Suh and Lee, 2003).

It has been argued that photoautotrophic production is the only method that is technically and economically feasible for large-scale production of algae biomass destined for phyto-chemical extraction (Borowitzka, 1997). Two main systems include open pond and closed photobioreactor production (Borowitzka, 1999), but, heterotrophic production systems have also been successfully used for algal biomass and metabolites (Chen, Zhang, and Guo, 1996; Miao and Wu, 2006). The technical viability of each system is influenced by intrinsic properties of the selected algae strain used, as well as climatic conditions, the costs of land and water, and the targeted end product (Borowitzka, 1992).

Phototrophic Production

Open Pond Production Systems   Raceway ponds are the most commonly used open pond production system (Jiménez *et al.*, 2003). They are typically made of closed-loop, oval-shaped recirculation channels (Figure 10.4), and are generally between 0.2 and 0.5 m deep, with integrated mixing and circulation methods to stabilise algae growth and productivity. Raceway ponds are usually built in concrete, but compacted earth-lined ponds with white plastic have also been used. In the continuous production cycle, algal broth and nutrients are introduced in front of the paddlewheel and are circulated through the loop to the harvest extraction point. The CO_2 requirement is met from surface air absorption, but aerators may also be installed to enhance absorption (Terry and Raymond, 1985).

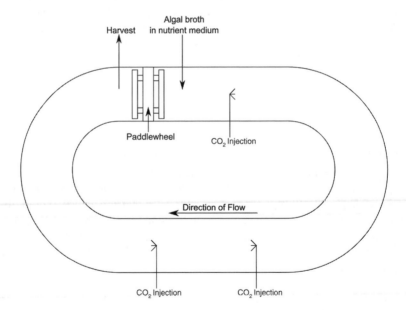

Figure 10.4   *Plan view of a raceway pond where algae broth is introduced after the paddlewheel (agitates to prevents sedimentation), and completes a cycle while being mechanically aerated with CO_2. Harvesting occurs before the paddlewheel, prior to recirculation. Reprinted with permission from Y. Chisti et al., 2007 © Elsevier (2007).*

Open pond systems require specific environments due to inherent threat of contamination and pollution from other algal species and protozoa (Pulz and Scheinbenbogan, 1998). Although monoculture cultivation is possible under extreme environmental culture conditions, only a small number of algae strains are suitable. The species *Chlorella* (adaptable to nutrient-rich media), *Dunaliella salina* (adaptable to high salinity) and *Spirulina* (adaptable to high alkalinity) are typical examples (Borowitzka, 1999). An example of large-scale monoculture cultivation is that of *Dunaliella salina* for β-carotene production in the extremely halophilic waters of Hutt-Lagoon, Western Australia (Pulz and Scheinbenbogan, 1998). In respect of biomass productivity, open pond systems are less efficient when compared with closed photobioreactors (Chisti, 2007). This can be attributed to several determining factors, including evaporation losses, temperature fluctuations in the growth media, CO_2 deficiency, inefficient mixing and limited lighting intensity (Chisti, 2007; Pulz, 2001; Setlik, Veladimir, and Malek, 1970; Ugwu, Aoyagi, and Uchiyama, 2008). Compared to closed photobioreactors, open pond production is the cheaper method of large-scale algal biomass production. For example, in 2008, the unit cost of producing *Dunaliella salina* in an open pond system was about €2.55 per kilogram of dry biomass (Tan, 2008).

Closed Photobioreactor Systems   Microalgae production in closed photobioreactors (PBRs) is designed to overcome some of the limitations of open pond production systems. For example, the pollution and contamination risks with open pond systems, in most part, preclude their use for high-value products directed to pharmaceutical and cosmetics industries (Ugwu, Aoyagi, and Uchiyama, 2008). Also, unlike open-pond production, PBRs permit culturing of single microalgae species for multiple harvesting cycles with significantly higher biomass productivity (Chisti, 2007). Closed systems include tubular, flat-plate and column PBRs. These systems are also more appropriate for sensitive strains since the closed configuration makes the control of potential contamination easier. However, the costs are substantially higher than open pond systems (Carvalho, Meireles, and Malcata, 2006).

Closed PBR systems consist of an array of straight glass or plastic tubes (Figure 10.5). The tubular array facilitates light absorption and can be aligned horizontally (Molina Grima *et al.*, 2001), vertically (Sánchez Mirón *et al.*, 1999), inclined (Ugwu, Ogbonna, and Tanaka, 2002) or helically (Watanabe and Saiki, 1997); the tubes are generally 0.1 m (or less) in diameter (Chisti, 2007). Algae cultures are re-circulated either with a mechanical pump or airlift system, the latter allowing CO_2 and O_2 to be exchanged between the liquid medium and the aeration gas, as well as providing a mechanism for mixing (Eriksen, 2008). Agitation and mixing may be required to enhance the gas exchange in the tubes. The largest closed photobioreactors are tubular, for example the 25 m^3 plant at Mera Pharmaceuticals, Hawaii (Olaizola, 2000) and the 700 m^3 plant in Klötze, Germany (Pulz, 2001).

Hybrid Algae Biomass Production Systems   The hybrid two-stage cultivation method combines distinct growth stages in PBRs and open pond production systems. The first stage is in a PBR where controllable conditions minimise contamination from other organisms and favour continuous cell division. The second production stage is aimed at exposing the cells to nutrient stresses, which enhances synthesis of the desired lipid product (Huntley and Redalje, 2007; Rodolfi *et al.*, 2008). The second stage is ideally suited to open pond systems, as the environmental stresses that stimulate production can occur naturally through the transfer of the culture from the PBR. Huntley and Redalje (2007) used such a two-stage system for the

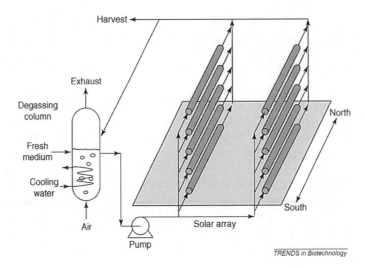

Figure 10.5   *An enclosed tubular photobioreactor showing the two main components, the airlift system and the solar receiver array. The degassing column allows for O_2/CO_2 exchange and provides an outlet to harvest the biomass. The solar receiver provides a platform for growth enhancement by giving a high surface area to volume ratio. Reprinted with permission from Y. Chisti et al., 2008 © Elsevier (2008).*

production of both lipids and astaxanthin from *Haematococcus pluvialis*, and achieved an annual average microbial lipid production rate >10 toe (tonnes oil equivalent) ha^{-1} per annum, with a maximum rate of 24 toe ha^{-1} per annum. They also demonstrated that under similar conditions, rates of up to 76 toe ha^{-1} per annum were feasible using species with higher photosynthetic efficiency and lipid content.

Heterotrophic Algae Production

In this process microalgae are grown on organic substrates such as glucose in stirred tank bioreactors or fermenters. In these cultures algae growth is independent of light energy, which allows for simpler scale-up possibilities than that of open ponds or closed PBRs, since smaller reactor surface to volume ratios may be used (Eriksen, 2008). Heterotrophic production systems provide a high degree of growth control and lower harvesting costs due to the higher cell densities achieved (Chen and Chen, 2006). The set-up costs are minimal, although the systems use more energy than the production of photosynthetic microalgae because the process cycle includes the initial production of organic carbon sources via the photosynthesis process (Chisti, 2007).

10.3.2   Macroalgae Biomass Production

Depending on the location and the expected scale of production, one of three methods may be used in commercial macroalgae/seaweed production: offshore, onshore and harvesting from natural stocks (Bird and Benson, 1987; Israel *et al.*, 2005; Martinez *et al.*, 2006; Rorrer and Cheney, 2004).

Offshore Seaweed Production

Offshore production involves the installation of rigs in marine environments that can support the growth of seaweeds. Seaweed species are selected for target products, (see Section 10.2), and high production potential. A key challenge is in the interactions between seaweed, and cultivation techniques and conditions (e.g., mooring of kelp plants, nutrition supply, seeding and biological interactions) (Bird and Ryther, 1990; Wheeler, 1987). The cultivation of *Porphyra* for Nori (edible *Porphyra* seaweed used as a wrap for sushi and onigiri) in Japan has been successful on offshore nets (McHugh, 2003). However, it is notable that the *Porphyra* life cycle is complex and was not fully understood until 1949 (Drew-Baker, 1949), thereby hampering attempts to scale up the process. Modern cultivation now incorporates an initial onshore seeding and growth period before the transfer of nets containing the small *Porhyra* blades out into the intertidal area and on rigs in the open ocean. The nets are exposed to the air for a couple of hours per day to control the growth of pest seaweeds. Initial harvesting is done after a 50 day growth period in the open sea and subsequently at 10–15 day intervals for up to 10–12 times annually.

Unsuccessful attempts have been made to develop seaweed farms in sub-tidal ocean environments (Bird and Benson, 1987). The main obstacle arises from poor control of growing conditions in the open ocean. However, there is evidence that production can be successfully located in intertidal zone enclosures, where the tidal regime exchanges seawater twice a day (Bravo *et al.*, 1992). Since water replacement occurs during high tide, the cost of water pumping and fertilisation normally involved in similar cultivation systems become irrelevant, but the scaling-up of such systems is still a challenge.

Onshore Seaweed Production

Onshore cultivation of seaweed in tanks cannot stand alone viably, due to the large production areas required and the very high installation costs. To improve performance, the use of aquaculture wastewater has been tested (Buschmann *et al.*, 1994), with observed successes in enhancing productivity (biomass production over 48 wet kg m^{-2} per annum). Integration of tank-based seaweed production with fish farming also enables the possible mitigation of the impact of fish waste and minimises algae cultivation costs. Such integration can enhance system profitability and ecological performance (Buschmann *et al.*, 2001). For example, the macroalga *Gracilaria lemaneiformis* can absorb and store large quantities of nitrogen and phosphorus, and produce large amounts of O$_2$, thereby effectively decreasing eutrophication, that is, the ageing process in water caused by nutrient enrichment which sustains the unwanted growth of plants and algae (Djodjic, Bergström, and Grant, 2005; Yang *et al.*, 2006). *G. lemaneiformis* also has a high productivity, and therefore has become one of the most important cultivated species in China. Matos *et al.* (2006) demonstrated the feasibility of cultivating valuable seaweeds economically using wastewaters from intensive and semi-intensive aquacultures, with the added advantage of bioremediating the effluent to allow for water re-circulation or direct discharge into the sea. Cascade production systems could enable onshore cultivation of species that are economically valuable, but less efficient biofilters. For example, in Portugal, *Gracilaria* is regarded as a most promising species for use as a biofilter, having a high nitrogen uptake efficiency and also good biomass yields (29 g DW m^{-2} per day). Therefore, a two-stage cascade system could be deployed for *Gracilaria* cultivation together with other species with commercial value, but less efficient nitrogen uptake, such as *Palmaria* and *Chondrus* (Matos *et al.*, 2006).

Biomass Yield from Natural Seaweed Stocks

Seaweeds are also produced through gathering/harvesting of natural stocks. However, the demand for seaweed-based products exceeds the natural supply of raw seaweed material, therefore, harvesting will ultimately lead to the depletion of the natural resource (Martinez *et al.*, 2006). Inclement weather conditions also make the production of seaweeds using natural stocks unpredictable. Cultivation methods should therefore be preferred since natural stocks are limited and may not sustain production of the desired products.

10.3.3 Phytochemical-Directed Algae Production Techniques

Carotenoid Production

An open pond system is the conventional method for commercial production of β-carotene. Minimal process control is required for source algal species such as *Dunaliella*, as it thrives at high salt concentrations thus reducing any contamination risk (Ben-Amotz, 1999). The system is operated in batch mode, with the required salinity achieved via evaporation of the growth medium, and nitrogen control to initiate the nutrient-limiting conditions that are necessary for β-carotene accumulation (Mojaat *et al.*, 2008). Open pond cultivation systems have low overall efficiency due to poor CO_2 assimilation, lack of process control and generally poor biomass production (García-González *et al.*, 2005). Currently, there are commercial β-carotene plants operating in Israel, China, USA and Australia (Ben-Amotz, 2004). In temperate regions, closed PBRs are used, allowing for more efficient production due to the higher degree of process control and low operational inputs, but they are more expensive to build and maintain (García-González *et al.*, 2005).

Numerous studies have investigated the cultivation of *H. pluvialis* for astaxanthin extraction (He, Duncan, and Barber, 2007; Issarapayup, Powtongsook, and Pavasant, 2009; Kobayashi *et al.*, 1997; Lorenz and Cysewski, 2000; Zhang *et al.*, 2009). A two-stage carotenoid production process was proposed to manage the interaction between cell growth and astaxanthin accumulation (Cysewski and Lorenz, 2004; Fábregas *et al.*, 2001), whereby biomass production and astaxanthin accumulation occur in sequence and in separate bioreactors (Aflalo *et al.*, 2007). In an industrial production context, the first stage occurs in the controllable conditions of a PBR, which prevents contamination by other organisms and favours continuous cell division. The second production stage in an open pond exposes the cells to nutrient stress, which enhances synthesis of astaxanthin (Brennan and Owende, 2010).

Phycobiliprotein Production

An outdoor raceway pond is preferred for commercial production of phycobiliprotein-directed algae biomass cultivation, for example, from *Spirulina platensis*. Although, *Spirulina* is generally regarded as a photoautotrophic organism, under certain culture conditions, it will assume mixotrophic and heterotrophic growth characteristics. Biomass harvesting is by filtration processes, such as gravity or horizontal vibratory screen, and vacuum bed filtering.

Subsequent spray or drum-drying yields *Spirulina* powder at 3–4% moisture content (Becker, 2004; Oliveira *et al.*, 2009).

10.3.4   Biorefinery Concept

The International Energy Agency's Bioenergy Task 42 on Biorefineries (IEA, 2007) defines biorefining as the sustainable processing of biomass into a spectrum of bio-based products (food, feed, chemicals) and bioenergy (biofuels, power and/or heat). Biorefining integrates multiple unit conversion processes that are analogues to petroleum refinery processes, depicted in Figure 10.6 (Subhadra, 2010; Taylor, 2008). The fundamental aim of a biorefinery is to deliver multiple products by depolymerising and deoxygenating the feedstock components, thereby maximising the value derived for the biomass feedstock (Cherubini, 2010). It is based on the concept that low-volume high-value outputs (proteins, carbohydrates, pigments, PUFAs, fertilisers and nutritional supplements) will increase the economic sustainability of the process and that high-volume low-value outputs (biofuels, bulk chemicals and fertilisers) will meet specific standard and production targets. There is an overall reduction in the unit's environmental footprint by reduced consumption of fossil fuel in the primary energy mix for electricity generation (Subhadra, 2010).

Available evidence indicates that multiple product extraction from the primary biomass production could be fundamental to successful commercial production of biofuels and co-products from algae (*viz.*, bulk chemicals, proteins and sugars) (Wijffels, Barbosa, and Eppink, 2010). An algal biorefinery process that is designed to maximise multiple product recovery from the biomass (including high-value phytochemicals), combined with process chain integration to include carbon capture and storage, and phycoremediation of wastewater, could enhance the viability and future sustainability of biofuel-directed microalgae production (Brennan and Owende, 2010).

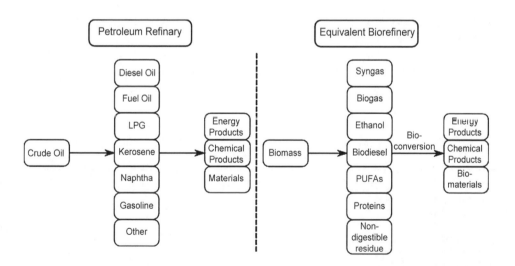

Figure 10.6   *Analogy of biorefinery concept for algal biomass with a petroleum refinery.*

Sustainable algal biorefineries need to minimise the consumption of non-renewable energy resources, while maximising the complete use of biomass (Cherubini, 2010). Key factors for biorefinery sustainability include: type of feedstock, conversion processes and type of products. Biorefinery sustainability should therefore be qualified by an evaluation of the system performance at a plant scale and environmental impact through life-cycle assessment (Wellisch *et al.*, 2010). A unique aspect of algae to other biorefinery feedstocks is the wide variety of species available (Subhadra, 2010), thus providing more potential products, but also resulting in the need for a large range of processing technologies, most of which are still at a pre-commercial stage (Cherubini, 2010). Algal biomass feedstock will experience seasonal changes in both quantity and quality; this may result in a change of biorefinery capacity towards products that can be derived from the biomass all year round and leave niche products to specialised producers.

10.4   Extraction Techniques for Algal Phytochemicals

Extraction processes are a major limitation to efficient recovery of phytochemicals and other high-value products from algal biomass (*viz.*, carotenoids, phycobiliproteins, phycocolloids and PUFAs). They are highly specific and are generally dictated by the end-product and desired quality standards (Mata, Martins, and Caetano, 2010; Richmond, Cheng-Wu, and Zarmi, 2003). Table 10.8 summarises the typical methods that have been applied, the choice of which generally depend on the strength of the microalgal cell wall and extracted product.

10.4.1   Pre-Treatment Processes

The nature of algal cell walls determines biodegradability, and therefore, strongly impacts on the extraction process (Sialve, Bernet, and Bernard, 2009). Cell disruption is necessary for most extraction processes as they depend upon the recovery of the intracellular contents of microalgae. Most cell disruption techniques have been adapted from applications for intracellular, non-photosynthetic bio-products (Middelberg, 1994), and have been successfully applied to microalgal biomass (Mendes-Pinto *et al.*, 2001). Specific disruption methods include the use of high-pressure cell homogenizers, bead mills, autoclaving and chemical action, for example, mixing with hydrochloric acid or sodium hydroxide.

Table 10.8   *Extraction methods available for various algal phytochemicals.*

Process Methods	Phytochemical Category				
	Carotenoids	PUFAs	Phycobiliproteins	Phycocolloids	Fertilisers
Pre-treatment	*	*	*	*	*
Extraction option:					
1. Solvent extraction	*	*		*	
2. Supercritical fluid (SCF) extraction	*	*			
3. Expanded bed adsorption chromatography (EBA)	*		*		
4. Pressurised liquid extraction (PLE)			*		

10.4.2   Solvent Extraction

Solvent extraction is the primary lipid extraction technique currently used for both macroalgae and microalgae, but the technique is laborious and environmentally unfriendly due to the excessive amounts of solvent required and waste generated (Denery *et al.*, 2004). Hexane is the preferred solvent for lipid recovery from microalgae, as it can recover >95% of lipid in the biomass (Muhs *et al.*, 2009). In solvent extraction, the biomass is extruded through an expeller or press, and a pulp is mixed with hexane which extracts the oil. The mixture is subsequently filtered, and the hexane/oil solution separated by distillation (Scott *et al.*, 2010). Sodium carbonate, sodium hydroxide, or lime are used in an alkaline extraction process to obtain various phycocolloids (e.g., alginate, and carrageenan) from macroalgae (Falshaw, Bixler, and Johndro, 2003; Vauchel *et al.*, 2009).

10.4.3   Supercritical Fluid Extraction

Supercritical fluid (SCF) extraction is one of the most promising, environmentally acceptable extraction techniques as it generates solvent-free extracts (Sajilata, Singhal, and Kamat, 2008), while avoiding the degradation of thermally labile compounds (Bruno *et al.*, 1993). Carbon dioxide is the most commonly used SCF solvent in food-directed applications, due in the most part to its relatively low cost, purity and safety in handling (Brunner, 2005). The SCF extraction process relies on exposure of the pre-treated algal biomass to the solvent fluid under supercritical conditions that effects dissolution of the target compound. The dissolved compound is subsequently separated from the solvent by a drop in solution pressure (Sahena *et al.*, 2009). SCF extraction is commonly used for the extraction of carotenoids and PUFAs from microalgae (Macías-Sánchez *et al.*, 2009; Mendes *et al.*, 1995; Mendes *et al.*, 2003; Sajilata, Singhal, and Kamat, 2008). A typical pilot plant (Figure 10.7) consists of a solvent pump to deliver fluid through the system, an extraction column (extract collection and solvent depressurisation) and specific separators for controlled fractionation of the extracted compounds.

10.4.4   Expanded Bed Adsorption Chromatography

Expanded bed adsorption chromatography (EBA) allows the capture of proteins from particle-containing feedstock with prior removal of particulates in a single operation. It is similar to column chromatography, but in expanded bed, a particulate adsorbent in a column is allowed to rise by applying upward flow, thus increasing the space between adsorbent particles, allowing cells and cell debris to pass through without blocking the bed (Hjorth, 1997). EBA achieves the extraction of phycobiliproteins in a single process, that is, without the need for feedstock pre-treatment or particle removal (Bermejo *et al.*, 2003). EBA columns allow higher flow rates than conventional packed bed chromatography, which can result in fluid flow between resin particles allowing the larger unwanted particles such as cellular debris to travel through the bed unhindered, while the target proteins are adsorbed onto the resin (McCann *et al.*, 2008). This method has been successfully used to purify R-phycoerythrin from the red alga *Polysiphonia urceolate* (Niu, Wang, and Tseng, 2006), B-phycoerythrin from the red microalga *Porphyridium cruentum* (Bermejo *et al.*, 2003) and C-phycocyanin from the cyanobacterium *Spirulina platensis* (Niu *et al.*, 2007).

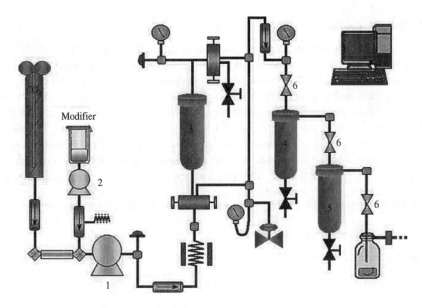

Figure 10.7   *Illustration of supercritical fluid extraction pilot plant equipped with two fractionation cells. The CO₂ pump (1) and modifier/solvent pump (2) pass the supercritical fluid through to the solid sample extraction cell (3) and onto the fractionation cells (4 and 5) via valves (6) in which the extract is collected and the solvent depressurised. Reprinted with permission from M. Herrero et al., 2006 © Elsevier (2006).*

10.4.5   Pressurised Liquid Extraction

Pressurized liquid extraction (PLE) combines high temperature and pressure with liquid solvents to achieve fast and efficient extraction of the target product (Carabias-Martínez *et al.*, 2005). Compared to other extraction processes, PLE is a relativity fast procedure that is readily automated, and has lower solvent consumption (Santoyo *et al.*, 2009). A basic PLE process (Figure 10.8) includes pre-heating of solvent to a set temperature and injection into a reaction vessel that is pre-loaded with the sample biomass. The extract compound is cooled in a temperature-controlled water bath to prevent possible degradation (Pitipanapong *et al.*, 2007).

PLE has been used for algal phytochemical recovery, such as: carotenoids from *Haematococcus pluvialis* (Jaime *et al.*, 2010) and *Dunaliella salina* (Denery *et al.*, 2004); phenolic acids from *Anabaena dolliolum*, *Spongiochloris spongiosa*, *Porphyra tenera* and *Undaria pinnatifida* (Onofrejová *et al.*, 2010), and fucoxanthin from *Eisenia bicyclis* (Shang *et al.*, 2011) amongst other compounds from other important species of algae.

10.4.6   Unit Process in Commercial Phytochemical Extraction

Carotenoid Extraction

The supercritical carbon dioxide (SC-CO₂) technique is now widely employed for β-carotene recovery from *Dunaliella* (Plaza *et al.*, 2009), due to the high diffusion capacity through complex cell membrane matrix, thereby alleviating the environmental burdens associated with other solvent extraction techniques (Section 10.4.2). Also, the low critical CO₂ temperature (31.1 °C)

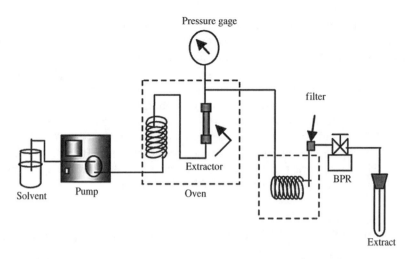

Figure 10.8   *Illustration of a laboratory PLE unit. The pump delivers the solvent, through a preheating coil installed in an oven, into the extracting vessel. The extract is cooled in a coil immersed in a water bath and collects in sample vials. Reprinted with permission from J. Pitipanapong et al., 2007 © Elsevier (2007).*

means that the system is operable at moderate temperature ($60\,°C$), therefore minimising thermal degradation of the product (Krichnavaruk *et al.*, 2008), In addition, CO_2 is easily separated from the extract compounds by depressurisation (Machmudah *et al.*, 2006). Compounds extracted by SC-CO_2 have high purity and good quality, and are therefore safe for nutraceutical and pharmaceutical applications (Mendes *et al.*, 2003).

PLE has also been used for β-carotene extraction. The automated technique uses minimal solvent, and involves sample processing in an oxygen- and light-free environment (Benthin, Danz, and Hamburger, 1999). PLE is a faster and easier technique to use when compared to SC-CO_2, with higher extraction yields obtained from a significantly lower amount of solvent (Jaime *et al.*, 2010). PLE has also been successfully used for β-carotene extraction from *Dunaliella* on a laboratory scale (Denery *et al.*, 2004; Herrero *et al.*, 2006a; Herrero, Cifuentes, and Ibañez, 2006; Jaime *et al.*, 2010).

Organic solvent extraction methods, SC-CO_2 and PLE, have also been used for extraction and purification of astaxanthin from *H. pluvialis* (Jaime *et al.*, 2010; Krichnavaruk *et al.*, 2008; Machmudah *et al.*, 2006). Traditionally, organic solvent extraction after cell-wall disruption was used (Sarada *et al.*, 2006), but the process was hindered by the inherent thick cell wall of *H. pluvialis*, necessitating excessive use of solvents and multiple extraction steps, thus making the process infeasible for large-scale astaxanthin production (Lim *et al.*, 2002; Mendes-Pinto *et al.*, 2001). The properties of SC-CO_2 enables more efficient cell wall diffusion (Krichnavaruk *et al.*, 2008), thereby overcoming the extraction limitations due to the cell wall thickness of *H. pluvialis*.

Phycobiliprotein Extraction

Phycobiliprotein extraction from species such as *Spirulina platensis* can be exceedingly difficult due to the heat sensitivity of the proteins (Oliveira *et al.*, 2008), the cyanobacteriums multilayered cell wall and small size (Herrero *et al.*, 2005), and the high number of unit operations required (Eriksen, 2008). Consequently, various extraction and purification protocols have been used. The basic process involves two steps: (1) intracellular material extraction by pre-treatment and purification using mainly chromatography, and involving reagent-activated cell wall disintegration, and precipitation of the phycobiliproteins (Bermejo, Ruiz, and Acien, 2007), (2) purification of phycobiliproteins is carried out by various chromatographic processes (e.g., adsorption, hydrophobic interaction, gel filtration, ion exchange) (Eriksen, 2008), or a two-phase aqueous extraction process (Patil and Raghavarao, 2007), or both (Soni *et al.*, 2006). Expanded bed adsorption (EBA) chromatography has also been used with the advantage of a low number of process steps producing a concentrated and partially pure product (Bermejo, Ruiz, and Acien, 2007).

10.5   Metabolic Engineering for Synthesis of Algae-derived Compounds

Genetic transformation is being carried out to engineer microalgae (high potential biofuel resource) as efficient cell factories for the expression of phytochemicals such as proteins, antibodies and PUFAs (León-Bañares *et al.*, 2004). Metabolic engineering of microalgae opens up the possibility of enhancing the production of target phytochemicals, and the potential discovery of new compounds with commercial applications (León-Bañares *et al.*, 2004). A comprehensive understanding of the regulatory and synthesis mechanisms of algae may enable enhanced production of phytochemicals (Beer *et al.*, 2009). Figure 10.9 illustrates a range of commercially important metabolic pathways where there is still a lack of understanding of the controlling mechanisms in eukaryotic algae (Scott *et al.*, 2010).

Currently, metabolic engineering is possible for only a few microalgae species which have a well-defined, mapped genome. These algae include *Chlamydomonas reinhardtii*, *Volvox carteri*, *Cyanidioschyzon merolae* and *Emiliania huxleyi*, and the diatoms *Phaeodactactylum tricornutum* and *Thalassiosira pseudonana*. Due to the growing interest in biofuel-directed algae production, it is expected that genome sequencing of other important strains will be undertaken to enable species-based optimisation of target compounds. Current, metabolic engineering includes the manipulation of culture conditions, nuclear and chloroplastic transformation, the expression of recombinant proteins, augmented fatty acid biosynthesis, trophic conversion and re-engineering light-harvesting antenna complexes (Rosenberg *et al.*, 2008).

10.5.1   Manipulation of Culture Conditions

One of the most effective methods for enhancing microalgae lipid accumulation is nitrogen limitation, which also results in a gradual change of lipid composition from free fatty acids to triacylglycerol (TAG) (Widjaja, Chien, and Ju, 2009). Lipid accumulation in microalgae occurs when nutrient depletion becomes the growth-limiting factor (typically nitrogen, but also silicate for diatoms). Under such conditions, cell proliferation is hindered, but assimilated carbon is converted to TAG and stored within existing cells (Meng *et al.*, 2009). Wu and Hsieh (2008) investigated the effects of salinity, nitrogen concentration and light intensity on lipid productivity, and recorded up to a 76% increase in lipid production under specific nitrogen-limiting growth conditions, by comparison with conventional batch processes.

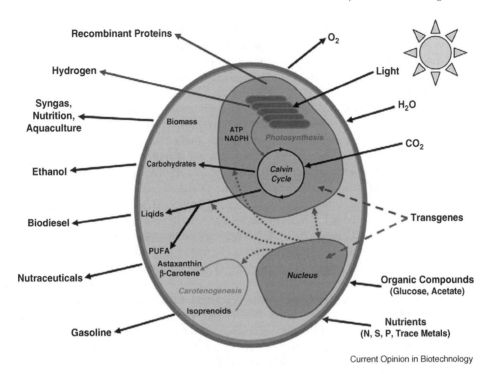

Recombinant Proteins

Hydrogen

Syngas,
Nutrition,
Aquaculture

Biomass

ATP
NADPH   *Photosynthesis*

O₂

Light

H₂O

CO₂

Ethanol

Carbohydrates

*Calvin
Cycle*

Biodiesel

Liqids

PUFA

Nutraceuticals

Astaxanthin
β-Carotene

Carotenogenesis

Nucleus

Transgenes

Organic Compounds
(Glucose, Acetate)

Isoprenoids

Gasoline

Nutrients
(N, S, P, Trace Metals)

Current Opinion in Biotechnology

Figure 10.9 *Microalgal biomass metabolic pathways with commercial importance. This figure depicts internal cellular pathways involved in the biosynthesis of various products. The chloroplast (depicted as the light-grey organelle absorbing light and CO₂) plays a direct role in the biosynthesis of protein and hydrogen (solid grey), the nucleus drives all other major biosynthesis mechanisms through metabolic regulation (dotted). Both of these organelles contain their own genome, and both offer the possibility of genetic engineering. Reprinted with permission from J.N. Rosenberg et al., 2007 © Elsevier (2007).*

However, experimental evidence suggests that algae growth-rate maximisation may be more important for overall enhancement of lipid production. For example, Weldy and Huesemann (2007) recorded higher lipid productivity ($0.46 \, g \, l^{-1}$ per day) under N-sufficient conditions and high light intensity when compared with N-deficient cultures of *Dunaliella salina* ($0.12 \, g \, l^{-1}$ per day).

10.5.2 Nuclear and Chloroplast Transformation

Integration of new genetic information into the nucleus results in recombinant protein expression and the possibility of manipulating the algal metabolism, while chloroplast transformation results in controlled site-directed recombination of constructs, leading to high expression levels due to high plastome copy number that are not subjected to gene silencing (León-Bañares *et al.*, 2004). The basis of gene transformation is to cause temporal permeability of the cell membrane, to enable transfer of DNA molecules while preserving cell viability. Transformation protocols have been developed for algal strains lacking cell walls (e.g., some strains of *C. reinhardtii*), which greatly increases the number of recoverable transformants. Otherwise, enzymatic removal of cell walls prior to transformation is required (Kumar *et al.*, 2004), different methods that have been developed to achieve this include particle bombardment, *Agrobacterium tumefaciens*-mediated transformation, agitation with glass beads and electroporation (exposure to electrical field pulses) (Potvin and Zhang, 2010).

Particle bombardment is a simple and highly reproducible transformation method, and remains the most effective technique for chloroplast transformation. It is consequently the most widely used method and has been successful applied to a range of algae species, including *C. reinhardtii* (El-Sheekh, 2000; Mayfield and Kindle, 1990), *Chlorella kessleri* (El-Sheekh, 1999), *Chlorella ellipsoidea* (Chen *et al.*, 1998), *Haematococcus pluvialis* (Teng *et al.*, 2002) and *Phaeodactylum tricornutum* (Zaslavskaia *et al.*, 2000).

A. tumefaciens-mediated transformation has mainly been used for engineering higher plant cells. It enables the transfer of T-DNA from *A. tumefaciens* tumour-inducing plasmid into the target genome (Potvin and Zhang, 2010). It has been shown to enhance the transfer frequency by 50-fold compared to glass bead transformation (Kumar *et al.*, 2004) and has been used with several algae species, including *C. reinhardtii* (Kumar *et al.*, 2004) and *H. pluvialis* (Kathiresan *et al.*, 2009).

Agitation of cell-wall-deficient microalgal cells with glass beads involves vortexing of cells in the presence of DNA, glass beads and polyethylene glycol (PEG), PEG being used to improve transformation efficiency (Kindle, 1998). Agitation with glass beads is a simpler, cheaper and more effective method than particle bombardment to transform nuclei, and is routinely used with cell-wall-deficient mutants (León-Bañares *et al.*, 2004).

Electroporation exposes cell walls to high-intensity electrical field pulses to induce molecular transport across the plasma membrane (Brown, Sprecher, and Keller, 1991). It has been successfully used for several species with biofuel and phytochemical potential, such as *Nannochloropsis oculata* (Chen *et al.*, 2008), *Chlorella* sp. (Wang *et al.*, 2007), *D. salina* (Feng *et al.*, 2009; Sun *et al.*, 2008; Wang *et al.*, 2007), *D. tertiolecta* (Walker *et al.*, 2005) and *C. reinhardtii* (Ladygin, 2004).

10.5.3   Expression of Recombinant Proteins

Recombinant protein expression in microalgae offers an alternative to conventional sources (i.e., plant cells) and is likely to be a key application of microalgae biotechnology. This is due in part to algae's enhanced protein-folding mechanisms (León-Bañares *et al.*, 2004), the presence of post translational modification systems (León-Bañares *et al.*, 2004), rapid plastid and nuclear transformation methods (Mayfield and Franklin, 2005) and the low costs of capitalisation and production compared to other protein expression platforms (Mayfield and Franklin, 2005).

10.5.4   Augmented Fatty Acid Biosynthesis

Microalgae are recognised as the main primary producers of fatty acids (Guschina and Harwood, 2006), and are also the source of many other valuable biological compounds such as alcohols, alkanes, aldehydes and ketones (Andreou, Brodhun, and Feussner, 2009). Fatty acid synthesis occurs within the plastid, before translocation to the cytoplasm for further assembly into diacylglycerides and triacylglycerides (Riekhof, Sears, and Benning, 2005). The complexity of the lipid metabolism in algae means that directed fatty acid biosynthesis is challenging. Manipulation of culture conditions, described in Section 10.5.1, has been used to direct and enhance lipid accumulation, but results have been variable and frequently inconsistent where a single product is preferred (Beer *et al.*, 2009). However, several strategies to enable fatty acid biosynthesis toward more consistent lipid profiles have been developed from research, namely (Beer *et al.*, 2009): over-expression of fatty acid biosynthetic enzymes; increasing the availability of precursor molecules, such as acetyl-CoA; down-regulation of fatty acid catabolism by inhibiting β-oxidation or lipase hydrolysis; alteration of saturation profiles through the introduction or regulation of desaturases; optimisation fatty acid chain length with thioesterases.

10.5.5 Trophic Conversion of Microalgae

The majority of microalgal species are obligate photoautotrophic organisms (see Section 10.1.1), while others are heterotrophic and have the natural ability to metabolise sugars. Through genetic engineering, some obligate photoautotrophs have been converted towards heterotrophy through the introduction of hexose transporters (Rosenberg *et al.*, 2008), in order to improve cell densities and secondary metabolite output. Examples of species that have undergone trophic conversion include *C. reinhardtii, P. tricornutum* and *V. carteri*.

10.5.6 Re-Engineering Light-Harvesting Antenna Complexes

The capture and conversion of solar energy are important factors in the development of biofuel- and/or phytochemical-directed production in algae (Johnson and Schmidt-Dannert, 2008). The absorption of solar energy via pigments present in the light-harvesting antenna complexes (LHCs) of microalgae effects photochemical reactions (Polle *et al.*, 2000), and channels the energy to two distinct photosystems (PSI and PSII). Microalgae species evolution has resulted in physiological and biochemical adaption for conditions of low light intensity (Rosenberg *et al.*, 2008). They are characterised by tightly stacked thylakoids and large LHCs that allow for efficient capture of solar energy. Under the artificial production conditions of PBRs, high light intensity can overwhelm the photosystems, causing reduction in photosynthetic efficiency and adverse cell stress (Long, Humphries, and Falkowski, 1994). Current research efforts have been directed at minimising the effects of photoinhibition by using cell cultures that have been re-engineered to develop smaller LHCs. For example, Beckmann *et al.* (2009) engineered a new PSII for *C. reinhardtii* to attain higher solar energy conversion efficiency and increased cell growth. The engineered strain (*T7*) was characterised by 10–17% antenna size reduction and up to a 50% increase in photosynthetic efficiency at high light intensity levels, resulting in the potential to achieve up to a 50% improvement in the mid growth phase and higher overall cell densities. RNA interference technology has also been used to suppress the expression of LHC proteins in *C. reinhardtii* (Mussgnug *et al.*, 2007). The resulting mutant strain contained less tightly stacked thylakoids and exhibited higher resistance to photoinhibition. Therefore, potential re-engineering of LHCs for wild algae strains offers potential for efficient conversion of solar energy to biomass, with positive implications for future biotechnology interventions towards development of efficient and sustainable algal resources.

10.6 Phytochemical Market Evolution

The global market for algal phytochemicals is still in its infancy, with mostly specialist companies developing specific, high-value products for niche markets (Table 10.9). The extent of the current phytochemical products market can be divided into two distinct areas: (1) high-value, low-volume products (proteins, PUFAs and carotenoids); (2) low-value, high-volume products (phycocolloids, feeds and fertilisers). The market value of the entire product range is still difficult to determine, but being organic, such products are generally considered as "premium" in comparison to synthetic alternatives, therefore commanding higher market prices. The associated companies, the majority of which are based in USA, tend to be focused on a specialist area; very few have linked their production strategy to biofuel-directed production systems.

At present, long-term development of algal products is somewhat hindered by sub-optimal performance of scaled-up algae production systems (Spolaore *et al.*, 2006), and specifically, the downstream processing techniques, raising questions relating to their economic competitiveness. However, as capacity increases and prices go down, new markets will emerge (Wijffels, Barbosa, and Eppink, 2010). Biorefining of algal biomass seem to be a rational pathway to enhancing the economic feasibility and future sustainability. It could be argued that realisation of the full inherent biofuel potential could result in enormous algal biomass production. It is envisaged that sustainable production can integrate the high-volume, low-value products and the low-volume, high-value products in the biorefinery concept. These could link to opportunities in phycoremediation for optimisation and full chain processes, and hence move towards sustainable production.

10.7 Conclusions

Multiple product extraction from biofuel-directed algal biomass production could be fundamental to process development and optimisation, and eventually the sustainability of algal biofuel production. It is therefore unlikely that algal biomass production processes will be developed for biofuels as the sole end-product. Biorefining will be key to maximisation of phytochemical recovery. Biorefining provides the basis for multiple algal biomass component extraction (*viz.*, carotenoids, phycobiliproteins, phycocolloids and PUFAs), and transformation to specific functional components. It has inherent opportunities for synergistic coupling of production process pathways with environmental impact mitigation associated with biowaste (e.g., biogas recovery from intermediate extraction waste and phycoremediation) and fossil fuel energy use (e.g., carbon sequestration and recycling). Therefore, development of the integrated production chain components including algal strain screening and growth optimisation, scale-up feasibility, metabolic engineering and holistic system design (including logistics and environmental impacts mitigation opportunities, e.g., phycoremediation) are important. Specific to phytochemical extraction, metabolic engineering of biofuel-directed microalgae production opens up the possibility of enhancing the production of target phytochemicals and even the discovery of new compounds for industrial and pharmaceutical applications.

Table 10.9 *Commercially valuable products derived from algae.*

Compound[a]	Product	Price range[b] (€ g^{-1})	Producer (Country)
Carotenoids	β-Carotene	0.3–3	AquaCarotene (USA), Cognis Nutrition and Health (Australia), Cyanotech (USA), Nikken Sohonsha Corporation (Japan), Tianjin Lantai Biotechnology (China), Parry Pharmaceuticals (India), Betatene (Australia), Western Biotechnology (Australia) Inner Mongolia Biological Engineering (China), Nature Beta Technologies (Israel)
	Astaxanthin	10	AlgaTechnologies (Israel), Bioreal (USA), Cyanotech (USA), Mera Pharmaceuticals (USA), Parry Pharmaceuticals (India), BioReal (Sweden), Fuji Health Science (USA) Cybercolors (Ireland)
Fatty Acids	EPA acid Oleic acid γ-Linolenic acid Palmitic acid Palmitoleic acid DHA fatty acid Arachidonic acid	60	BlueBiotech International (Germany), Spectra Stable Isotopes (USA), Martek Biosciences (USA)
Proteins	Phycobiliproteins	15 000	BlueBiotech International (Germany), Cyanotech (USA),
	Recombinant proteins	—	Rincon Pharmaceuticals (USA)
	Heavy isotope labelled metabolites	1000–20 000	Spectra Stable Isotopes (USA),
Phycocolloids	Agar	—	—
	Alginates	0.001	Arramara Teo. (Ireland)
	Carrageenans	—	FMC, USA
Feed	Whole-cell dietary supplements	0.05	BlueBiotech International (Germany), Cyanotech (USA), Earthrise Nutritionals (USA), Phycotransgenics (USA)
	Whole-cell aquaculture feed	0.07	BlueBiotech International (Germany), Aquatic Eco-Systems (USA), Coastal BioMarine (USA), Reed Mariculture (USA)
Phycosupplements	Fertiliser	0.001	Arramara Teo. (Ireland)
	Biochar	—	—

[a] By the overlapping nature of the ranges considered, the figures provided were adopted from multiple sources including: Plaza *et al.* (2009), U.S. DOE (2010), Spolaore *et al.* (2006), Pulz *et al.* (2004), Amin Hejazi and Wijffels (2004), Lorenz *et al.* (2000), Ratledge (2004), Rosenberg *et al.* (2008).
[b] Figures are considered current at time of stated publication, and have not been updated to represent market changes. Price ranges for products have in some cases been converted into €, but not adjusted for inflation.

Acknowledgement

The contribution of this chapter was made possible under the Charles Parsons Energy Research Programme of Science Foundation Ireland (SFI), Grant No. 6C/CP/E001.

References

Abdulqader, G., Barsanti, L., and Tredici, M.R. (2000) Harvest of *Arthrospira platensis* from Lake Kossorom (Chad) and its household usage among the Kanembu. *Journal of Applied Phycology*, **12**, 493–498.

Aflalo, C., Meshulam, Y., Zarka, A., and Boussiba, S. (2007) On the relative efficiency of two- vs. one-stage production of astaxanthin by the green alga *Haematococcus pluvialis*. *Biotechnology and Bioengineering*, **98**, 300–305.

Andrade, L.R., Salgado, L.T., Farina, M. *et al.* (2004) Ultrastructure of acidic polysaccharides from the cell walls of brown algae. *Journal of Structural Biology*, **145**, 216–225.

Andreou, A., Brodhun, F., and Feussner, I. (2009) Biosynthesis of oxylipins in non-mammals. *Progress in Lipid Research*, **48**, 148–170.

Balat, M. and Balat, H. (2010) Progress in biodiesel processing. *Applied Energy*, **87**, 1815–1835.

Barringgton, K., Chopin, T., and Robinson, S. (2009) Integrated multi-trophic aquaculture (IMTA) in marine temperate waters, in *Integrated Mariculture. A Global Review* (ed. D. Sato), Food and Agriculture Organization, Rome, Fisheries and Aquaculture Technical Paper, No. 529, Rome.

Barsanti, L. and Gualtieri, P. (2006) *Algae: Anatomy, Biochemistry, and Biotechnology*, Taylor & Francis, Boca Raton, FL.

Becker, E.W. (1994) *Microalgae: Biotechnology and Microbiology*, Cambridge University Press, Cambridge.

Becker, E.W. (2004) Microalgae in human and animal nutrition, in *Handbook of Microalgae Culture: Biotechnology and Applied Phycology* (ed. A. Richmond), Wiley-Blackwell, Oxford.

Beckmann, J., Lehr, F., Finazzi, G. *et al.* (2009) Improvement of light to biomass conversion by de-regulation of light-harvesting protein translation in *Chlamydomonas reinhardtii*. *Journal of Biotechnology*, **142**, 70–77.

Beer, L.L., Boyd, E.S., Peters, J.W., and Posewitz, M.C. (2009) Engineering algae for biohydrogen and biofuel production. *Current Opinion in Biotechnology*, **20**, 264–271.

Ben-Amotz, A. (1999) Production of β-carotene from Dunaliella, in *Chemicals from Microalgae* (ed. Z. Cohen), Taylor & Francis, London.

Ben-Amotz, A. (2004) Industrial production of microalgal cell-mass and secondary products - major industrial species Dunaliella, in *Handbook of Microalgae Culture: Biotechnology and Applied Phycology* (ed. A. Richmond), Wiley-Blackwell, Oxford.

Benthin, B., Danz, H., and Hamburger, M. (1999) Pressurized liquid extraction of medicinal plants. *Journal of Chromatography. A*, **837**, 211–219.

Bermejo, R., Alvárez-Pez, J.M., Acién Fernández, F.G., and Molina Grima, E. (2002) Recovery of pure B-phycoerythrin from the microalga *Porphyridium cruentum*. *Journal of Biotechnology*, **93**, 73–85.

Bermejo, R., Gabriel Acién, F., Ibáñez, M.J. *et al.* (2003) Preparative purification of B-phycoerythrin from the microalga *Porphyridium cruentum* by expanded-bed adsorption chromatography. *Journal of Chromatography. B*, **790**, 317–325.

Bermejo, R., Ruiz, E., and Acien, F.G. (2007) Recovery of B-phycoerythrin using expanded bed adsorption chromatography: Scale-up of the process. *Enzyme and Microbial Technology*, **40**, 927–933.

Bermejo, R., Talavera, E.M., and Alvarez-Pez, J.M. (2001) Chromatographic purification and characterization of B-phycoerythrin from *Porphyridium cruentum*: Semipreparative high-performance liquid chromatographic separation and characterization of its subunits. *Journal of Chromatography. A*, **917**, 135–145.

Bilanovic, D., Andargatchew, A., Kroeger, T., and Shelef, G. (2009) Freshwater and marine microalgae sequestering of CO_2 at different C and N concentrations - Response surface methodology analysis. *Energy Convers Manage*, **50**, 262–267.

Bird, K.T. (1988) Agar production and quality from *Gracilaria* sp. strain G-16: Effects of environmental factors. *Botanica Marina*, **31**, 33–39.

Bird, K.T. and Benson, P.H. (1987) *Seaweed Cultivation for Renewable Resources*, Elsevier, Amsterdam.

Bird, K.T. and Ryther, J.H. (1990) Cultivation of *Gracilaria verrucosa* (Gracilariales, Rhodophyta) strain G-16 for agar. *Hydrobiologia*, **204–205**, 347–351.

Bird, M.I., Wurster, C.M., de Paula Silva, P.H. *et al.* (2011) Algal biochar - production and properties. *Bioresource Technology*, **102**, 1886–1891.

Borowitzka, M.A. (1992) Algal biotechnology products and processes - matching science and economics. *Journal of Applied Phycology*, **4**, 267–279.

Borowitzka, M.A. (1997) Microalgae for aquaculture: Opportunities and constraints. *Journal of Applied Phycology*, **9**, 393–401.

Borowitzka, M.A. (1999) Commercial production of microalgae: Ponds, tanks, tubes and fermenters. *Journal of Biotechnology*, **70**, 313–321.

Boussiba, S., Bing, W., Yuan, J.P. *et al.* (1999) Changes in pigments profile in the green alga *Haeamtococcus pluvialis* exposed to environmental stresses. *Biotechnology Letters*, **21**, 601–604.

Bravo, A., Buschmann, A.H., Valenzuela, M.E. *et al.* (1992) Evaluation of artificial intertidal enclosures for *Gracilaria* farming in Southern Chile. *Aquacultural Engineering*, **11**, 203–216.

Brennan, L. and Owende, P. (2010) Biofuels from microalgae – A review of technologies for production, processing, and extractions of biofuels and co-products. *Renewable & Sustainable Energy Reviews*, **14**, 557–577.

Brown, L.E., Sprecher, S.L., and Keller, L.R. (1991) Introduction of exogenous DNA into *Chlamydomonas reinhardtii* by electroporation. *Molecular and Cellular Biology*, **11**, 2328–2332.

Brown, M.R., Mular, M., Miller, I. *et al.* (1999) The vitamin content of microalgae used in aquaculture. *Journal of Applied Phycology*, **11**, 247–255.

Brunner, G. (2005) Supercritical fluids: technology and application to food processing. *Journal of Food Engineering*, **67**, 21–33.

Bruno, T., Castro, C.A.N., Hamel, J.F.P., and Palavra, A.M.F. (1993) Supercritical fluid extraction of biological products, in *Recovery Processes for Biological Materials* (eds J.F. Kennedy and J.M.S. Cabral), John Wiley & Sons, Chichester.

Buschmann, A.H., Correa, J.A., Westermeier, R. *et al.* (2001) Red algal farming in Chile: A review. *Aquaculture (Amsterdam, Netherlands)*, **194**, 203–220.

Buschmann, A.H., Mora, O.A., Gomez, P. *et al.* (1994) *Gracilaria chilensis* outdoor tank cultivation in Chile: Use of land-based salmon culture effluents. *Aquacultural Engineering*, **13**, 283–300.

Carabias-Martínez, R., Rodríguez-Gonzalo, E., Revilla-Ruiz, P., and Hernández-Méndez, J. (2005) Pressurized liquid extraction in the analysis of food and biological samples. *Journal of Chromatography. A*, **1089**, 1–17.

Cardozo, K.H.M., Guaratini, T., Barros, M.P. *et al.* (2007) Metabolites from algae with economical impact. *Comparative Biochemistry and Physiology C-Pharmacology Toxicology & Endocrinology*, **146**, 60–78.

Carvalho, A.P., Meireles, L.A., and Malcata, F.X. (2006) Microalgal reactors: A review of enclosed system designs and performances. *Biotechnology Progress*, **22**, 1490–1506.

Çelekli, A. and Dönmez, G. (2006) Effect of pH, light intensity, salt and nitrogen concentrations on growth and β-carotene accumulation by a new isolate of *Dunaliella* sp. *World Journal of Microbiology & Biotechnology*, **22**, 183–189.

Chen, F., Zhang, Y., and Guo, S. (1996) Growth and phycocyanin formation of *Spirulina platensis* in photoheterotrophic culture. *Biotechnology Letters*, **18**, 603–608.

Chen, G.Q. and Chen, F. (2006) Growing phototrophic cells without light. *Biotechnology Letters*, **28**, 607–616.

Chen, H.L., Li, S.S., Huang, R., and Tsai, H.J. (2008) Conditional production of a functional fish growth hormone in the transgenic line of *Nannochloropsis oculata* (Eusticmatophyceae). *Journal of Phycology*, **44**, 768–776.

Chen, Y., Li, W.B., Bai, Q.H., and Sun, Y.R. (1998) Study on transient expression of gus gene in *Chlorelia ellipsoidea* (Chlorophyta) by using biolistic particle delivery system. *Chinese Journal of Oceanology and Limnology*, **16**, 47–49.

Cherubini, F. (2010) The biorefinery concept: Using biomass instead of oil for producing energy and chemicals. *Energy Convers Manage*, **51**, 1412–1421.

Chisti, Y. (2007) Biodiesel from microalgae. *Biotechnology Advances*, **25**, 294–306.

Chisti, Y. (2008) Biodiesel from microalgae beats bioethanol. *Trends in Biotechnology*, **26**, 126–131.

Chiu, S.-Y., Kao, C.-Y., Tsai, M.-T. *et al.* (2009) Lipid accumulation and CO_2 utilization of *Nannochloropsis oculata* in response to CO_2 aeration. *Bioresource Technology*, **100**, 833–838.

Chuntapa, B., Powtongsook, S., and Menasveta, P. (2003) Water quality control using *Spirulina platensis* in shrimp culture tanks. *Aquaculture (Amsterdam, Netherlands)*, **220**, 355–366.

Cox, P.A., Banack, S.A., Murch, S.J. *et al.* (2005) Diverse taxa of cyanobacteria produce (-N-methylamino-L-alanine, a neurotoxic amino acid. *Proceedings of the National Academy of Sciences of the United States of America*, **102**, 5074–5078.

Cysewski, G.R. and Lorenz, R.T. (2004) Industrial production of microalgal cell-mass and secondary products—species of high potential Haematococcus, in *Handbook of Microalgal Culture: Biotechnology and Applied Phycology* (ed. A. Richmond), Wiley-Blackwell, Oxford.

Del Campo, J., García-González, M., and Guerrero, M. (2007) Outdoor cultivation of microalgae for carotenoid production: Current state and perspectives. *Applied Microbiology and Biotechnology*, **74**, 1163–1174.

Demmig-Adams, B. and Adams, W.W. (2002) Antioxidants in photosynthesis and human nutrition. *Science*, **298**, 2149–2153.

Denery, J.R., Dragull, K., Tang, C.S., and Li, Q.X. (2004) Pressurized fluid extraction of carotenoids from *Haematococcus pluvialis* and *Dunaliella salina* and kavalactones from *Piper methysticum*. *Analytica Chimica Acta*, **501**, 175–181.

Djodjic, F., Bergström, L., and Grant, C. (2005) Phosphorus management in balanced agricultural systems. *Soil Use Manage*, **21**, 94–101.

Drew-Baker, K.M. (1949) Conchocelis-Phase in the Life-History of *Porphyra umbilicalis* (L.). *Nature*, **164**, 748–749.

Durairatnam, M. (1987) Studies of the yield of agar, gel strength and quality of agar of *Gracilaria edulis* (Gmel.) Silva from Brazil. *Hydrobiologia*, **151/152**, 509–512.

El-Sheekh, M.M. (1999) Stable transformation of the intact cells of *Chlorella kessleri* with high velocity microprojectiles. *Biologia Plantarum*, **42**, 209–216.

El-Sheekh, M.M. (2000) Stable chloroplast transformation *Chlamydomonas reinhardtii*; using microprojectile bombardment. *Folia Microbiologica*, **45**, 496–504.

Eriksen, N. (2008) The technology of microalgal culturing. *Biotechnology Letters*, **30**, 1525–1536.

Eriksen, N. (2008) Production of phycocyanin—a pigment with applications in biology, biotechnology, foods and medicine. *Applied Microbiology and Biotechnology*, **80**, 1–14.

Fábregas, J., Otero, A., Maseda, A., and Domínguez, A. (2001) Two-stage cultures for the production of astaxanthin from *Haematococcus pluvialis*. *Journal of Biotechnology*, **89**, 65–71.

Fajardo, A.R., Cerdán, L.E., Medina, A.R. *et al.* (2007) Lipid extraction from the microalga *Phaeodactylum tricornutum*. *European Journal of Lipid Science and Technology*, **109**, 120–126.

Falkowski, P.G. and Raven, J.A. (1997) *Aquatic Photosynthesis*, Blackwater Science, London.

Falshaw, R., Bixler, H.J., and Johndro, K. (2003) Structure and performance of commercial κ-2 carrageenan extracts. Part III. Structure analysis and performance in two dairy applications of extracts from the New Zealand red seaweed, Gigartina atropurpurea. *Food Hydrocolloids*, **17**, 129–139.

FAO (2008) *The State of World Fisheries and Aquaculture 2008*, Food and Agriculture Organization of the United Nations, Rome.

Farrar, W.V. (1966) Tecuitlatl: A glimpse of aztec food technology. *Nature*, **211**, 341–342.

Fazeli, M.R., Tofighi, H., Samadi, N., and Jamalifar, H. (2006) Effects of salinity on β-carotene production by *Dunaliella tertiolecta* DCCBC26 isolated from the Urmia salt lake, north of Iran. *Bioresource Technology*, **97**, 2453–2456.

Feng, S., Xue, L., Liu, H., and Lu, P. (2009) Improvement of efficiency of genetic transformation for *Dunaliella salina* by glass beads method. *Molecular Biology Reports*, **36**, 1433–1439.

Fenton, W.S., Hibbeln, J., and Knable, M. (2000) Essential fatty acids, lipid membrane abnormalities, and the diagnosis and treatment of schizophrenia. *Biological Psychiatry*, **47**, 8–21.

Fowles, M. (2007) Black carbon sequestration as an alternative to bioenergy. *Biomass Bioenergy*, **31**, 426–432.

García-González, M., Moreno, J., Manzano, J.C. *et al.* (2005) Production of *Dunaliella salina* biomass rich in 9-cis-β-carotene and lutein in a closed tubular photobioreactor. *Journal of Biotechnology*, **115**, 81–90.

García, F., Freile-Pelegrín, Y., and Robledo, D. (2007) Physiological characterization of *Dunaliella* sp. (Chlorophyta, Volvocales) from Yucatan, Mexico. *Bioresource Technology*, **98**, 1359–1365.

Graham, L.E., Graham, J.M., and Wilcox, L.W. (2009) *Algae*, 2nd edn, Pearson Education, Inc., California.

Griffiths, M. and Harrison, S. (2009) Lipid productivity as a key characteristic for choosing algal species for biodiesel production. *Journal of Applied Phycology*, **21**, 493–507.

Guerin, M., Huntley, M.E., and Olaizola, M. (2003) *Haematococcus* astaxanthin: applications for human health and nutrition. *Trends in Biotechnology*, **21**, 210–216.

Guschina, I.A. and Harwood, J.L. (2006) Lipids and lipid metabolism in eukaryotic algae. *Progress in Lipid Research*, **45**, 160–186.

Hagen, C., Siegmund, S., and Braune, W. (2002) Ultrastructural and chemical changes in the cell wall of *Haematococcus pluvialis* (Volvocales, Chlorophyta) during aplanospore formation. *European Journal of Phycology*, **37**, 217–226.

Harrington, G.W., Beach, D.H., Dunham, J.E., and HolzJJr., G.G. (1970) The polyunsaturated fatty acids of marine dinoflagellates. *The Journal of Protozoology*, **17**, 213–219.

He, M.L., Hollwich, W., and Rambeck, W.A. (2002) Supplementation of algae to the diet of pigs: a new possibility to improve the iodine content in the meat. *Journal of Animal Physiology and Animal Nutrition-Zeitschrift fur Tierphysiologie Tierernahrung und Futtermittelkunde*, **86** (3–4), 97–104.

He, P., Duncan, J., and Barber, J. (2007) Astaxanthin accumulation in the green alga *Haematococcus pluvialis*: Effects of cultivation parameters. *Journal of Integrative Plant Biology*, **49**, 447–451.

Hejazi, M.A. and Wijffels, R.H. (2003) Effect of light intensity on β-carotene production and extraction by *Dunaliella salina* in two-phase bioreactors. *Biomolecular Engineering*, **20**, 171–175.

Hejazi, M.A. and Wijffels, R.H. (2004) Milking of microalgae. *Trends in Biotechnology*, **22**, 189–194.

Herrero, M., Cifuentes, A., and Ibañez, E. (2006) Sub- and supercritical fluid extraction of functional ingredients from different natural sources: Plants, food by-products, algae and microalgae: A review. *Food Chemistry*, **98**, 136–148.

Herrero, M., Jaime, L., Martín-Álvarez, P.J. *et al.* (2006) Optimization of the extraction of antioxidants from *Dunaliella salina* microalga by pressurized liquids. *Journal of Agricultural and Food Chemistry*, **54**, 5597–5603.

Herrero, M., Simó, C., Ibáñez, E., and Cifuentes, A. (2005) Capillary electrophoresis-mass spectrometry of *Spirulina platensis* proteins obtained by pressurized liquid extraction. *Electrophoresis*, **26**, 4215–4224.

Hjorth, R. (1997) Expanded-bed adsorption in industrial bioprocessing: Recent developments. *Trends in Biotechnology*, **15**, 230–235.

Hosseini Tafreshi, A. and Shariati, M. (2009) *Dunaliella* biotechnology: Methods and applications. *Journal of Applied Microbiology*, **107**, 14–35.

Hu, C., Li, M., Li, J. *et al.* (2008) Variation of lipid and fatty acid compositions of the marine microalga *Pavlova viridis* (Prymnesiophyceae) under laboratory and outdoor culture conditions. *World Journal of Microbiology & Biotechnology*, **24**, 1209–1214.

Huntley, M. and Redalje, D. (2007) CO_2 mitigation and renewable oil from photosynthetic microbes: A new appraisal. *Mitigation and Adaptation Strategies for Global Change*, **12**, 573–608.

IEA (2007) IEA Bioenergy Task 42 - Countries Report Final, [Available at http://www.biorefinery.nl/fileadmin/biorefinery/docs/CountryReportsIEABioenergyTask42Final170809.pdf. Access date: 18th January 2011].

Israel, A., Gavrieli, J., Glazer, A., and Friedlander, M. (2005) Utilization of flue gas from a power plant for tank cultivation of the red seaweed *Gracilaria cornea*. *Aquaculture (Amsterdam, Netherlands)*, **249**, 311–316.

Issarapayup, K., Powtongsook, S., and Pavasant, P. (2009) Flat panel airlift photobioreactors for cultivation of vegetative cells of microalga *Haematococcus pluvialis*. *Journal of Biotechnology*, **142**, 227–232.

Jaime, L., Rodríguez-Meizoso, I., Cifuentes, A. *et al.* (2010) Pressurized liquids as an alternative process to antioxidant carotenoids' extraction from *Haematococcus pluvialis* microalgae. *LWT–Food Science and Technology*, **43**, 105–112.

Janssen, M., Tramper, J., Mur, L.R., and Wijffels, R.H. (2003) Enclosed outdoor photobioreactors: Light regime, photosynthetic efficiency, scale-up, and future prospects. *Biotechnology and Bioengineering*, **81**, 193–210.

Jiang, Y., Chen, F., and Liang, S.Z. (1999) Production potential of docosahexaenoic acid by the heterotrophic marine dinoflagellate *Crypthecodinium cohnii*. *Process Biochemistry*, **34**, 633–637.

Jiménez, C., Cossío, B.R., Labella, D., and Xavier Niell, F. (2003) The feasibility of industrial production of *Spirulina (Arthrospira)* in Southern Spain. *Aquaculture (Amsterdam, Netherlands)*, **217**, 179–190.

Jin, E. and Melis, A. (2003) Microalgal biotechnology: Carotenoid production by the green algae *Dunaliella salina*. *Biotechnology and Bioprocess Engineering*, **8**, 331–337.

Johnson, E.A. and An, G.H. (1991) Astaxanthin from microbial sources. *Critical Reviews in Biotechnology*, **11**, 297–326.

Johnson, E.T. and Schmidt-Dannert, C. (2008) Light-energy conversion in engineered microorganisms. *Trends in Biotechnology*, **26**, 682–689.

Kathiresan, S., Chandrashekar, A., Ravishankar, G.A., and Sarada, R. (2009) *Agrobacterium*-mediated transformation in the green alga *Haematococcus pluvialis* (Chlorophyceae volvocales). *Journal of Phycology*, **45**, 642–649.

Khan, S.A., Rashmi, Hussain, M.Z. *et al.* (2009) Prospects of biodiesel production from microalgae in India. *Renewable & Sustainable Energy Reviews*, **13** 2361–2372.

Kindle, K.L. (1998) Nuclear transformation: technology and applications, in *The Molecular Biology of Chloroplasts and Mitochondria in Chlamydomonas* (eds J.D. Rochaix, M. Goldschmidt-Clermont, and S. Merchant), Kluwer Academic Publishers, Dordrecht.

Kobayashi, M., Kakizono, T., and Nagai, S. (1991) Astaxanthin production by a green alga *Haematococcus pluvialis* accompanied with morphological changes in acetate media. *Journal of Fermentation and Bioengineering*, **71**, 335–339.

Kobayashi, M., Kurimura, Y., Kakizono, T. *et al.* (1997) Morphological changes in the life cycle of the green alga *Haematococcus pluvialis*. *Journal of Fermentation and Bioengineering*, **84**, 94–97.

Krichnavaruk, S., Shotipruk, A., Goto, M., and Pavasant, P. (2008) Supercritical carbon dioxide extraction of astaxanthin from *Haematococcus pluvialis* with vegetable oils as co-solvent. *Bioresource Technology*, **99**, 5556–5560.

Kumar, A., Ergas, S., Yuan, X. *et al.* (2010) Enhanced CO_2 fixation and biofuel production via microalgae: Recent developments and future directions. *Trends in Biotechnology*, **28**, 371–380.

Kumar, S.V., Misquitta, R.W., Reddy, V.S. *et al.* (2004) Genetic transformation of the green alga - *Chlamydomonas reinhardtii* by *Agrobacterium tumefaciens*. *Plant Science (Shannon, Ireland)*, **166**, 731–738.

Ladygin, V.G. (2004) Efficient transformation of mutant cells of *Chlamydomonas reinhardtii* by electroporation. *Process Biochemistry (Barking, London, England)*, **39**, 1685–1691.

Lardon, L., Hellias, A., Sialve, B. *et al.* (2009) Life-cycle assessment of biodiesel production from microalgae. *Environmental Science & Technology*, **43**, 6475–6481.

Lavens, P. and Sorgeloos, P. (1996) Manual on the production and use of live food for aquaculture in: FAO Fisheries Technical Paper No. 361, Food and Agriculture Organization of the United Nations, Rome, p. 296.

Lee, R.E. (1980) *Phycology*, Cambridge University Press, New York.

Lee, Y.K. and Zhang, D.H. (1999) Production of astaxanthin by Haematococcus, in *Chemicals from Microalgae* (ed. Z. Cohen), Taylor & Francis, London.

Lehmann, J. (2007) A handful of carbon. *Nature*, **447**, 143–144.

León-Bañares, R., González-Ballester, D., Galván, A., and Fernández, E. (2004) Transgenic microalgae as green cell-factories. *Trends in Biotechnology*, **22**, 45–52.

Lignell, A. and Pedersén, M. (1989) Agar composition as a function of morphology and growth rate. Studies on some morphological strains of Gracilaria secundata and Gracilaria verrucosa (Rhodophyta). *Botanica Marina*, **32**, 219–227.

Lim, G.B., Lee, S.Y., Lee, E.K. *et al.* (2002) Separation of astaxanthin from red yeast *Phaffia rhodozyma* by supercritical carbon dioxide extraction. *Biochemical Engineering Journal*, **11**, 181–187.

Long, S.P., Humphries, S., and Falkowski, P.G. (1994) Photoinhibition of photosynthesis in nature. *Annual Review of Plant Physiology and Plant Molecular Biology*, **45**, 633–662.

Lorenz, R.T. and Cysewski, G.R. (2000) Commercial potential for *Haematococcus* microalgae as a natural source of astaxanthin. *Trends in Biotechnology*, **18**, 160–167.

Lourenço, S.O., Barbarino, E., De-Paula, J.C. *et al.* (2002) Amino acid composition, protein content and calculation of nitrogen-to-protein conversion factors for 19 tropical seaweeds. *Phycological Research*, **50**, 233–241.

Lyons, H. (2000) National seaweed forum report, in *Marine and Natural Resources* (ed. E. Nic Dhonncha), Irish Marine Institute, Dublin.

Machmudah, S., Shotipruk, A., Goto, M. *et al.* (2006) Extraction of astaxanthin from *Haematococcus pluvialis* using supercritical CO_2 and ethanol as entrainer. *Industrial & Engineering Chemistry Research*, **45**, 3652–3657.

Macías-Sánchez, M.D., Serrano, C.M., Rodríguez, M.R., and Martínez de la Ossa, E. (2009) Kinetics of the supercritical fluid extraction of carotenoids from microalgae with CO_2 and ethanol as cosolvent. *Chemical Engineering Journal*, **150**, 104–113.

Mansour, M.P., Frampton, D.M.F., Nichols, P.D. *et al.* (2005) Lipid and fatty acid yield of nine stationary-phase microalgae: Applications and unusual C_{24}-C_{28} polyunsaturated fatty acids. *Journal of Applied Phycology*, **17**, 287–300.

Marinho-Soriano, E. and Bourret, E. (2003) Effects of season on the yield and quality of agar from *Gracilaria* species (Gracilariaceae, Rhodophyta). *Bioresource Technology*, **90**, 329–333.

Marinho-Soriano, E., Bourret, E., de Casabianca, M.L., and Maury, L. (1999) Agar from the reproductive and vegetative stages of *Gracilaria bursa-pastoris*. *Bioresource Technology*, **67**, 1–5.

Martinez, B., Viejo, R.M., Rico, J.M. *et al.* (2006) Open sea cultivation of *Palmaria palmata* (Rhodophyta) on the northern Spanish coast. *Aquaculture (Amsterdam, Netherlands)*, **254**, 376–387.

Mata, T.M., Martins, A.A., and Caetano, N.S. (2010) Microalgae for biodiesel production and other applications: A review. *Renewable & Sustainable Energy Reviews*, **14**, 217–232.

Matos, J., Costa, S., Rodrigues, A. *et al.* (2006) Experimental integrated aquaculture of fish and red seaweeds in Northern Portugal. *Aquaculture (Amsterdam, Netherlands)*, **252**, 31–42.

Mayfield, S.P. and Franklin, S.E. (2005) Expression of human antibodies in eukaryotic micro-algae. *Vaccine*, **23**, 1828–1832.

Mayfield, S.P. and Kindle, K.L. (1990) Stable nuclear transformation of *Chlamydomonas reinhardtii* by using a *C. reinhardtii* gene as the selectable marker. *Proceedings of the National Academy of Sciences of the United States of America*, **87**, 2087–2091.

Mayne, S.T., Janerich, D.T., Greenwald, P. *et al.* (1994) Dietary β-carotene and lung cancer risk in U.S. nonsmokers. *Journal of the National Cancer Institute*, **86**, 33–38.

McCann, K.B., Gomme, P.T., Wu, J., and Bertolini, J. (2008) Evaluation of expanded bed adsorption chromatography for extraction of prothrombin complex from Cohn Supernatant I. *Biologicals*, **36**, 227–233.

McHugh, D.J. (1991) Worldwide distribution of commercial sources of seaweeds including *Gelidium*. *Hydrobiologia*, **221**, 19–21.

McHugh, D.J. (2003) A Guide to the Seaweed Industry*in FAO Fisheries Technical Paper, N. 441*, Food and Agriculture Organization of the United Nations, Rome.

McHugh, D.J. (2002) Prospects for seaweed production in developing countries*in FAO Fisheries Circulars* Food and Agriculture Organization of the United Nations, Rome.

Mendes-Pinto, M.M., Raposo, M.F.J., Bowen, J. *et al.* (2001) Evaluation of different cell disruption processes on encysted cells of *Haematococcus pluvialis*: effects on astaxanthin recovery and implications for bio-availability. *Journal of Applied Phycology*, **13**, 19–24.

Mendes, R.L., Fernandes, H.L., Coelho, J. *et al.* (1995) Supercritical CO_2 extraction of carotenoids and other lipids from *Chlorella vulgaris*. *Food Chemistry*, **53**, 99–103.

Mendes, R.L., Nobre, B.P., Cardoso, M.T. *et al.* (2003) Supercritical carbon dioxide extraction of compounds with pharmaceutical importance from microalgae. *Inorganica Chimica Acta*, **356**, 328–334.

Meng, X., Yang, J., Xu, X. *et al.* (2009) Biodiesel production from oleaginous microorganisms. *Renewable Energy*, **34**, 1–5.

Miao, X. and Wu, Q. (2006) Biodiesel production from heterotrophic microalgal oil. *Bioresource Technology*, **97**, 841–846.

Michaud, D.S., Feskanich, D., Rimm, E.B. *et al.* (2000) Intake of specific carotenoids and risk of lung cancer in 2 prospective US cohorts. *The American Journal of Clinical Nutrition*, **72**, 990–997.

Middelberg, A.P.J. (1994) The release of intracellular bioproducts, in *Bioseparation and Bioprocessing: A Handbook* (ed. G. Subramanian), Wiley-VCH Verlag GrmbH, Weinheim.

Mogedas, B., Casal, C., Forján, E., and Vílchez, C. (2009) β-Carotene production enhancement by UV-A radiation in *Dunaliella bardawil* cultivated in laboratory reactors. *Journal of Bioscience and Bioengineering*, **108**, 47–51.

Mojaat, M., Pruvost, J., Foucault, A., and Legrand, J. (2008) Effect of organic carbon sources and Fe^{2+} ions on growth and β-carotene accumulation by *Dunaliella salina*. *Biochemical Engineering Journal*, **39**, 177–184.

Molina Grima, E., Fernandez, J., Acién Fernández, F.G., and Chisti, Y. (2001) Tubular photobioreactor design for algal cultures. *Journal of Biotechnology*, **92**, 113–131.

Molina Grima, E., Medina, A., Giménez, A., and González, M. (1996) Gram-scale purification of eicosapentaenoic acid (EPA, 20:5n-3) from wet *Phaeodactylum tricornutum* UTEX 640 biomass. *Journal of Applied Phycology*, **8**, 359–367.

Muhs, J., Viamajala, S., Heydon, B. *et al.* (2009) *A Summary of Opportunities, Challenges, and Research Needs - Algae Biofuels & Carbon Recycling*, Utah State University, Logan.

Muller-Feuga, A. (2000) The role of microalgae in aquaculture: situation and trends. *Journal of Applied Phycology*, **12**, 527–534.

Mussgnug, J.H., Thomas-Hall, S., Rupprecht, J. *et al.* (2007) Engineering photosynthetic light capture: Impacts on improved solar energy to biomass conversion. *Plant Biotechnology Journal*, **5**, 802–814.

Naik, P.S., Chanemougasoundharam, A., Paul Khurana, S.M., and Kalloo, G. (2003) Genetic manipulation of carotenoid pathway in higher plants. *Current Science*, **85**, 1423–1430.

Niu, J.F., Wang, G.C., Lin, X.Z., and Zhou, B.-C. (2007) Large-scale recovery of C-phycocyanin from *Spirulina platensis* using expanded bed adsorption chromatography. *Journal of Chromatography. B*, **850**, 267–276.

Niu, J.F., Wang, G.C., and Tseng, C.K. (2006) Method for large-scale isolation and purification of R-phycoerythrin from red alga *Polysiphonia urceolata* Grev. *Protein Expres Purif*, **49**, 23–31.

Olaizola, M. (2000) Commercial production of astaxanthin from *Haematococcus pluvialis* using 25,000-liter outdoor photobioreactors. *Journal of Applied Phycology*, **12**, 499–506.

Oliveira, E.G., Rosa, G.S., Moraes, M.A., and Pinto, L.A.A. (2008) Phycocyanin content of *Spirulina platensis* dried in spouted bed and thin layer. *Journal of Food Process Engineering*, **31**, 34–50.

Oliveira, E.G., Rosa, G.S., Moraes, M.A., and Pinto, L.A.A. (2009) Characterization of thin layer drying of *Spirulina platensis* utilizing perpendicular air flow. *Bioresource Technology*, **100**, 1297–1303.

Onofrejová, L., Vasícková, J., Klejdus, B. *et al.* (2010) Bioactive phenols in algae: The application of pressurized-liquid and solid-phase extraction techniques. *Journal of Pharmaceutical and Biomedical Analysis*, **51**, 464–470.

Patil, G. and Raghavarao, K.S. (2007) Aqueous two phase extraction for purification of C-phycocyanin. *Biochemical Engineering Journal*, **34**, 156–164.

Peet, M. (2004) Nutrition and schizophrenia: beyond omega-3 fatty acids, *Prostaglandins, Leukotrienes Essent. Fatty Acids*, **70**, 417–422.

Phadwal, K. and Singh, P.K. (2003) Effect of nutrient depletion on β-carotene and glycerol accumulation in two strains of *Dunaliella* sp. *Bioresource Technology*, **90**, 55–58.

Pitipanapong, J., Chitprasert, S., Goto, M. *et al.* (2007) New approach for extraction of charantin from *Momordica charantia* with pressurized liquid extraction, *Sep. Purification Technology*, **52**, 416–422.

Plaza, M., Herrero, M., Cifuentes, A., and Ibáñez, E. (2009) Innovative natural functional ingredients from microalgae. *Journal of Agricultural and Food Chemistry*, **57**, 7159–7170.

Polle, J.E.W., Benemann, J.R., Tanaka, A., and Melis, A. (2000) Photosynthetic apparatus organization and function in the wild type and a chlorophyll *b*-less mutant of *Chlamydomonas reinhardtii*. Dependence on carbon source. *Planta*, **211**, 335–344.

Potvin, G. and Zhang, Z. (2010) Strategies for high-level recombinant protein expression in transgenic microalgae: A review. *Biotechnology Advances*, **28**, 910–918.

Pulz, O. (2001) Photobioreactors: Production systems for phototrophic microorganisms. *Applied and Environmental Microbiology*, **57**, 287–293.

Pulz, O. and Gross, W. (2004) Valuable products from biotechnology of microalgae. *Applied Microbiology and Biotechnology*, **65**, 635–648.

Pulz, O. and Scheinbenbogan, K. (1998) Photobioreactors: Design and performance with respect to light energy input. *Advances in Biochemical Engineering/Biotechnology*, **59** 123–152.

Ratledge, C. (2004) Fatty acid biosynthesis in microorganisms being used for single cell oil production. *Biochimie*, **86**, 807–815.

Reijnders, L. (2009) Are forestation, bio-char and landfilled biomass adequate offsets for the climate effects of burning fossil fuels? *Energy Policy*, **37**, 2839–2841.

Reijnders, L. and Huijbregts, M.A.J. (2008) Biogenic greenhouse gas emissions linked to the life cycles of biodiesel derived from European rapeseed and Brazilian soybeans. *Journal of Cleaner Production*, **16**, 1943–1948.

Reitan, K.I., Rainuzzo, J.R., Øie, G., and Olsen, Y. (1997) A review of the nutritional effects of algae in marine fish larvae. *Aquaculture (Amsterdam, Netherlands)*, **155**, 207–221.

Richmond, A., Cheng-Wu, Z., and Zarmi, Y. (2003) Efficient use of strong light for high photosynthetic productivity: interrelationships between the optical path, the optimal population density and cell-growth inhibition. *Biomolecular Engineering*, **20**, 229–236.

Riekhof, W.R., Sears, B.B., and Benning, C. (2005) Annotation of genes involved in glycerolipid biosynthesis in *Chlamydomonas reinhardtii*: Discovery of the betaine lipid synthase BTA1Cr. *Eukaryotic Cell*, **4**, 242–252.

Riley, H. (2002) Effects of algal fibre and perlite on physical properties of various soils and on potato nutrition and quality on a gravelly loam soil in southern Norway. *Acta Agriculturae Scandinavica Section B-Soil and Plant Science*, **52**, 86–95.

Rodolfi, L., Zittelli, G.C., Barsanti, L. *et al.* (2003) Growth medium recycling in *Nannochloropsis* sp. mass cultivation. *Biomolecular Engineering*, **20**, 243–248.

Rodolfi, L., Zittelli, G.C., Bassi, N. *et al.* (2008) Microalgae for oil: Strain selection, induction of lipid synthesis and outdoor mass cultivation in a low-cost photobioreactor. *Biotechnology and Bioengineering*, **102**, 100–112.

Rorrer, G.L. and Cheney, D.P. (2004) Bioprocess engineering of cell and tissue cultures for marine seaweeds. *Aquacultural Engineering*, **32**, 11–41.

Rosenberg, J.N., Oyler, G.A., Wilkinson, L., and Betenbaugh, M.J. (2008) A green light for engineered algae: Redirecting metabolism to fuel a biotechnology revolution. *Current Opinion in Biotechnology*, **19**, 430–436.

Rubio-Rodríguez, N., Beltrán, S., Jaime, I. *et al.* (2010) Production of omega-3 polyunsaturated fatty acid concentrates: A review. *Innovative Food Science & Emerging Technologies*, **11**, 1–12.

Sahena, F., Zaidul, I.S.M., Jinap, S. *et al.* (2009) Application of supercritical CO_2 in lipid extraction - A review. *Journal of Food Engineering*, **95**, 240–253.

Sajilata, M.G., Singhal, R.S., and Kamat, M.Y. (2008) Supercritical CO_2 extraction of γ-linolenic acid (GLA) from *Spirulina platensis* ARM 740 using response surface methodology. *Journal of Food Engineering*, **84**, 321–326.

Sánchez Mirón, A., Contreras Gomez, A., Garca Camacho, F. *et al.* (1999) Comparative evaluation of compact photobioreactors for large-scale monoculture of microalgae. *Journal of Biotechnology*, **70**, 249–270.

Santoyo, S., Rodríguez-Meizoso, I., Cifuentes, A. *et al.* (2009) Green processes based on the extraction with pressurized fluids to obtain potent antimicrobials from *Haematococcus pluvialis* microalgae. *LWT–Food Science and Technology*, **42**, 1213–1218.

Sarada, R., Vidhyavathi, R., Usha, D., and Ravishankar, G.A. (2006) An efficient method for extraction of astaxanthin from green alga *Haematococcus pluvialis*. *Journal of Agricultural and Food Chemistry*, **54**, 7585–7588.

Schaeffer, D.J. and Krylov, V.S. (2000) Anti-HIV activity of extracts and compounds from algae and cyanobacteria. *Ecotoxicology and Environmental Safety*, **45**, 208–227.

Scott, S.A., Davey, M.P., Dennis, J.S. *et al.* (2010) Biodiesel from algae: Challenges and prospects. *Current Opinion in Biotechnology*, **21**, 277–286.

Setlik, I., Veladimir, S., and Malek, I. (1970) Dual purpose open circulation units for large scale culture of algae in temperate zones. I. Basic design considerations and scheme of a pilot plant. *Algol Stud (Trebon)*, **1**, 111–164

Shaish, A., Harari, A., Hananshvili, L. *et al.* (2006) 9-cis β-carotene-rich powder of the alga *Dunaliella bardawil* increases plasma HDL-cholesterol in fibrate-treated patients. *Atherosclerosis*, **189**, 215–221.

Shang, Y.F., Kim, S.M., Lee, W.J., and Um, B.H. (2011) Pressurized liquid method for fucoxanthin extraction from *Eisenia bicyclis* (Kjellman) Setchell. *Journal of Bioscience and Bioengineering*, **111**, 237–241.

Sialve, B., Bernet, N., and Bernard, O. (2009) Anaerobic digestion of microalgae as a necessary step to make microalgal biodiesel sustainable. *Biotechnology Advances*, **27**, 409–416.

Sidhu, K.S. (2003) Health benefits and potential risks related to consumption of fish or fish oil. *Regulatory Toxicology and Pharmacology*, **38**, 336–344.

Soni, B., Kalavadia, B., Trivedi, U., and Madamwar, D. (2006) Extraction, purification and characterization of phycocyanin from *Oscillatoria quadripunctulata* - Isolated from the rocky shores of Bet-Dwarka, Gujarat, India. *Process Biochemistry (Barking, London, England)*, **41**, 2017–2023.

Spoehr, H.A. and Milner, H.A. (1949) The chemical composition of *Chlorella*: Effect of environmental conditions. *Plant Physiology*, **24**, 120–149.

Spolaore, P., Joannis-Cassan, C., Duran, E., and Isambert, A. (2006) Commercial applications of microalgae. *Journal of Bioscience and Bioengineering*, **101**, 87–96.

Subhadra, B.G. (2010) Sustainability of algal biofuel production using integrated renewable energy park (IREP) and algal biorefinery approach. *Energy Policy*, **38**, 5892–5901.

Suh, I.S. and Lee, C.G. (2003) Photobioreactor engineering: Design and performance. *Biotechnology and Bioprocess Engineering*, **8**, 313–321.

Sun, G., Zhang, X., Sui, Z., and Mao, Y. (2008) Inhibition of *pds* gene expression via the RNA interference approach in *Dunaliella salina* (Chlorophyta). *Journal of Marine Biotechnology*, **10**, 219–226.

Suresh Kumar, G., Deepa, T., Sushma, S. *et al.* (2003) Lycopene attenuates oxidative stress induced experimental cataract development: An *in vitro* and *in vivo* study. *Nutrition*, **19**, 794–799.

Tan, H.H. (2008) Algae-to-biodiesel at least five to 10 years away *Energy Current: News for the Business of Energy,* Singapore, [Available at http://www.energycurrent.com/index.php?id=3&storyid=14415 Access date: 12th January 2009].

Tanvetyanon, T. and Bepler, G. (2008) β-carotene in multivitamins and the possible risk of lung cancer among smokers versus former smokers: A meta-analysis and evaluation of national brands. *Cancer*, **113**, 150–157.

Taylor, G. (2008) Biofuels and the biorefinery concept. *Energy Policy*, **36**, 4406–4409.

Teng, C., Qin, S., Liu, J. *et al.* (2002) Transient expression of *lacZ* in bombarded unicellular green alga *Haematococcus pluvialis*. *Journal of Applied Phycology*, **14**, 497–500.

Terry, K.L. and Raymond, L.P. (1985) System design for the autotrophic production of microalgae. *Enzyme and Microbial Technology*, **7**, 474–487.

Thompson, G.A. (1996) Lipids and membrane function in green algae. *Biochim Biophys Acta, Lipids Lipid Metab*, **1302**, 17–45.

Tisdale, M.J. (1999) Wasting in cancer. *The Journal of Nutrition*, **129**, 243S–246.

US DOE (2010) National Algal Biofuels Technology Roadmap, US Department of Energy, Office of Energy Efficiency and Renewable Energy, Biomass Program, Maryland.

Ugwu, C.U., Aoyagi, H., and Uchiyama, H. (2008) Photobioreactors for mass cultivation of algae. *Bioresource Technology*, **99**, 4021–4028.

Ugwu, C.U., Ogbonna, J., and Tanaka, H. (2002) Improvement of mass transfer characteristics and productivities of inclined tubular photobioreactors by installation of internal static mixers. *Applied Microbiology and Biotechnology*, **58**, 600–607.

van de Velde, F., Antipova, A.S., Rollema, H.S. *et al.* (2005) The structure of κ/ι-hybrid carrageenans II. Coil-helix transition as a function of chain composition. *Carbohydrate Research*, **340**, 1113–1129.

van de Velde, F., Pereira, L., and Rollema, H.S. (2004) The revised NMR chemical shift data of carrageenans. *Carbohydrate Research*, **339**, 2309–2313.

van Poppel, G. (1993) Carotenoids and cancer: An update with emphasis on human intervention studies. *European Journal of Cancer (Oxford, England: 1990)*, **29**, 1335–1344.

Vauchel, P., Leroux, K., Kaas, R. *et al.* (2009) Kinetics modeling of alginate alkaline extraction from *Laminaria digitata*. *Bioresource Technology*, **100**, 1291–1296.

Villanueva, R.D., Montaño, N.E., Romero, J.B. *et al.* (1999) Seasonal variations in the yield, gelling Properties, and chemical composition of agars from *Gracilaria eucheumoides* and *Gelidiella acerosa* (Rhodophyta) from the Philippines. *Botanica Marina*, **42**, 175–182.

Viskari, P.J. and Colyer, C.L. (2003) Rapid extraction of phycobiliproteins from cultured cyanobacteria samples. *Analytical Biochemistry*, **319**, 263–271.

Waldenstedt, L., Inborr, J., Hansson, I., and Elwinger, K. (2003) Effects of astaxanthin-rich algal meal (*Haematococcus pluvalis*) on growth performance, caecal campylobacter and clostridial counts and tissue astaxanthin concentration of broiler chickens. *Animal Feed Science and Technology*, **108**, 119–132.

Walker, T., Purton, S., Becker, D., and Collet, C. (2005) Microalgae as bioreactors. *Plant Cell Reports*, **24**, 629–641.

Wang, B., Li, Y., Wu, N., and Lan, C. (2008) CO_2 bio-mitigation using microalgae. *Applied Microbiology and Biotechnology*, **79**, 707–718.

Wang, C., Wang, Y., Su, Q., and Gao, X. (2007) Transient expression of the GUS gene in a unicellular marine green alga, *Chlorella* sp. MACC/C95, via electroporation. *Biotechnology and Bioprocess Engineering*, **12**, 180–183.

Wang, T., Xue, L., Hou, W. *et al.* (2007) Increased expression of transgene in stably transformed cells of *Dunaliella salina* by matrix attachment regions. *Applied Microbiology and Biotechnology*, **76**, 651–657.

Ward, O.P. and Singh, A. (2005) Omega-3/6 fatty acids: Alternative sources of production. *Process Biochemistry (Barking, London, England)*, **40**, 3627–3652.

Watanabe, Y. and Saiki, H. (1997) Development of a photobioreactor incorporating *Chlorella* sp. for removal of CO_2 in stack gas. *Energy Convers Manage*, **38**, S499–S503.

Weldy, C.S. and Huesemann, M. (2007) Lipid production by *Dunaliella salina* in batch culture: Effects of nitrogen limitation and light intensity. *U.S. Department of Energy Journal of Undergraduate Research*, **7**, 115–122.

Wellisch, M., Jungmeier, G., Karbowski, A. *et al.* (2010) Biorefinery systems – potential contributors to sustainable innovation. *Biofpr*, **4**, 275–286.

Wen, Z.-Y. and Chen, F. (2003) Heterotrophic production of eicosapentaenoic acid by microalgae. *Biotechnology Advances*, **21**, 273–294.

Wen, Z.Y. and Chen, F. (2001) A perfusion-cell bleeding culture strategy for enhancing the productivity of eicosapentaenoic acid by *Nitzschia laevis*. *Applied Microbiology and Biotechnology*, **57**, 316–322.

Wheeler, J.N. (1987) Oceanic farming of macrocystis: The problems and non-problems, in *Seaweed Cultivation for Renewable Resources (Developments in Aquaculture and Fisheries Science)* (eds K.T. Bird and P.H. Benson), Elsevier Science Ltd., Amsterdam.

Widjaja, A., Chien, C.C., and Ju, Y.H. (2009) Study of increasing lipid production from fresh water microalgae *Chlorella vulgaris*. *Journal of the Taiwan Institute of Chemical Engineers*, **40**, 13–20.

Wijffels, R.H. and Barbosa, M.J. (2010) An Outlook on Microalgal Biofuels. *Science*, **329**, 796–799.

Wijffels, R.H., Barbosa, M.J., and Eppink, M.H.M. (2010) Microalgae for the production of bulk chemicals and biofuels. *Biofpr*, **4**, 287–295.

Wu, W.T. and Hsieh, C.H. (2008) Cultivation of microalgae for optimal oil production. *Journal of Biotechnology*, **136**, S521–S521.

Yang, Y.F., Fei, X.G., Song, J.M. *et al.* (2006) Growth of *Gracilaria lemaneiformis* under different cultivation conditions and its effects on nutrient removal in Chinese coastal waters. *Aquaculture (Amsterdam, Netherlands)*, **254**, 248–255.

Yaphe, W. (1984) Properties of *Gracilaria* agars. *Hydrobiologia*, **116/117**, 171–186.

Ye, Z.W., Jiang, J.G., and Wu, G.H. (2008) Biosynthesis and regulation of carotenoids in *Dunaliella*: Progresses and prospects. *Biotechnology Advances*, **26**, 352–360.

Zaslavskaia, L.A., Lippmeier, J.C., Kroth, P.G. *et al.* (2000) Transformation of the diatom *Phaeodactylum tricornutum* (Bacillariophyceae) with a variety of selectable marker and reporter genes. *Journal of Phycology*, **36**, 379–386.

Zhang, B.Y., Geng, Y.H., Li, Z.K. *et al.* (2009) Production of astaxanthin from *Haematococcus* in open pond by two-stage growth one-step process. *Aquaculture (Amsterdam, Netherlands)*, **295**, 275–281.

Zhou, G., Sun, Y., Xin, H. *et al.* (2004) In vivo antitumor and immunomodulation activities of different molecular weight λ-carrageenans from *Chondrus ocellatus*. *Pharmacological Research*, **50**, 47–53.

Zhou, G., Xin, H., Sheng, W. *et al.* (2005) In vivo growth-inhibition of S180 tumor by mixture of 5-Fu and low molecular λ-carrageenan from *Chondrus ocellatus*. *Pharmacological Research*, **51**, 153–157.

Zilinskas Braun, G. and Zilinskas Braun, B. (1974) Light absorption, emission and photosynthesis, in *Algal Physiology and Biochemistry* (ed. W.D.P. Stewart), Blackwell Scientific Publications, Oxford.

11

New Bioactive Natural Products from Canadian Boreal Forest

François Simard, André Pichette and Jean Legault

*Laboratoire d'Analyse et de Séparation des Essences Végétales (LASEVE),
Département des Sciences Fondamentales, Université du Québec à Chicoutimi,
Chicoutimi, Québec, Canada*

11.1   Introduction

From an economical point of view, the Canadian forest represents one of the most important natural resources of this country. In fact, the forest sector is one of Canada's largest employers, providing about 340 000 jobs [1] mainly in rural communities and all across the country. Since 2002, the Canadian forest sector has been experiencing a great deal of difficulty maintaining its profitability. These problems can mostly be explained by the countervailing duties imposed by the United States on the Canadian softwood industry, but also by the reduction in cutting rights and the growing competition in world markets [1–3]. This situation has led to the closure of several wood processing plants and job losses in the last few years. To overcome these difficulties, the forest sector has been forced to review its resource exploitation. In an effort to diversify the production of this sector, the Forest Product Association of Canada (FPAC) has identified different bio products (biofuels, biomaterials and biochemicals) that can be produced from wood fibre and that might strengthen this industry and boost employment in the future [2]. Producing biofuel and biochemicals from forest biomass would also help decrease Canada's dependence on fossil fuels, together with decreasing greenhouse gas production. The FPAC have proposed to integrate the production of these bioproducts at existing wood processing plants [4,5] by modifying them into better-tailored biorefineries.

Biorefinery Co-Products: Phytochemicals, Primary Metabolites and Value-Added Biomass Processing, First Edition.
Edited by Chantal Bergeron, Danielle Julie Carrier and Shri Ramaswamy.
© 2012 John Wiley & Sons, Ltd. Published 2012 by John Wiley & Sons, Ltd.

Being an important supplier of forest biomass [6] and possessing important infrastructures, Canada is in an ideal situation for the development of forest-based biorefineries [6].

A biorefinery is similar to a petroleum refinery; the major difference is that biomass is used instead of fossil fuels as a feedstock for biofuels, biochemicals and biomaterials production. Forest-based biorefineries must be able to process different biomass feedstocks, such as logging and pulping residues. The forest biomass is mainly composed of cellulose, hemicellulose and lignin that complicate its conversion into bioproducts. In fact, cellulose and hemicellulose conversion into free sugars used in the production of most biofuels and biochemicals is still a process in development. Moreover, cheaper solutions already exist for biofuel, such as ethanol production from corn [7–9]. Biofuels and biochemicals are very attractive products. Unfortunately, the high cost of lignocellulosic biomass processing, combined with relatively low fossil fuels costs, makes their production from forest biomass still unprofitable. To increase profitability, it has been proposed to include the production of high-value/low-volume co-products into biorefineries [8]. The development of such high-value co-products from forest biomass is therefore essential for the implantation of forest-based biorefineries in Canada.

Bioactive natural products represent potentially one of the most interesting high-value co-products and their extraction could become the initial processing step for future forest-based biorefineries [10]. Natural products are compounds biosynthesized from the metabolism of living organism. These products can be readily extracted with organic solvents or water under room temperature and atmospheric conditions [7]. Therefore, natural products are often referred to as "extractables" or "extractable bioproducts". In plant organisms, many natural products are known to play a protective role against pathogens, other plants, animals and UV radiation [11]. For several centuries, Native people have been taking advantage of these defensive properties by using plants as medicines [12]. In fact, natural products are known to possess a wide range of beneficial effects on human health [13] and the discovery of many modern drugs results from the identification of bioactive natural products [14]. Thanks to these discoveries, many bioactive natural products or plant extracts have become feedstock for the pharmaceutical industry [15,16]. One of the most well-known and commercially successful natural product is paclitaxel. Paclitaxel was first isolated in 1969 by the US National Cancer Institute (NCI) from the bark of *Taxus brevifolia* [17], a coniferous tree growing on the pacific coast of North America [18]. It was found to inhibit cancer cell division thanks to a unique mechanism involving the cell's microtubules [17]. Taxol®, the commercial drug containing paclitaxel, became one of the most popular anticancer drug worldwide, with sales reaching $1,5 billion in 1999, and today's total market remains over $1 billion per year [19]. During the development of paclitaxel as a commercial drug, its availability rapidly became a problem. Wild-harvested *T. brevifolia* biomass was not sufficient to provide the quantities needed. To overcome this hurdle, a semisynthesis of paclitaxel using baccatin III as a starting point has been developed [20]. Baccatin III can be isolated in significant yield from the foliage of *Taxus baccata,* a European species. In Canada, a few companies (Chatam Biotec Ltd., Biolyse Pharma and Active Botanicals Co. Ltd.) are producing paclitaxel from *Taxus canadensis* [21], a slow-growth shrub found throughout the Atlantic and Quebec provinces of Canada [16]. The paclitaxel example illustrates the great potential of a bioactive natural product to evolve into a high-value pharmaceutical bioproduct. However, the commercial use of bioactive natural products is not limited to the pharmaceutical industry. Plant extracts and isolated natural products are also used in manufacturing of cosmeceuticals [22–26] and nutraceuticals [27,28]. There are also potential markets to be developed, such as plant extractables that can be used as pesticides [29,30], antiparasitic agents for animal husbandry [31–33] and food preservatives [34,35]. Most of these uses require relatively small volumes of bioactive natural products, but would yield a larger income than biofuels or commodity chemicals extracted from the biomass. Consequently, bioactive natural products represent an excellent opportunity to facilitate the implantation of forest-based biorefineries in Canada.

Canada's First Nations have acquired a rich knowledge of how to use boreal forest plant species as food, medicine and materials [36–38]. These ethno-pharmacological skills clearly highlight the potential of developing new bioactive natural products from the boreal forest of Canada. However, these skills remain largely empirical and, combined with the fact that scientific studies examining their traditional uses are scarce, the development of high-value co-products derived from a biorefinery setting is somewhat limited. Over the last 10 years, our research group has undertaken the task of gathering information about the chemical composition and the bioactivity of plant species from the Canadian boreal forest. The work was focused on the identification of new bioactive natural products and the development of these natural products into bioproducts. In this chapter, the work carried out in our laboratory will be presented in two main reviews: (i) bioactive compounds isolated from widespread and commercial tree species currently used by the forest industry sector. These species, namely *Betula papyrifera*, *Larix laricina*, *Populus tremuloides* and *Abies balsamea*, are potential feedstock for the future Canadian forest-based biorefineries and their bioactive extractables are potential high-value co-products; (ii) synthesis of derivatives from bioactive natural products, which is an attractive step in the development of bioactive natural products into biopharmaceuticals.

11.2 Identification of New Bioactive Natural Products from Canadian Boreal Forest

11.2.1 Selection of Plant Species and Bio Guided Isolation Process

The first step to discovering new bioactive compounds from the Canadian boreal forest consists in the selection of species suspected of containing bioactive natural products. In addition to chemo-taxonomic knowledge (data available on closely related species), ethno-pharmacological knowledge is a rich source of information for plant species screening. Reports on the folk medicine developed by the First Nations from the North American boreal forest territory are available [36–41] and are remarkable tools for plant selection for screening of their different bioactivities. The next step is the evaluation of the extract from the selected plant species for bioactivity. Plant extracts are complex mixtures of highly diversified compounds which generally are not good candidates for modern high-throughput screening assays [42]. For example, tannin complexes with numerous proteins can lead to false positives in purified protein bioassays [43]. Development of bioassays applicable to plant extracts is therefore critical to assess bioactivity. It must be pointed out that in our research group *in vitro* cell-based bioassays have been efficient for the discovery of new antibacterial [44] and cytotoxic compounds [45–50].

Once the bioassay is well established, bioguided isolation is a straightforward and very efficient strategy for the identification of the compounds responsible for the plant extract bioactivity [51]. During this process, the plant extract is roughly fractionated in a limited number of fractions using simple and cheap separation methods, such as liquid–liquid extraction or column chromatography. The obtained fractions are assayed using the biological target, to identify the fraction that maintains the bioactivity detected earlier in the extract.

Depending on the complexity of the active fraction, a second fractionation can be performed before isolation. The latter is carried out using fine chromatographic techniques such as high-performance liquid chromatography (HPLC) or high-speed counter-current chromatography (HSCCC). Finally, isolated compound structures are characterized using spectroscopic methods (IR, NMR and MS spectroscopy) and their bioactivity is assayed. Recent progress in the analytical tools available for the analysis of plant extracts, as well as chromatographic fractions, has facilitated the identification of ubiquitous and known compounds early in the fractionation process, thus significantly accelerating the bioguided isolation process. This progress is reviewed elsewhere [52,53]. In this section, a review will be presented concerning the screening work that has been done over the last 10 years in our research group on Canadian boreal forest commercial species to discover new bioactive natural products.

11.2.2   Diarylheptanoids from the Inner Bark *Betula Papyrifera*

Betula papyrifera Marsh. is widely distributed all across Canada following the northern limit of tree growth from Newfoundland to British Columbia [18]. Over the years this plant has gained huge industrial importance in Canada, mainly for the production of paper pulp. This production yields large quantities of waste bark which is currently used as a cheap fuel at paper mills to save energy [54]. An attractive avenue to add value to this waste is its use as a component of growing media [55] or for particle board production [56]. Combined with these new applications, the extraction of bioactive natural products as high-value co-products could significantly enhance the profitability of paper mills, making them efficient biorefineries. *B. papyrifera* bark is divided into inner bark and outer bark. Lupane triterpene extractibles from the outer bark have been extensively studied for their use in cancer and HIV treatments. Recent developments in the chemistry and the bioactivity of these extractibles have been reviewed [54]. On the other hand, hardly any scientific information is available on the chemical composition and the bioactivity of the extractibles from *B. papyrifera* inner bark. Studies on the inner bark of other species of the *Betula* genus revealed that the main extractibles are phenolics, mainly arylbutanoids, diaryl-heptanoids, lignans and phenolic glycoside [57–61]. This information, combined with the fact that inner bark biomass is more abundant than the outer bark prompted us to investigate inner bark extractibles of *B. papyrifera* to identify new bioactive natural products.

The investigation was performed as follows: the inner bark was first extracted using a hydro-alcoholic mixture and the concentrated aqueous residue was submitted to liquid–liquid extraction using ethyl acetate (EtOAc). The dried EtOAc phase was assayed by a multi-well *in vitro* cell culture method using human lung carcinoma (A549), human colorectal adenocarcinoma (DLD-1) and human normal skin fibroblast (WS1). The EtOAC phase was found to exert significant cytotoxic activity against A549 and DLD-1 so bioguided fractionation was initiated. The fractionation steps were executed first by using an open Diaion® column, followed by silica gel column chromatography. From the second fractionation step, three fractions exhibited interesting *in vitro* cytotoxic activities against the cancer cell lines. These fractions were purified using silica gel column chromatography, polyamide flash column chromatogaphy and preparative HPLC to afford 10 phenolic compounds of the following types: diarylheptanoid glycosides (**1–4**), a diarylheptanoid (**5**), a lignan (**6**), flavonoids (**7–8**) and chavicol glycosides (**9–10**) [47], see Figure 11.1. Amongst them, papyriferoside A (**1**) had never been isolated before, while the other nine products (**2–10**) had never been reported from *B. papyrifera*. The *in vitro* cytotoxic activities of all the isolated compounds were assessed and platyphylloside (**4**) exhibited the strongest cytotoxicity with IC_{50} values

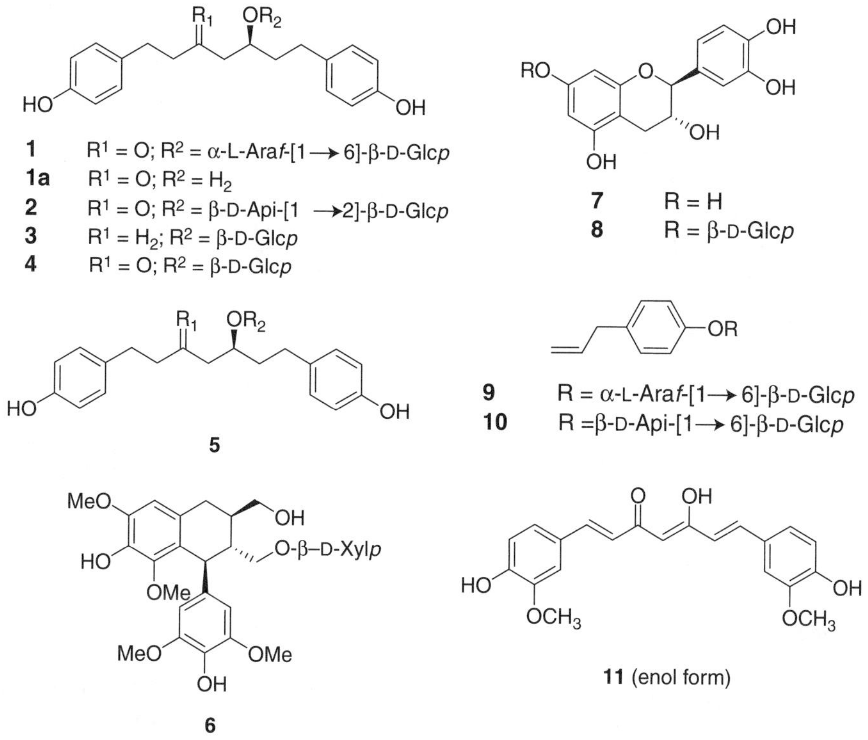

Figure 11.1   *Phenolics compounds from B.* papyrifera *inner bark and curcumin.*

of 13.8 ± 0.5 and 12.6 ± 0.3 µM against A549 and DLD-1 cells, respectively. The new compound papyriferoside A (**1**) also showed significant cytotoxicity, with IC_{50} values of 30 ± 2 and 20.8 ± 0.8 µM against A549 and DLD-1 cells, respectively.

Several studies have demonstrated the cytotoxic properties of diarylheptanoids [62–65] such as curcumin (**11**), a well-known diarylheptanoid from the Indian spice turmeric (*Curcuma longa*). Curcumin (**11**) has been extensively studied for its use in the treatment of different cancers [66] and other inflammatory-related diseases [67], and has now reached clinical II trials [68]. These results suggest that diarylheptanoids from *B. papyrifera* inner bark might be good candidates for evaluation for cancer chemotherapy. Since platyphylloside (**4**) can be isolated from *B. papyrifera* inner bark with a good yield [47], it could also represent an excellent starting point for the semi-synthesis of new anticancer diarylheptanoid derivatives. For all these reasons, birch waste bark is a promising source of anticancer diarylheptanoids for development as pharmaceutical bioproducts.

11.2.3   Labdane Diterpenes from Larix Laricina

Larix laricina is another industrially important tree species widely used for paper production [18]. Since the bark from *L. laricina* had never been evaluated for its cytotoxicity before, the potential of this biomass was assessed in our laboratory for anticancer agents. A bark extract obtained with $CHCl_3$ as the extraction solvent showed significant *in vitro* cytotoxic activity against two human cancer cell lines (A549 and DLD-1) [69]. Fractionation of this

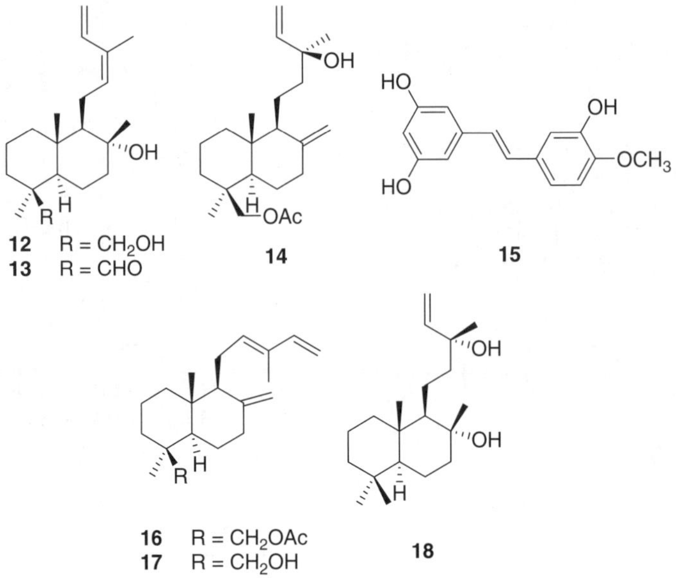

Figure 11.2 *Compounds isolated from* L. laricina *bark and other bioactive labdane diterpenes.*

extract using silica gel column chromatography, followed by preparative HPLC purification of active fractions led to the isolation of two new labdane diterpenes (**12** and **13**, Figure 11.2), a known labdane diterpene (**14**) and rhapontigenin (**15**). The labdane diterpenes **13** and **14** showed moderate cytotoxicity against the two cancer cell lines (A549 and DLD-1) but also against the normal skin fibroblast cell line WS1 with IC_{50} values ranging from 32 to 69 μM. Compound **12** showed an interesting selectivity by being cytotoxic to the DLD-1 cancer cell line ($IC_{50} = 37 \pm 3$ μM), while being inactive against the A549 cancer cell line and the WS1 normal cell line ($IC_{50} > 200$ μM) [69]. Recently, structurally related labdane diterpenes **16** and **17** isolated from the *Pinus massoniana* resin showed strong cytotoxicity, with IC_{50} values ranging from 0.38 to 19.62 μM against A431 and A549 cancer cell lines [70]. Sclareol (**18**), another related labdane diterpene has been studied for its effect on human leukaemia and breast cancer cells [71] growth, as well as its ability to decrease *in vivo* tumour cell growth rates [72]. These results show the potential of labdane diterpenes for their use in cancer chemotherapy. The high yield obtained in our studies for the isolation of the labdane diterpene **12** from *L. laricina* bark [69] could contribute to making this widespread species a good raw material for the production of bioactive labdane diterpenes from biomass.

11.2.4 Phenolic Compounds from *Populus Tremuloïdes* Buds

Poplar trees are well known for their phenolic compounds content which includes flavonoids [73,74], hydroxycinnamic acid derivatives [75–77] and phenolic glycosides [78,79]. Interest in the chemical composition of the extractibles from the genus *Populus* is, in part, due to thousands of years old historical use of the resinous hive product propolis. In fact, propolis from North America, Asia and Europe is produced from exudate collected by honeybees from poplar tree buds [80]. Propolis is known to possess a broad spectrum of biological activities which include antitumor and antimicrobial activities [80,81]. Its chemical composition is determined by the tree species from which the bees collected the exudate [82]. It has been demonstrated that one of the main sources of exudate from the boreal region of Canada is *P. tremuloïdes* [83]. Since *P. tremuloïdes* is also one of the most widespread trees in North America [18], a bud extract was included in our screening work to identify new bioactive natural products.

Buds were extracted using an EtOH–H$_2$O mixture and the aqueous concentrated residue was submitted to liquid–liquid extraction using hexane and H$_2$O saturated *n*-butanol. The *n*-butanol phase was dried and submitted to an open Diaion® column followed by further fractionation using an open silica gel column. Fractions were investigated for their *in vitro* cytotoxic activity (against A549, DLD-1 and WS1 cell lines) and antibacterial activity (against *Staphylococcus aureus* and *Escherichia coli*). One fraction exhibited weak cytotoxicity against the three cell lines and was purified using semi-preparative HPLC to afford 10 phenolic compounds [49]. Amongst these, the two related hydroxycinnamic acids **19** and **20** (Figure 11.3) were moderately cytotoxic (IC$_{50}$ values ranging from 19 to 51 μM) against the cancer cell lines used. An earlier report showed that hydroxycinnamic acid **20** along with 20 other synthetic derivatives were strongly cytotoxic against the highly liver-metastatic murine colon 26-L5 carcinoma cell line [84]. Additionally, other simple hydroxycinnamic acid derivatives such as artepillin C (**21**), baccharin (**22**) and drupanin (**23**) have been shown to

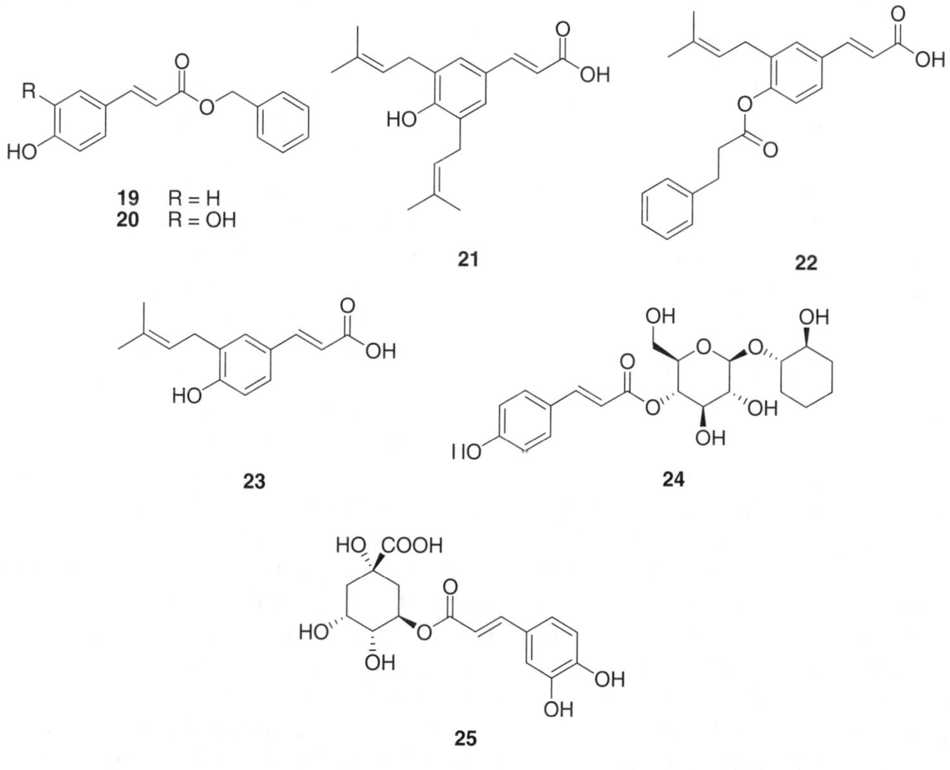

Figure 11.3   *Hydroxycinnamic acid derivatives from* Populus *spp and propolis.*

induce apoptosis in SW480 and HL60 cancer cell lines [85]. Overall, these results suggest that hydroxycinnamic acid derivatives are potential bioproducts for cancer treatment.

The new hydroxycinnamic acid derivative obtained from this work, compound **24**, did not exhibit good abilities for cancer chemotherapy. However, it will be interesting to evaluate its potential as a dietary supplement. In fact, quinic acid ester and the glucose ester of hydroxycinnamic acid structurally related to compound **24** are well-known dietary phenolic compounds found in many fruits and vegetables [86]. Moreover, chlorogenic acid (**25**), the most common ester of hydroxycinnamic acid possesses antioxidant properties [87,88] that have been correlated with the prevention of different cancers [79–81], as well as cardiovascular diseases [87,89]. Evaluation of the antioxidant activity of compound **24** is currently progressing in our laboratory to assess its potential as a dietary supplement from the buds of *P. tremuloïdes*.

11.2.5   Sesquiterpenes from *Abies Balsamea*

Essential oils are already well-established bioproducts used in perfumes and make-up products, in sanitary products, in dentistry, in agriculture, and as food preservatives and additives, as well as natural remedies [90]. Most of these applications are based on bactericidal, virucidal and fungicidal properties, as well as on essential oil fragrances [90]. *A. balsamea* foliage produces an important essential oil [91] mainly employed in the perfumery industry, but little is known about its biological properties. Since the chemical composition of this essential oil is well known [35,85,86] our investigations were focused on the cytotoxicity and antibacterial activities of both the oil and its components [35,37] to characterize useful bioactive compounds.

Using several solid tumour cell lines. including MCF-7, A549, DLD-1, M4BEU and CT-26, *A. balsamea* essential oil showed significant cytotoxic activity with GI_{50} values ranging from 0.76 mg ml^{-1} to 1.7 mg ml^{-1}. GC-MS analysis revealed that this product was constituted of about 96% monoterpenes along with some sesquiterpenes already described in the literature [92,93]. From the 17 compounds identified in this analysis, only the sesquiterpene, α-humulene (**26**; Figure 11.4) was found to be cytotoxic against human breast adenocarcinoma cells MCF-7 with a GI_{50} value of 73 ± 2 μM. α-Humulene (**26**) was also tested on the other solid tumour cell lines (A549, DLD-1, M4BEU and CT-26) and was found to be active against all of them, with GI_{50} values ranging from 53 to 73 μM. Further investigations showed that α-humulene (**26**) and *A. balsamea* essential oil induce intracellular glutathione (GSH) depletion and increased ROS production that might be responsible for their cytotoxicity [46]. Surprisingly, the concentration of α-humulene (**26**) was not adequate to explain the cytotoxicity of *A. balsamea* oil. In fact, another compound present in the oil might be acting in synergy with α-humulene (**26**) to promote its cytotoxic activity. β-Caryophyllene (**27**) appeared to be a good candidate since it was reported that it promoted 5-fluorouracil absorption across human skin [94]. Promotion of skin absorption suggests that β-caryophyllene (**27**) could increase intracellular accumulation of anticancer agents, thereby potentiating their cytotoxicity. Furthermore, β-caryophyllene (**27**) did not show any cytotoxicity at a concentration of 250 μM against the MCF-7 solid tumour cell line. As it turns out, the use of a combination of 10 μg ml^{-1} of β-caryophyllene (**27**) with 32 μg ml^{-1} of α-humulene (**26**) inhibited the growth of MCF-7 solid tumour cell line by 75 ± 6%, which is significantly more than α-humulene (**26**) alone at the same concentration (50 ± 6% inhibition). To corroborate these potential properties, β-caryophyllene (**27**) was also used in combination with paclitaxel (**28**),

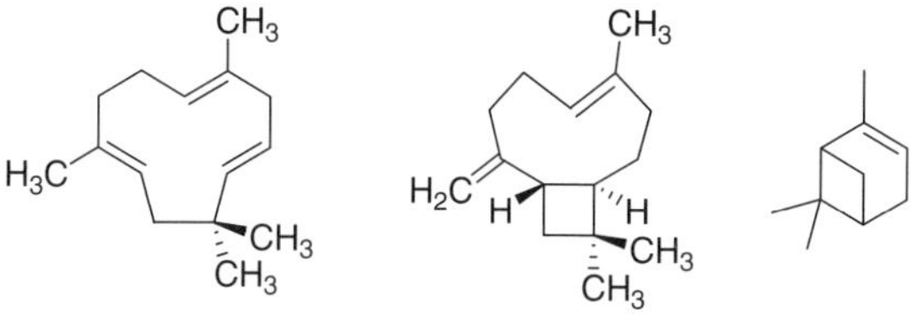

Figure 11.4   *α-humulene (**26**), β-caryophyllene (**27**), paclitaxel (**28**) and α-pinene (**29**).*

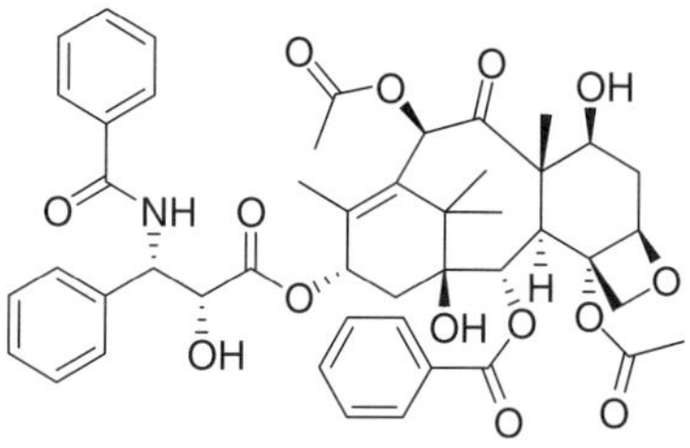

an antitumor agent clinically used to treat breast, ovarian and lung cancers. β-Caryophyllene (**27**) significantly increased the growth inhibition of paclitaxel (**28**) against MCF-7 and L-929 cancer cell lines and the highest potentiating effect was obtained against DLD-1 cancer cell line, with an increase in growth inhibition of about 189%. Thanks to its low toxicity, β-caryophyllene (**27**) is a promising molecule in cancer chemotherapy by allowing reduction of the quantity of paclitaxel (**28**) or other anticancer agent necessary for the treatment, thereby reducing unwanted side effects.

Antibacterial activity is one of the most common biological properties of essential oils [90]. However, no scientific study regarding the antibacterial activity of *A. balsamea* essential oil had been performed. This task was undertaken by our laboratory using one gram-positive (*Staphylococcus aureus*) and one gram-negative (*Escherichia coli*) bacterial strain. The essential oil was inactive against *E. coli*, while the minimum inhibitory concentration (MIC) was determined to be 56 μg ml^{-1} against *S. aureus* [49]. To identify the compounds responsible for the antibacterial activity, all the compounds identified by GC-MS analysis were assayed against *E. coli* and *S. aureus*. Surprisingly, α-pinene (**29**; Figure 11.5) was found to be active against both bacterial strains with an MIC value of 13.6 μg ml^{-1}, while the essential oil was inactive against *E. coli*. The antibacterial activity of α-pinene (**29**) against *S. aureus* was already known [95]. The most active compounds against *S. aureus* were β-caryophyllene (**27**) and α-humulene (**26**), with MIC values of 5.1 and 2.6 μg ml^{-1}, respectively. The antibacterial activity of these two sesquiterpenes (**26** and **27**) had not been evaluated before. The results of this work established the antibacterial properties of essential oil from *A. balsamea* foliage pinpointing its potential industrial application. Additionally, the antibacterial activity of β-caryophyllene (**27**) and α-humulene (**26**) was demonstrated for the first time. Further work should determine if these sesquiterpenes (**26** and **27**) could be used as promising antibacterial forest bioproducts in cosmetic, food or pharmaceutical applications.

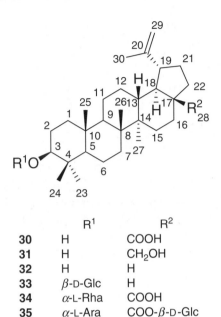

	R¹	R²
30	H	COOH
31	H	CH₂OH
32	H	H
33	β-D-Glc	H
34	α-L-Rha	COOH
35	α-L-Ara	COO-β-D-Glc
36	α-L-Ara	COO-α-L-Ara
37	α-L-Ara	CH₂ O-α-L-Ara

Figure 11.5   Lupane-type triterpenes, saponins and non-natural saponins.

11.3   Chemical Modification of Bioactive Natural Products from the Canadian Boreal Forest

Derivatization is an important part of the development of bioactive natural products into viable bioproducts. This is especially true regarding pharmaceuticals, where 23% of the drugs approved between 1981 and 2002 in the United States were derivatives of natural products [96,97]. Chemical or enzymatic synthesis of derivatives from natural products may have different purposes. Availability could be one criterion as many bioactive natural products are obtained in small yield from the biomass. Therefore, the semi-synthesis of a bioactive natural product from a more abundant structurally related compound may provide the targeted quantity for development or commercialization. A good example is the cytotoxic compound betulinic acid (**30**), which can be obtained with a 2.8% yield or lower from *Betula* spp. outer bark [54]. To pursue the development of betulinic acid (**30**) as an anticancer agent, a simple two-step semi-synthesis has been developed using betulin (**31**) as the starting product [91,92]. Betulin (**31**) can be isolated at a much higher yield, around 17%, from the outer bark of *Betula* spp. [54]. Derivative synthesis may be also related to pharmacokinetic properties (absorption, distribution, metabolism and elimination) [98]. For example, if a given bioactive natural product possesses a low water solubility (high partition coefficient), functionalities can be added to the compound to increase its solubility. Finally, derivative synthesis can also be used to determine structure activity relationships (SARs). Defining SARs results in the determination of important functionalities associated with the action mechanism of the bioactive natural product and may lead to improved bioactivity. In the following section, a review of the glycosidation of bioactive natural products from Canadian boreal forest will be covered. This strategy has been developed in our facilities to promote new bioproduct development. Many classes of bioactive natural products such as terpenes, alkaloids and flavonoids appear as glycosides in plants [99]. The glycosidic residue can be directly involved in the action mechanism of the bioactive compound or modulate its pharmacokinetic properties [99]. It is generally accepted that glycosides are more hydrophilic than their aglycon, aqueous solubility being an important factor in pharmacokinetics [98].

11.3.1   Glycosidation of Triterpenoids from Outer Bark of *B. papyrifera*

Lupane triterpenes are readily available in good yield from *B. papyrifera* outer bark. Betulin (**31**), which is the most abundant triterpenoid from the outer bark [54], can be used as a starting point for a simple two-step semi-synthesis of betulinic acid (**30**) [91,92], which can also be isolated from the bark, but with a much smaller yield. Betulinic acid (**30**) is well known for its *in vitro* cytotoxic activity against a broad panel of human cancer cells [95–97], for its *in vivo* tumour growth inhibition [100] and for its non-toxicity in normal cells [101], making betulinic acid a very promising anticancer agent. However, an important limitation to clinical development is its weak water solubility in media such as blood serum and the polar solvents used for bioassays [102]. In our laboratory, the chosen strategy to increase betulinic acid's hydro solubility was the synthesis of glycoside derivatives.

To evaluate properly the effect of a sugar moiety addition upon the cytotoxicity of lupane triterpenes; betulin (**31**), betulinic acid (**30**) and lupeol (**32**) were used for glycosidation with D-glucose, L-rhamnose and D-arabinose. The importance of the COOH group at C-28 position for the cytotoxicity of betulinic acid (**30**) has been established elsewhere [103], therefore glycosidation was performed selectively at the C-3 position. Glycosidation was also performed at C-3 with lupeol (**32**) which bears only one hydroxyl group, while both C-3 and C-28 glycosides of betulin (**31**) were prepared. The strategy used for glycosidation starts with the preparation of sugars for coupling. Sugar hydroxyl groups were completely protected with benzoyl groups, followed by selective deprotection in anomeric position and by activation at the same position using a trichloroacetimidate group. Betulinic acid and betulin also needed to be selectively protected to avoid the formation of a mixture of C-3 and C-28 glycosides. Betulin was selectively protected by an acetate group at the C-28 position because of the higher reactivity at that site. To obtain acetate at position C-3 of betulin, both hydroxyl groups (C-3 and C-28 positions) were acetylated, followed by a selective deprotection at C-28 position. In the case of betulinic acid, the COOH group at C-28 position was protected using an allyl group which was chosen because of a more efficient regeneration of the acid function after glycosidation. Glycosidation reactions were carried out between the activated sugars and triterpenes having only one free hydroxyl group. Removal of the protecting groups (benzoyl, acetate and allyl) gave 15 different glycosides which were submitted to cytotoxicity bioassays against three cancer cell lines (A549, DLD-1and B16-F1) and one healthy cell line (WS1) [104]. The addition of a D-glucose moiety to lupeol (**32**), which was weakly cytotoxic against the tested cell lines (IC_{50} values ranging from 63 to 165 μM), substantially enhanced its cytotoxicity by about 7- to 12-fold with IC_{50} values ranging from 14 to 15 μM. Additions of L-rhamnose and D-arabinose to lupeol led to less improvement in cytotoxicity. As far as betulin

(**31**) was concerned, addition of a D-glucose moiety at C-3 provoked a complete loss of cytotoxicity, while C-3 α-L-rhamnoside and C-3 α-L-arabinoside were weakly cytotoxic (IC_{50}values ranging from 18 to 63 µM). Glycosidation on the C-28 primary alcohol of betulin (**31**) completely suppressed cytotoxicity. As expected, the higher cytotoxic glycosides obtained in this study were those having betulinic acid as aglycone. The most effective sugar moiety added at the C-3 position of betulinic acid was L-rhamnose (compound **34**) having IC_{50}values against cancer cell lines ranging from 2.6 to 3.9 µM. Furthermore, C-3 betulinic acid glycosides were the only compounds exhibiting different cytotoxicities toward cancer cell lines in comparison to healthy ones. This selectivity toward cancer cells was increased by up to 12-fold in the case of the C-3 α-L-rhamnoside of betulinic acid (**34**) which is the most potent glycoside obtained in this study. These positive results prompted us to investigate in depth the cytotoxic potential of glycosylated lupane triterpenes by undertaking the preparation of bidesmosides [105].

Lupane triterpene bidesmosides consist of the lupane-type triterpenoid aglycone bearing two sugar moieties usually at the C-3 and C-28 positions. Although bidesmosides of the lupan-type are scarce in nature, 3β-O-(α-L-arabinopyranosyl)lup-20(29)-ene-28-O-β-D-glucopyra-nosyl ester (**35**) isolated from *Schefflera rotundifolia* has been shown to exhibit noticeable cytotoxicity against J774. A1, HEK-293 and WEHI-164 cell lines. Therefore, our study began with the semi-synthesis of the natural bidesmoside **35** starting from betulinic acid (**30**). To obtain significant information about the SAR of lupane-type bidesmosides, seven other non-natural lupane bidesmosides bearing D-glucose, L-arabinose and L-rhamnose sugar moieties were also prepared, starting from betulin (**31**) and betulinic acid (**30**). The syntheses were carried out by a combination of Schmidt's procedure and phase transfer conditions using fully benzoylated trichloroacetimidate and sugar bromide donors [105]. All bidesmosides were assayed against A549, DLD-1, MCF7 and PC-3 human cancer cell lines. These studies showed that the most active bidesmosides were those bearing L-rhamnose sugar moieties. Betulinic acid (**30**) and betulin (**31**) aglycones bearing L-rhamnose moieties at both the C-3 and the C-28 positions (compounds **36** and **37**) were highly cytotoxic against cancer cell lines, with IC_{50} values ranging from 6.0 to 7.2 and 1.7 to 1.9 µM, respectively. These results clearly highlighted the beneficial effect of the addition of L-rhamnose moieties that has also been observed in the cytotoxicity of solasodine steroidal glycosides [106]. It was suggested that certain types of cancer cells may have protein receptors that recognize L-rhamnose moieties and help deliver the anticancer agent directly to the tumour [105]. It is also interesting to note that most betulinic acid bidesmosides showed similar or higher cytotoxicity than betulinic acid (**30**), even though the COOH group at C-28 was not free. As mentioned above, the importance of this COOH group for betulinic acid cytotoxicity has been demonstrated in the literature [103]. In summary, our work on the synthesis of glycosides [104] and bidesmosides [104] clearly revealed the potentiating effect of adding L-rhamnose moieties to lupane-type triterpenes. Glycoside **34** and bidesmosides **36** and **37** prepared in these two studies showed strong cytotoxicity, making them very attractive as clinical anticancer agents. To further investigate this potential, haemolytic activity evaluation of a library of natural and semi-synthetic glycosides and bidesmosides including compounds **34**, **36** and **37** were performed.

A major drawback for the clinical development of glycosides and bidesmosides as antitumour agents is the toxicity induced in most animals by red blood cell haemolysis. No study on the haemolytic activity of glycosides and bidesmosides having betulin and betulinic acid as aglycones was available. Therefore, this task was undertaken in our laboratory to better evaluate their clinical potential. The results obtained in this study showed that glycosides and bidesmosides of the lupane family possess much weaker haemolytic activity than most other aglycone families, including the oleanane family [107]. The highly potent glycoside **34** and bidesmoside **37** exhibited no haemolytic activity making them promising anticancer agents for studies of tumour growth inhibition in animals. Work is currently in progress in our laboratory to elucidate the mechanism of action of these two compounds, as well as to evaluate their efficiency on animal models. We do believe that these two compounds represent excellent prospects to become valuable bioproducts. Process development beginning with the extraction of betulin followed by semi-synthesis of glycosides derivatives **34** and **37** would be an excellent approach to add value to waste bark generated by industrial uses of *B. papyrifera*.

11.4   Conclusion

In this chapter, studies by our research group over the past 10 years, aiming at demonstrating the potential of natural products to become valuable bioproducts were covered. Over the years, cytotoxic activities on cancer cell lines of diarylheptanoids **1** and **4**, labdane diterpenes **12**, **13** and **14**, hydroxycinnamic acids **20** and **21** and sesquiterpene **26** all isolated from commercial species of Canadian boreal forest were demonstrated. A non-toxic sesquiterpene (**28**) possessing a high potentiating effect on anticancer agents used clinically, such as paclitaxel, was also identified. Some of these compounds represent very attractive opportunities for the development of biopharmaceutical products used in the treatment of cancer. Our pursuit of new bioactive natural products from commercial species of the Canadian boreal forest biomass also led us to screen for antibacterial activity. Essential oil from the foliage of *A. balsamea*, along with two of its components (sesquiterpenes **26** and **27**) showed significant *in vitro* antibacterial activity on *S. aureus*, making them potential bioproducts for the cosmetic, food or pharmaceutical industries. Synthesis of derivatives has also been used in our laboratory to modulate the pharmacokinetics of bioactive natural products and ultimately increase their potential to be developed as biopharmaceuticals. Glycosidation of anticancer lupane triterpenoids has allowed us to identify compounds **34**, **36** and **37** as potential clinical anticancer agents. Overall, we do believe that these results clearly show the potential of natural products to become profitable co-products that would facilitate the implantation of forest-based biorefineries in Canada.

While these results represent an excellent starting point to promote the concept of integrating bioactive natural product extraction in biorefineries, there is still some work to be done before biorefineries use this type of process. During the course of this screening work, our research group have put together a large library of natural and semi-synthetic compounds obtained from the Canadian boreal forest biomass. This library is a precious tool to identify new valuable bioproducts for the discovery of cosmeceuticals, nutraceuticals, pesticides, food preservatives and antiparasitic agents. Since the library contains isolated and characterized natural products, as well as derivatives, positive results on an assay lead directly to the active principle. Future use of this library could rapidly lead to the development of new bioproducts.

An urgent need for broader development of natural products as bioproducts is the evaluation of potential markets. Natural products have a wide range of potential commercial uses [16–28], but little is known about the potential return they can generate. An accurate evaluation of the potential profit associated with bioactive natural products would guide investors before committing themselves to production. Finally, green extraction, isolation and transformation processes need to be developed to limit the environmental impacts of bioactive natural product processing. Since biorefineries are seen as a way to reduce greenhouse gas emission, processes integrated into the facility must have a minimal ecological footprint. Such green processes could facilitate the introduction of bioactive natural products into Canadian forest-based biorefineries.

References

1. Smith, M. and Parkins, J.R. (2011) Community response to forestry transition in rural Canada: Analysis of media and census data for six case study communities in New Brunswick and British Columbia, p. 108.
2. Forest Products Association of Canada (2011) The New Face of the Canadian Forest Industry.
3. Lapointe, M., Dunn, K., Tremblay-Coté, N. *et al.* (2006) *Looking-Ahead: A 10-Year Outlook for the Canadian Labour Market (2006–2015).* Human Resources and Social Development Canada, Strategic Policy Research Directorate, p. 82.
4. Chambost, B.V., Mcnutt, J., and Stuart, P.R. (2008) Guided tour: implementing the forest biorefinery (FBR) at existing pulp and paper mills. *Pulp and Paper Canada*, **109** (7–8), 19–27.
5. Hämäläinen, S., Näyhä, A., and Pesonen, H.-L. (2011) Forest biorefineries - a business opportunity for the finnish forest cluster. *Journal of Cleaner Production*, **19**, 1884–1891.
6. Mabee, W., Gregg, D., and Saddler, J. (2005) Assessing the emerging biorefinery sector in Canada. *Applied Biochemistry and Biotechnology*, **123** (1), 765–778.
7. Amidon, T.E., Wood, C.D., Shupe, A.M. *et al.* (2008) Biorefinery: conversion of woody biomass to chemicals, energy and materials. *Journal of Biobased Materials and Bioenergy*, **2**, 100–120.
8. Fernando, S., Adhikari, S., Chandrapal, C., and Murali, N. (2006) Biorefineries: current status, challenges, and future direction. *Energy & Fuels*, **20** (4), 1727–1737.
9. Zhang, Y.H. (2008) Reviving the carbohydrate economy via multi-product lignocellulose biorefineries. *Journal of Industrial Microbiology & Biotechnology*, **35** (5), 367–375.
10. Clark, J.H., Deswarte, F.E.I., and Farmer, T.J. (2009) The integration of green chemistry into future biorefineries. *Biofuels, Bioproducts and Biorefining*, **3** (1), 72–90.
11. Bourgaud, F., Gravot, A., Milesi, S., and Gontier, E. (2001) Production of plant secondary metabolites: a historical perspective. *Plant Science*, **161** (5), 839–851.
12. Saklani, A. and Kutty, S.K. (2008) Plant-derived compounds in clinical trials. *Drug Discovery Today*, **13** (3–4), 161–171.
13. Raskin, I., Ribnicky, D.M., Komarnytsky, S. *et al.* (2002) Plants and human health in the twenty-first century. *Trends in Biotechnology*, **20** (12), 522–531.
14. Harvey, A.L. (2008) Natural products in drug discovery. *Drug Discovery Today*, **13** (19–20), 894–901.
15. Fowler, M.W. (2006) Plants, medicines and man. *Journal of the Science of Food and Agriculture*, **86** (12), 1797–1804.
16. Wetzel, S., Duchesne, L.C., and Laporte, M.F. (2006) Pharmaceuticals from the forest, in *Bioproducts from Canada's Forests*, Springer, pp. 131–146.
17. Guenard, D., Gueritte-Voegelein, F., and Potier, P. (1993) Taxol and taxotere: discovery, chemistry, and structure-activity relationships. *Accounts of Chemical Research*, **26** (4), 160–167.
18. Burns, R.M. and Honkala, B.H. (1990) *Silvics of North America: 1.Conifers 2. Hardwoods*, U.S. Department of Agriculture, Forest Service, p. 877.
19. Malik, S., Cusidó, R.M., Mirjalili, M.H. *et al.* (2011) Production of the anticancer drug taxol in Taxus baccata suspension cultures: A review. *Process Biochemistry*, **46** (1), 23–34.
20. Baloglu, E. and Kingston, D.G.I. (1999) A new semisynthesis of paclitaxel from baccatin III. *Journal of Natural Products*, **62** (7), 1068–1071.

21. Bourgault, S., Boulfroy, A., Verstraete, A., and Plante, A. (2008) Portrait socio-économique de la région de la Capitale-Nationale.
22. Thornfeldt, C. (2005) Cosmeceuticals containing herbs: fact, fiction, and future. *Dermatologic Surgery*, **31** (7), 873–880.
23. Baumann, L.S. (2007) Less-known botanical cosmeceuticals. *Dermatologic Therapy*, **20** (5), 330–342.
24. Davis, S.C. and Perez, R. (2009) Cosmeceuticals and natural products: wound healing. *Clinics in Dermatology*, **27** (5), 502–506.
25. Epstein, H. (2009) Cosmeceuticals and polyphenols. *Clinics in Dermatology*, **27** (5), 475–478.
26. Arct, J. and Pytkowska, K. (2008) Flavonoids as components of biologically active cosmeceuticals. *Clinics in Dermatology*, **26** (4), 347–357.
27. Wetzel, S., Duchesne, L., and Laporte, M.F. (2006) Nutraceuticals from the Forest, in *Bioproducts from Canada's Forests*, Springer, pp. 113–130.
28. Espín, J.C., García-Conesa, M.T., and Tomás-Barberán, F.A. (2007) Nutraceuticals: facts and fiction. *Phytochemistry*, **68** (22–24), 2986–3008.
29. Copping, L.G. and Duke, S.O. (2007) Natural products that have been used commercially as crop protection agents. *Pest Management Science*, **63**, 524–554.
30. Dayan, F.E., Cantrell, C.L., and Duke, S.O. (2009) Natural products in crop protection. *Bioorganic & Medicinal Chemistry*, **17** (12), 4022–4034.
31. Zahir, A., Rahuman, A., Kamaraj, C. *et al.* (2009) Laboratory determination of efficacy of indigenous plant extracts for parasites control. *Parasitology Research*, **105** (2), 453–461.
32. Zahir, A., Rahuman, A., Bagavan, A. *et al.* (2010) Evaluation of botanical extracts against *Haemaphysalis bispinosa* Neumann and *Hippobosca maculata* Leach. *Parasitology Research*, **107** (3), 585–592.
33. Naidoo, V., McGaw, L.J., Bisschop, S.P.R. *et al.* (2008) The value of plant extracts with antioxidant activity in attenuating coccidiosis in broiler chickens. *Veterinary Parasitology*, **153** (3–4), 214–219.
34. Tripathi, P. and Dubey, N.K. (2004) Exploitation of natural products as an alternative strategy to control postharvest fungal rotting of fruit and vegetables. *Postharvest Biology and Technology*, **32** (3), 235–245.
35. Balasundram, N., Sundram, K., and Samman, S. (2006) Phenolic compounds in plants and agri-industrial by-products: Antioxidant activity, occurrence, and potential uses. *Food Chemistry*, **99** (1), 191–203.
36. Marles, R.J., Clavelle, C., Monteleone, L. *et al.* (1999) *Aboriginal Plant Use in Canada's Northwest Boreal Forest*, University of Washington Press, p. 368.
37. Moermann, D.E. (1998) *Native American Ethnobotany*, Timber Press, Inc., p. 927.
38. Black, M.J. (1980) *Algonquin Ethnobotany: An Interpretation of Aboriginal Adaptation in Southwestern Quebec*, vol. **65** National Museum of Canada, p. 243.
39. Arnason, T., Hebda, R.J., and Johns, T. (1981) Use of plants for food and medicine by Native Peoples of eastern Canada. *Revue Canadienne De Botanique*, **59**, 2189–2325.
40. Black, P.L., Arnason, J.T., and Cuerrier, A. (2008) Medicinal plants used by the Inuit of Qikiqtaaluk (Baffin Island, Nunavut). *Botany*, **86** (2), 157–163.
41. Ritch-Krc, E.M., Thomas, S., Turner, N.J., and Towers, G.H.N. (1996) Carrier herbal medicine: traditional and contemporary plant use. *Journal of Ethnopharmacology*, **52** (2), 85–94.
42. Koehn, F.E. and Carter, G.T. (2005) The evolving role of natural products in drug discovery. *Nature Reviews Drug Discovery*, **4** (3), 206–220.
43. Potterat, O. and Hamburger, M. (2008) Drug discovery and development with plant-derived compounds, in *Natural Compounds as Drugs, Volume I*, vol. **65**, Birkhäuser, Basel, pp. 46–118.
44. Pichette, A., Larouche, P.-L., Lebrun, M., and Legault, J. (2006) Composition and antibacterial activity of *Abies balsamea* essential oil. *Phytotherapy Research*, **20** (5), 371–373.
45. Dufour, D., Pichette, A., Mshvildadze, V. *et al.* (2007) Antioxidant, anti-inflammatory and anticancer activities of methanolic extracts from *Ledum groenlandicum* Retzius. *Journal of Ethnopharmacology*, **111** (1), 22–28.

46. Legault, J., Dahl, W., Debiton, E. *et al.* (2003) Antitumor activity of balsam fir oil: production of reactive oxygen species induced by α-humulene as possible mechanism of action. *Planta Medica*, **69** (05), 402–407.

47. Mshvildadze, V., Legault, J., Lavoie, S. *et al.* (2007) Anticancer diarylheptanoid glycosides from the inner bark of *Betula papyrifera*. *Phytochemistry*, **68** (20), 2531–2536.

48. Pichette, A., eacute Lavoie, S., Morin, P. *et al.* (2006) New labdane diterpenes from the stem bark of *Larix laricina*. *Chemical & Pharmaceutical Bulletin*, **54** (10), 1429–1432.

49. Pichette, A., Eftekhari, A., Georges, P. *et al.* (2010) Cytotoxic phenolic compounds in leaf buds of *Populus tremuloides*. *Canadian journal of chemistry*, **88**, 104–110.

50. Simard, F., Legault, J., Lavoie, S. *et al.* (2008) Isolation and identification of cytotoxic compounds from the wood of *Pinus resinosa*. *Phytotherapy Research*, **22** (7), 919–922.

51. Pieters, L. and Vlietinck, A.J. (2005) Bioguided isolation of pharmacologically active plant components, still a valuable strategy for the finding of new lead compounds? *Journal of Ethnopharmacology*, **100** (1–2), 57–60.

52. Koehn, F.E. (2008) High impact technologies for natural products screening, in *Natural Compounds as Drugs Volume I*, vol. **65**, Birkhäuser, Basel, pp. 175–210.

53. Lang, G., Mayhudin, N.A., Mitova, M.I. *et al.* (2008) Evolving trends in the dereplication of natural product extracts: new methodology for rapid, small-scale investigation of natural product extracts. *Journal of Natural Products*, **71** (9), 1595–1599.

54. Krasutsky, P.A. (2006) Birch bark research and development. *Natural Product Reports*, **23** (6), 919–942.

55. Naasz, R., Caron, J., Legault, J., and Pichette, A. (2009) Efficiency factors for bark substrates: biostability, aeration, or phytotoxicity. *Soil Science Society of America Journal*, **73** (3), 780–791.

56. Pedieu, R., Riedl, B., and Pichette, A. (2009) Properties of mixed particleboards based on white birch (*Betula papyrifera*) inner bark particles and reinforced with wood fibres. *European Journal of Wood and Wood Products*, **67** (1), 95–101.

57. Smite, E., Lundgren, L.N., and Andersson, R. (1993) Arylbutanoid and diarylheptanoid glycosides from inner bark of *Betula pendula*. *Phytochemistry*, **32** (2), 365–369.

58. Smite, E., Pan, H., and Lundgren, L.N. (1995) Lignan glycosides from inner bark of *Betula pendula*. *Phytochemistry*, **40** (1), 341–343.

59. Pan, H. and Lundgren, L.N. (1994) Rhododendrol glycosides and phenyl glucoside esters from inner bark of *Betula pubescens*. *Phytochemistry*, **36** (1), 79–83.

60. Santamour, J.F.S. and Lundgren, L.N. (1996) Distribution and inheritance of platyphylloside in *Betula*. *Biochemical Systematics and Ecology*, **24** (2), 145–156.

61. Santamour, F.S. and Lundgren, L.N. (1997) Rhododendrin in *Betula*: a reappraisal. *Biochemical Systematics and Ecology*, **25** (4), 335–341.

62. Tian, Z., An, N., Zhou, B. *et al.* (2009) Cytotoxic diarylheptanoid induces cell cycle arrest and apoptosis via increasing ATF3 and stabilizing p53 in SH-SY5Y cells. *Cancer Chemotherapy and Pharmacology*, **63** (6), 1131–1139.

63. Wohlmuth, H., Deseo, M.A., Brushett, D.J. *et al.* (2010) Diarylheptanoid from *Pleuranthodium racemigerum* with in vitro prostaglandin E2 inhibitory and cytotoxic activity. *Journal of Natural Products*, **73** (4), 743–746.

64. Choi, S., Kim, K., Kwon, J. *et al.* (2008) Cytotoxic activities of diarylheptanoids from *Alnus japonica*. *Archives of Pharmacal Research*, **31** (10), 1287–1289.

65. Ali, M.S., Banskota, A.H., Tezuka, Y. *et al.* (2001) Antiproliferative activity of diarylheptanoids from the seeds of *Alpinia blepharocalyx*. *Biological & Pharmaceutical Bulletin*, **24** (5), 525–528.

66. Bar-Sela, G., Epelbaum, R., and Schaffer, M. (2011) Curcumin as an anti-cancer agent: review of the gap between basic and clinical applications. *Current Medicinal Chemistry*, **17** (3), 1–11.

67. Aggarwal, B.B. and Harikumar, K.B. (2009) Potential therapeutic effects of curcumin, the anti-inflammatory agent, against neurodegenerative, cardiovascular, pulmonary, metabolic, autoimmune and neoplastic diseases. *The International Journal of Biochemistry & Cell Biology*, **41** (1), 40–59.

68. Dhillon, N., Aggarwal, B.B., Newman, R.A. *et al.* (2008) Phase II trial of curcumin in patients with advanced pancreatic cancer. *Clinical Cancer Research*, **14** (14), 4491–4499.

69. Pichette, A., Lavoie, S., Morin, P. *et al.* (2006) New labdane diterpenes from the stem bark of *Larix laricina*. *Chemical & Pharmaceutical Bulletin*, **54** (10), 1429–1432.

70. Yang, N.-Y., Liu, L., Tao, W.-W *et al.* (2010) Diterpenoids from *Pinus massoniana* resin and their cytotoxicity against A431 and A549 cells. *Phytochemistry*, **71** (13), 1528–1533.

71. Dimas, K., Papadaki, M., Tsimplouli, C. *et al.* (2006) Labd-14-ene-8,13-diol (sclareol) induces cell cycle arrest and apoptosis in human breast cancer cells and enhances the activity of anticancer drugs. *Biomedecine & Pharmacotherapy*, **60** (3), 127–133.

72. Noori, S., Hassan, Z.M., Mohammadi, M. *et al.* (2010) Sclareol modulates the Treg intra-tumoral infiltrated cell and inhibits tumor growth in vivo. *Cellular Immunology*, **263** (2), 148–153.

73. Kurkin, V.A., Zapesochnaya, G.G., and Braslavskii, V.B. (1990) Flavonoids of the buds of *Populus balsamifera*. *Chemistry of Natural Compounds*, **26** (2), 224–225.

74. Komoda, Y. (1989) Isolation of flavonoids from Populus nigra as oe^4-3-ketosteroid (5.a) reductase inhibitors. *Chemical & Pharmaceutical Bulletin*, **37** (11), 3128–3130.

75. Isidorov, V. (2003) GC-MS analysis of compounds extracted from buds of *Populus balsamifera* and Populus nigra. *Zeitschrift für Naturforschung*, **58** (5–6), 355–360.

76. Greenaway, W., Scaysbrook, T., and Whatley, F.R. (1988) Phenolic analysis of bud exudate of *Populus lasiocarpa* by GC/MS. *Phytochemistry*, **27** (11), 3513–3515.

77. Greenaway, W., Gümüsdere, I., and Whatley, F.R. (1991) Analysis of phenolics of bud exudate of *Populus euphratica* by GC-MS. *Phytochemistry*, **30** (6), 1883–1885.

78. Ogawa, Y., Oku, H., Iwaoka, E. *et al.* (2006) Allergy-preventive phenolic glycosides from *Populus sieboldii*. *Journal of Natural Products*, **69** (8), 1215–1217.

79. Zhang, X., Thuong, P.T., Min, B.-S *et al.* (2006) Phenolic glycosides with antioxidant activity from the stem bark of *Populus davidiana*. *Journal of Natural Products*, **69** (9), 1370–1373.

80. Banskota, A.H., Tezuka, Y., and Kadota, S. (2001) Recent progress in pharmacological research of propolis. *Phytotherapy Research*, **15** (7), 561–571.

81. Ghedira, K., Goetz, P., and Le Jeune, R. (2009) Propolis. *Phytothérapie*, **7** (2), 100–105.

82. Vardar-Ünlü, G., Silici, S., and Ünlü, M. (2008) Composition and in vitro antimicrobial activity of Populus buds and poplar-type propolis. *World Journal of Microbiology and Biotechnology*, **24** (7), 1011–1017.

83. Christov, R., Trusheva, B., Popova, M *et al.* (2005) Chemical composition of propolis from Canada, its antiradical activity and plant origin. *Natural Product Research*, **19** (7), 673–678.

84. Nagaoka, T., Banskota, A.H., Tezuka, Y. *et al.* (2002) Selective antiproliferative activity of caffeic acid phenethyl ester analogues on highly liver-Metastatic murine colon 26-L5 carcinoma cell line. *Bioorganic & Medicinal Chemistry*, **10** (10), 3351–3359.

85. Akao, Y., Maruyama, H., Matsumoto, K. *et al.* (2003) Cell growth inhibitory effect of cinnamic acid derivatives from propolis on human tumor cell lines. *Biological & Pharmaceutical Bulletin*, **26** (7), 1057–1059.

86. Winter, M. and Herrmann, K. (1986) Esters and glucosides of hydroxycinnamic acids in vegetables. *Journal of Agricultural and Food Chemistry*, **34** (4), 616–620.

87. Lafay, S., Gil-Izquierdo, A., Manach, C. *et al.* (2006) Chlorogenic acid is absorbed in its intact form in the stomach of rats. *The Journal of Nutrition*, **136** (5), 1192–1197.

88. Tsychiya, T., Suzuki, O., and Igarashi, D. (1996) Protective effects of chlorogenic acid on paraquat-induced oxidative stress in rats. *Bioscience, Biotechnology, and Biochemistry*, **60** (5), 765–768.

89. Suzuki, A., Kagawa, D., Ochiai, R. *et al.* (2002) Green coffee bean extract and its metabolites have a hypotensive effect in spontaneously hypertensive rats. *Hypertension Research*, **25** (1), 99–107.

90. Bakkali, F., Averbeck, S., Averbeck, D., and Idaomar, M. (2008) Biological effects of essential oils - A review. *Food and Chemical Toxicology*, **46** (2), 446–475.

91. Berger, R.G., Hüsnü, K., Başer, C., and Demirci, F. (2007) *Chemistry of essential oils*, in *Flavours and Fragrances*, Springer Berlin Heidelberg, pp. 43–86.
92. Simard, S., Hachey, J.M., and Collin, G.J. (1988) The variations of essential oil composition during the extraction process. The case of *Thuja occidentalis L.* and *Abies balsamea (L.)* Mill. *Journal of Wood Chemistry and Technology*, **8** (4), 561–573.
93. von Rudolph, E. and Grant, A.M. (1982) Seasonal variation of the terpenes of the leaves, buds and twigs of alpine and balsam firs. *Canadian Journal of Botany*, **60**, 2682–2685.
94. Cornwell, P.A. and Barry, B.W. (1994) Sesquiterpene components of volatile oils as skin penetration enhancers for the hydrophilic permeant 5-fluorouracil. *Journal of Pharmacy and Pharmacology*, **46** (4), 261–269.
95. Raman, A., Weir, U., and Bloomfield, S.F. (1995) Antimicrobial effects of tea-tree oil and its major components on *Staphylococcus aureus*, *Staph. epidermidis* and *Propionibacterium acnes*. *Letters in Applied Microbiology*, **21** (4), 242–245.
96. Butler, M.S. (2005) Natural products to drugs: natural product derived compounds in clinical trials. *Natural Product Reports*, **22** (2), 162–195.
97. Newman, D.J., Cragg, G.M., and Snader, K.M. (2003) Natural products as sources of new drugs over the period 1981–2002. *Journal of Natural Products*, **66** (7), 1022–1037.
98. Rautio, J., Kumpulainen, H., Heimbach, T. *et al.* (2008) Prodrugs: design and clinical applications. *Nature Reviews Drug Discovery*, **7** (3), 255–270.
99. Kren, V. and Martinkova, L. (2001) Glycosides in medicine: "The Role of Glycosidic Residue in Biological Activity". *Current Medicinal Chemistry*, **8** (11), 1303.
100. Pisha, E., Chai, H., Lee, I.-S. *et al.* (1995) Discovery of betulinic acid as a selective inhibitor of human melanoma that functions by induction of apoptosis. *Nature Medicine*, **1** (10), 1046–1051.
101. Zuco, V., Supino, R., Righetti, S.C. *et al.* (2002) Selective cytotoxicity of betulinic acid on tumor cell lines, but not on normal cells. *Cancer Letters*, **175** (1), 17–25.
102. Cichewicz, R.H., and Kouzi, S.A. (2004) Chemistry, biological activity, and chemotherapeutic potential of betulinic acid for the prevention and treatment of cancer and HIV infection. *Medicinal Research Reviews*, **24** (1), 90–114.
103. Kim, D.S.H.L., Pezzuto, J.M., and Pisha, E. (1998) Synthesis of betulinic acid derivatives with activity against human melanoma. *Bioorganic & Medicinal Chemistry Letters*, **8** (13), 1707–1712.
104. Gauthier, C., Legault, J., Lebrun, M. *et al.* (2006) Glycosidation of lupane-type triterpenoids as potent in vitro cytotoxic agents. *Bioorganic & Medicinal Chemistry*, **14** (19), 6713–6725.
105. Gauthier, C., Legault, J., Lavoie, S. *et al.* (2008) Synthesis and cytotoxicity of bidesmosidic betulin and betulinic acid saponins. *Journal of Natural Products*, **72** (1), 72–81.
106. Chang, L.-C., Tsai, T.-R., Wang, J.-J. *et al.* (1998) The rhamnose moiety of solamargine plays a crucial role in triggering cell death by apoptosis. *Biochemical and Biophysical Research Communications*, **242** (1), 21–25.
107. Gauthier, C., Legault, J., Girard-Lalancette, K. *et al.* (2009) Haemolytic activity, cytotoxicity and membrane cell permeabilization of semi-synthetic and natural lupane- and oleanane-type saponins. *Bioorganic & Medicinal Chemistry*, **17** (5), 2002–2008.

12

Pressurized Fluid Extraction and Analysis of Bioactive Compounds in Birch Bark

Michelle Co[1] and Charlotta Turner[2]

[1] *Department of Physical and Analytical Chemistry, Uppsala University, Uppsala, Sweden*
[2] *Department of Chemistry, Centre for Analysis and Synthesis, Lund University, Lund, Sweden*

12.1 Introduction

Birch is a deciduous tree which belongs to the family of *Betulaceae* and grows mostly in boreal and temperate zones. The trees are tolerant to temperatures below zero; for example, it was found that birch can withstand temperatures as low as $-70\,°C$. Hence, many birch species are well adapted to northern climates. The two major species of birch are Eurasian and North American. The phytochemicals in birch vary by the geographical location, climate and naturally by the species [1,2]. In Europe, the most common birch species are silver birch (*Betulapendula*) and downy birch (*Betulapubescens*), whereas in North America yellow birch (*Betulaalleghaniensis*) and paper birch (*Betulapapyrifera*) are more common. Birch is excellent as a resource for energy production since it has fast growth rates, and higher density and lower moisture content than many of the other common forestry species (e.g., spruce, pine and alder) [3]. Birch bark is a by-product of the forestry industry. The amount of bark produced per annum is estimated to be in the thousands of tonnes [4]. Commonly, the main function of bark is to produce energy through incineration. The produced energy is either recycled back into the processes or sold as district heating to nearby communities.

Biorefinery Co-Products: Phytochemicals, Primary Metabolites and Value-Added Biomass Processing, First Edition.
Edited by Chantal Bergeron, Danielle Julie Carrier and Shri Ramaswamy.
© 2012 John Wiley & Sons, Ltd. Published 2012 by John Wiley & Sons, Ltd.

Sweden is the world's second largest producer of forest processed products (e.g., paper, pulp and sawn timber); thus the Swedish forest industry is vital for the national economy. In Sweden, the total commercial production area of forest is vast, approximately 21.5 Mha [4], predominately in the northern part of the country. Consequently, the amount of by-products generated by the forest industries are important and are currently used for the production of relatively low-value wood products (e.g., chipboard and plywood), chemicals and bioenergy [4]. In Sweden, the 2008 total energy consumption was approximately 612 TWh, of which 20% (i.e., 122 TWh) of the consumed energy was produced from biomass, 31.6% from crude oils and oil products, and 29.9% from nuclear power. Thus, the production of bioenergy from forestry by-products is of both environmental and economic importance. The forest industry consumes large amounts of energy, of which 90% of the energy is produced from biomass [4,5]. Furthermore, the demand for biomass-based energy and energy from other green sources will increase due to political decisions and regulation. For instance, the German government has recently decided to phase out nuclear facilities by 2022. A similar decision is expected to take place in Sweden, which would result in price increases for forestry products and by-products. New competitors in South America and Asia have enticed the Swedish forest industry to develop new, more valuable and specialized products that are derived from wood. Viscous and speciality cellulose for the textile and medical products industries, as well as cellulose composites for the car industry are examples of higher-value products that are being developed. The production of ethanol from wood bark and other wood processing by-products is already a commercial business in several of the forest industries in Sweden. There is also interest in identifying new uses for by-products, such as bark, or the production of higher-value products, without putting the core ethanol mission aside.

For example, plant sterols are extracted from tall oil, a by-product from chemical paper pulp processing [6–8]. Tall oil is found in pine trees as a mixture of compounds, such as, turpentine, and is used as a resin in different industries, including mining, paper manufacture, paint manufacture and synthetic rubber manufacture. In Finland, several companies (e.g., Stora Enso, Rakennustoimisto Rasto Oy, Finnforest, Raute) formed a joint non-profit wood product industry venture in 1988, Finnish Wood Research Oy (FWR) [9], where the main focus is to develop new innovative products and production technologies, as well as to implement research results in industrial-scale production. The FWR research projects for 2010–2013 are: (i) energy- and material-efficient wood construction, (ii) new customer-orientated product and production technologies, (iii) recycling of wood products and the recovery of materials and (iv) development of bioenergy and biochemical business for wood product industry companies. Starting from bark obtained from a wood processing plant, the extraction process, using ethanol as a solvent, will not decrease the energy value of the by-product, and the bark can still be incinerated for energy production as it is currently done. This concept has been thoroughly discussed in a recent review article by Ekman *et al.* [10].

It is known that birch bark contains a variety of bioactive compounds. Therefore, it is worthwhile to extract these compounds in order to increase the value of the bark. Thus, the already strained forestry industry could, with this novel approach, strengthen their competitiveness. Birch bark contains a wide assortment of compounds, ranging from small molecules, such as terpenoids, carbohydrates, flavonoids, steroids and tannins, to larger molecules, such as celluloses and lignins. Nevertheless, the focus here will be on the small molecules (< 5000 Da), which are often referred to as bioactive compounds, given the fact that they show biological activity in living tissues. These properties could be antibacterial, anti-inflammatory or providing protection against cardiovascular diseases and ageing-related disorders [11–16]. The extracted compounds from birch bark could be further purified to be of use in the pharmaceutical and the nutraceutical industries. Many of the extracted compounds exhibit antioxidant capacity and could therefore be used in the food and oil industry to prevent food rancidity and oxidation of industrial oils. The extraction process does not significantly affect the bark, since only a negligible amount of substance is usually extracted and thus would not disturb the incineration process for energy production.

There are many conventional techniques, for example, Soxhlet and solid–liquid extraction (SLE) for extracting bioactive compounds from birch bark. However, the environmental awareness of today encourages the industry to develop or improve new or already established techniques/methods towards less energy requirement and avoidance or minimization of the use of hazardous chemicals. Two rather newer techniques are pressurized fluid extraction (PFE) and supercritical fluid extraction (SFE). PFE is an extraction technique based on elevated pressure and temperature. The extraction efficiency is in principle improved at elevated temperature and the elevated pressure used in PFE is mainly to maintain the extraction solvent in a liquid state. It is important to understand that the temperature used in PFE is below the critical temperature of the solvent. SFE is an extraction technique that differs markedly from PFE. The conditions of SFE are that the temperature and pressure used are above the critical values of the solvent. There are several fluid options in SFE but the most common one is carbon dioxide ($SC-CO_2$).

In this chapter, PFE using hot water and ethanol as extraction solvents of different compounds from birch bark will be discussed, in addition to examples of $SC-CO_2$ extraction. Furthermore, qualitative and quantitative analysis of birch bark will be presented, especially regarding antioxidant determination. Analytical methodologies will be described for assessing the chemical mechanism of the birch-extracted bioactive compounds towards reactive oxygen species (ROS) and reactive nitrogen species (RNS), which often are the cause behind food deterioration and human diseases such as atherosclerosis, diabetes mellitus, chronic inflammation, neurodegenerative diseases and many types of cancers [17,18].

12.2 Qualitative Analysis of Birch Bark

12.2.1 Antioxidant Assays

There are many simple and reliable *in vitro* analytical methodologies for determining the antioxidant capacity of compounds in different food and biological samples. However, up to now, in the field of antioxidant research, there is yet no agreement upon a standard methodology for determining the antioxidant capacity of a sample. According to Prior *et al.* [19], a standard assay should fulfil certain criteria, such as: (i) measurement of the chemistry actually occurring in potential applications, (ii) utilization of biologically relevant molecules, (iii) technically simple, (iv) with a defined endpoint and chemical mechanism, (v) readily available instrumentation, (vi) good repeatability and reproducibility, (vii) adaptable for assay of both hydrophilic and lipophilic antioxidants and (viii) adaptable to high-throughput analysis. However, antioxidant capacity measurements *in vitro* still bear no or only little similarity to *in vivo* measurements. Subsequently, the results obtained from *in vitro* measurement might not have any implications at all *in vivo*, but are rather just an indication of antioxidant power.

The current assays used for antioxidant determination are in general divided into two categories, competitive assay and non-competitive assay. A competitive assay involves an antioxidant, a synthetic free radical generator and an oxidizable molecular probe (substrate), of which changes in concentration can be measured by using a fluorescence or ultraviolet (UV) detector. A non-competitive assay, on the other hand, only involves the oxidizable molecular probe, and an antioxidant [20]. In addition, the different antioxidant assays are also defined as hydrogen atom transfer (HAT) and electron transfer (ET) depending on their reaction mechanisms. Examples of HAT assays are the oxygen radical absorbance capacity (ORAC) [21], the radical-trapping antioxidant parameter (TRAP) [22] and the Crocin bleaching assay [23]. As for ET-based assays, the most well-known ones are the Folin–Ciocalteu reagent (FCR) assay [24,25], the Trolox equivalent antioxidant capacity (TEAC) assay [26,27], the ferric ion reducing antioxidant power (FRAP) assay [28], and the 2,2-diphenyl-1-picrylhydrazyl (DPPH) method [29]. The DPPH assay is the most commonly used method for assessing antioxidants in tree bark and other plant materials. Therefore, its merits and disadvantages are described more thoroughly below.

Nevertheless, common for the above-mentioned antioxidant assays are that an absorbance/emission change, either UV/Vis or fluorescence, is determined over a certain time period. The steady-state or end-point time for the different assays widely differs, ranging from five minutes to several hours. The wide range in testing times for different assays results in data that is difficult to compare. Furthermore, some antioxidants are slow in their reaction and need hours to reach the end-point, while others react instantaneously. Finally, some of the assays work only at basic or acidic conditions, which have crucial effects on the reducing capacity of antioxidants. Antioxidants in acidic conditions usually show lower reducing capacity than in basic conditions. This is mainly due to hydrogen protonation and dissociation. For example, the FRAP assay requires acidic conditions, the opposite is true for the FCR assay, which is carried out in basic conditions, and the TEAC assay is performed in neutral conditions. Furthermore, some assays work well for hydrophilic antioxidants, but do not produce optimum results when using lipophilic antioxidants, and vice versa.

However, new adjustments and developments to the ORAC assay enable the use of both hydrophilic and lipophilic chain-breaking antioxidants, in which the antioxidant inhibition of peroxyl radical induced oxidation is determined [24]. ORAC is a HAT-based assay. In the ORAC assay, the peroxyl radical is normally generated by an azo radical initiator, 2,2′-azobis (2-amidino-propane) dihydrochloride (AAPH), which also is commonly use in different TRAP assays [30]. Cao and Prior developed the ORAC assay in 1998 [31], the principle of the assay is prevention of peroxyl-radical-induced oxidation. The effectiveness of the antioxidants in the ORAC assay is given by loss of fluorescence intensity when peroxyl-radical-induced oxidation is inhibited. For example, the ORAC assay was used in determining whether roasting affects the total phenol content and antioxidant activity of cashew nuts, kernel and testa [32]. The roasting was conducted at two temperatures, 70 °C for 6 h and 130 °C for 33 min. Cashew testa processed at the higher temperature contained the highest total phenol content, while cashew kernel had the lowest. Similar to ORAC, the TRAP assay is also based on HAT, in which the oxygen consumption during a controlled induced lipid peroxidation reaction is measured. Since its development by Wayner *et al.* in 1985 [33], the assay has been widely used for total antioxidant capacity in plasma and serum.

A common denominator of the presented assays is that they usually report the obtained result as Trolox equivalents (TE). Trolox is the trade name of 6-hydroxyl- 2,5,7,8,-

tetramethylchroman-2-carboxylic acid, which is a water-soluble vitamin E analogue. For instance, the TEAC assay is a Trolox-based assay, where the scavenging of the long-lived radical anion, 2,2'-azinobis 3-ethylbenzothiazoline-6-sulfonicacid (ABTS$^\bullet$+) by an antioxidant is measured by a spectrophotometer, since ABTS$^\bullet$+ is intensely coloured. The obtained result is then expressed as TE. Another antioxidant equivalent also commonly used is gallic acid equivalent (GAE). For instance, in a study of yellow birch bark, FCR was used to determine the total phenol content of a crude extract, which was obtained by ultrasonication-assisted extraction and maceration [34], and the FCR results were then expressed as GAE. Different parts of the yellow birch bark tissue were studied; twigs, wood, foliage, inner bark and outer bark. The obtained results from the maceration process displayed the highest total phenol content in the inner bark, with 313 mg GEA, followed by 240 mg GEA in the wood, 170 mg GEA in the outer bark and 70 mg GEA in the foliage. The total phenol content was lowest in twigs, with 58 mg GEA. Extracts obtained from ultrasonication-assisted extraction showed the same consistency, with the highest phenol amount in the inner bark. Kähkönen *et al.* [35] also investigated the total phenolic content in silver birch by using FCR. The total phenolic content of leaf and phloem, as well as bark was compared and they reported that the highest phenolic content was in phloem, which is located in the innermost layer of the bark. The phloem distributes organic nutrients to all parts of the plant, thus it makes sense that highest phenol content was found there.

A variety of antioxidant assays are listed in Table 12.1.

DPPH is a scavenging assay that is fairly simple to execute and is based on the use of organic nitrogen radicals (Figure 12.1). DPPH dissolves in organic solvent (e.g., methanol or ethanol) at ambient temperature and becomes a deep purple colour. A change in colour from deep purple to faded yellow is observed when the DPPH is reduced by the antioxidant. The decrease in absorbance is monitored at 515 nm, which is the absorbance maximum of DPPH. The principle of the DPPH assay is based on the fact that the DPPH concentration is proportional to the antioxidant concentration. Therefore, the total antioxidant capacity is expressed as an EC_{50} value, which is the amount of antioxidant that will reduce the initial DPPH concentration by 50%. Thus, a low EC_{50} value corresponds to a high total antioxidant capacity. One of the shortcomings of the DPPH assay is that DPPH acts both as a probe (substrate) and as an oxidant. There is no competition between the DPPH and other radicals (a non-competitive assay, as described above). Therefore, the DPPH assay cannot be translated to living systems because of

Table 12.1   A summary of antioxidant assays and their recent applications.

Antioxidant Assay	Reaction mechanism	Recent applications	References
ORAC	HAT	Beers, bioavailabilty of flavonoids in pecan nuts, jalapeno and serrano peppers	[36–38]
TRAP	HAT	Anthocyanins, blood plasma, vegetables	[39–41]
Crocin bleaching	HAT	Alkannins and shikonins, structure-activity relationship of phenolic compounds, essential oils	[42–44]
FCR	ET	Birch bark, aromatic alkaloids, fruit and vegetables waste, quercetin	[34,35,45,46]
TEAC	ET	Herbs, medicinal plants, human serum lipids	[47–49]
FRAP	ET	Peptides, herbal medicine, fruit extracts	[50–52]
DPPH	ET	Essential oils, phenols in solvents, Maillard reaction products	[53–55]

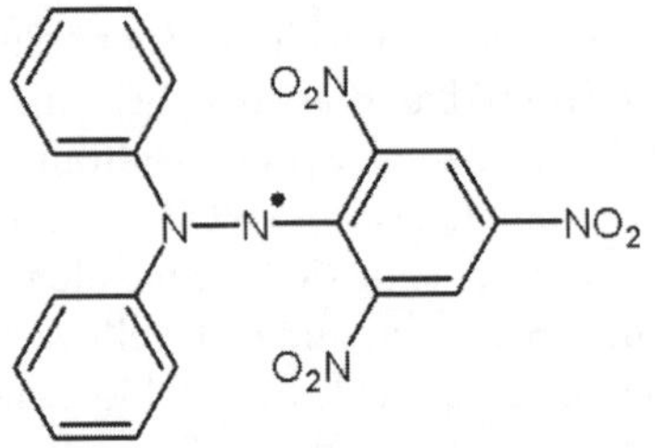

Figure 12.1   *The molecular structure of 2,2-diphenyl-1-picrylhydrazyl, Mw = 394.32 g mol⁻¹.*

the competition stemming from the various substrates. In addition, the DPPH molecule is a stable nitrogen radical, which is usually not present naturally in biological systems and has no similarity to the highly reactive peroxyl radicals involved in lipid peroxidation processes encountered inside the human body or in higher plants. Thus, antioxidants that are highly reactive with peroxyl radicals may be slow in reacting with DPPH or in the worst-case scenario, not react at all. Furthermore, the DPPH assay is limited by steric accessibility, thus small molecules have better access to the DPPH molecules than larger ones [19,56,57].

Despite the shortcomings of the DPPH assay, it is by far the most-used assay for screening/ measuring the antioxidant capacity of pure and complex samples. The DPPH radical is stable, commercially available and inexpensive; the data are obtained with a widely available spectrophotometer. Brand-Williams and co-workers [57] introduced the traditional DPPH assay, where 25 mg DPPH is dissolved in 100 ml methanol, with a reaction time of approximately 30 min, which in many cases is not sufficient for larger molecules (e.g., antioxidants) to reach a steady state [19,56]. Brand-Williams and co-workers plotted the remaining proportion of DPPH as a function of different concentrations of antioxidants, expressed as the number of moles of antioxidant/number of moles of DPPH, for calculating the EC$_{50}$ value. This approach is ideal if the identity of the antioxidant is known; however, when working with an unknown antioxidant, estimation of the mole ratio becomes problematic. Thus, a modified Brand-Williams method was used to measure the antioxidant capacity of the inner and outer layer of birch bark extracts, which contained unknown antioxidants [58]. The modified method called for a reaction time of four hours to ensure that steady state was reached without risking degradation. The EC$_{50}$ values were determined by using an external calibration curve of different concentrations of DPPH. The antioxidant capacity was found highest in the extracts produced from the outer layer of birch and finely ground particles (Table 12.2).

Table 12.2   *Effect of sample particle size on DPPH antioxidant capacity of extracts from the inner and outer layers of birch bark.*

Experiment # (n = 3)	Birch layer type	Particle size	EC$_{50}$ value (µg bark/µg DPPH)	RSD (%)
1	Inner	Squares (1 × 1 cm)	57	8
2	Inner	Finely ground (<1 mm)	37	6
3	Outer	Long slices (1 × 10 cm)	9	9
4	Outer	Finely ground (<1 mm)	5	15

The extracts were obtained by pressurized fluid extraction using ethanol at 130 °C, 50 bar and an extraction time of 3 × 5 min [58]. Reproduced by permission of The Royal Society of Chemistry (RSC).

In another study on yellow birch twigs, García-Pérez *et al.* [59] used the FCR assay to determine the total phenol content and the DPPH assay for antioxidant capacity determination. The yellow birch twig extracts were produced by SLE, firstly with 70% aqueous acetone and followed by fractionation resulting in three fractions: (i) *tert*-butyl-methyl ether fraction, (ii) ethyl acetate fraction and (iii) aqueous fraction. The results showed that the ethyl acetate and aqueous fractions displayed the highest antioxidant capacity, as well as the highest phenolic content. Results showed that the antioxidant capacity was correlated to the phenolic content. However, it is important to note that the antioxidant activity is not always correlated to a high phenolic content because the FCR assay does not give details of the identification of the compounds present in the extracts [19]. Therefore, it is risky to predict the antioxidant capacity on the basis of the total phenolic content as measured by FCR.

12.2.2 Antimicrobial Activity

There are several different methods to determine the antimicrobial activity of a sample, but the requirements of the different methods are similar: (i) test bacteria, (ii) indicator solution to ascertain the bacterial growth, (iii) growth medium and (iv) controls. The most general methods fall into three categories that consist of bioautographic, diffusion and dilution methods [60,61]. For qualitative assessment of the antimicrobial activity, the bioautographic and diffusion methods are preferred to the dilution method because they yield the minimal inhibitory concentration (MIC) [62,63]. Briefly, various concentrations of the test extract are incubated together with the test bacteria in a growth medium at a certain temperature and duration (e.g., 24 or 48 h). After the incubation period, the antimicrobial activity is assessed by either measuring the bacterial growth zones or the population of bacteria grown. The sample is deemed to display antimicrobial properties if the growth of the bacteria is inhibited compared to that of the controls.

Recently, Salin *et al.* [64] screened betulin and its derivatives against *Chlamydia pneumoniae* (a gram-negative bacterium), which is involved in asthma, atherosclerosis and lung cancer disease mechanisms. Betulin and 32 other betulin derivatives were incubated together with Hodgkin's lymphoma cells (HL-cells), which were infected by a *C. pneumoniae* strain, at 37 °C for 70 h under conditions of 5% CO_2 and 95% humidity. The results showed that betulin and the majority of the tested derivatives inhibited *C. pneumoniae* growth. For example, betulin showed a 53% inhibition and three other betulin derivatives showed inhibition of *C. pneumoniae* above 80% at a concentration of 1 µM. Five derivatives of betulin showed an inhibition close to 100% and were subsequently tested in dose response experiments, using traditional immunofluorescence labelling. Four out of the five derivatives showed detectable inclusions, and one, betulindioxime, completely eradicated the *C. pneumoniae* strain from the HL-cells. Betulindioxime resulted in 80% inhibition with a MIC of 1 µM at a concentration of 500 nM. Salin *et al.* [59] reported a relationship between the structure–activity and the pharmaco-logical profiles of the tested betulin derivatives, which could help further drug development.

12.2.3 Antitumour Activity

There are many ways to determine drug-induced cytotoxicity and cell proliferation such as fluorescence-based and colorimetric-based assays [65]. The sulforhodamine B (SRB) assay is considered to be a simple and reliable method for fast screening of potentially active compounds due to its high sensitivity (comparable with several fluorescence assays) and linear correlation to number of cells and cellular proteins [66]. Furthermore, the SRB assay has a colorimetric endpoint and is stable and non-destructive. The SRB molecule consists of two sulfonic groups and is an anionic aminoxanthene protein dye that is bright pink in colour. After incubation of tumour cells and substrate, SRB is added for staining purposes; SRB binds to the living cells. The cell growth is determined by the optical density (OD) of SRB, measured at 570 nm.

Drag *et al.* [67] used the SRB assay to test birch bark extract, betulin and betulinic acid against human gastric carcinoma and pancreatic carcinoma drug-sensitive and drug-resistant cell lines [67]. The concentrations tested for the three substances were 0, 2, 4 and 8 µg ml^{-1} with incubation for five days prior to SRB staining. The obtained results from Drag *et al.* showed that crude birch bark extract, as well as purified betulin and betulinic acid, exhibited cytotoxic activity toward the investigated tumour cells, with betulinic acid the most effective substance, having an IC$_{50}$ value of 6.16 µM for gastric carcinoma cells and an IC$_{50}$ value of 7.42 µM for pancreatic carcinoma cells. Birch bark extracts had IC$_{50}$ values of 11.66 and 25.26 µM, respectively. Betulin was the least effective substance against the tested cell lines with IC$_{50}$ values of 18.74 and 21.09 µM, respectively. The IC$_{50}$ values of birch bark extracts and betulin for pancreatic carcinoma cells differed slightly. Nevertheless, the IC$_{50}$ values of birch bark extracts for drug-sensitive and drug-resistant cells differed markedly, with birch bark extracts having an IC$_{50}$ value of 9.07 µM and betulin having a value of 20.62 µM.

Another study of betulinic acid aimed at primary paediatric acute leukaemia cells showed marked apoptosis at a betulinic acid concentration of 10 µg ml^{-1}. In addition, comparison between the efficiency of betulinic acid and conventional cytotoxic drugs demonstrated that betulinic acid was more potent than the standard drugs [68]. In spite of the positive results, one has to remember that *in vitro* and *in vivo* tests may differ considerably due to competitive effects and other uncontrolled mechanisms in biological systems.

The MTT assay, MTT is 3-(4,5-dimethylthiazol-2-yl)-2,5-diphenyl tetrazolium bromide, is another appropriate assay for cytotoxicity testing [69]. MTT is a dye which is yellow coloured and due to its high water solubility it crosses both plasma and mitochondrial membranes. MTT is a tetrazolium salt that can be reduced to a formazan (see Figure 12.2), which is bright blue in colour. Hence, the optical measurement of the formed formazan compound is correlated with the number of cells present at the end of the assay period. For instance, the cytotoxic effect of betulinic acid against melanoma, neuroectodermal and malignant brain tumour cell lines was studied using the MTT assay [70]. Zuco *et al.* tested nine different human tumour lines: two clones from a subcutaneous melanoma metastasis (Me665/2/21 and Me665/2/60); three ovarian carcinomas (A2780, OVCAR-5 and IGROV-1); one cervix carcinoma (A431); two lung carcinoma (H460) and (POGB) and one selected doxorubicin-resistant subline (POGB/DX). Within a very narrow range of doses of betulinic acid (1.5–4.5 mg ml^{-1}), the

Tetrazolium
(yellow)

Reduction

Formazan
(bright blue)

Figure 12.2   *The reduction of a tetrazolium salt to a formazan in the MTT assay – a colour change from yellow to bright blue is observed.*

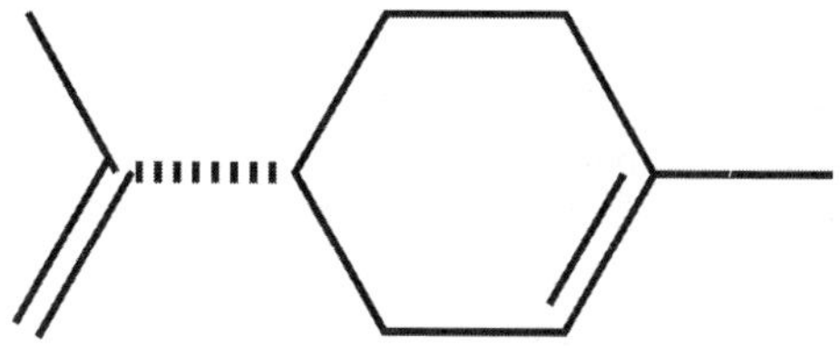

Figure 12.3 *Limonene.*

in vitro results showed that betulinic acid inhibited the growth in all neoplastic cell lines. Furthermore, *in vivo* tests with betulinic acid showed some indication of antitumour activity on ovarian carcinoma IGROV-1 xenografts [70]. Betulinic acid was administered to infected mice by intraperitoneal injections (i.p.) at a dose of 100 mg kg^{-1} of body weight. The results of the *in vivo* test showed that mice given betulinic acid survived significantly longer than control mice: 16 ± 1.03 days for controls and 22 ± 2.59 days for mice receiving betulinic acid.

12.3 Quantitative Analysis of Bioactive Compounds in Birch

12.3.1 Terpenoids

The plant kingdom produces a vast amount of diverse small organic compounds (<5000 Da). These compounds have apparent function for the growth and development of the plant and are called primary metabolites. Those organic molecules that are not involved in the basic growth and development of the plants are called secondary metabolites, which are known in the literature as "natural products".

The largest group of secondary metabolites consists of terpenes, which are also known as terpenoids or isoprenoids. The building blocks of terpenes are isoprenes. In nature, one terpene often encountered is limonene, a monoterpene containing 10 carbon atoms (Figure 12.3). Triterpenes, also common in nature, contain 30 carbon atoms, and are often found in different tree species. For example, betulin is abundant in birch bark (Figure 12.4).

Recently, studies have reported that terpenes exhibit a broad range of biological activities, such as antimicrobial, antifungal, antiviral, antihyperglycemic, anti-inflammatory and

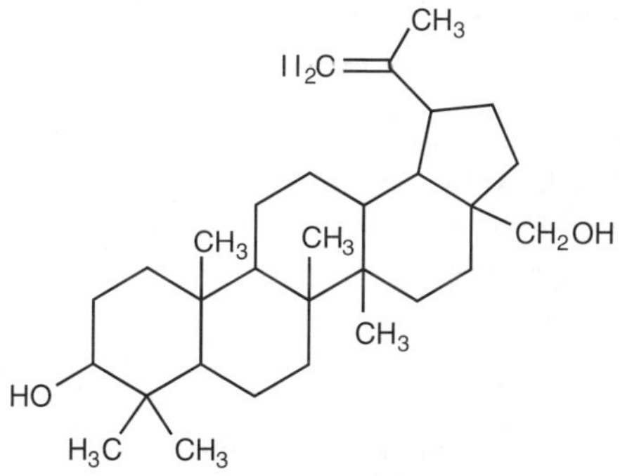

Figure 12.4 *The molecular structure of betulin, Mw = 442.72 mol g^{-1}.*

antiparasitic [71–73]. Many of the terpenes are highly volatile and are analyzed by gas chromatography (GC) with either a flame ionization detector (FID) or mass spectrometric detection (MS). For example, mulberry bark and leaves contain the terpenoid β-sitosterol, which has been extracted by SFE with SC-CO$_2$ and Soxhlet extraction with 96% ethanol or *n*-hexane [74]. The amount of β-sitosterol extracted from mulberry bark was reported to be the highest using *n*-hexane in a Soxhlet extraction. The concentration of β-sitosterol in mulberry bark extracts was determined as 103.85 ± 1.43 g/100 g dry weight of plant, using both GC-FID and GC-MS, without any derivatization and with 5α-cholestan-3-one as the internal standard. The best SFE method yielded 31.84 ± 1.27 g/100 g dry weight, at 400 bar and for 60 min. Headspace solid phase micro-extraction coupled to GC-MS (HS-SPME-GC–MS) has also proven to be useful to extract compounds from birch samples; several volatile sesqui-terpenoid and phenol derivatives were detected [75].

Terpenoids can also be analyzed using high-performance liquid chromatography (HPLC) coupled to UV/VIS or diode array detection (DAD), if reference samples are available. For instance, in a study of birch bark, betulin (lup-20(29)-ene-3,28-diol) (Figure 12.4) was identified and quantified using HPLC with UV detection at 209 nm [58]. As stated above, betulin is a bioactive compound that displays antimicrobial and antitumour activity.

The structure of betulin is shown in Figure 12.4 and is a good example of a triterpene [2,76]. An external calibration curve was conducted using pure betulin as standard to quantify the extracted betulin from birch bark. The extracted betulin was determined to be 26% of the total dry weight of the outer bark [58]. The results obtained by Co *et al.* [58] corresponded well with the literature. The betulin content in birch varies from 10–35% of the total dry weight of the outer bark. The variation in betulin concentration can be attributed to geographical location, growth rate, climate and differences in birch species. Betulinic acid is a derivative of betulin and it is a minor constituent of birch bark. However, in terms of cytotoxic and apoptosis activity, betulinic acid has been shown to be more active than betulin [70,77–80]. Thus, the interest in extracting betulin from birch trees mainly lies in that it can easily be converted to betulinic acid [81,82].

Terpenoids are difficult to ionize with the available atmospheric pressure ionization (API) technique, electrospray, which is often used in LC-MS, hence GC-MS is more useful for analyzing terpenoids. The drawback in using GC-MS is mainly that derivatization is required, which sometimes leads to loss of the compound. Water-soluble terpenoids, on the other hand, can be analyzed by LC-MS, which is rather straightforward since the plant extract is in general in aqueous solution. The optimization of a suitable ionization interface in MS is often also the critical step in LC-MS-based analysis. Rhourri-Frih *et al.* [83] compared two different API techniques: atmospheric pressure chemical ionization (APCI) and atmospheric pressure photon ionisation (APPI), for analyzing triterpenes (e.g., betulin and betulinic acid) from birch bark [83]. Both APCI and APPI could successfully detect triterpenes, with APPI in the positive mode resulting in the best sensitivity, whereas APCI in the negative mode was more sensitive regarding acidic triterpenes.

12.3.2   Carbohydrates

Carbohydrates, also known as saccharides, are divided into several groups based on their molecular weights. The smallest saccharide is a monosaccharide, followed by a disaccharide and then an oligosaccharide; the largest saccharides are called polysaccharides, usually containing more than 10 monosaccharide residues. A rough upper limit of polysaccharides is usually around 3000 monosaccharide residues, except for certain larger varieties, such as cellulose, which can contain from 7000 to 15 000 monosaccharide residues. Polysaccharides can be neutral or acidic. The common name for carbohydrates is sugars. They serve as energy storage deposits and as building blocks for larger molecules such as hemicellulose, cellulose, pectin and lignin. The carbohydrate content of hardwood (wood from angiosperm trees) and softwood (wood from conifers trees) varies. Moreover, the carbohydrate content varies within the tree, depending on if the original sampling location is sapwood or heartwood. Sustainability aspects in the forest industry have invigorated research on polysaccharides and lignin to produce new hybrid materials. Together biomass-derived polysaccharides blended with synthetic polymers could result in novel products that could be used in the food and pharmaceutical industries. This in turn would lead to a demand for sensitive and robust analytical tools and methodologies.

The most common analytical technique to separate underivatised carbohydrates is ion chromatography (IC) with an anion-exchange column, using an alkali-hydroxide- or alkali-acetate-based mobile phase [84]. Other columns such as hilic, porous graphitic carbon and amide have in recent years proved to be efficient in separating carbohydrates [85–88]. The hydrophilic interaction of these columns allows the use of aqueous organic mobile phases at neutral pH that enable mass spectrometric detection. However, carbohydrates are not chromophores, thus UV detection is not feasible. Detectors that can be used are: (i) refractive index (RI) [89], (ii) evaporative light scattering (ELSD) [90], (iii) integrated pulsed amperometric (IPAD) [91], (iv) charged aerosol (CAD) [92], and (v) mass spectrometric (MS) [93].

Anion-exchange columns are still frequently used for carbohydrate separation despite the advances in column technology. The alkali mobile phase used to achieve good separation is unfortunately not compatible with MS, especially using an API interface. The low volatility and high conductivity of the alkali mobile phase requires a desalting step prior to MS detection. This can be achieved by, for example, a cation exchange membrane [93]. An online desalting IC system using a RI detector removed the salt from the carbohydrate by using an H^+ ion-exclusion column, and a Pb^{2+} ion-exclusion column [89]. This online valve-switching technique is extremely useful when coupling different columns in parallel. Cheng *et al.* [89] used this as an online system for analyzing cellulosic biomass and the limits of detection (LOD) and limit of quantification (LOQ) for cellobiose, glucose, xylose, galactose, mannose and arabinose varied between 0.12–4.88 and 0.40–16.3 g ml^{-1}, respectively. In addition, cellulose was found to be the principal component in the cellulosic biomass and hemicellulose appeared in minor amounts. This speculation is based on the highest amount of glucose (1736 µg ml^{-1}) and xylose (168 µg ml^{-1}) present in the studied cellulosic biomass.

As mentioned above, LC using an ion-exchange column is the most common methodology for analyzing carbohydrates. However, GC is also used to analyze certain carbohydrates due to its high efficiency. The critical criterion for using GC is that the compound has to be sufficiently volatile, hence the carbohydrates have to undergo derivatization prior to GC analysis. For many years, acetylation and the use of trifluoroacetates were the most popular methods for derivatizing carbohydrates, but today, various silanization methods are more commonly used, for instance using trimethylchlorosilane or hexamethyltrisilazine [94,95].

12.3.3 Flavonoids

Polyphenols are made up of multiples of phenol substructures. Specifically, flavonoids containing three phenol rings belong to the polyphenolic group. Flavonoids are commonly found in various concentrations throughout the plant kingdom, for example in flowers, trees, fruits, nuts and vegetables [96]. The role of flavonoids is possibly involving plant defences, for example against stress, infection or herbivore attack [97]. There is a large body of research devoted to the identification of flavonoids in plants and their role in human nutrition [98,99].

Due to the structural variation of flavonoids, they can be into four subfamilies: (i) flavonols, (ii) flavones, (iii) flavanols and (iv) isoflavones. The functional groups attached to the flavonoid are in many cases hydroxyl groups, methoxy groups and sugar moieties, for example rutinose, galactose, xylose, glucose and rhamnose [96,100]. A flavonoid without the sugar moiety is a flavonoid aglycone. Many of the reported flavonoids (i.e., quercetin and catechin) exhibit antioxidant capacities. The antioxidant capability of a flavonoid depends on its structural arrangement and the number and position of the hydroxyl groups.

Birch bark, like other tree barks, contains various types of flavonoids. Hiltunen *et al.* [1] studied silver birch bark and isolated one flavonoid and three phenolic compounds [1]. First they extracted the birch bark using a Soxhlet with methanol as solvent. The isolated flavonoid was identified on the basis of its ^{13}C and ^{1}H NMR chemical shifts as (+)-catechin-7-*O*-β-D-xylopyranoside. The other phenolic compounds were 3,4,5-trimethoxyphenyl-β-D-apiofuranosyl-β-D-glucopyranoside and 3,4,5-trimethoxyphenyl-β-D-glucopyranoside. The presence of catechin in birch bark was also confirmed in another study, where liquid chromatography was coupled to MS detection using electrospray ionization [101]. Mammela [101] also identified other compounds, for example betuloside, platyphylloside and 5-hydroxyplatyphyllone. In addition, losses of 132 u and 162 u were observed in the obtained mass spectrum, probably corresponding to pentose (e.g., xylose, arabinose ribose and lyxose) and hexose (e.g., glucose, mannose, galactose and allose) monosaccharides, respectively.

Proanthocyanidins belong to the flavanol family. "Soluble proanthocyanindins" simply implies proanthocyanidins that are water-soluble, which have been reported to be the cause in reddening and ripening of plum skin. In wood, Barry *et al.* reported that the colour of wood tends to change in connection with fungal infection, in which soluble proanthocyanidins may be involved [102]. A study on birch-bark-derived soluble proanthocyanidins reported that the concentration depends on growing site, sampling season and location in trunk [103]. The soluble proanthocyanidins were extracted by solid–liquid extraction using 95% acetone at room temperature and the quantification of the soluble proanthocyanidins was determined using acid solution, which converted the soluble proanthocyanidins to coloured anthocyanidins [104].

12.4   High-Performance Liquid Chromatography with Diode Array, Electrochemical and Mass Spectrometric Detection of Antioxidants

As discussed above regarding antioxidant assays, there are no perfect antioxidant assays; each of the presented assays had its limitations. The capacity of an antioxidant is related to its ability to be reduced, hence it is tightly correlated to its redox potential. Blasco *et al.* [105] used electrochemistry to determine the antioxidant capacity based on the facts that: (i) all antioxidants exhibit electroactivity, (ii) direct measurements can be made without the need for reactive species, (iii) many antioxidant assays are conceptual, based on electron transfer reactions, (iv) the antioxidant capacity is correlated to the oxidation potential, (v) the obtained current is proportional to the antioxidant amount and the functional groups, (vi) this method is cheap and reliable compared to antioxidant assays, (vii) the response is not dependent on the optical path length or sample turbidity and (viii) the synergistic effect of the antioxidants can be avoided using a separation step prior to the electrochemical measurements.

Direct electrochemical measurements with controlled potential can be conducted in batch and flow modes. The electrochemical batch approaches are, for example, cyclic voltammetry (CV) and differential-pulse voltammetry (DPV). Both techniques have been reported for determination of the antioxidant capacity in many natural antioxidants derived from the plant kingdom [106–108]. One of the restrictions of CV and DPV is in measuring individual antioxidants in a complex sample. Consequently, electrochemical flow approaches using amperometric or coulorimetric detection coupled to an analytical column are more suitable for that purpose [109,110].

Zettersten *et al.* [111] developed an online system, where an analytical column was coupled to three different detectors: (i) DAD, (ii) electrochemical detection (ECD) that was amperometric-based, comprising a flow-cell with glassy carbon electrodes and (iii) tandem mass spectrometric (MS/MS) detection. The schematic setup of the system is shown in Figure 12.5. The online system is extremely selective and informative, since compounds, which exhibit chromphoric as well as electroactivity can be monitored. In addition, the identity of the unknown antioxidants can be elucidated using MS/MS. The selectivity of the ECD is related to the set potential, and can be adjusted to screen different types of antioxidants. Zettesten *et al.* determined that setting the amperometric detection at +0.4 V vs. Ag/AgCl gave a satisfactory correlation with the spectrophotometric-based DPPH assay.

The same online system was used in the study of the degradation of birch bark antioxidants extracted at elevated temperature and pressure [112]. Figure 12.6 shows which of the birch bark compounds display chromophoric and electroactive characteristics. At a retention time of

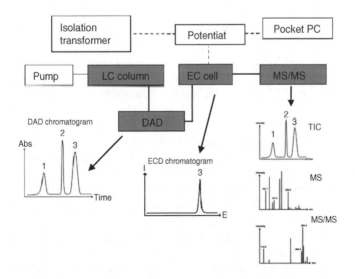

Figure 12.5   *A schematic picture of an HPLC-DAD-ECD-MS/MS system.*

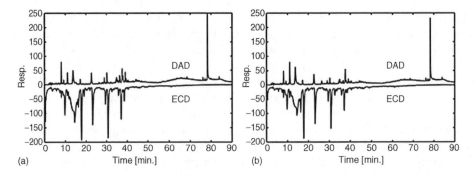

Figure 12.6 *Overlaid chromatograms of DAD monitored at 280 nm and ECD with an applied redox potential of +0.4 V vs. Ag/AgCl. (a) Birch bark extracted at 180 °C for 5 min and (b) birch bark extracted at 180 °C for 5 min followed by hydrothermal treatment at 180 °C for five minutes.*

79 min, the separated compounds showed chromophoric activity without displaying electro-activity at all investigated potentials. The advantage of using this system is that rapid characterization of unknown antioxidant compounds can be obtained. Catechin, which is a flavonoid commonly found in birch trees, was detected when studying the obtained DAD and MS/MS spectrum [112]. At elevated temperatures, catechin was converted into its different diastereoisomers. Separation coupled to DAD, ECD and MS/MS is a powerful technique, which could rapidly screen potent antioxidants, simultaneously elucidating their structure through DAD and MS/MS analysis.

12.5   Extraction of Bioactive Compounds

12.5.1   Conventional Solid Liquid Extraction (SLE)

The simplest way of conducting SLE is by leaching, which in principle consists of immersing the solid sample in a solvent for a certain length of time. SLE can also be conducted together with a shaking device or by more advanced apparatus, for example, Soxhlet. Nonetheless, SLE is a straightforward and simple technique, which is simple to perform and with no need for sophisticated instrumentation. Lengthy extraction time and dilution of sample are the draw-backs of SLE. Despite this, SLE is still the most used extraction technique for extracting bioactive compounds from diverse plant materials [1]. The solvents frequently used in SLE are methanol, propanol and ethanol. Chloroform and other non-sustainable organic solvents are sometimes used to enhance the extraction selectivity.

SLE, with water and ethanol as solvents, was used and compared with SFE and PFE for the extraction of antioxidants from spruce bark [113]. SLE, using water and ethanol at ambient temperature in which the macerated biomass was kept overnight, was not as efficient as PFE at 80, 130 and 180 °C, using the same solvents. SFE using pure CO_2, as well as a mixture of CO_2 and ethanol, resulted in the lowest solid yields with the extracts displaying the lowest antioxidant activity. The extracted antioxidants were identified as stilbenes (astringin, iso-rhapontin and piceid) using NMR and MS [113]. In another study, SLE using different solvents: (i) dichloromethane, (ii) ethyl acetate, (iii) acetone, (iv) chloroform, (v) methanol and (vi) 95% ethanol (aqueous solution, v/v), were tested for extracting betulin and betulinic acid from birch bark [114]. The extraction was conducted at ambient temperature for two hours, and repeated twice. The obtained extracts were analyzed using HPLC-UV at 210 nm. The results showed that 95% ethanol was the best solvent for betulin extraction, followed by chloroform, ethyl acetate and acetone. The same trend was also observed for betulinic acid.

12.5.2 Supercritical Fluid Extraction (SFE)

A pure component can exist in several different states, for example as liquid, solid, gas and supercritical fluid. The physical state of a pure component depends mainly on temperature and pressure. Thus, a pure component is considered as a supercritical fluid at temperatures and pressures above its critical values (i.e., T_c and P_c). The physical properties of a supercritical fluid are both liquid-like and gas-like, with a density close to a liquid, a gas-like viscosity and diffusivity somewhere between a liquid and a gas [115,116]. Furthermore, the surface tension of a supercritical fluid is zero, which enables proper wetting and efficient penetration into the sample matrix. These properties of the supercritical fluid facilitate faster mass transport, enabling efficient extraction [117]. In addition, the solvent strength and selectivity can easily be tuned by changing the temperature, pressure and the choice of modifier (e.g., ethanol, methanol or water) [116]. SC-CO_2 ($T_c = 30.9\,°C$ and $P_c = 73.8$ bar) is the most widely used supercritical fluid [118]. It is readily available in high purity at a low cost. In addition, SC-CO_2 is easily removed by simple expansion or in many cases recycled back into the process by a gas compressor. A schematic of SFE equipment is shown in Figure 12.7.

SC-CO_2 is a non-polar solvent similar to hexane or heptane, and hence only suitable for extracting non-polar or slightly polar compounds. Moreover, the solvation power of SC-CO_2 is related to the molecular weight of the extracted compound. Roughly described, SC-CO_2 has good solvation power for small molecules and poor for large ones. An exception is, for instance, fluorinated polymers that are highly soluble in SC-CO_2 despite their large molecular weight. Oxygenated organic compounds are readily soluble in SC-CO_2, whereas free fatty acids and water have low solubilities in SC-CO_2. Compounds that are insoluble in SC-CO_2 are proteins, polysaccharides and mineral salts. Sometimes, adding a small amount of a polar modifier can shift the polarity of SC-CO_2. The added modifier can also improve the yield of SC-CO_2 extraction [119,120]. European white birch (*Betulapendula*) is a rich source of triterpenes, especially betulin and its derivative, betulinic acid. Pure SC-CO_2 as well as SC-CO_2 containing 5% ethanol were demonstrated to effectively extract betulin from birch bark [121]. The temperature used in both methods was $40\,°C$, with a pressure of 450 bar for the former method, while a lower pressure was used (300 bar) in the latter. The betulin yield was 64% higher using SC-CO_2 containing 5% ethanol compared to pure CO_2. Furthermore, lupeol and β-sitosterol were also extracted with SC-CO_2 containing ethanol.

12.5.3 Pressurized Fluid Extraction (PFE)

Richter introduced PFE in 1995 [122] as an extraction technique mainly applicable to extract polycyclic aromatic hydrocarbons (PAHs), polychlorinated biphenyl (PCB) congeners and chlorinated pesticides from sediments [123,124]. Today, PFE is an attractive technique for extracting phytochemicals from diverse plant materials [125,126]. PFE has proven to be an extraction technique, which can be more effective than or comparable to conventional extraction techniques (e.g., Soxhlet and SLE) [125,127–129]. The principle of PFE is simple:

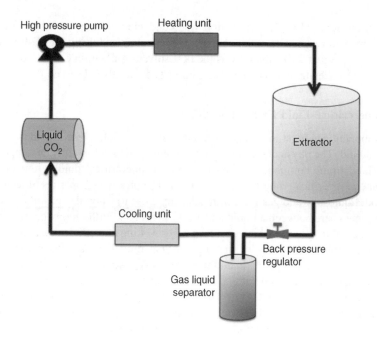

Figure 12.7   *A schematic picture of SFE where the used CO$_2$ is recycled back into the unit. The SFE set-up comprises liquid CO$_2$, high pressure pump, heating unit, extractor, back pressure regulator, gas liquid separator and a cooling unit.*

the solute is extracted out from the sample matrix using a liquid solvent at elevated temperature and pressure. The function of the applied pressure in PFE is mainly to maintain the extraction solvent in a liquid state during the extraction [130]. A schematic of PFE equipment is shown in Figure 12.8.

Extraction at elevated temperature affects the physico-chemical properties of the extraction solvent in such a way that the density of the solvent decreases with increasing temperature. In addition, the solvent viscosity decreases with increasing temperature. This is mainly due to the breaking of the intermolecular interaction between the solvent molecules. Furthermore, the selectivity of PFE is correlated with the temperature. For instance, the temperature affects the dielectric constant of the extraction solvent. This phenomenon is especially extensive for water, with its strong hydrogen bonding properties. At ambient temperature, water is a polar solvent, which is suitable for extracting polar compounds. At temperatures around 200 °C, water has a dielectric constant of around 35, which is a remarkable reduction from 80 at ambient temperature [131]. The low dielectric constant of water at elevated temperature is because of the weakening of its hydrogen-bonding property. Water becomes disordered when energy is added and this leads to disruption of the hydrogen bonds between water molecules. Consequently, water at a temperature of around 200 °C becomes less polar and has a dielectric constant comparable with methanol and ethanol at ambient temperature [124,131].

Water is probably the most environmentally friendly solvent and can be used in PFE for extracting less polar compounds from biomass because the extraction step can be integrated into the pre-treatment unit operation. Extraction of betulin from birch bark is an interesting

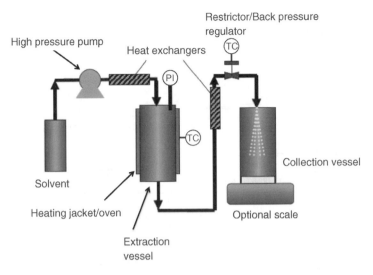

Figure 12.8 *A schematic picture of PFE where the collection vessel is put on a scale to determine the amount of obtained extract. PI: pressure indicator, TC: temperature control.*

example. In a previous study by Co *et al.* [58], PFE was optimized using a response surface design to determine the best condition for extracting high yields of betulin from birch bark. The solvents used in the study were ethanol and water. Betulin was effectively extracted using ethanol, and the optimized conditions were determined to be 120 °C, 50 bar and 15 minutes duration, as shown in Figure 12.9.

Interestingly, betulin could not be extracted by water, even at the highest temperature tested, 180 °C. In the same study, Co *et al.* [58] used solubility parameter plots to explain why water at elevated temperatures was not optimum to extract betulin from birch bark, as shown in Figure 12.10. The solubility parameters of betulin, water and ethanol were calculated by

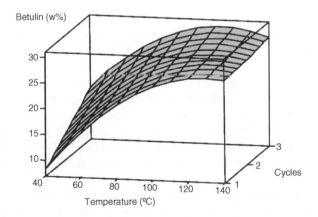

Figure 12.9 *A response surface plot of betulin yield vs. the extraction temperature and extraction time using ethanol as a solvent. Reproduced by permission of The Royal Society of Chemistry (RSC).*

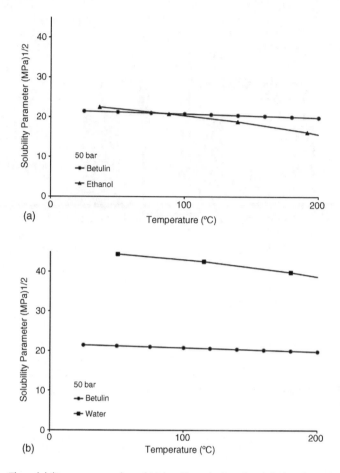

Figure 12.10   *The solubility parameter plots of (a) betulin and ethanol and (b) betulin and water at 50 bar.*

combining Hansen's solubility parameter with Joback's group contribution and Fedor's solubility parameter.

Figure 12.10 shows the solubility parameter plots of betulin and the tested extraction solvents, ethanol and water, respectively. Two components have mutual miscibility if their curves are close and the differences between the curves do not exceed four units. In Figure 12.10a, the curves of betulin and ethanol are close, which indicates mutual miscibility. The opposite is observed for the curves of betulin and water (see Figure 12.10b); the curves are more than four units away from each other. Hence, the developed solubility parameter supports the experimental result that water, even at the highest temperature tested, could not extract betulin from birch bark.

Betulin, as mentioned earlier, is a bioactive compound that has functional activity towards bacteria, fungi and viruses. Despite this, betulin is not an antioxidant using the DPPH assay [58]. This can be explained by examining the molecular structure of betulin (Figure 12.4): betulin does not have a conjugated ring system to which methyl groups are

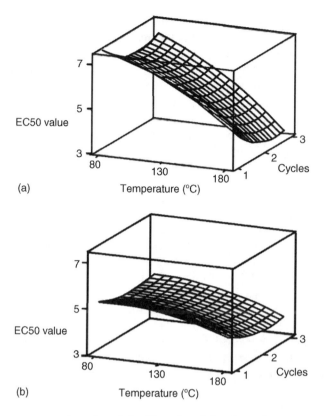

(a)

(b)

Figure 12.11 *The response surface plots of the EC$_{50}$ value of the obtained extracts using (a) ethanol and (b) water, respectively. One cycle corresponds to a five minute static extraction.*

attached. In general, antioxidants are electron-rich molecules with conjugated ring system and hydroxyl groups.

Co *et al.* extracted molecules other than betulin that displayed antioxidant capacity from birch bark, using PFE. The optimal conditions using ethanol and water, respectively, were also investigated by a response surface design, comprising three levels and 13 middle points. The antioxidant capacity of the extract was determined by the DPPH assay, and the antioxidant capacity was expressed as an EC$_{50}$ value, as described above. Two response surface plots, as depicted in Figure 12.11, were assembled. A general trend was noticed for PFE using ethanol and water, respectively. The highest antioxidant capacity was obtained with extracts that had been prepared at the highest temperature tested, which was 180 °C. In addition, a statistical test with a 95% confidence interval showed that the extraction temperature significantly affected the antioxidant capacity, while the extraction time did not.

Willför *et al.* determined some polysaccharides in 11 industrially important tree species, and the birch tree was amongst these [132]. Different wood samples were extracted with PFE using first hexane to obtain lipophilic compounds and then an acetone:water (95:5 v/v) mixture for hydrophilic compounds. It is important to note that the acetone:water extracts contained monomeric and dimeric sugar units, and no hemicelluloses. Different treatments were applied to the extracts in order to obtain celluloses, hemicelluloses and pectins, prior to silylation for

GC-analysis. For example, acid hydrolysis of the extracts was conducted to obtain celluloses, and acid methanolysis to obtain hemicelluolse and pectins (i.e., non-cellulosic sugars). The results obtained by Willför *et al.* displayed that the predominant polysaccharide was cellulose, with content between 60 to 80% (e.g., 331–515 mg g^{-1}) of the dry wood for all 11 species. For birch, the cellulose content was approximately 415 mg g^{-1} dry wood. The non-cellulosic sugar found in the investigated species was between 15–25% (e.g., 198–378 mg g^{-1}), with birch in the high end, around 250 mg g^{-1} dry wood. The dominant sugar unit in the non-cellulosic sugar was xylose, while glucose and mannose were in much smaller amounts. The high amount of polysaccharides in different wood species could be recovered and be of use in the biorefinery concept. For example, the recovered polysaccharide could be used in papermaking to improve the quality of paper, or as polymer in food and pharmaceutical applications.

Finally, xylose is an interesting sugar species with pharmaceutical applications, for example, for its antiproliferative activity. The xylose is linked to naphthalene to form naphthoxylosides, which attack the cancer cells and save the healthy cells [133]. The first trial performed by Mani *et al.* showed that treatment with this xyloside at a pharmacologically relevant dose reduced the average tumour load in severe combined immunodeficiency mice (SCID-mice) by 70–97% [134]. Furthermore, Nilsson *et al.* investigated the antiproliferative activity of two series of peracetylated mono- or bis-xylosylated naphthoxylosides and compared the antiproliferative effects with an unprotected analogue for normal human foetal lung fibroblasts cells (HFL-1) and human bladder carcinoma cells (T24) [135]. Their data showed that the antiproliferative activity increased upon peracetylation, but selectivity for T24 cells was not noticed. Nonetheless, all tested compounds succeeded in priming glycosaminoglycan chains, which means that they were capable of entering the cells.

12.6   Discussion and Future Perspectives

The interest in bioactive compounds from plant material is increasing due to environmental awareness. Extracting these bioactive compounds from biomass that is destined for biorefineries is attractive because this maximizes the use of renewable resources. Birch bark is a rich source of bioactive compounds with historical anecdotal uses, such as material that was used to combat diseases. For instance, birch bark was found in the stomach of the oldest human found, Ötzi the Iceman, and there are theories around the reason for ingesting birch bark, as it is antiviral, and perhaps quite nutritious too. It is therefore worthwhile to extract these bioactive compounds from birch bark, since the production of bark is in the thousands of tonnes per annum [4]. Up-to-date, analytical scale PFE has proved successful in rapidly extracting betulin and other bioactive compounds from birch barks without using any environmentally hazardous solvents. Hot water extraction of birch bark gave extracts that contained a variety of compounds with antioxidant capacity. Identification and characterization of the bioactivity of these compounds is required in order to use them wisely and optimally. PFE scale-up has so far not been investigated, which would be of major interest, especially if placed on-site at a debarking process where the bark is easily accessible. Separation and purification of the obtained extracts would be another interesting development of PFE of bioactive compounds from birch bark. For instance, betulin is a crystalline compound, which preferably could be separated from co-extracted compounds through simple crystallization or precipitation. Finally, advanced analytical separation and detection are important for the development of future high-technology biorefinery concepts. With more knowledge about what valuable compounds can be found in a biomass material such as birch bark, as well as their concentrations, it would become possible to calculate if a larger-scale process is feasible in terms of economic and environmental advantages.

Acknowledgements

FORMAS (Swedish Research Council for or Environment, Agricultural Sciences and Spatial Planning, 209-2006-1346, 229-2009-1527), SSF (Swedish Foundation for Strategic Research 2005:0073/13), VR (Swedish Research Council, 2006-4084), STINT (Swedish Foundation for International Cooperation in Research and Higher Education, YR2009-7015), Oscar & LiliLamm's Foundation, Innovationsbron and SIDA (Swedish International Development Cooperation Agency, SWE-2005-473) are acknowledged for financial support. Skutskär Stora Enso is greatfully acknowledged for collecting and providing birch bark in two of our larger-scale studies. Last but not least, Monica Waldebäck, Per Sjöberg, Leif Nyholm and Camilla Zettersten along with former M.Sc. thesis students, Sandra Wende, Pirjo Koskela and Peter Eklund-Åkergren, are thanked for all their interesting contributions, with critical discussions and experimental data.

References

1. Hiltunen, E., Pakkanen, T.T., and Alvila, L. (2004) Phenolic extractives from wood of birch (Betula pendula). *Holzforschung*, **58**, 326–329.
2. Krasutsky, P.A. (2006) Birch bark research and development. *Natural Product Report*, **23**, 919–942.
3. Uri, V., Lohmus, K., Kiviste, A., and Aosaar, J. (2009) The dynamics of biomass production in relation to foliar and root traits in a grey alder (Alnus incana (L.) Moench) plantation on abandoned agricultural land. *Forestry*, **82**, 61–74.
4. Skogsstyrelsen (2011) http://www.skogsstyrelsen.se/Myndigheten/Statistik/Skogsstatistisk-Arsbok/.
5. Energimyndigheten (2010) Energiläget 2010, Statens Energimyndighet, 1403–1892.
6. Chemical, A. (2011) http://www.arizonachemical.com/.
7. Maki-Arvela, P., Holmbom, B., Salmi, T., and Murzin, D.Y. (2007) Recent progress in synthesis of fine and specialty chemicals from wood and other biomass by heterogeneous catalytic processes. *Catalysis Reviews-Science and Engineering*, **49**, 197–340.
8. Ryan, E., Galvin, K., O'Connor, T.P. *et al.* (2006) Fatty acid profile, tocopherol, squalene and phytosterol content of brazil, pecan, pine, pistachio and cashew nuts. *International Journal of Food Sciences and Nutrition*, **57**, 219–228.
9. Oy, F.W.R. (2011) http://www.fwr.fi/en.
10. Ekman, A., Campos, M., Lindahl, S. *et al.* (2011) Value addition in bioresource utilization by sustainable technologies in new biorefinery concepts. *Submitted to Energy & Environmental Science*.
11. Beking, K. and Vieira, A. (2010) Flavonoid intake and disability-adjusted life years due to Alzheimer's and related dementias: a population-based study involving twenty-three developed countries. *Public Health Nutrition*, **13**, 1403–1409.
12. Choi, A.Y., Choi, J.H., Lee, J.Y. *et al.* (2010) Apigenin protects HT22 murine hippocampal neuronal cells against endoplasmic reticulum stress-induced apoptosis. *Neurochemistry International*, **57**, 143–152.
13. Gonzalez-Gallego, J., Garcia-Mediavilla, M.V., Sanchez-Campos, S., and Tunon, M.J. (2010) Fruit polyphenols, immunity and inflammation. *British Journal of Nutrition*, **104**, S15–S27.

14. Obrenovich, M.E., Nair, N.G., Beyaz, A. *et al.* (2010) The role of polyphenolic antioxidants in health, disease, and aging. *Rejuvenation Research*, **13**, 631–643.
15. Halliwell, B. (2001) Role of free radicals in the neurodegenerative diseases - Therapeutic implications for antioxidant treatment. *Drugs & Aging*, **18**, 685–716.
16. Loizzo, M.R., Tundis, R., and Menichini, F. (2008) Natural products and their derivatives as cholinesterase inhibitors in the treatment of neurodegenerative disorders: An update. *Current Medicinal Chemistry*, **15**, 1209–1228.
17. Valko, M., Leibfritz, D., Moncol, J. *et al.* (2007) Free radicals and antioxidants in normal physiological functions and human disease. *International Journal of Biochemistry & Cell Biology*, **39**, 44–84.
18. Droge, W. (2002) Free radicals in the physiological control of cell function. *Physiological Reviews*, **82**, 47–95.
19. Prior, R.L., Wu, X., and Schaich, K. (2005) Standardized methods for the determination of antioxidant capacity and phenolics in foods and dietary supplements. *Journal of Agricultural and Food Chemistry*, **53**, 4290–4302.
20. Magalhaes, L.M., Segundo, M.A., Reis, S., and Lima, J. (2008) Methodological aspects about in vitro evaluation of antioxidant properties. *Analytica Chimica Acta*, **613**, 1–19.
21. Cao, G.H., Alessio, H.M., and Cutler, R.G. (1993) Oxygen-radical absorbency capacity assay for antioxidants. *Free Radical Biology and Medicine*, **14**, 303–311.
22. Ghiselli, A., Serafini, M., Natella, F., and Scaccini, C. (2000) Total antioxidant capacity as a tool to assess redox status: critical view and experimental data. *Free Radical Biology and Medicine*, **29**, 1106–1114.
23. Bors, W., Michel, C., and Saran, M. (1984) Inhibition of bleaching of the carotenoid crocin - a rapid test for quantifying antioxidant activity. *Biochimica Et Biophysica Acta*, **796**, 312–319.
24. Folin, O. and Ciocalteu, V. (1927) On tyrosine and tryptophane determinations in proteins. *Journal of Biological Chemistry*, **73**, 627–650.
25. Singleton, V.L., Orthofer, R., and Lamuela-Raventos, R.M. (1999) *Oxidants and Antioxidants, Pt A (Methods in Enzymology), Analysis of Total Phenols and other Oxidation Substrates and Antioxidants by Means of Folin-Ciocalteu Reagent*, Academic Press Inc., San Diego, pp. 152–178.
26. Re, R., Pellegrini, N., Proteggente, A. *et al.* (1999) Antioxidant activity applying an improved ABTS radical cation decolorization assay. *Free Radical Biology and Medicine*, **26**, 1231–1237.
27. Miller, N.J., Rice-Evans, C., Davies, M.J. *et al.* (1993) A novel method for measuring antioxidant capacity and its application to monitoring the antioxidant status in premature neonates. *Clinical Science*, **84**, 407–412.
28. Benzie, I.F.F. and Strain, J.J. (1999) *Oxidants and Antioxidants, Pt A (Methods in Enzymology), Ferric Reducing Antioxidant Power Assay: Direct Measure of Total Antioxidant Activity of Biological Fluids and Modified Version for Simultaneous Measurement of Total Antioxidant Power and Ascorbic Acid Concentration*, Academic Press Inc., San Diego, pp. 15–27.
29. Huang, D.J., Ou, B.X., and Prior, R.L. (2005) The chemistry behind antioxidant capacity assays. *Journal of Agricultural and Food Chemistry*, **53**, 1841–1856.
30. Somogyi, A., Rosta, K., Pusztai, P. *et al.* (2007) Antioxidant measurements. *Physiological Measurement*, **28**, R41–R55.
31. Cao, G.H. and Prior, R.L. (1998) Comparison of different analytical methods for assessing total antioxidant capacity of human serum. *Clinical Chemistry*, **44**, 1309–1315.
32. Chandrasekara, N. and Shahidi, F. (2011) Effect of roasting on phenolic content and antioxidant activities of whole cashew nuts, kernels, and testa. *Journal of Agricultural and Food Chemistry*, **59**, 5006–5014.
33. Wayner, D.D.M., Burton, G.W., Ingold, K.U., and Locke, S. (1985) Quantitative measurement of the total peroxyl radical trapping antioxidant capability of human blood plasma by controlled peroxidation - the important contribution made by plasma proteins. *Febs Letters*, **187**, 33–37.
34. Diouf, P.N., Stevanovic, T., and Boutin, Y. (2009) The effect of extraction process on polyphenol content, triterpene composition and bioactivity of yellow birch (Betula alleghaniensis Britton) extracts. *Industrial Crops and Products*, **30**, 297–303.

35. Kahkonen, M.P., Hopia, A.I., Vuorela, H.J. *et al.* (1999) Antioxidant activity of plant extracts containing phenolic compounds. *Journal of Agricultural and Food Chemistry*, **47**, 3954–3962.

36. Granato, D., Branco, G.F., Faria, J.D.F., and Cruz, A.G. (2011) Characterization of Brazilian lager and brown ale beers based on color, phenolic compounds, and antioxidant activity using chemometrics. *Journal of the Science of Food and Agriculture*, **91**, 563–571.

37. Hudthagosol, C., Haddad, E.H., McCarthy, K. *et al.* (2011) Pecans acutely increase plasma postprandial antioxidant capacity and catechins and decrease LDL oxidation in humans. *Journal of Nutrition*, **141**, 56–62.

38. Alvarez-Parrilla, E., de la Rosa, L.A., Amarowicz, R., and Shahidi, F. (2011) Antioxidant activity of fresh and processed jalapeno and serrano Peppers. *Journal of Agricultural and Food Chemistry*, **59**, 163–173.

39. Denev, P., Ciz, M., Ambrozova, G. *et al.* (2010) Solid-phase extraction of berries' anthocyanins and evaluation of their antioxidative properties. *Food Chemistry*, **123**, 1055–1061.

40. Venturini, D., Simao, A.N.C., Barbosa, D.S. *et al.* (2010) Increased oxidative stress, decreased total antioxidant capacity, and iron overload in untreated patients with chronic hepatitis C. *Digestive Diseases and Sciences*, **55**, 1120–1127.

41. Ciz, M., Cizova, H., Denev, P. *et al.* (2010) Different methods for control and comparison of the antioxidant properties of vegetables. *Food Control*, **21**, 518–523.

42. Ordoudi, S.A., Tsermentseli, S.K., Nenadis, N. *et al.* (2011) Structure-radical scavenging activity relationship of alkannin/shikonin derivatives. *Food Chemistry*, **124**, 171–176.

43. Ordoudi, S.A. and Tsimidou, M.Z. (2006) Crocin bleaching assay (CBA) in structure-radical scavenging activity studies of selected phenolic compounds. *Journal of Agricultural and Food Chemistry*, **54**, 9347–9356.

44. Rodriguez, S.A. and Murray, A.P. (2010) Antioxidant activity and chemical composition of essential oil from atriplex undulata. *Natural Product Communications*, **5**, 1841–1844.

45. Utkina, N.K. (2009) Antioxidant activity of aromatic alkakoids from marine sponges Aaptos aaptos and Hyrtios SP. *Chemistry of Natural Compounds*, **45**, 849–853.

46. Zielinska, D., Wiczkowski, W., and Piskula, M.K. (2008) Determination of the relative contribution of quercetin and its glucosides to the antioxidant capacity of onion by cyclic voltammetry and spectrophotometric methods. *Journal of Agricultural and Food Chemistry*, **56**, 3524–3531.

47. Yi, W.G. and Wetzstein, H.Y. (2011) Effects of drying and extraction conditions on the biochemical activity of selected herbs. *Hortscience*, **46**, 70–73.

48. Gan, R.Y., Xu, X.R., Song, F.L. *et al.* (2010) Antioxidant activity and total phenolic content of medicinal plants associated with prevention and treatment of cardiovascular and cerebrovascular diseases. *Journal of Medicinal Plants Research*, **4**, 2438–2444.

49. Pinchuk, I., Shoval, H., Bor, A. *et al.* (2011) Ranking antioxidants based on their effect on human serum lipids peroxidation. *Chemistry and Physics of Lipids*, **164**, 42–48.

50. Di Bernardini, R., Rai, D.K., Bolton, D. *et al.* (2011) Isolation, purification and characterization of antioxidant peptidic fractions from a bovine liver sarcoplasmic protein thermolysin hydrolyzate. *Peptides*, **32**, 388–400.

51. Boyle, S.P., Doolan, P.J., Andrews, C.E., and Reid, R.G. (2011) Evaluation of quality control strategies in Scutellaria herbal medicines. *Journal of Pharmaceutical and Biomedical Analysis*, **54**, 951–957.

52. Wang, R.J. and Hu, M.L. (2011) Antioxidant capacities of fruit extracts of five mulberry genotypes with different assays and principle components analysis. *International Journal of Food Properties*, **14**, 1–8.

53. Frassinetti, S., Caltavuturo, L., Cini, M. *et al.* (2011) Antibacterial and antioxidant activity of essential oils from citrus spp. *Journal of Essential Oil Research*, **23**, 27–31.

54. Gaikwad, P., Barik, A., Priyadarsini, K.I., and Rao, B.S.M. (2010) Antioxidant activities of phenols in different solvents using DPPH assay. *Research on Chemical Intermediates*, **36**, 1065–1072.

55. Hwang, I.G., Kim, H.Y., Woo, K.S. *et al.* (2011) Biological activities of Maillard reaction products (MRPs) in a sugar-amino acid model system. *Food Chemistry*, **126**, 221–227.

56. Koleva, II., van Beek, T.A., Linssen, J.P.H. *et al.* (2002) Screening of plant extracts for antioxidant activity: a comparative study on three testing methods. *Phytochemical Analysis*, **13**, 8–17.

57. Brand-Williams, W., Cuvelier, M.E., and Berset, C. (1995) Use of a free radical method to evaluate antioxidant activity. *Food Science Technology (N. Y.)*, **28**, 25–30.

58. Co, M., Koskela, P., Eklund-Åkergren, P. *et al.* (2009) Pressurized liquid extraction of betulin and antioxidants from birch bark. *Green Chemistry*, **11**, 668–674.

59. García-Pérez, M.-E., Diouf Papa, N., and Stevanovic, T. (2008) Comparative study of antioxidant capacity of yellow birch twigs extracts at ambient and high temperatures. *Food Chemistry*, **107**, 344–351.

60. Valgas, C., de Souza, S.M., Smania, E.F.A., and Smania, A. (2007) Screening methods to determine antibacterial activity of natural products. *Brazilian Journal of Microbiology*, **38**, 369–380.

61. Rios, J.L., Recio, M.C., and Villar, A. (1988) Screening methods for natural products with antimicrobial activity: A review of the literature. *Journal of Ethnopharmacology*, **23**, 127–149.

62. Lambert, R.J.W. and Pearson, J. (2000) Susceptibility testing: accurate and reproducible minimum inhibitory concentration (MIC) and non-inhibitory concentration (NIC) values. *Journal of Applied Microbiology*, **88**, 784–790.

63. van Vuuren, S.F. (2008) Antimicrobial activity of South African medicinal plants. *Journal of Ethnopharmacology*, **119**, 462–472.

64. Salin, O., Alakurtti, S., Pohjala, L. *et al.* (2010) Inhibitory effect of the natural product betulin and its derivatives against the intracellular bacterium Chlamydia pneumoniae. *Biochemical Pharmacology*, **80**, 1141–1151.

65. Davis, E.M. (1988) Protein assays - a review of common techniques. *American Biotechnology Laboratory*, **6**, 28–37.

66. Wieland, V. (2005) *Chemosensitivity, Sulforhodamine B Assay and Chemosensitivity*, Humana Press, Totowa, N.

67. Drag, M., Surowiak, P., Drag-Zalesinska, M. *et al.* (2009) Comparision of the cytotoxic effects of birch bark extract, betulin and betulinic acid towards human gastric carcinoma and pancreatic carcinoma drug-sensitive and drug-resistant cell lines. *Molecules*, **14**, 1639–1651.

68. Ehrhardt, H., Fulda, S., Fuhrer, M. *et al.* (2004) Betulinic acid-induced apoptosis in leukemia cells. *Leukemia*, **18**, 1406–1412.

69. Burton, J.D. (2005) *Chemosensitivity, The MTT Assay to Evaluate Chemosensitivity*, Human Press.

70. Zuco, V., Supino, R., Righetti, S.C. *et al.* (2002) Selective cytotoxicity of betulinic acid on tumor cell lines, but not on normal cells. *Cancer Letters*, **175**, 17–25.

71. Yang, X.W., Li, S.M., Shen, Y.H., and Zhang, W.D. (2008) Phytochemical and biological studies of Abies species. *Chemistry & Biodiversity*, **5**, 56–81.

72. Lewinsohn, E., Gijzen, M., and Croteau, R.B. (1992) Regulation of monoterpene biosynthesis in conifer defense. *Acs Symposium Series*, **497**, 8–17.

73. Trapp, S. and Croteau, R. (2001) Defensive resin biosynthesis in conifers. *Annual Review of Plant Physiology and Plant Molecular Biology*, **52**, 689–724.

74. Boszormenyi, A., Szarka, S., Hethelyi, E. *et al.* (2009) Triterpenes in traditional and supercritical-fluid extracts of morus alba leaf and stem bark. *Acta Chromatographica*, **21**, 659–669.

75. Regert, A., Alexandre, V., Thomas, N., and Lattuati-Derieux, A. (2006) Molecular characterisation of birch bark tar by headspace solid-phase microextraction gas chromatography-mass spectrometry: A new way for identifying archaeological glues. *Journal of Chromatography A*, **1101**, 245–253.

76. Jäger, S., Laszczyk, M., and Scheffler, A. (2008) A preliminary pharmacokinetic study of betulin, the main pentacyclic triterpene from extract of outer bark of birch (Betulae alba cortex). *Molecules*, **13**, 3224–3235.

77. Yi, J.E., Obminska-Mrukowicz, B., Yuan, L.Y., and Yuan, H. (2010) Immunomodulatory effects of betulinic acid from the bark of white birch on mice. *Journal of Veterinary Science*, **11**, 305–313.

78. Fujioka, T., Kashiwada, Y., Kilkuskie, R.E. *et al.* (1994) Anti-aids agents - 11 betulinic acid and platanic acid as anti-HIV principles from syzigium-claviflorum, and the anti-HIV activity of structurally related triterpenoids. *Journal of Natural Products*, **57**, 243–247.

79. Dorr, C.R., Yemets, S., Kolomitsyna, O. *et al.* (2011) Triterpene derivatives that inhibit human immunodeficiency virus type 1 replication. *Bioorganic & Medicinal Chemistry Letters*, **21**, 542–545.

80. Yogeeswari, P. and Sriram, D. (2005) Betulinic acid and its derivatives: A review on their biological properties. *Current Medicinal Chemistry*, **12**, 657–666.

81. Liu, J., Fu, M.L., and Chen, Q.H. (2011) Biotransformation optimization of betulin into betulinic acid production catalysed by cultured Armillaria luteo-virens Sacc ZJUQH100-6 cells. *Journal of Applied Microbiology*, **110**, 90–97.

82. Csuk, R., Schmuck, K., and Schafer, R. (2006) A practical synthesis of betulinic acid. *Tetrahedron Letters*, **47**, 8769–8770.

83. Rhourri-Frih, B., Chaimbault, P., Claude, B. *et al.* (2009) Analysis of pentacyclic triterpenes by LC-MS. A comparative study between APCI and APPI. *Journal of Mass Spectrometry*, **44**, 71–80.

84. Lee, Y.C. (1990) High-performance anion-exchange chromatography for carbohydrate analysis. *Analytical Biochemistry*, **189**, 151–162.

85. Churms, S.C. (1996) Recent progress in carbohydrate separation by high-performance liquid chromatography based on hydrophilic interaction. *Journal of Chromatography A*, **720**, 75–91.

86. Ikegami, T., Tomomatsu, K., Takubo, H. *et al.* (2008) Separation efficiencies in hydrophilic interaction chromatography. *Journal of Chromatography A*, **1184**, 474–503.

87. Ikegami, T., Horie, K., Saad, N. *et al.* (2008) Highly efficient analysis of underivatized carbohydrates using monolithic-silica-based capillary hydrophilic interaction (HILIC) HPLC. *Analytical and Bioanalytical Chemistry*, **391**, 2533–2542.

88. Antonio, C., Pinheiro, C., Chaves, M.M. *et al.* (2008) Analysis of carbohydrates in Lupinus albus stems on imposition of water deficit, using porous graphitic carbon liquid chromatography-electrospray ionization mass spectrometry. *Journal of Chromatography A*, **1187**, 111–118.

89. Cheng, C., Chen, C.S., and Hsieh, P.H. (2010) On-line desalting and carbohydrate analysis for immobilized enzyme hydrolysis of waste cellulosic biomass by column-switching high-performance liquid chromatography. *Journal of Chromatography A*, **1217**, 2104–2110.

90. Ganzera, M. and Stuppner, H. (2005) Evaporative light scattering detection (ELSD) for the analysis of natural products. *Current Pharmaceutical Analysis*, **1**, 135–144.

91. Wilson, R., Cataldo, A., and Andersen, C.P. (1995) Determination of total nonstructural carbohydrates in tree species by high-performance anion-exchange chromatography with pulsed amperometric detection. *Canadian Journal of Forest Research-Revue Canadienne De Recherche Forestiere*, **25**, 2022–2028.

92. Vehovec, T. and Obreza, A. (2010) Review of operating principle and applications of the charged aerosol detector. *Journal of Chromatography A*, **1217**, 1549–1556.

93. Bruggink, C., Maurer, R., Herrmann, H. *et al.* (2005) Analysis of carbohydrates by anion exchange chromatography and mass spectrometry. *Journal of Chromatography A*, **1085**, 104–109.

94. Scott, R. (2009) *Encyclopedia of Chromatography*, 3rd edn (Print Version), Carbohydrates, CRC Press.

95. Gao, X.B., Yang, J.H., Huang, F. *et al.* (2003) Progresses of derivatization techniques for analyses of carbohydrates. *Analytical Letters*, **36**, 1281–1310.

96. Rice-Evans, C.A., Miller, N.J., and Paganga, G. (1996) Structure-antioxidant activity relationships of flavonoids and phenolic acids. *Free Radical Biology and Medicine*, **20**, 933–956.

97. Mayer, A.M. (2006) Polyphenol oxidases in plants and fungi: Going places? A review. *Phytochemistry*, **67**, 2318–2331.

98. Houghton, P. and Howes, M.-J. (2005) Natural products and derivatives affecting neurotransmission relevant to Alzheimer's and Parkinson's Disease. *Neurosignals*, **14**, 6–22.

99. Scalbert, A., Manach, C., Morand, C. *et al.* (2005) Dietary polyphenols and the prevention of diseases. *Critical Reviews in Food Science and Nutrition*, **45**, 287–306.

100. Herrmann, K. (1988) On the occurrence of flavonol glycosides in vegetables. *Zeitschrift Fur Lebensmittel-Untersuchung Und-Forschung*, **186**, 1–5.

101. Mammela, P. (2001) Phenolics in selected European hardwood species by liquid chromatography-electrospray ionisation mass spectrometry. *Analyst*, **126**, 1535–1538.

102. Barry, K.M., Davies, N.W., and Mohammed, C.L. (2002) Effect of season and different fungi on phenolics in response to xylem wounding and inoculation in Eucalyptus nitens. *Forest Pathology*, **32**, 163–178.

103. Luostarinen, K. and Mottonen, V. (2004) Effect of growing site, sampling date, wood location in trunk and drying method on concentration of soluble proanthocyanidins in Betula pendula wood, with special reference to wood colour. *Scandinavian Journal of Forest Research*, **19**, 234–240.

104. Porter, L.J., Hrstich, L.N., and Chan, B.G. (1986) The conversion of procyanidins and prodelphinidins to cyanidin and delphinidin. *Phytochemistry*, **25**, 223–230.

105. Blasco, A.J., González Crevillén, A., González, M.C., and Escarpa, A. (2007) Direct electrochemical sensing and detection of natural antioxidants and antioxidant capacity in vitro systems. *Electroanalysis*, **19**, 2275–2286.

106. Ragubeer, N., Beukes, D.R., and Limson, J.L. (2010) Critical assessment of voltammetry for rapid screening of antioxidants in marine algae. *Food Chemistry*, **121**, 227–232.

107. Corduneanu, O., Janeiro, P., and Brett, A.M.O. (2006) On the electrochemical oxidation of resveratrol. *Electroanalysis*, **18**, 757–762.

108. Chevion, S., Roberts, M.A., and Chevion, M. (2000) The use of cyclic voltammetry for the evaluation of antioxidant capacity. *Free Radical Biology and Medicine*, **28**, 860–870.

109. Zielinska, D., Nagels, L., and Piskula, M.K. (2008) Determination of quercetin and its glucosides in onion by electrochemical methods. *Analytica Chimica Acta*, **617**, 22–31.

110. Aaby, K., Hvattum, E., and Skrede, G. (2004) Analysis of flavonoids and other phenolic compounds using high-performance liquid chromatography with coulometric array detection: Relationship to antioxidant activity. *Journal of Agricultural and Food Chemistry*, **52**, 4595–4603.

111. Zettersten, C., Co, M., Wende, S. *et al.* (2009) Identification and characterization of polyphenolic antioxidants using on-line liquid chromatography, electrochemistry, and electrospray ionization tandem mass spectrometry. *Analytical Chemistry*, **81**, 8968–8977.

112. Co, M., Zettersten, C., Nyholm, L. *et al.* (2011) Degradation effects in the extraction of antioxidants from birch bark using water at elevated temperature and pressure. *Analytica Chimica Acta*. In Press, Corrected Proof. doi: 10.1016/j.aca.2011.04.038

113. Co, M., Fagerlund, A., Engman, L. *et al.* (2011) Extraction of antioxidants from spruce (Picea abies) bark using eco-friendly solvents. *Phytochemical Analysis*. In Press, Corrected Proof. doi: 10.1002/pca.1316

114. Zhao, G.L., Yan, W.D., and Cao, D. (2007) Simultaneous determination of betulin and betulinic acid in white birch bark using RP-HPLC. *Journal of Pharmaceutical and Biomedical Analysis*, **43**, 959–962.

115. Bowadt, S. and Hawthorne, S.B. (1995) Supercritical fuid extraction in environmental analysis. *Journal of Chromatography A*, **703**, 549–571.

116. Hawthorne, S.B. (1990) Analytical-scale supercritical fluid extraction. *Analytical Chemistry*, **62**, A633-&.

117. Chrastil, J. (1982) Solubility of solids and liquids in supercritical gases. *Journal of Physical Chemistry*, **86**, 3016–3021.

118. Turner, C., King, J.W., and Mathiasson, L. (2001) Supercritical fluid extraction and chromatography for fat-soluble vitamin analysis. *Journal of Chromatography A*, **936**, 215–237.

119. Lang, Q. and Wai, C.M. (2001) Supercritical fluid extraction in herbal and natural product studies - a practical review. *Talanta*, **53**, 771–782.

120. Langenfeld, J.J., Hawthorne, S.B., Miller, D.J., and Pawliszyn, J. (1994) Role of modifiers for analytical-scale supercritical fluid extraction of environmental samples. *Analytical Chemistry*, **66**, 909–916.

121. Felfoldi-Gava, A., Simandi, B., Plander, S. *et al.* (2009) Betulaceae and platanaceae plants as alternative sources of selected lupane-type triterpenes - Their composition profile and betulin content. *Acta Chromatographica*, **21**, 671–681.

122. Richter, B.E., Jones, B.A., Ezzell, J.L. *et al.* (1996) Accelerated solvent extraction: a technique for sample preparation. *Analytical Chemistry*, **68**, 1033–1039.

123. Schantz, M.M., Nichols, J.J., and Wise, S.A. (1997) Evaluation of pressurized fluid extraction for the extraction of environmental matrix reference materials. *Analytical Chemistry*, **69**, 4210–4219.

124. Hawthorne, S.B., Yang, Y., and Miller, D.J. (1994) Extraction of organic pollutants from environmental solids with subcritical and supercritical water. *Analytical Chemistry*, **66**, 2912–2920.

125. Williams, F.B., Sander, L.C., Wise, S.A., and Girard, J. (2006) Development and evaluation of methods for determination of naphthodianthrones and flavonoids in St. John's wort. *Journal of Chromatography A*, **1115**, 93–102.

126. da Costa, C.T., Margolis, S.A., Benner, B.A., and Horton, D. (1999) Comparison of methods for extraction of flavanones and xanthones from the root bark of the osage orange tree using liquid chromatography. *Journal of Chromatography A*, **831**, 167–178.

127. Schantz, M.M. (2006) Pressurized liquid extraction in environmental analysis. *Analytical and Bioanalytical Chemistry*, **386**, 1043–1047.

128. Ong, E.S. and Len, S.M. (2003) Pressurized hot water extraction of berberine, baicalein and glycyrrhizin in medicinal plants. *Analytica Chimica Acta*, **482**, 81–89.

129. Hyotylainen, T., Hartonen, K., Saynajoki, S., and Riekkola, M.L. (2001) Pressurised hot-water extraction of brominated flame retardants in sediment samples. *Chromatographia*, **53**, 301–305.

130. Turner, C. and Waldebäck, M. (2010) in *Separation, Extraction and Concentration Processes in the Food, Beverage and Nutraceutical Industries, Principles of Pressurized Fluid Extraction and Environmental, Food and Agricultural Applications* (ed. S. Rizvi), Woodhead Publishing Limited Incorporating Chandos, Publishing.

131. Yang, Y., Belghazi, M., Lagadec, A. *et al.* (1998) Elution of organic solutes from different polarity sorbents using subcritical water. *Journal of Chromatography A*, **810**, 149–159.

132. Willfor, S., Sundberg, A., Pranovich, A., and Holmbom, B. (2005) Polysaccharides in some industrially important hardwood species. *Wood Science and Technology*, **39**, 601–617.

133. Jacobsson, M. and Ellervik, U. (2002) Synthesis of naphthoxylosides on solid support. *Tetrahedron Letters*, **43**, 6549–6552.

134. Mani, K., Belting, M., Ellervik, U. *et al.* (2004) Tumor attenuation by 2(6-hydroxynaphthyl)-beta-D-xylopyranoside requires priming of heparan sulfate and nuclear targeting of the products. *Glycobiology*, **14**, 387–397.

135. Nilsson, U., Jacobsson, M., Johnsson, R. *et al.* (2009) Antiproliferative effects of peracetylated naphthoxylosides. *Bioorganic & Medicinal Chemistry Letters*, **19**, 1763–1766.

13

Adding Value to the Integrated Forest Biorefinery with Co-Products from Hemicellulose-Rich Pre-Pulping Extract

Abigail S. Engelberth[1] and G. Peter van Walsum[2]

[1] *Laboratory of Renewable Resources Engineering, Department of Agricultural and Biological Engineering, Potter Engineering Center, Purdue University, West Lafayette, Indiana, USA*
[2] *Forest Bioproducts Research Institute, Department of Chemical and Biological Engineering, University of Maine, Orono, Maine, USA*

13.1   Introduction

The only source for sustainable and renewable organic carbon for use in chemicals and transportation fuels is plant biomass (Huber, Iborra, and Corma, 2006). Forestry biomass potential in the US is around 368 million dry tons annually (Perlack *et al.*, 2005), with around 108 million tons used for pulp production (Ragauskas *et al.*, 2006a). Combining existing pulp production with new technologies for production of biofuels and other bioproducts can leverage existing biomass collection methods, conversion infrastructure and technical know-how to advance the development of new bio-based products. The idea of an integrated biorefinery is to optimize the use of all fractions of biomass for the production of biofuels, bioenergy and biomaterials (Ragauskas *et al.*, 2006b).

In an integrated forest biorefinery (IFBR), the three major components of wood can be allocated to different uses that make best use of the component characteristics: the cellulose

Biorefinery Co-Products: Phytochemicals, Primary Metabolites and Value-Added Biomass Processing, First Edition.
Edited by Chantal Bergeron, Danielle Julie Carrier and Shri Ramaswamy.
© 2012 John Wiley & Sons, Ltd. Published 2012 by John Wiley & Sons, Ltd.

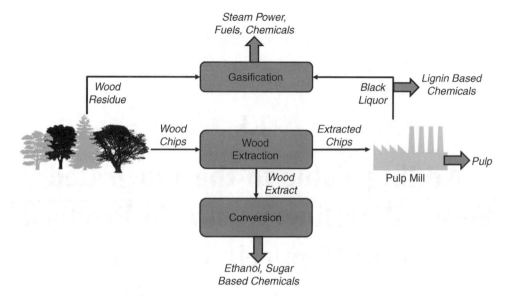

Figure 13.1   *Concept for an integrated forest biorefinery (IFBR) which would use all portions of woody biomass in a useful manner (Marinova et al., 2009). Adapted with permission from M. Marinova et al., 2009 © Elsevier (2009).*

would be used for the production of pulp, the hemicellulose would act as a precursor to sugar-based chemicals and the lignin could be used for production of high-value products such as carbon fibers, chemicals or polymers, or simply relegated to boiler fuel (van Heiningen, 2006; Marinova *et al.*, 2009). In current pulp mill operations, the cellulose is allocated to pulp, but most of the hemicellulose is allocated to boiler fuel, where it delivers low net value. In an IFBR, some of the hemicellulose could be removed prior to pulping, thus enabling the possibility of adding greater value to this stream. With current extraction technologies and pulp production, around 14 million tons of hemicellulose could be recovered annually (Ragauskas *et al.*, 2006a). Figure 13.1 is a depiction of the integration of the IFBR concept with woody biomass used in pulping. This particular design envisions using pre-pulping extraction and gasification to derive a greater variety of materials from the starting materials, all the while maintaining pulp production. An IFBR would take advantage of the "know-how" of the pulping industry to effectively transport and process lignocellulosic biomass (Kautto *et al.*, 2010).

13.1.1   Why Hemicellulose

In the traditional Kraft pulping process, wood chips are cooked in a solution of sodium hydroxide and sodium sulfide (white liquor) which dissolves the lignin and hemicellulose from the chips. These dissolved components are separated from the cellulose fiber as black liquor, which is evaporated to reduce moisture content and burnt in a recovery boiler to generate steam and recycle the pulping chemicals (van Heiningen, 2006). Roughly 52 million dry tons of pulping liquors are produced annually (Perlack *et al.*, 2005). The heating value of hemicellulose ($13.6\,\mathrm{MJ\,kg^{-1}}$) is low compared to lignin ($27\,\mathrm{MJ\,kg^{-1}}$), and hemicellulose could command a higher value if it were used to produce higher-value chemicals (van Heiningen, 2006). The recovery of hemicellulose prior to pulping could allow for an increase in the production of pulp, since most mills are bottlenecked by the capacity of the capital intensive recovery boiler. By diverting some of the hemicellulose from the black liquor, this frees some of the boiler capacity for increased pulp production (Ragauskas *et al.*, 2006a). The pre-extraction of hemicellulose from wood prior to pulping has been in practice since the early 1930s for the case of dissolving pulps (Ragauskas *et al.*, 2006a) and for Kraft pulp, recent advances have been made in the pre-extraction process to recover the hemicellulose without degrading the pulp quality. For example, pre-pulping extraction with green liquor has been shown to result in minimal reduction of pulp yield. (Mao *et al.*, 2008; Tunc and van Heiningen, 2008; Walton *et al.*, 2010c; Yoon, Cullinan, and Krishnagopalan, 2010).

13.1.2 Increased Revenue

One aim of the IFBR is to increase the revenue of the pulp mill by diversifying the product portfolio of the pulp industry (Hytonen and Stuart, 2009). The North American pulping industry has been in a period of decreased profitability due to competition from faster-growing species, reduced demand for newsprint, and higher raw material and energy costs (Marinova *et al.*, 2009; Helmerius *et al.*, 2010). One way of adding revenue is to extract and convert the hemicellulose into a more valuable product. Hemicelluloses are heterogenous biopolymers made of pentoses (C5), hexoses (C6) and sugar acids (Saha, 2003). These sugars can be depolymerized into monomers which then undergo further treatment to produce higher-value commodity chemicals or alcohols.

13.1.3 Hemicellulose Possibilities

Possible products that can be derived from hemicellulose include fermentation products, such as alcohols (ethanol, butanol), organic acids (acetic acid, lactic acid), microbial or algal lipids and commodity organic chemicals. Chemical conversions can also be applied to hemicellulose to produce hydrocarbons such as tridecane.

13.2 Hemicellulose Recovery

Hemicellulose recovery from hardwood chips prior to pulping has been demonstrated both in the lab and at an industrial scale. Hemicellulose pre-extraction using either hot water or an alkali-based solvent is discussed in this chapter. The extraction of hemicellulose prior to pulping must not degrade or diminish either the quality or the quantity of the pulp (Al-Dajani and Tschirner, 2010; Yoon, Cullinan, and Krishnagopalan, 2010). In keeping with the concept of an IFBR, the pre-extraction should also be performed in such a way that the lignin is minimally removed so that it can be recovered after pulping to create valuable products like carbon fiber or polyurethane (van Heiningen, 2006; Stoutenburg *et al.*, 2008).

13.2.1 Integration of Hemicellulose Extraction with the Current Pulping Process

The extraction of hemicellulose is integrated into the current pulping process prior to Kraft pulping. The wood chips are subjected to either hot water or an alkali solution to remove hemicellulose. The hemicellulose-rich solution is processed into higher-value chemicals and the remaining chips resume processing in standard pulping operations. The extraction of hemicellulose must be performed with a solvent that is compatible with the pulping process because some solvent will transfer into the pulping unit (van Heiningen, 2006). Because of this constraint, only water and pulp-compatible inorganic (alkali) solutions will be discussed as possible extraction solvents. Generally, softwoods use water as the pre-extraction solvent since their major hemicellulose component is acetyl-galactomannan (20 w/w %), which has been shown to degrade in alkali conditions. Hardwoods are better suited for extraction using alkali conditions since the major hemicellulose component of hardwood is glucoronoxylan (15–30 w/w %) and the xylan is degraded in the acidic conditions encountered during water extraction (van Heiningen, 2006).

A possible integration scheme for the extraction and recovery of hemicellulose with a current Kraft pulp mill is shown in Figure 13.2. This proposed scheme assumes a hardwood feed and makes use of green liquor to generate an alkali extraction solution. The wood chips are sent first to the extraction unit and then through the Kraft pulping process (Mao *et al.*, 2008). The proposed scheme shown in Figure 13.2 is an example of a biorefinery producing two products: acetic acid and ethanol. The acetic acid is recovered from the hemicellulose-rich stream using liquid–liquid extraction and the ethanol is produced by fermentation downstream of the acetic acid removal. This configuration could also be used to produce fermentation products other than ethanol, such as butanol, or an organic acid such as lactic acid.

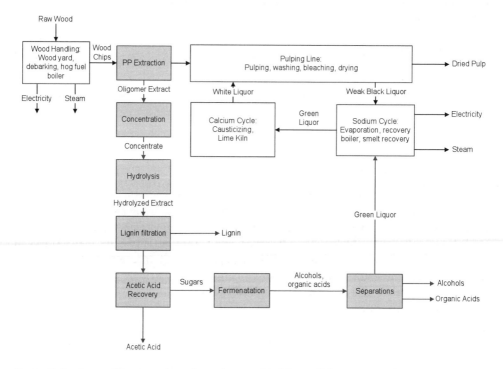

Figure 13.2   *Proposed pre-extraction scheme for removal of hemicellulose prior to pulping and subsequent recovery of acetic acid and ethanol from extracted hemicellulose. The current process steps are in white and the additional steps required for the recovery of acids and alcohols are shown in grey.*

Hot Water Extraction

Water extraction of hemicellulose is also termed autohydrolysis because at elevated temperatures acetic acid is released from the acetylated polysaccharides in the wood (Brasch and Free, 1965). The resulting hydronium ions lower the pH of the solution to around 3 or 4, and act as a catalyst for the depolymerization of hemicellulose (Garrote, Dominguez, and Parajo, 1999). The effect of the decrease in pH on the extract is seen by the accumulation of sugar dehydration products that lead to the formation of 5-hydroxymethylfurfural (HMF) or furfural from hexose or pentose sugars, respectively (Chheda, Roman-Leshkov, and Dumesic, 2007). Due to the facile degradation of xylan at moderately low pH, hardwood has generally been designated for alkali-based hemicellulose extraction because of its high xylose content (van Heiningen, 2006).

Extraction of hemicellulose from softwood prior to pulping using hot water has been successfully demonstrated (Yoon, Macewan, and van Heiningen, 2008; Yoon, Cullinan, and Krishnagopalan, 2010). The major hemicellulose component of softwood is acetyl-galacto-glucomannan, which has been shown to degrade in alkali conditions via the peeling reaction (Wigell, Brelid, and Theliander, 2007a; Wigell, Brelid, and Thellander, 2007b). In order for hemicellulose to remain in a usable form and to avoid degradation products, softwood should ideally use hot water as the extraction solvent. The extraction of hemicellulose from loblolly pine was performed using pressurized water heated to 170 °C for up to 80 minutes. A preliminary study by Yoon, Macewan, and van Heiningen (2008) found that 170 °C was ideal for hemicellulose recovery and temperatures above 170 °C have been shown to produce low-molecular-mass degradation products (Sjostrom, 1991). The resulting liquid showed hemicellulose recovery, as measured by monomeric sugars, amounting to 10.91% on oven-dry wood (ODW) for the 80 minute extraction. The effect of the pre-extraction on the wood chips was measured against untreated chips in the Kraft process. The weight loss encountered during the pre-extraction showed an effect on the pulp quality (Yoon and van Heiningen, 2008), but this effect was counteracted with the addition of sodium borohydride in a mild alkaline sodium sulfide solution to the Kraft process (Yoon, Cullinan, and Krishnagopalan, 2010). This addition resulted in a faster refining response using a PFI (Norwegian Paper and Fibre Research Institute)-type beating mill and the pulp showed no significant change in the properties of the resulting paper (Yoon, Cullinan, and Krishnagopalan, 2010).

Tunc and van Heiningen (2008) examined the extraction of hemicellulose from a mixture of southern hardwood chips. The extraction was performed at a lower temperature of 150 °C in order to counteract the formation of resinous deposits of lignin (Tunc and van Heiningen, 2008). The water extraction of hardwood was performed in order to test the ability of an accelerated solvent extractor (ASE) to screen for lignocellulosic autohyrolysis products. The results of this study demonstrated that the ASE was a suitable screening tool and that it is possible to extract xylan from southern hardwood (Tunc and van Heiningen, 2008).

Alkali-Based Extraction

Hardwoods are better suited for alkali-based extraction since they are comprised mostly of xylose oligomers, and neutral pH will prevent the formation of degradation and by-products from the hemicellulose. The acidity of water extraction may also degrade the cellulose and decrease the quality of the pulp (Yoon and van Heiningen, 2008). Hardwood alkali extraction has been the most extensively studied. Pre-extraction of aspen hardwood chips using strong alkali conditions (1.04 to 2.08 M NaOH) resulted in the recovery of 40–50 kg of hemicellulose per tonne of wood, with no effect on the pulp yield (Al-Dajani and Tschirner, 2008).

In order to create near-neutral extraction conditions, the amount of alkali is decreased, but the temperature is increased (Mao *et al.*, 2008; Walton, van Walsum, and van Heiningen, 2009; Mao *et al.*, 2010; Walton *et al.*, 2010c). The near-neutral conditions aid in the prevention of oligosaccharide degradation to hydroxyl acids (Walton *et al.*, 2010c) and allow for the release of hemicellulose and some lignin, while keeping the cellulose intact. One major advantage of the near-neutral process is that the yield and properties of the brownstock pulp are essentially the same as found in the traditional Kraft process. A second advantage is that the demand or load to the lime kiln will be reduced because the green-liquor-extracted wood chips require less white liquor (Mao *et al.*, 2010). Off-loading the lime kiln reduces fuel consumption, which is attractive when the kiln is fired using expensive fuel oil. The near-neutral extraction uses the already present green liquor stream, which is comprised mostly of Na_2CO_3 and Na_2S, and some white liquor, mostly NaOH and Na_2S (Mao *et al.*, 2008). The near-neutral process uses a charge of green liquor of 3% (as NaO_2 on dry wood) counted as total titratable alkali (TTA) (Mao *et al.*, 2008). Extraction using 2% green liquor was able to maintain the pH at near-neutral conditions, while reducing the release of acetic acid (Walton *et al.*, 2010c). In order to maintain the quality of the pulp, anthraquinone (AQ) is added to the extraction process to act as a pulping catalyst (Mao *et al.*, 2008). It has been demonstrated that a near-neutral pre-extraction performed at 160 °C for 110 minutes with the addition of 0.05% AQ followed by 12% effective alkali (EA) modified Kraft cooking produces the same yield of pulp as untreated wood that has undergone 15% EA Kraft cooking; both trials used a sulfidity of 30% (Mao *et al.*, 2008). The EA is a measure of the alkalinity of the extraction solution in terms of the sodium salts present (NaOH + $\frac{1}{2}Na_2S$). The results from Mao *et al.*, displayed in Figure 13.3, shows the effect of the near neutral extraction or no pre-extraction on the pulp yield (Mao *et al.*, 2008). The measure used for comparison is the kappa number, which is a measure of the residual lignin content of the pulp. The kappa number is determined from the volume (mL) of 0.1N potassium permanganate solution that is consumed by one gram of dried pulp (TAPPI, 1993). The kappa number is a measure of the relative hardness, bleachability or the degree of delignification of pulp and can be used during the pulping process as a control for the amount of bleaching the pulp will require based on the desired whiteness of the final product. The graph shows that the yield of the modified Kraft pulp is within experimental error of the control Kraft pulp. These results are very promising for the near-neutral extraction of hemicellulose prior to pulping.

Alkali extraction has also been carried out on some softwood samples. Extraction of hemicellulose from southern pine was performed using green liquor and it was found that the recovery of xylose was greater than that of mannose (Yoon and van Heiningen, 2010). This result supports the reported degradation of mannose in alkali conditions. The amount of dissolved lignin increased, while the hemicellulose recovered decreased, when the amount of green liquor in the extraction solvent increased. The amount of green liquor used did not seem to have an effect on the cellulose content.

Comparison of Water and Alkali Pre-Extraction of Birch

In order to better understand the effect of extraction solvent used for hemicellulose recovery, a comparison of extraction solvent on the effect of hemicellulose removal, pulp quality and

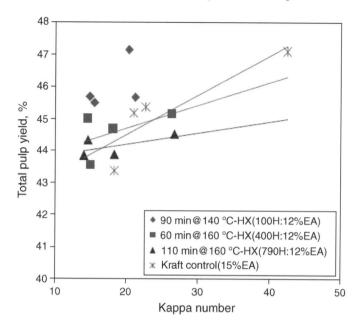

Figure 13.3 *Pulp yield versus kappa number for the near-neutral extraction of hemicellulose from hardwood chips followed by a modified Kraft pulping, compared to the traditional Kraft pulping process (Mao et al., 2008). Reprinted with permission from Mao, H. et al., 2008 © American Scientific Publishers (2008).*

fermentability of the hemicellulose removed was performed on silver birch (*Betula pendula*) (Helmerius *et al.*, 2010). The study varied the extraction time, temperature and EA of the pre-extraction solution. The EA of the solvent was varied from 0 to 7% by changing the amount of green liquor in the extract. The values measured for comparison were the pulp yield, the amount of xylan recovered – as xylose – and the final pH of the extraction solution. From an initial screening, five conditions were selected for testing the pulping quality of the chips post extraction. The pulping test was performed against a control of untreated chips and the EA for all pulping tests was set so that the EA charge was effectively the same for all conditions (i.e., the amount of liquor added for pulping the alkali was decreased, since there was some liquor already present in the chips). The results found that the amount of hemicellulose extracted was greater for hot water than for alkali pre-extraction, but that the quality of the pulp was less for the hot-water pre-treatment. Hemicellulose was shown to play a role in the tearing strength and flexibility of the resulting paper since it adds to the bonding strength of the fibers. The amorphous nature of hemicellulose promotes fiber–fiber bonding and increases the durability of the paper (Karlsson, 2006). The recovered hemicellulose was tested to see how well it performed during fermentation. The hot-water-extracted hemicellulose was converted to succinate more easily and to a greater extent than the alkali-extracted hemicellulose. The study demonstrated that the pre-extraction step may involve trade-offs based on the desired product. The two pre-extraction pathways proposed for either hot-water- or alkali-based extraction are shown in Figure 13.4 (Helmerius *et al.*, 2010). The hot-water-extraction technique may also be better for incorporation into a dissolving pulp mill. Hot water is able to extract more hemicellulose as xylan, which has been shown to decrease the quality of the dissolving pulp since it has the ability to precipitate into the cellulose microfibrils (Li *et al.*, 2010).

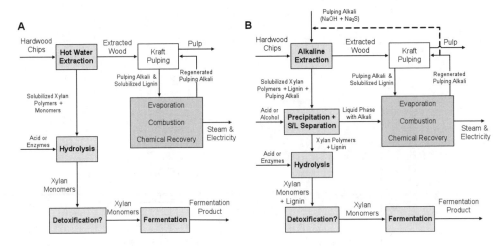

Figure 13.4   *Two schemes for the recovery of hemicellulose prior to pulping (A) Hot-water extraction and (B) alkali extraction (Helmerius et al., 2010). Reprinted with permission from Helmerius, J. et al., 2010 © Elsevier (2010).*

13.2.2   Applications of Hot-Water Extraction

Wood composites, such as orientated strand board (OSB) or particle board can be constructed of wood that has been pre-extracted prior to compression into the final composite product. When using hot water only, the extraction will remove primarily hemicellulose, leaving the strong cellulose fibers and adhesive lignin to be included in the final material. The absence of hemicellulose, which is a hygroscopic polymer, from the OSB would aid in decreasing the swelling and shrinking of the resulting board since the moisture bonding sites would be lessened (Sattler *et al.*, 2008). Research on the composite products produced from such extracted wood have demonstrated higher strength to weight ratios, reduced water absorption and increased fire resistance compared to composite made from conventional wood (Paredes *et al.*, 2008). Paredes *et al.* also found that when the hemicellulose was removed there was a reduction in the emission of both methanol and acetaldehyde during the board pressing process (Paredes *et al.*, 2010). In addition, since the hemicellulose that has been extracted from the wood is the component most easily attacked by microorganisms, composites made from extracted wood are also more resistant to microbial attack. The extraction of hemicellulose decreases the decay of white rot fungi on OSB, though it was found to have no effect on the decay of brown rot fungi (Howell, Paredes, and Jellison, 2009). Pre-extraction on loblolly pine using water at 140 °C has been shown to be the best extraction temperature to recover hemicellulose while maintaining the desired properties of OSB (Sattler *et al.*, 2008). For hardwoods, the recommended pre-extraction should use a severity factor (SF) of 3.54 in order to maintain the desired properties of the OSB (Paredes *et al.*, 2009). The severity factor is a measure that accounts for the temperature, time and chemical quantities used for the conversion of lignocellulosic components (Dang and Nguyen, 2006).

Because hot-water extraction removes primarily the hemicellulose of wood, which has the lowest energy content of the major wood components, such extraction could also be used prior to combustion of wood. Pre-extracted hemicellulose could command higher value as a fermented product and the extracted wood, with reduced hemicellulose, can be more easily dewatered prior to combustion than untreated wood.

13.3 Hemicellulose Conversion

Hemicelluloses extracted from wood chips can be converted to products of higher value than boiler fuel. Commodity chemicals that can be produced through fermentation, such as lactic acid, or liquid transportation fuels that have increasingly high value are both possible products that can be derived from hemicellulose extracts. The hemicelluloses in hardwood extracts consist primarily of xylan oligomers, which are converted to other products by either chemical or biological means. Conversion by fermentation requires hydrolysis of the oligomeric sugars to fermentable monomers, which can be accomplished through acid or enzymatic hydrolysis. Resulting monomer sugars can then be converted to fermentation products, though xylose, the main monomer from hardwood hemicellulose, converts less readily than glucose. Other challenges for conversion of hemicellulose extracts by fermentation are the dilute concentration of sugars in raw extract, which results in high separation costs for the dilute products, and the presence of microbial inhibitors in the extract, which can prevent efficient fermentation.

13.3.1 Hydrolysis of Hemicellulose Oligomers

Soluble oligomers have been hydrolyzed to monomer sugars by both acid and enzymatic catalysis. Um and van Walsum (2009) showed that hydrolysis using sulfuric acid could achieve a 84% conversion of extracted oligomers to monomeric sugars with relatively low production of degradation products such as furfural. This was achieved at combined SFs ranging from 1.4 to 1.8, including a wide range of temperatures, reaction durations and acid levels. Acid hydrolysis of extract has also been demonstrated at larger scales in stirred batch reactors. Um and van Walsum (2010) showed that enzymes, particularly in optimized mixtures, could produce up to 84–92% yield of monomer sugars at either separate hydrolysis and fermentation (SHF) or simultaneous saccharification and fermentation (SSF) conditions over a reaction time of 72 hours. In general, enzymatic hydrolysis provides benefits of higher overall yields, due to lack of acid-induced degradation, and the reaction conditions are milder than those of acid hydrolysis. Additionally, acid hydrolysis typically produces a large waste stream of gypsum from neutralization of the sulfuric acid. The drawbacks of enzymatic hydrolysis are that the reaction is slower, thus requiring longer residence times and larger equipment, and the cost of enzyme can be prohibitive.

13.3.2 Fermentation to Alcohols

Both ethanol and butanol can be produced by fermenting organisms. These two alcohols are proven fuels and both have been produced at commercial scale through fermentation of various sugar and starch feedstocks. Fermentation to alcohols takes place in anaerobic conditions, which results in relatively slow fermentation, but high retention of carbon from the original sugar to the product fuel, since no oxygen is present to oxidize carbon to CO_2.

Fermentation of Hemicellulose Extract to Ethanol

The principle monosaccharide in aqueous hemicellulose extracts is xylose, which is relatively difficult to convert to ethanol at high yield. Several organisms are known to convert the five-carbon sugar, but typically they do so with co-production of acetic acid or other fermentation products, thus lowering the yield of ethanol. Several organisms have been genetically modified to convert xylose to ethanol with improved yields and selectivity. One of the earlier successful organisms which received a fair amount of attention is *Escherichia coli* KO11 (Takahashi *et al.*, 1999; Walton, van Heiningen, and van Walsum, 2010a). Walton, van Heiningen, and van Walsum (2010a) demonstrated that *E. coli* KO11 could convert all of the hemicellulose sugars, xylose, glucose, mannose, arabinose and galactose, to ethanol, and was relatively resilient to extract inhibitors such as acetic acid and the high salt levels of sodium salts. Takahashi *et al.* (1999) found that *E. coli* KO11 is more easily inhibited by acetic acid when growing on xylose than when growing on glucose. Greater sensitivity to inhibitors while growing on xylose is typical of organisms that have been genetically modified to ferment xylose to ethanol. Inhibition by acetate is more severe at lower pH. The optimum fermentation pH in the presence of acetic acid was found to be 7, where pH 6 is optimum for cultures which were not grown in acetic acid (Takahashi *et al.*, 1999). *E. coli* is generally more tolerant of furfural and 5-hydroxymethyl furfural than other organisms (Klinke, Thomsen, and Ahring, 2004). Inhibition by components in the extract becomes especially problematic when it is necessary to concentrate the sugars in the extract so as to increase product concentration. For reasonable economics of distillation, ethanol product concentrations should be at least $50\,g\,L^{-1}$ (Galbe *et al.*, 2007) which, in turn, requires of the order of $100\,g\,L^{-1}$ of sugars. Near neutral pH hemicellulose extracts, which can be produced without diminishing pulp quantity, normally produce less than $20\,g\,L^{-1}$ sugars. Thus, concentrating such extracts runs the risk of also concentrating the inhibitors to intolerable levels.

Walton, van Walsum, and van Heiningen (2009) demonstrated fermentation of concentrated green liquor to ethanol. Concentration was accomplished by two means: evaporation, which concentrates most inhibitors, as well as sugars, and ultrafiltration, which can concentrate sugar oligomers, but not smaller components such as acetic acid and salts. Figure 13.5 shows the results of fermenting hemicellulose extract at three consistencies: dilute hemicellulose extract at 3.7 wt%, and evaporated extracts at 5.4 wt% and 9.8 wt% solids. The two lower-concentration solutions were both readily fermentable, producing 1.2 and $2.1\,g\,L^{-1}$ of ethanol, respectively. These concentrations of ethanol are too dilute for economic product recovery, but did represent approximately 85% of the theoretical yields based on sugar consumption. The hemicellulose extract at 9.8 wt% solids was not initially fermentable, but fermentation was observed when using a large inoculum of cells derived from the twofold evaporated extract at 5.4 wt% solids.

Figure 13.6 shows results of fermenting extracts that had been concentrated using ultrafiltration. The extract was produced from 4% green liquor extraction and concentrated using tangential flow filtration with membranes of 0.22 μm and 10 kD, operated in series. The concentration resulted in fivefold xylose concentration, but also in threefold lignin concentration. Acetic acid did not increase, and sodium, though not measured, was not expected to increase. A long lag was observed, but after seven days all of the sugars were consumed, and from the 18.4 g L^{-1} sugars, a concentration of $10\,g\,L^{-1}$ ethanol was produced, indicating likely incomplete quantification of initial sugars present.

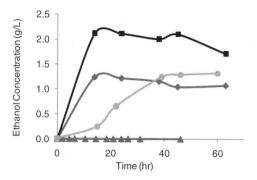

Figure 13.5   *Ethanol production at varying extract evaporation levels. Fermentation conducted by E. coli K011 at 37°C, pH 7 in a 3 L bioreactor. 3.7 wt% solids extract (◆); 5.4 wt% solids extract (■); 9.8 wt% solids extract, no strain adaptation (▲), 9.8 wt% solids extract, adapted with large inoculum from 5.4 wt% in log growth (●) (Walton et al., 2010b). Reprinted with kind permission form Walton, S. L. et al., 2006 © Springer Science + Business Media (2006).*

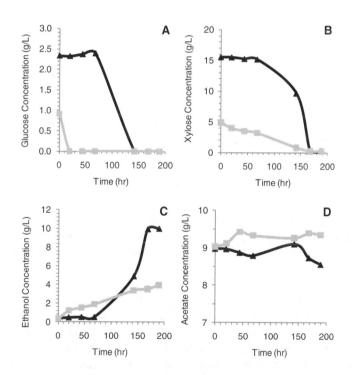

Figure 13.6   *Fermentation by E. coli KO11 of ultra-filtered extracts. Feedstocks for fermentation produced from 4% green liquor extracts, 800 H-Factor. Retentate from 0.22 µm filtration, sugars concentrated fivefold (▲), permeate from 0.22 µm filtration, filtered with 10 kD ultrafiltration, concentrated fourfold (■) (Walton, van Walsum, and van Heiningen, 2009). Reprinted with kind permission form Walton, S. L. et al., 2006 © Springer Science + Business Media (2006).*

Fermentation of Hemicellulose Extract to Butanol

Butanol is considered by many to be a superior automotive fuel to ethanol (Gheshlaghi *et al.*, 2009). Butanol has a higher energy density than ethanol and also more easily blends with gasoline, making it more compatible with the petroleum distribution systems (Atsumi, Hanai, and Liao, 2008; Irimescu, 2010). Like ethanol, butanol has a high octane rating, but lower vapour pressure, which also improves its blending qualities. Butanol production through fermentation is possible using certain bacteria, many of which are capable of consuming wood-derived sugars. In most cases, the bacteria also produce acetone and ethanol, and thus the fermentation is referred to as ABE (acetone, butanol, ethanol) fermentation. Although butanol has superior fuel properties compared to ethanol, it is also more toxic to the fermenting organism, and thus is difficult to produce at high concentrations (Garcia *et al.*, 2011).

Work is currently underway to produce butanol through fermentation of wood extracts, though the greater sensitivity of the butanol fermentation organisms to inhibition by the extract is challenging for this application.

13.3.3   Conversion of Extracts to Triacylglycerides (TAGs)

Conversion of triacylglycerides (TAGs) to biodiesel through transesterification with methanol has been demonstrated and is currently a commercial technology. Most biodiesel is produced from edible plant oils such as palm or soy bean oil, thus production of biodiesel competes directly with food production. TAGs can also be produced by fermenting sugar with lipid-accumulating algae or bacteria, which accumulate lipids as an energy storage medium (Waltermann *et al.*, 2000). If the sugars are derived from wood extracts, this represents another route for producing liquid fuels from wood extracts. Extraction of the TAG from the algae or microbes and subsequent chemical upgrading can turn these oils into suitable transportation fuels with high energy densities.

Production of TAGs Using Rhodococcus Opacus

Rhodococcus opacus is one bacterium that has been found to accumulate TAG as an energy storage medium. TAG yields on microbial biomass have been reported to be as high as 76% (MacEachran, Prophete, and Sinskey, 2010) and growth on glucose has resulted in total a product concentration of $77.6 \, g \, L^{-1}$ (Kurosawa *et al.*, 2010). Generally, the bacteria will begin to accumulate TAG when its cellular multiplication has stopped as a result of nitrogen limitation, but it still has carbon sources available for consumption. In this situation, rather than multiply in numbers, the individual cells will accumulate intracellular TAG from consumption of the available carbon source. *R. opacus* is an aerobic organism, thus it requires a high rate of oxygen transfer into the bioreactor. Respiration allows for rapid cell production, though much of the carbon is lost to respiration. Thus, the carbon retention will be poorer than for anaerobic fermentation to alcohols, and the energy demands for oxygen transfer will reduce the energy efficiency of the overall process. To its benefit, aerobic growth is more rapid than anaerobic growth, thus smaller fermentation vessels will be required for consumption of the available feed extracts. Recently, *R. opacus* was reported to have been genetically modified to consume xylose as well as glucose (Stauffer, 2010).

Extraction of TAG from Microbial Biomass

The lipids produced by microorganisms for energy storage are held within the cell membranes. Thus, to harvest the TAG, the cells must be ruptured to allow collection of the valuable oil. Although many methods are available for rupturing the cells and extracting the oils, all require large amounts of energy or powerful solvents, both of which raise operating costs to prohibitive levels (Voss and Steinbuchel, 2001).

Upgrading TAG

Biodiesel is produced from triglycerides through transesterification with methanol, producing long-chain esters that perform well in diesel engines. For more demanding fuel applications, such as jet fuel, complete conversion to hydrocarbon is required. Conversion of triglycerides to hydrocarbons has been demonstrated through catalytic hydrotreatment, resulting in a mixture of hydrocarbons (Durrett, Benning, and Ohlrogge, 2008; Duan and Savage, 2011). Thus, by this route of microbial fermentation to TAG and hydrogen upgrading, high-quality fuels could be produced from wood extracts.

13.3.4 Hemicellulose Upgrading Via the Carboxylate Platform

The carboxylate platform is an approach to producing biofuels and bioproducts that is analogous to the sugar and thermochemical conversion platforms, wherein biomass is first converted to an intermediate product, such as sugar or syngas, and the intermediate is then upgraded to the desired product (Holtzapple and Granda, 2009). In the carboxylate platform, the intermediate products are carboxylic acids, and these are then upgraded to a variety of value-added products. Biodiesel could be considered a carboxylate platform process, since the intermediate material is long-chain fatty acids that are chemically upgraded to diesel-compatible fuel. When applied to cellulosic biomass, the carboxylate platform usually includes a fermentation step to convert carbohydrates, and other anaerobically biodegradable materials in the biomass, to carboxylic acids.

One configuration of the carboxylate platform has been termed the MixAlco™ process, a technology developed at Texas A&M University and currently being commercialized by Terrabon, LLC. The mixed alcohol, or MixAlco™, process is a patented technology in which a biodegradable material is converted by acidogenic fermentation with subsequent upgrading to mixed alcohols (Holtzapple *et al.*, 1999). Recent work has also further upgraded the mixed alcohols to hydrocarbon fuels, such as gasoline, realizing a yield of 70 gallons of gasoline from one tonne of municipal solid waste (Terrabon, 2011). Some of the biomaterials that have been studied as feedstocks for the process include; chicken and dairy manure, municipal solid waste, sewage sludge, aloe vera, molasses, glycerol, office paper, corn stover, sugarcane bagasse, and recently, wood extracts (Ross and Holtzapple, 2001; Thanakoses, Black, and Holtzapple, 2003; Blackman and van Walsum, 2009; Forrest, Sierra, and Holtzapple, 2010; Fu and Holtzapple, 2010b; Fu and Holtzapple, 2010a; Rughoonundun *et al.*, 2010). Figure 13.7 shows a simplified flow diagram of the MixAlco™ process as applied to conversion of dairy manure (Blackman and van Walsum, 2009).

The flexibility of feedstock and the numerous possibilities for chemical production are attractive advantages of the MixAlco™ process. The diversity in feedstock selection affords a

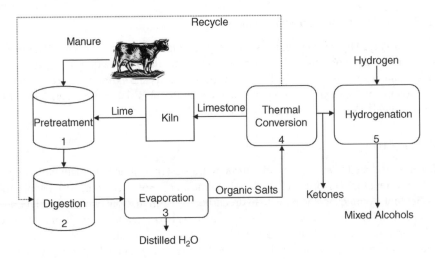

Figure 13.7   *Example configuration of the carboxylate platform converting dairy manure to mixed alcohols. Unit operations include pretreatment, digestion, drying, thermal conversion (deoxygenation) and hydrogenation (Blackman and van Walsum, 2009). Application to a wood biorefinery would replace the alkali pre-treatment with pre-pulping extraction. Reprinted with permission from Blackman, E. D. et al., 2009 © American Institute of Chemical Engineers & John Wiley & Sons (2009).*

great advantage for fermentation of the hemicellulose that is recovered prior to pulping. Alkali-extracted hemicellulose is advantageous, since fermentation is performed with the addition of a buffering agent to raise the pH. When solid lignocellulosic biomass is used as the feedstock, the process begins with a pre-treatment step to remove the lignin, thus enhancing the reactivity of the material (Granda *et al.*, 2009). The pre-pulping extraction acts as the pre-treatment step and the recovered hemicellulose is initially subjected to fermentation to convert the hemicellulose to carboxylate salts. The salts are then dried and sent along to thermal deoxygenation where they are converted into ketones. The ketones are then subjected to hydrogenation to produce an array of alcohols, primarily secondary alcohols. Further hydrogen treatment can also be applied to the secondary alcohols to produce saturated hydrocarbons of various chain lengths.

Fermentation

Fermentation to carboxylate salts is considered a "sour" fermentation because the aim is to produce acids instead of alcohols or methane. The fermentation works as a consolidated bioprocessing reactor, and includes in one vessel the different steps of anaerobic decomposition, including: hydrolysis, acidogenesis and acetogenesis. For the carboxylate platform, an important aim is to avoid methanogenesis, which volatilizes the organic acids as carbon dioxide and methane. The acids produced in the fermentation are neutralized with the addition of an alkaline buffer, usually calcium carbonate or ammonium bicarbonate, in order to produce carboxylate salts (Pham, Holtzapple, and El-Halwagi, 2010).

Use of a mixed culture of microorganisms is one of the novel features of the MixAlco™ approach to the carboxylate platform. Mixed cultures are able to hydrolyze and consume all of the biodegradable materials and they are tolerant of inhibitors; in fact they often consume compounds that are known inhibitors of some pure cultures, such as furans, and they operate

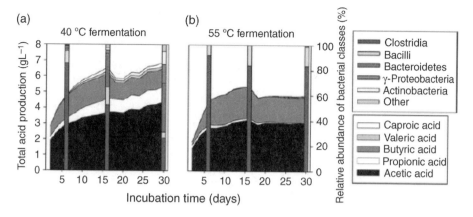

Figure 13.8   *Acid production at 40 °C (a) and 55 °C (b) in the fermentation of sorghum using a marine sediment (Hollister et al., 2010). The change of bacterial classes as the fermentation progresses is shown at three different points in the fermentation as a bar graph. The acid accumulation during fermentation is shown as an additive-area line graph. These graphs demonstrate how the acid production changes at different temperatures because of the evolving bacterial presence. Reprinted with kind permission form Hollister, E. B. et al., 2010 © Springer Science + Business Media (2010).*

with no risk of contamination. The microorganisms used for fermentation are able to digest fats and proteins, as well as carbohydrates (Pham, Holtzapple, and El-Halwagi, 2010). Halophillic organisms, such as those found in marine sediments, have been shown to afford higher acid production than terrestrial inoculum (Thanakoses, 2002; Fu and Holtzapple, 2010a). The acids produced during fermentation differ in chain length depending on the duration of incubation, since the microorganisms present evolve as the fermentation advances. During the first part of fermentation, the broth is rich in *Clostridia* and acetic acid is most abundantly produced. As the fermentation progresses, the bacterial community changes and longer-chain acids are produced at higher rates. The temperature of the fermentation also plays a key role in the types of acid produced: in general, lower temperature gives rise to longer-chain acids. Figure 13.8 displays the results of a study on the acid production and bacterial analysis of a marine sediment inoculum on the fermentation of sorghum. The graphs indicate that the lower temperature (40 °C) fermentation not only has a wider variety of bacteria present but also that more longer chain acids are produced than at 55 °C (Hollister *et al.*, 2010).

Fermentation of a hemicellulose-rich pre-pulping green liquor extract using a marine sediment from Rockland, ME at 37 and 55 °C was performed over a period of 21 days (Baddam and van Walsum, 2009). The fermentation used ammonium bicarbonate as the buffer. Some results from this fermentation are shown in Figure 13.9 and align well with the acid production results from Hollister *et al.* (2010). The production of longer-chain acids occurred later in the fermentation period and was greater for the lower-temperature fermentation. Acetic acid is initially present in the fermentation broth because it is released from the extraction as an acetyl group from the bound xylan.

Dewatering

Once the carboxylate salts are formed, the next step in the process is to dry the salts. A high-temperature vapour compression unit has been developed as a more economical technique to

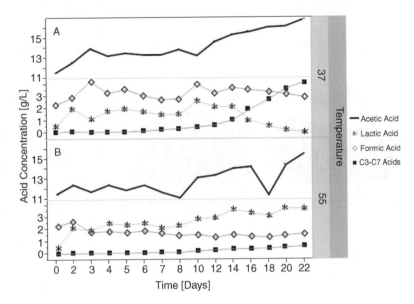

Figure 13.9   *Results of fermentation of green liquor extracts from hardwood, prior to pulping, at (A) 37 °C and (B) 55 °C. Acetic acid is initially present in the fermentation broth since it is released from the bound xylan in the wood chips (Baddam and van Walsum, 2009). The major product of fermentation is acetic acid and is shown on a different scale than the other acids produced so that the acetic acid production line does not fully diminish the production lines of the other acids. The C3–C7 denotation refers to acids with a three- to seven-carbon backbone. Reprinted from Baddam, R. et al., 2009 © American Institute of Chemical Engineers (2009).*

dry the carboxylate salts. The design makes used of a novel latent-heat exchanger coated with a hydrophobic monolayer of polytetra fluoroethylene (PTFE), in order to increase heat transfer through drop-wise condensation (Lara, Noyes, and Holtzapple, 2008). The fermentation broth is heated to around 172 °C and then enters the latent heat exchanger where the heat from vapour condensing in an adjacent stage vaporizes the water in the broth (Pham, Holtzapple, and El-Halwagi, 2010). The vapour is then compressed to a higher temperature and pressure so that it easily condenses and this is then sent back through the exchanger. The power system of the water-removal process uses a combined cycle and takes advantage of injecting liquid water into the compressor unit. By injecting liquid water, the compression is nearly isothermal, which minimizes the energy consumption of the compressor (Lara, Noyes, and Holtzapple, 2008). The normal operation temperature of a vapour compression unit is around 80 °C and by operating at an elevated temperature (172 °C) the size of the exchanger and compressor is decreased, since the temperature differential is decreased. The smaller size allows for lower capital and operating costs for this unit. The dewatering process, known as AdVE™, is a technology that has been patented by Terrabon LLC. This technology is applicable to the desalination of brackish water and the dewatering of carboxylate salts produced during acidogenic fermentation.

Thermal Deoxygenation

Once the water has been removed from the carboxylate salts, the next step is conversion of the salts into ketones. The decomposition of calcium carboxylate salts into ketones occurs at temperatures above 300 °C (Ardagh *et al.*, 1924) and more specifically at 430 °C (Holtzapple *et al.*, 1999). The ketones formed during this process can rapidly degrade, so they must be removed quickly once they are formed. During thermal decomposition, calcium carbonate is regenerated and can be used in a recycle stream to the fermentation. The conversion of calcium carboxylate salts to ketones is very high, up to 99% (Yeh, 2002). The thermal deoxygenation step has been successfully demonstrated in a carboxylate platform if the reactor is kept under vacuum in order to remove the ketones as soon as they are produced (Pham, Holtzapple, and El-Halwagi, 2010).

Hydrogenation to Alcohols

The ketones formed during thermal deoxygenation are converted into alcohols by the addition of hydrogen. A mixture of ketones that would be present in a MixAlco™ fermentation were evaluated using a Raney-nickel catalyst and it was determined that the reactions followed a Langmuir model and were first order with respect to hydrogen pressure (Chang *et al.*, 2000). The conversion for the MixedAlco™ platform has been demonstrated in a series of three continuously stirred tank reactors under high pressure (55 bar) and in isothermal conditions (130 °C) (Pham, Holtzapple, and El-Halwagi, 2010). The conversion yields for the third reactor in the series are up to 98% (Holtzapple *et al.*, 1999).

13.3.5   Conversion to Tridecane

Carbohydrates in wood extracts can be converted through chemical means to alkanes. A four-step process for converting xylan oligomers from hot-water wood extracts has been demonstrated (Xing *et al.*, 2010). The four steps to convert xylan oligomers to tridecane are: (1) acid hydrolysis and biphasic dehydration, (2) aldol condensation, (3) low-temperature hydrogenation and (4) high-temperature hydrodeoxygenation. Acid hydrolysis is carried out using sulfuric acid, which hydrolyzes the oligomers to xylose monomers, and continues to dehydrate the xylose to furfural. Since furfural is relatively hydrophobic, a biphasic reaction contacting an organic phase to the aqueous reaction phase can extract the furfural as it is produced and drive the reaction forward, generating high yields and selectivity for furfural. Transfer of the furfural to the organic phase is enhanced with the addition of salt to the aqueous phase, which could make this technology a good fit with green liquor extraction, since it has high salt content in the extract. The Aldol condensation is carried out with the addition of acetone at a 1:2 molar ratio with furfural. This produces the unsaturated C-13 compound $C_{13}H_{10}O_3$, which takes the form of two furan rings on either end of 3-pentanone. In the third step this highly unsaturated compound is stabilized with the addition of 7 moles H_2 per mole of $C_{13}H_{10}O_3$ over Ru/C catalyst under mild conditions to produce $C_{13}H_{24}O_3$. The final step is a hydrodeoxygenation, to produce linear tridecane $C_{13}H_{28}$. Xing *et al.* (Xing *et al.*, 2010) used 4 wt% Pt/SiO$_2$-Al$_2$O$_3$ catalyst which combines acidic SiO$_2$-Al$_2$O$_3$ to catalyze the dehydration of H-dimers to form C=C bonds and water, and Pt to hydrogenate the C=C bonds.

The overall stoichiometric reaction for these four steps can be written as:

$$2C_5H_{10}O_5 + C_3H_6O + 12H_2 \rightarrow C_{13}H_{28} + 11H_2O$$

Theoretically, 1.0 kg of tridecane can be produced from 1.6 kg xylan oligomers, 0.3 kg of acetone and 0.13 kg of H$_2$.

13.3.6 Fermentation to Commodity Chemicals

Although fuels provide a large market with little chance of saturation by biomass-derived products, commodity chemicals such as lactic acid provide higher potential revenue for utilization of the same resources. Fermentation of wood extracts to lactic acid has been demonstrated on extracts derived from hardwoods and Siberian larch (Walton *et al.*, 2010b). Other fermentation-derived chemicals, such as itaconic acid or succinic acid may also be candidates for production from hemicelluloses.

Fermentation to Lactic Acid

Walton *et al.* (2010b) demonstrated conversion of wood extract to lactic acid using the bacterium *Bacillus coagulans MXL-9*, which is capable of fermenting xylose and other wood-derived sugars. This organism was found to have a high tolerance for wood extract inhibitors such as acetic acid and high levels of sodium. With an extended lag time, suggesting that acclimation to inhibitors was achieved; fermentations with *B. coagulans* were capable of withstanding 30 g L^{-1} acetic acid and 20 g L^{-1} of sodium to produce lactic acid from xylose at high yields. Fermentation of hot-water extracts from Siberian larch, which contained mainly galactose and arabinose was also successful. Figure 13.10 shows the fermentation results of hot-water larch extract. Interestingly, the furfural present in the hydrolysate appears to be consumed by *B. coagulans* more readily than the major sugar, galactose. The ability to consume all the wood-derived sugars, the strong resistance to inhibitors and the high yield of lactic acid fermentations (often > 80%) suggests that production of this commodity chemical could be one of the more promising near-term applications of extract fermentation.

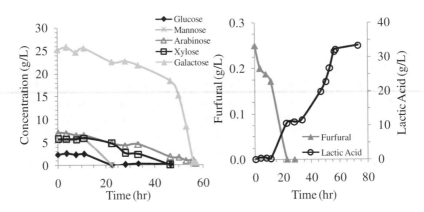

Figure 13.10   *Conversion of hot-water extract derived from Siberian Larch to lactic acid by B. coagulans MXL-9 (Walton et al., 2010b). Reprinted with kind permission form Walton, S. L. et al., 2010 © Springer Science + Business Media (2010).*

Fermentation to Itaconic Acid

Itaconic acid is a non-toxic, biodegradable five-carbon acid with the formula $C_5H_6O_4$ that can be blended with other monomers for the production of polymer plastics and gels. Experiments are currently underway to produce itaconic acid from wood extracts.

13.4 Process Economics

13.4.1 Integrating Extraction into an Existing Mill

Energy Costs from Black Liquor Reduction

Pre-pulping extraction preferentially removes hemicellulose from the wood chips entering the plant. Depending on the pH, the extraction can also remove considerable amounts of lignin and organic acids from the wood chips. For the most part, relatively little cellulose is extracted. The extracted organic acids and hemicellulose are the components most likely to be utilized for conversion to new products, while the lignin is likely to pass through the conversion process and be returned to the recovery boiler. Thus, the black liquor produced from pre-extracted wood chips will have reduced volume and energy content due to the reduced amount of hemicellulose and organic acids being burnt. This loss of fuel energy is a real cost to the mill and must be compensated. To estimate the value of this lost energy, it can be assumed that the mill will increase its steam production to pre-extraction levels through increased consumption of biomass-quality wood, which is less expensive than pulp-quality wood. Because the components diverted away from the black liquor stream (hemicellulose) have lower average heat of combustion than regular wood (which still contains high-energy lignin), the quantity of biomass required to replace the lost fuel value is slightly less than that extracted from the wood.

13.4.2 Energy Cost for Extraction

Pre-pulping extract is carried out separately from the primary Kraft pulping and is carried out at high temperature. Most of the heat from the extraction vessel is removed with the extract, but a large portion of the heat, contained in the extracted chips and entrained liquor, is sent on into the primary pulping digester. Thus, the real heat demand for the extraction depends on the degree of heat integration achieved through the processing of the extract. For most applications, evaporation of the extract immediately after the extraction is desirable in order to increase the concentration of sugars and organic acids in the dilute extract. Since the extract is produced at high temperature, and most processing is carried out at lower temperatures, the sensible heat in the extract can be used to evaporate part of the water. Use of multi-effect or vapour recompression evaporation and other heat integration steps can extend the efficiency of the evaporation to the extent that little additional steam is required to achieve considerable volume reduction. Further downstream processing of the extract, which may involve agitated fermentation, liquid–liquid extraction, gas stripping, distillation or thermal conversion will add additional power and steam requirements to the biorefinery operation. Partly balancing these demands, integration of the pre-pulping extraction will reduce some steam loads in the main pulp mill, for example reduced volume through the pulping digester, and black liquor evaporation reduces steam demands somewhat. Mao *et al.* (2008) reported that for a

1000-tonne (of product) per day pulp mill, net energy demands for the extraction step and associated reductions in pulp mill steam demand resulted in an energy requirement of 6.6 tons steam per ton of hemicellulose sugars extracted. These calculations assumed heat savings in the pulping process, but no use of multi-effect or vapour recompression evaporation to concentrate the extract.

Cost Estimate for Extracted Sugars

Determining an appropriate cost for extracted sugars is not a straightforward calculation, since there is so much integration with the host mill. However, a preliminary assessment can be made by taking into account the cost of an extraction vessel and associated plumbing and machinery, the cost of lost energy content in the black liquor and the net energy cost of the extraction step. Downstream processing of the extract, which varies greatly from one application to the next, is excluded from this analysis. For an extraction system sized for a 1000 ton per day pulp mill, the extraction vessel was reported by Mao *et al.* to be of the order of $12 000 000 in 2008 dollars (Mao *et al.*, 2008). This estimate is likely too low, but taken as a starting point, and given an extracted hemicellulose recovery of 50 tons per day, results in a capital cost per annual kg of sugar production of the order of $0.88. Accounting for energy costs for the extraction step, applicable energy savings in the pulping process and some additional labor, Mao's numbers suggest an operating cost of about $0.11 kg^{-1} of extracted sugars. This is comparable to other sources of sugar. For example, at $5.50 per bushel ($\sim$ $0.10/lb) and 67% starch content, the unrefined starch in corn has a basic cost of about $0.32 kg^{-1}. In the spring of 2011, world sugar prices were of the order of $0.50 kg^{-1} (ISO, 2011).

13.5   Conclusion

Pre-pulping extraction has been shown to be a viable source of soluble hemicellulose that can be produced in parallel with existing chemical pulp production. Extraction with hot water or dilute alkaline aqueous solutions has been shown to be effective for softwoods and hardwoods, and if extraction is carried out under near-neutral pH conditions, Kraft pulp yield and quality from hardwood can be maintained. Extracted hemicellulose has been converted via fermentation and chemical processes to a variety of commodity chemicals and fuels, including ethanol, lactic acid, mixed carboxylate salts and tridecane. Current work is continuing to develop more effective extraction methods and other applications of this sustainable source of convertible carbohydrates.

References

Al-Dajani, W.W. and Tschirner, U.W. (2008) Pre-extraction of hemicelluloses and subsequent Kraft pulping Part I: Alkaline extraction. *Tappi Journal*, **7** (6), 3–8.

Al-Dajani, W.W. and Tschirner, U.W. (2010) Pre-extraction of hemicelluloses and subsequent ASA and ASAM pulping: Comparison of autohydrolysis and alkaline extraction. *Holzforschung*, **64** (4), 411–416.

Ardagh, E.G.R., Barbour, A.D., McClellan, G.E., and McBride, E.W. (1924) Distillation of acetate of lime. *Industrial and Engineering Chemistry*, **16**, 1133–1139.

Atsumi, S., Hanai, T., and Liao, J.C. (2008) Non-fermentative pathways for synthesis of branched-chain higher alcohols as biofuels. *Nature*, **451** (7174), 86- U13.

Baddam, R. and van Walsum, G.P. (2009) Anaerobic fermentation of hemicellulose present in pre-pulping extracts of northern hardwoods to carboxylic acids. AIChE Annual Meeting, Nashville TN.

Blackman, E.D. and van Walsum, G.P. (2009) Production of renewable bioproducts and reduction of phosphate pollution through the lime pretreatment and acidogenic digestion of dairy manure. *Environmental Progress & Sustainable Energy*, **28** (1), 121–133.

Brasch, D.J. and Free, K.W. (1965) Prehydrolysis-Kraft pulping of *Pinus radiata* grown in New Zealand. *Tappi*, **48** (4), 245–248.

Chang, N.S., Aldrett, S., Holtzapple, M.T., and Davison, R.R. (2000) Kinetic studies of ketone hydrogenation over Raney-nickel catalyst. *Chemical Engineering Science*, **55** (23), 5721–5732.

Chheda, J.N., Roman-Leshkov, Y., and Dumesic, J.A. (2007) Production of 5-hydroxymethylfurfural and furfural by dehydration of biomass-derived mono- and poly-saccharides. *Green Chemistry*, **9** (4), 342–350.

Dang, V. and Nguyen, K.L. (2006) Characterisation of the heterogeneous alkaline pulping kinetics of hemp woody core. *Bioresource Technology*, **97** (12), 1353–1359.

Duan, P. and Savage, P.E. (2011) Hydrothermal liquefaction of a microalga with heterogeneous catalysts. *Industrial & Engineering Chemistry Research*, **50** (1), 52–61.

Durrett, T.P., Benning, C., and Ohlrogge, J. (2008) Plant triacylglycerols as feedstocks for the production of biofuels. *The Plant Journal*, **54** (4), 593–607.

Forrest, A.K., Sierra, R., and Holtzapple, M.T. (2010) Suitability of pineapple, *Aloe vera,* molasses, glycerol, and office paper as substrates in the MixAlco process™. *Biomass & Bioenergy*, **34** (8), 1195–1200.

Fu, Z.H. and Holtzapple, M.T. (2010a) Consolidated bioprocessing of sugarcane bagasse and chicken manure to ammonium carboxylates by a mixed culture of marine microorganisms. *Bioresource Technology*, **101** (8), 2825–2836.

Fu, Z.H. and Holtzapple, M.T. (2010b) Fermentation of sugarcane bagasse and chicken manure to calcium carboxylates under thermophilic conditions. *Applied Biochemistry and Biotechnology*, **162** (2), 561–578.

Galbe, M., Sassner, P., Wingren, A., and Zacchi, G. (2007) Process engineering economics of bioethanol production. *Biofuels.*, **108**, 303–327.

Garcia, V., Pakkila, J., Ojamo, H. *et al.* (2011) Challenges in biobutanol production: How to improve the efficiency? *Renewable & Sustainable Energy Reviews*, **15** (2), 964–980.

Garrote, G., Dominguez, H., and Parajo, J.C. (1999) Mild autohydrolysis: an environmentally friendly technology for xylooligosaccharide production from wood. *Journal of Chemical Technology and Biotechnology*, **74** (11), 1101–1109.

Gheshlaghi, R., Scharer, J.M., Moo-Young, M., and Chou, C.P. (2009) Metabolic pathways of clostridia for producing butanol. *Biotechnology Advances*, **27** (6), 764–781.

Granda, C.B., Holtzapple, M.T., Luce, G. *et al.* (2009) Carboxylate platform: the MixAlco™ process Part 2: Process economics. *Applied Biochemistry and Biotechnology*, **156** (1–3), 537–554.

Helmerius, J., von Walter, J.V., Rova, U. *et al.* (2010) Impact of hemicellulose pre-extraction for bioconversion on birch Kraft pulp properties. *Bioresource Technology*, **101** (15), 5996–6005.

Hollister, E.B., Forrest, A.K., Wilkinson, H.H. *et al.* (2010) Structure and dynamics of the microbial communities underlying the carboxylate platform for biofuel production. *Applied Microbiology and Biotechnology*, **88** (1), 389–399.

Holtzapple, M.T., Davison, R.R., Ross, M.K. *et al.* (1999) Biomass conversion to mixed alcohol fuels using the MixAlco™ process. *Applied Biochemistry and Biotechnology*, **77–9**, 609–631.

Holtzapple, M.T. and Granda, C.B. (2009) Carboxylate platform: the MixAlco™ process Part 1: Comparison of three biomass conversion platforms. *Applied Biochemistry and Biotechnology*, **156** (1–3), 525–536.

Howell, C., Paredes, J.J., and Jellison, J. (2009) Decay resistance properties of hot water extracted oriented strandboard. *Wood and Fiber Science*, **41** (2), 201–208.

Huber, G.W., Iborra, S., and Corma, A. (2006) Synthesis of transportation fuels from biomass: Chemistry, catalysts, and engineering. *Chemical Reviews*, **106** (9), 4044–4098.

Hytonen, E. and Stuart, P.R. (2009) Integrating bioethanol production into an Integrated Kraft pulp and paper mill: Techno-economic assessment. *Pulp & Paper-Canada*, **110** (5), 25–32.

Irimescu, A. (2010) Study of cold start air-fuel mixture parameters for spark ignition engines fueled with gasoline-isobutanol blends. *International Communications in Heat and Mass Transfer*, **37** (9), 1203–1207.

ISO International Sugar Organization (2011) Retrieved 03/11/11, available at http://www.isosugar.org/Prices.aspx.

Karlsson, H. (2006) *Fibre Guide — Fibre Analysis and Process Applications in the Pulp and Paper Industry. A Handbook*, AB Lorentzen & Wettre, Kista, Sweden.

Kautto, J., Henricson, K., Sixta, H. *et al.* (2010) Effects of integrating a bioethanol production process to a Kraft pulp mill. *Nordic Pulp & Paper Research Journal*, **25** (2), 233–242.

Klinke, H.B., Thomsen, A.B., and Ahring, B.K. (2004) Inhibition of ethanol-producing yeast and bacteria by degradation products produced during pre-treatment of biomass. *Applied Microbiology and Biotechnology*, **66** (1), 10–26.

Kurosawa, K., Boccazzi, P., de Almeida, N.M., and Sinskey, A.J. (2010) High-cell-density batch fermentation of Rhodococcus opacus PD630 using a high glucose concentration for triacylglycerol production. *Journal of Biotechnology*, **147** (3–4), 212–218.

Lara, J.R., Noyes, G., and Holtzapple, M.T. (2008) An investigation of high operating temperatures in mechanical vapor-compression desalination. *Desalination*, **227** (1–3), 217–232.

Li, H.M., Saeed, A., Jahan, M.S. *et al.* (2010) Hemicellulose removal from hardwood chips in the pre-hydrolysis step of the Kraft-based dissolving pulp production process. *Journal of Wood Chemistry and Technology*, **30** (1), 48–60.

MacEachran, D.P., Prophete, M.E., and Sinskey, A.J. (2010) The *Rhodococcus opacus* PD630 heparin-binding hemagglutinin homolog TadA mediates lipid body formation. *Applied and Environmental Microbiology*, **76** (21), 7217–7225.

Mao, H., Genco, J.M., Yoon, S.H. *et al.* (2008) Technical economic evaluation of a hardwood biorefinery using the "near-neutral" hemicellulose pre-extraction process. *Journal of Biobased Materials and Bioenergy*, **2** (2), 177–185.

Mao, H.B., Genco, J.M., van Heiningen, A., and Pendse, H. (2010) Kraft mill biorefinery to produce acetic acid and ethanol: Technical economic analysis. *Bioresources*, **5** (2), 525–544.

Marinova, M., Mateos-Espejel, E., Jemaa, N., and Paris, J. (2009) Addressing the increased energy demand of a Kraft mill biorefinery: the hemicellulose extraction case. *Chemical Engineering Research & Design*, **87** (9A), 1269–1275.

Paredes, J.J., Jara, R., Shaler, S.M., and van Heiningen, A. (2008) Influence of hot water extraction on the physical and mechanical behavior of OSB. *Forest Products Journal*, **58** (12), 56–62.

Paredes, J.J., Mills, R., Shaler, S.M. *et al.* (2009) Surface characterization of red maple strands after hot water extraction. *Wood and Fiber Science*, **41** (1), 38–50.

Paredes, J.J., Shaler, S.M., Edgar, R., and Cole, B. (2010) Selected volatile organic compound emissions and performance of oriented strandboard from extracted southern pine. *Wood and Fiber Science*, **42** (4), 429–438.

Perlack, R.D., Wright, L.L., Turhollow, A.F. *et al.* (2005) Biomass as feedstock for a bioenergy and bioproducts industry: the technical feasibility of a billion-ton annual supply. Oak Ridge National Laboratory.

Pham, V., Holtzapple, M., and El-Halwagi, M. (2010) Techno-economic analysis of biomass to fuel conversion via the MixAlco™ process. *Journal of Industrial Microbiology & Biotechnology*, **37**, 1–12.

Ragauskas, A.J., Nagy, M., Kim, D.H. *et al.* (2006a) From wood to fuels: Integrating biofuels and pulp production. *Industrial Biotechnology*, **2** (1), 55–65.

Ragauskas, A.J., Williams, C.K., Davison, B.H. *et al.* (2006b) The path forward for biofuels and biomaterials. *Science*, **311** (5760), 484–489.

Ross, M.K. and Holtzapple, M.T. (2001) Laboratory method for high-solids countercurrent fermentations. *Applied Biochemistry and Biotechnology*, **94** (2), 111–126.

Rughoonundun, H., Granda, C., Mohee, R., and Holtzapple, M.T. (2010) Effect of thermochemical pretreatment on sewage sludge and its impact on carboxylic acids production. *Waste Management*, **30** (8–9), 1614–1621.

Saha, B.C. (2003) Hemicellulose bioconversion. *Journal of Industrial Microbiology & Biotechnology*, **30** (5), 279–291.

Sattler, C., Labbe, N., Harper, D. *et al.* (2008) Effects of hot water extraction on physical and chemical characteristics of oriented strand board (OSB) wood flakes. *Clean-Soil Air Water*, **36** (8), 674–681.

Sjostrom, E. (1991) Carbohydrate degradation products from alkaline treatment of biomass. *Biomass & Bioenergy*, **1** (1), 61–64.

Stauffer, N.W. (2010) Engineering fat-making bacteria: A road to plentiful biodiesel MIT Energy Initiative, available at http://web.mit.edu/mitei/research/spotlights/fat-bacteria.html.

Stoutenburg, R.M., Perrotta, J.A., Amidon, T.E., and Nakas, J.P. (2008) Ethanol production from a membrane purified hemicellulosic hydrolysate derived from sugar maple by *pichia stipitis* NRRL Y-7124. *Bioresources*, **3** (4), 1349–1358.

Takahashi, C.M., Takahashi, D.F., Carvalhal, M.L.C., and Alterthum, F. (1999) Effects of acetate on the growth and fermentation performance of *Escherichia coli* KO11. *Applied Biochemistry and Biotechnology*, **81** (3), 193–203.

TAPPI (1993) Kappa number of pulp T 236 cm-85.

Terrabon (2011) Terrabon Inc. reports successful results of economical green gasoline production through the use of innovative renewable fuel catalyst technologies, available at http://terrabon.com/release_20110126.html.

Thanakoses, P. (2002) Conversion of bagasse and corn stover to mixed carboxylic acids using a mixed culture of mesophilic microorganisms, in *Chemical Engineering*, Texas A&M, **PhD.**

Thanakoses, P., Black, A.S., and Holtzapple, M.T. (2003) Fermentation of corn stover to carboxylic acids. *Biotechnology and Bioengineering*, **83** (2), 191–200.

Tunc, M.S. and van Heiningen, A.R.P. (2008) Hemicellulose extraction of mixed southern hardwood with water at 150 °C: Effect of time. *Industrial & Engineering Chemistry Research*, **47** (18), 7031–7037.

Um, B.H. and van Walsum, G.P. (2009) Acid hydrolysis of hemicellulose in green liquor pre-pulping extract of mixed northern hardwoods. *Applied Biochemistry and Biotechnology*, **153** (1–2), 127–138.

Um, B.H. and van Walsum, G.P. (2010) Evaluation of enzyme mixtures in releasing fermentable sugars from pre-pulping extracts of mixed northeast hardwoods. *Applied Biochemistry and Biotechnology*, **161** (1–8), 432–447.

van Heiningen, A. (2006) Converting a Kraft pulp mill into an integrated forest biorefinery. *Pulp & Paper-Canada*, **107** (6), 38–43.

Voss, I. and Steinbuchel, A. (2001) High cell density cultivation of *Rhodococcus opacus* for lipid production at a pilot-plant scale. *Applied Microbiology and Biotechnology*, **55** (5), 547–555.

Waltermann, M., Luftmann, H., Baumeister, D. *et al.* (2000) *Rhodococcus opacus* strain PD630 as a new source of high-value single-cell oil? Isolation and characterization of triacylglycerols and other storage lipids. *Microbiology-Sgm*, **146**, 1143–1149.

Walton, S., van Heiningen, A., and van Walsum, P. (2010a) Inhibition effects on fermentation of hardwood extracted hemicelluloses by acetic acid and sodium. *Bioresource Technology*, **101** (6), 1935–1940.

Walton, S., van Walsum, G.P., and van Heiningen, A. (2009) Fermentation of near-neutral pH extracted hemicellulose derived from northern hardwood. 8th World Congress on Chemical Engineering, Montreal.

Walton, S.L., Bischoff, K.M., van Heiningen, A.R.P., and van Walsum, G.P. (2010b) Production of lactic acid from hemicellulose extracts by *Bacillus coagulans* MXL-9. *Journal of Industrial Microbiology & Biotechnology*, **37** (8), 823–830.

Walton, S.L., Hutto, D., Genco, J.M. *et al.* (2010c) Pre-extraction of hemicelluloses from hardwood chips using an alkaline wood pulping solution followed by Kraft pulping of the extracted wood chips. *Industrial & Engineering Chemistry Research*, **49** (24), 12638–12645.

Wigell, A., Brelid, H., and Theliander, H. (2007a) Degradation/dissolution of softwood hemicellulose during alkaline cooking at different temperatures and alkali concentrations. *Nordic Pulp & Paper Research Journal*, **22** (4), 488–494.

Wigell, A., Brelid, H., and Thellander, H. (2007b) Kinetic modelling of (galacto)glucomannan degradation during alkaline cooking of softwood. *Nordic Pulp & Paper Research Journal*, **22** (4), 495–499.

Xing, R., Subrahmanyam, A.V., Olcay, H. *et al.* (2010) Production of jet and diesel fuel range alkanes from waste hemicellulose-derived aqueous solutions. *Green Chemistry*, **12** (11), 1933–1946.

Yeh, H. (2002) *Conversion of Carboxylate Salts to Carboxylic Acids Via Reactive Distillation*, Texas A&M, MS.

Yoon, S.H., Cullinan, H.T., and Krishnagopalan, G.A. (2010) Reductive modification of alkaline pulping of southern pine, integrated with hydrothermal pre-extraction of hemicelluloses. *Industrial & Engineering Chemistry Research*, **49** (13), 5969–5976.

Yoon, S.H., Macewan, K., and van Heiningen, A. (2008) Hot-water pre-extraction from loblolly pine (*Pinus taeda*) in an integrated forest products biorefinery. *Tappi Journal*, **7** (6), 27–32.

Yoon, S.H. and van Heiningen, A. (2008) Kraft pulping and papermaking properties of hot-water pre-extracted loblolly pine in an integrated forest products biorefinery. *Tappi Journal*, **7** (7), 22–27.

Yoon, S.H. and van Heiningen, A. (2010) Green liquor extraction of hemicelluloses from southern pine in an integrated forest biorefinery. *Journal of Industrial and Engineering Chemistry*, **16** (1), 74–80.

14

Pyrolysis Bio-Oils from Temperate Forests: Fuels, Phytochemicals and Bioproducts

Mamdouh Abou-Zaid[1] and Ian M. Scott[2]

[1] *Canadian Forest Service, Great Lakes Forestry Service, Natural Resources Canada, Sault Ste. Marie, Ontario, Canada*
[2] *Agriculture and Agri-Food Canada, London, Ontario, Canada*

14.1 Introduction

Pyrolysis is the conversion of organic materials under high temperatures between 350 to 800 °C and inert atmosphere into useful fuels and chemical feedstock (Fu *et al.*, 2010; Ucar and Ozkan, 2008). High temperatures convert the chemical backbone of plant matter, cellulose, hemicellulose, lignin and organic extractives into CO, H_2 and water (Ladisch *et al.*, 2010). The pyrolysis process can lead to the production of charcoal, condensable organic liquids or biofuels and non-condensable gases. In the context of this chapter, the term biomass refers to "living and recently living biological material which can be used as fuel or for industrial production, and commonly refers to plant matter grown for use as biofuels" (BIMAT, 2011). Biomass includes specifically grown herbaceous and woody resources, including their residues, as well as forestry and agricultural residues. The types of available feedstocks specifically include short-rotation forestry and herbaceous crops, residues from both herbaceous and forestry harvests and other sources of biomass, including cooking oils, food processing and green waste (Murphy *et al.*, 2011).

Biorefinery Co-Products: Phytochemicals, Primary Metabolites and Value-Added Biomass Processing, First Edition.
Edited by Chantal Bergeron, Danielle Julie Carrier and Shri Ramaswamy.

In the Northern hemisphere, forestry residues offer a large quantity of biomass for energy and fuel products, and for the separation of valuable chemicals. The successful extraction of value-added chemicals and bioproducts will directly influence the cost and use of the fuels produced from pyrolysis bio-oil.

14.2   Overview of Forest Feedstock

14.2.1   Residues

The potential benefits of isolating phytochemicals from Canadian forestry operations are important, considering the amount of material available. It is estimated that almost 2% of the carbon photosynthesized by plants is converted into phytochemicals. Plants respond to environmental stresses and insect/microbial infestations by increasing the synthesis of such phytochemicals as phenolics, flavonoids, lignans, terpenoids and phytosterols. In trees, the phytochemicals are most concentrated in needles, leaves, roots, bark, compression wood and knots. Some of these phytochemicals provide natural defenses against forest pests and have been found to have medicinal benefits. The pharmaceutical industry has successfully developed many drugs from phytochemicals, including acetylsalicylic acid (aspirin) from willow and paclitaxel (taxol) from the western yew tree and Canadian yew, respectively. Today, 10 of the 20 top-selling drugs worldwide are either natural products or have their structures derived from them.

Nutraceuticals are higher-risk products because of a lack of formal regulatory mechanisms, which leaves liability with the primary manufacturer. However, the functional foods/dietary supplements field is well developed and there is some Canadian experience in taking such products to market. There is an opportunity for the development of new and significant sources of nutraceuticals which could give a competitive advantage to some pulp and paper mills. For example, the discovery of large amounts of hydroxymatairesinol in Norway spruce *Picea abies* knots resulted in a by-product now manufactured in Finland as HMR-lignan. HMR-lignan acts as an antioxidant and free-radical scavenger.

Some traditional phytochemical industries, such as natural rubber and plant tannin industries are well developed, but have declined to some extent after reaching their peaks in the 1950s. The decline was mainly due to replacement by inexpensive petroleum-derived products. However, recent petroleum price increases and public environmental concerns create a unique opportunity for a renaissance in the forest industry. The potential opportunities for using these extractives from bark, branches and foliage include a range of bioactive products, for example flavonoids, lignans, terpenoids and alkaloids are recommended as drugs and nutraceuticals. Natural antioxidants and radical-scavenging agents such as flavones, flavanones, flavonols and anthocyanins are promoted for food and health-related products. Fine chemicals include turpentine and essential oils, and adhesives and polymer materials can be manufactured from rosin, tannic acids and lignin for use as valuable household products. Similarly, natural waxes can be processed into dyestuff, flavours and food additives. In agriculture, auxin, abscisic acid and cytokinin, can provide natural plant-growth regulators, while phenolic compounds may be useful natural pesticides for agriculture and horticulture. Cellulose and leaf protein can be made into feedstuff and pet food additives. Traditional phytochemicals, such as natural rubber, tannic acid, rosin and turpentine can be made into new products, whereas bioethanol, biodiesel and biogas may provide a large source of bioenergy.

As a bioenergy and bioprocessing hub, the integrated pulp mill biorefinery is an ideal location for the extraction and processing of phytochemicals from Canada's forest biomass.

14.2.2 Phytochemistry and Distribution of Feedstock

The boreal forest is the largest forest in Canada and is a natural factory for the production of enormous numbers of organic substances with a variety of roles and bioactivities. Each species has a unique profile of phytochemicals, with characteristic classes of substances presented as many related derivatives, a phenomenon known as phytochemical diversity and redundancy. Their biosynthesis, the metabolic sequences leading to the production of various classes of natural products, is interconnected. Naturally occurring compounds are identified as products of primary metabolism, for example, carbohydrates, nucleosides, amino acids, fatty acids and lipids, or as products of secondary metabolism, for example, phenolics, terpenoids, steroids and alkaloids (Harborne *et al.*, 1999). In forestry, there is an increasing interest in the array of biochemicals present in the foliage of trees, an array that exists in various combinations and concentrations, some of which provide a means of natural defence against pathogens and defoliating insects. Because of the abundance of conifers (Pinaceae, Taxaceae and Cupressaceae) and heath family (Ericaceae) plants in boreal forests, many of the compounds of particular interest are phenolic compounds (aromatic compounds with hydroxyl substituents, including phenolic acids, phenylpropanoids and flavonoids), which are hydrophilic substances; their common origin is the aromatic precursor shikimic acid. Amongst other activities, these flavonoid compounds have notable antioxidant activity with beneficial health effects for humans (Rice-Evans and Packer, 2003). The evidence for flavonoids as cardioprotective and chemopreventive agents is steadily accumulating. Classes of phenolic compounds found in northern deciduous and coniferous species include (Figure 14.1): (a) *phenolic acids*: for example, protocatechuic acid, gallic acid, ellagic acid; (b) *phenylpropanoids*: for example, *p*-coumaric acid, caffeic acid, ferulic acid; (c) *flavan-3-ols*: for example, (+)-catechin, (−)-epicatechin, (+)-gallocatechin; (d) *flavanones*: for example, pinocembrin, naringenin, eriodictyol; (e) *dihydroflavonols*: for example, taxifolin, taxifolin glycosides; (f) *flavones*: for example, apigenin; and (g) *flavonol glycosides*: for example, kaempferol and quercetin glycosides. Phenolic glycosides (e.g., 1-*O*-galloyl-β-D-glucoside, quercetin 3-*O*-β-D-glucoside) are located in the vacuoles throughout the plant, while aglycones (gallic acid, quercetin) (more lipophilic forme) are found in or on the cuticular waxes.

14.2.3 Bioactivities and Applications in Forestry

Plant–Insect Interactions

Phytochemicals are present in tree foliage in various combinations and concentrations, some of which provide natural defenses against defoliating insects. The study carried out by Abou-Zaid *et al.* (2000) illustrates the role of biochemicals from deciduous leaves and coniferous needles in plant–insect interactions as host resistance factors and as novel sources of control agents for forest pests. For example, the feeding behaviour of the gypsy moth spans across four local species of pine (*Pinus resinosa* Ait., red pine; *P. strobus* L., white pine; *P. banksiana* Lamb., jack pine and *P. sylvestris* L., scotch pine) and is highly variable. It has been shown that phenolic compounds present in the needles of the four pine species are responsible for the variations in the feeding behaviour of the gypsy moth (Beninger and Abou-Zaid, 1997). Also, it was shown that red maple is highly resistant to the forest tent caterpillar, while sugar maple is not. Several studies (Abou-Zaid *et al.*, 2001; Nicol *et al.*, 1997) have determined that ethyl *m*-digallate is the major resistance factor in red maple. It has both the highest insect antifeedant activity and the highest concentration of any compound in red maple, but is not present in sugar maple, suggesting it is the major resistance factor in red maple leaves. Thus, susceptibility (or resistance) to defoliation by the gypsy moth and forest tent caterpillar may be, at least in part, determined by the phenolic composition of the needles and leaves of the host trees. These baseline correlations provide the basis for future research in developing tools and methods for enhanced protection of timber trees, as well as providing insights into novel biotechnological means for improving resistance of susceptible species.

Applications of Forest Natural Products in Health Promotion and Disease Prevention

Although biological activity in forest species is thought to have evolved in plant-pest co-evolution, the First Nations Peoples of Canada clearly had a good understanding of

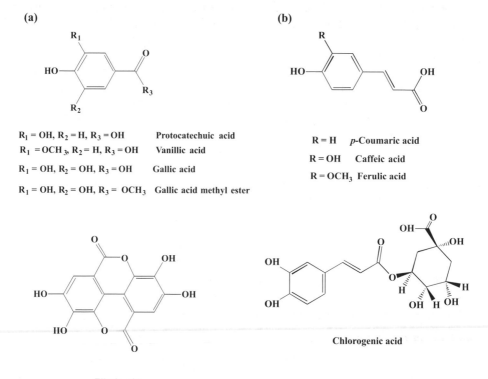

(a)

R₁ = OH, R₂ = H, R₃ = OH   Protocatechuic acid
R₁ = OCH₃, R₂ = H, R₃ = OH   Vanillic acid
R₁ = OH, R₂ = OH, R₃ = OH   Gallic acid
R₁ = OH, R₂ = OH, R₃ = OCH₃   Gallic acid methyl ester

(b)

R = H   *p*-Coumaric acid
R = OH   Caffeic acid
R = OCH₃ Ferulic acid

Chlorogenic acid

Ellagic acid

Figure 14.1   *Structures of phenolic compounds found in northern deciduous and coniferous species: phenolic acids (a) for example, protocatechuic acid, vanillic acid, gallic acid, gallic acid methyl ester, ellagic acid; phenyl-propanoids (b) for example, p-coumaric acid, caffeic acid, ferulic acid, chlorogenic acid; flavan-3-ols (c) for example, (+)-catechin, (−)-epicatechin, (+)-gallocatechin; flavanones (d) for example, pinocembrin, naringenin, naringin, eriodictyol, hesperitin; dihydroflavonols (e) for example, taxifolin, taxifolin glycosides; flavones (f) for example, apigenin, apigenin glycosides; flavonol and phenolic glycosides (g) for example, kaempferol and quercetin glycosides, quercetin 3-O-β-ᴅ-glucoside.*

(c)

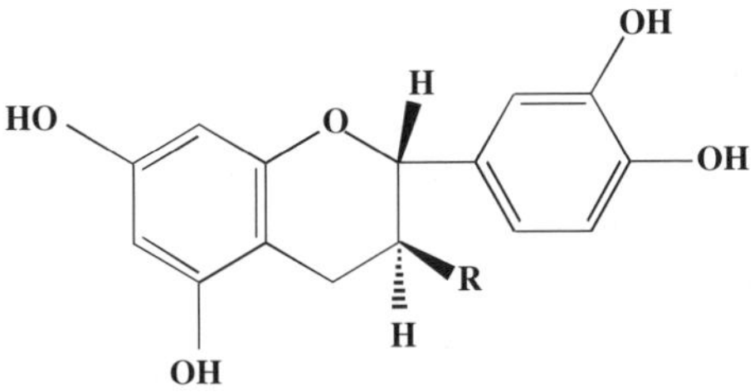

R = H (+) -Catechin

R = OH (+) -Gallocatechin

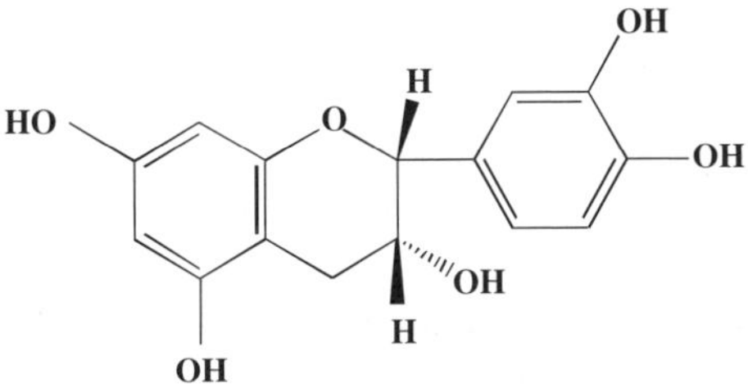

(–)-Epicatechin

(d)

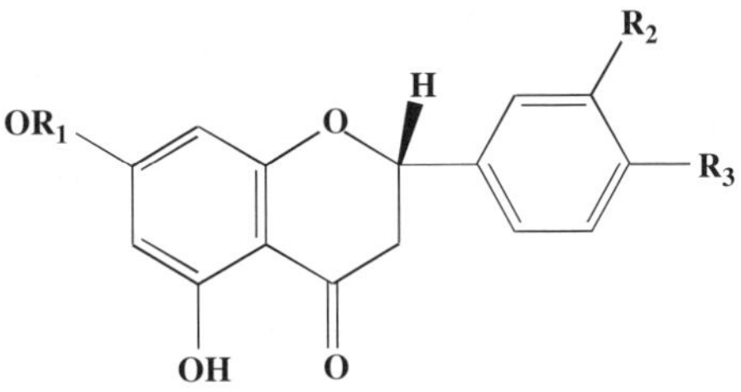

$R_1 = R_2 = R_3 = H$ Pinocembrin

$R_1 = R_2 = H, R_3 = OH$ Naringenin

$R_1 = $ neohesperidoside, $R_2 = H, R_3 = OH$ Naringin

$R_1 = H, R_2 = R_3 = OH$ Eriodictyol

$R_1 = H, R_2 = OH$ $R_3 = OCH_3$ Hesperetin

(e)

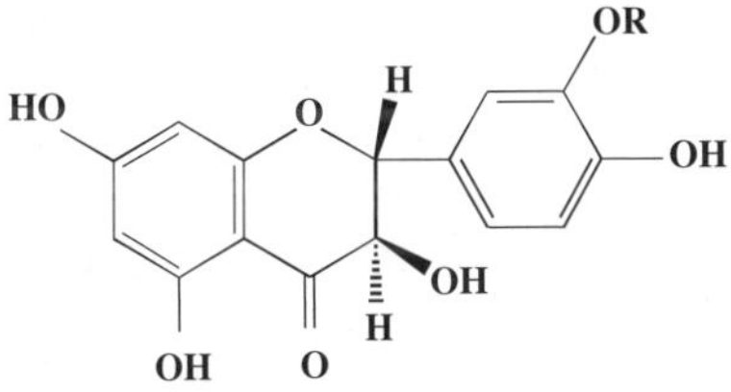

R = H   Taxifolin

R = glucosyl Taxifolin 3'-*O*-glucoside

(f)

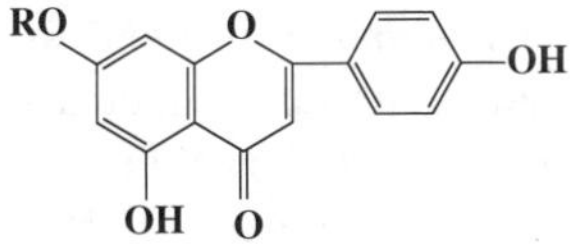

R = H   Apgienin

R = glucosly Apigenin 7-*O*-glucoside

(g)

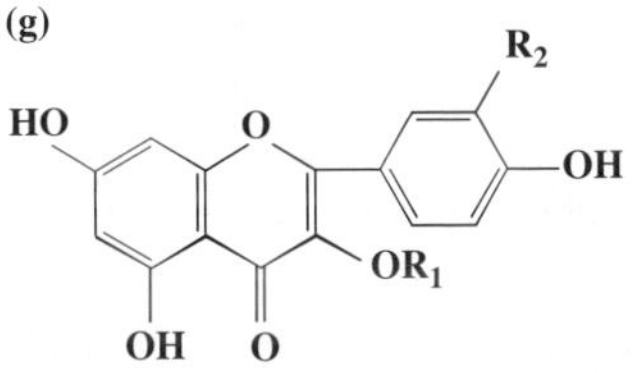

$R_1 = $ glucosyl, $R_2 = H$ Kaempferol 3-*O*-β-D-glucoside

$R_1 = $ galactosyl, $R_2 = H$ Kaempferol 3-*O*-β-D-galactoside

$R_1 = $ rhamnosyl, $R_2 = H$ Kaempferol 3-*O*-β-L-rhamnoside

$R_1 = $ rhamnoglucosyl, $R_2 = H$ Kaempferol 3-*O*-rhamnoglucoside

$R_1 = $ glucosyl, $R_2 = OH$ Quercetin 3-*O*-β-D-glucoside

$R_1 = $ galactosyl, $R_2 = OH$ Quercetin 3-*O*-β-D-galactoside

$R_1 = $ rhamnosyl, $R_2 = OH$ Quercetin 3-*O*-β-L-rhamnoside

$R_1 = $ rhamnoglucosyl, $R_2 = OH$ Quercetin 3-*O*-rhamnoglucoside

Figure 14.1   *(Continued)*

which species were active and which could be applied in traditional medicine. Research on non-forest timber products (NTFP), such as traditional native plants used by the First Nations Peoples of eastern Canada (Jones, 2000; Jones *et al.*, 2000), shows the importance of the First Nations traditional science in understanding native biodiversity. A review of the historic ethnobotanies showed that over 400 species of forest plants were used for 1800 medicinal applications by First Nations Peoples in eastern Canada (Arnason, Hebda, and Johns, 1981). The second aspect was to call attention to the distinct possibility that research on some of these plants could provide scientific validation of health benefits and renew the use of culturally used flora in northern communities. Ultimately, northern plants may be valuable sources of raw materials for the natural-products industry (Rice-Evans and Packer, 2003).

A logical first step is to review indigenous knowledge of the eastern Canadian flora using a quantitative ethnobotany approach (Jones *et al.*, 2000), and to concentrate first on those species for which there was a high consensus by the native peoples for use. For example, research results indicate that a highly significant correlation exists between the frequency of medicinal usage by the First Nations Peoples and the antifungal activity of the plant extract, based on laboratory bioassays (Jones *et al.*, 2000). In particular, pipsissewa, *Chimaphila umbellata*, was exceptionally active against a panel of eight medically important fungi, including azole-resistant forms of *Candida albicans*. The active principles included the light-activated anthraquinone, chimaphilin and the terpene, arbutin. For standardization of this plant product as a "natural health product", validated methods were developed to quantify the phytochemical markers of this plant. Another study showed that many traditionally used bark medicines had a broad spectrum of antibacterial activity, including activity against antibiotic-resistant strains (Omar *et al.*, 2000). Exceptional activity was found in black cherry bark, *Prunus serotina*, which was traditionally used for coughs and sore throats. Bioassay-guided fractionation led to the isolation of substantial amounts of naringenin, eriodictyol and other flavonoids as active principles. It has been shown that the use of several plants for the relief of diabetic symptoms, such as diabetic sores, has, in part, a pharmacological basis (McCune and Johns, 2002). In particular, foliage of evergreen species such as Pinaceae and Ericaceae has exceptional levels of antioxidants which are useful in wound healing. This may be related to their need to prevent leaf tissue photo-oxidation in winter. Through collaborative research with Haddad's group at the University of Montreal, we have found that several plants used as traditional medicines by the Cree of Northern Quebec have direct hypoglycaemia effects (Spoor *et al.*, 2006).

While traditionally used plants should be directed to prove benefits to indigenous communities, other approaches to research can also provide several key economic benefits for large forest industries. For example, raw materials from timber residues and natural products from minor species could increase economic success in the various forest-related sectors and industries, increase employment and provide opportunities within a broad spectrum of disciplines. Unfortunately, except for a large Canadian investment in lignin research, the forest industry has had minimal efforts in natural products. Increasing the use of native plant sources, both by using existing and developing biotechnology-based procedures, will benefit natural-health-product industries and healthcare technologies. As the potential is large, industry needs to develop ways of benefit sharing with native people who have provided much of the fundamental knowledge for natural product development.

14.3 Pyrolysis Technology

The pyrolysis process can be divided into three stages – (i) moisture evaporation ($< 200\,°C$), (ii) main devolatization (200 to 500 °C) and (iii) continuous slight devolatilization ($> 500\,°C$) (Fu *et al.*, 2010). Fast pyrolysis is the rapid heating of biomass particles with a short residence time in the reactor and is characterized by reaction temperatures around 500 °C, short residence times of less than 2 s and rapid cooling of vapours (Bridgwater, 2006). This temperature range and fast heating increases the yield of the liquid products, but the biomass must be dried to less than 10% water, and ground to less than 2 mm particle size to ensure a rapid reaction (Briens, Piskorz, and Berruti, 2008). The types of pyrolysis reactors available include fluidized bed and ablative; however, only the former process is used commercially at present. In the fluidized-bed reactor, the fluidized particles used are an inert material, for example sand, that can provide high heat transfer to the injected biomass particles. In most commercial fluidized "bubbling" beds the vapour residence time is typically between 5 and 10 s. Biomass particles in the mm size range will remain in the bed until they have reacted, but the liquid yield will be less when smaller particle sizes are used (Briens, Piskorz, and Berruti, 2008). A second type of fluidized reactor with a "circulating" fluidized bed allows for the circulation of a solid, often sand, which is heated by combustion of gaseous or solid byproducts of pyrolysis. After the product vapours are released, the sand flows back to a regular fluidized bed to be reheated. Circulating-bed processes have higher temperatures and shorter residence times compared to the bubbling bed, therefore requiring smaller- particle-size biomass (Briens, Piskorz, and Berruti, 2008). Regardless of the reactor type, it is possible to obtain similar liquid yields with variable vapour residence times, as long as the temperature is reduced for the higher residence time. This is important since most commercial pyrolyzers run at longer residence times, since they are easier to construct and will require less energy to run (Briens, Piskorz, and Berruti, 2008).

A less common pyrolysis is the ablative process where biomass particles contact hot metal surfaces that are externally heated. The biomass particles can be circulated by either a gas stream or by mechanical means to keep the particles moving against the heated wall. The advantage of this process is that large particles can achieve fast pyrolysis, but currently heat transfer through a heated wall makes scale-up difficult (Briens, Piskorz, and Berruti, 2008).

14.4 Prospects for Fuel Production

The first generation of biofuels is based on the conversion of existing agricultural commodity crops (Murphy *et al.*, 2011). Bio-oils can provide fuel for boilers, automobile and truck engines, and turbines or diesel generators to produce electricity (Bridgwater, 2006). However, due to water content and oxygenated compounds in bio-oil, it has been estimated that the heating value is approximately 50% of the heating value of diesel fuel (Briens, Piskorz, and Berruti, 2008). In most cases bio-oil cannot be used as a sole fuel, requiring combination with either an alcohol or diesel fuel. An advantage of bio-oil used as fuels is the lower production of air pollutants such as sulfur and nitrogen oxides, with the exception of carbon monoxide.

Bio-oil produced by pyrolysis is considered a "micro-emulsion" since it is an aqueous solution of holocellulose decomposition products that acts to stabilize the lignin macromolecules through hydrogen bonding (Bridgwater, 2006). It can tolerate the addition of water up to a limit, after which there will be phase separation.

Bio-oils can be upgraded either through hydrotreating or catalytic cracking to remove oxygen from the bio-oil molecules. Due to negative effects, such as contamination and increased gas yields, neither is as attractive as co-processing with fossil fuels (Briens, Piskorz, and Berruti, 2008). The promising second-generation biofuels, referred to as 2G, will be developed from lignocellulosic material through thermo-chemical processes into advanced biofuels (Murphy *et al.*, 2011). The 2G fuels are considered more challenging to produce compared to 1G, since the latter, including vegetable oils, starches and sugars, are often converted directly without pre-treatment. 2G fuels provide a 50% reduction or better in life-cycle greenhouse gas (GHG) emissions during the entire product life. Dedicated cellulosic crops can provide increased gross energy yields, but with reduced energy and life-cycle GHG emissions and input costs relative to first-generation crops.

14.5   Chemicals in the Bio-Oil

Hundreds of compounds have been identified in bio-oil as the basic compounds of biomass (Venderbosch and Prins, 2010). Separation of bio-oil by fractionation into aqueous and oily phases can produce aqueous compounds for meat browning and smoky flavour, and non-polar phenols that can be used as resins in plywood. Glycoaldehydes and sugars such as levoglucosan (Figure 14.2a and b) are created from the pyrolysis of celluloses and starches (Briens, Piskorz, and Berruti, 2008). The lignin component produces phenols, guaiacols and eugenols (Figure 14.2c and d) and the cellulose and hemicellulose yield sugars, acetaldehydes and formic acid. Between 40 and 50% of the bio-oil identity has been determined, what remains are the larger, less severely cracked or de-/re-polymerized compounds to be identified (Venderbosch and Prins, 2010). Higher pyrolysis temperatures are associated with greater cracking of most plant compounds. Faster vapour residence times or lower temperatures may allow for chemical separation from selected biomass. For example, the bio-oil produced from the pyrolysis of tobacco leaves still contained the alkaloid nicotine (Briens, Piskorz, and Berruti, 2008), while mustard straw bio-oil pyrolyzed at 300 °C contained the glycoalkaloid sinigrin (Scott *et al.*, unpublished). Thermo-chemical technology that can efficiently extract existing valuable phytochemicals from bio-oils is a desirable process and one that would make large-scale natural product extraction from biomass economical. It can be speculated that compounds which are more volatile might be removed from fluidized biomass through a distillation process. Currently under development at the University of Western Ontario, London Ontario, is a mechanized fluid reactor (MFR) that can increase heating in a step-wise manner to allow for pyrolysis to take place from low to high temperatures. This pyrolysis process is similar to batch distillation, where the biomass is exposed to a steadily increasing temperature. Temperature control to adjust the heating rate will allow for easier separation, purer products and protection of thermally sensitive products. The reactor functions through intense heat transfer to the biomass which is continually mixed. The temperature range is from ambient to 700 °C, with a gas residence time of 1 to 20 s. To date the MFR has provided temperature cuts of tomato plant waste that displays antioxidant activity, indicating that value-added compounds could be obtained from a pyrolysis-based process.

Using thermogravimetric analysis coupled with Fourier transform infrared (FTIR) analysis, it is possible to measure the evolution patterns of gas products during pyrolysis (Fu *et al.*, 2010). The release of CO_2 and CO was mainly due to the cracking and reforming of

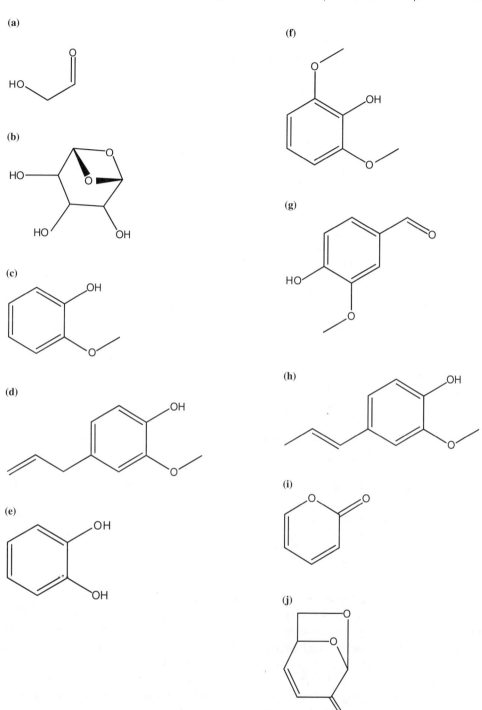

Figure 14.2   *Structures of compounds isolated from pyrolysis bio-oil: Glycolaldehyde (a); Levoglucan (b); Guaiacol (c); Eugenol (d); Catechol (e); Syringol (f); Vanillin (g); Isoeugenol (h); 2-pyrone (i) and Levoglucosenone (j).*

carbonyl, carboxyl and ether groups, while CH_4 was based on the cracking of methoxyl groups and formaldehyde was formed by C–C fragmentation of $-CH_2OH$ groups. HCN is the major N-containing product formed. The evolution patterns for CO, CH_4, ethane, acetone and HCN were similar between different biomass (straw and stalk from cotton and corn, respectively); but differences existed between the size of the HCN peak (cotton was smallest).

14.6   Valuable Chemical Recovery Process

A biorefinery that can convert biomass into fuels and chemicals is the most efficient approach to extract energy, fuel and valuable chemicals from a large amount of agricultural biomass (Briens, Piskorz, and Berruti, 2008; Demirbas, 2009). The major products that can be obtained from thermal conversion processes of biomass fall in to the categories of fuels (additives, alcohols, charcoal and gas) and chemicals (activated carbon, ethanol, fertilizers, fine chemicals, food additives, hydrogen, methane, methanol and resins) (Bridgwater, 2006). Integrating processing systems to yield multiple marketable products in order to increase the value of the raw materials is not a recent idea (Broder and Barrier, 1988). In the 1950s, the Tennessee Valley Authority (TVA) evaluated over 30 cellulosic feedstocks to determine the potential for producing higher-value products Hemicellulose and cellulose from alfalfa stem, corn stover, sugarcane bagasse and oak wood were converted to sugars, mainly xylose and glucose (Stinson, 1981 in Broder and Barrier, 1988). From the hydrolyzate sugars, products such as ethanol, lactic acid, furfural or citric acid were obtained. The solid lignin that remains was burnt to make steam and electricity, but also used to make adhesives.

Rapid thermal processing (RTP™) (fast pyrolysis) is a rapid destructive distillation process that breaks the chemical bonds by using thermal energy (Wampler, 2007). The use of pyrolysis as a tool for extracting non-volatile phytochemicals from forest biomass is a relatively new development and requires further investigation. Bio-based products or bioproducts are the spectrum of materials, chemicals, fuels and energy that can be derived from renewable bioresources as part of an integrated biorefiney.

The latter will be separated from the bio-oil, a dark brown, free-flowing organic liquid composed of highly oxygenated compounds (Demirbas, 2009). The chemical mixture includes water, guaiacol, catechol, syringol, vanillin (Figure 14.2e, f and g), furancarboxaldehyde, isoeugenol, pyrone (Figure 14.2h and i), acetic acid, formic acid and other carboxylic acids.

14.6.1   Sugars

The pyrolysis of the cellulose and hemicellulose components of biomass produces sugars, acetaldehyde and formic acids. Several sugars, including levoglucosan, levoglucosenone (Figure 14.2j) and hydroxyl-acetaldehyde (HAA) are found in bio-oil. These compounds are detectable by GC, and may be present, in the case of levoglucosenone, at 24 wt% in bio-oil from selected biomass (Venderbosch and Prins, 2010). Levoglucosan is a valuable product that is suitable for further fermentation. Glucosenone can be used in the synthesis of antibiotics and pheromones, rare sugars, butenolide, immunosuppresive agents and whisky lactones (Venderbosch and Prins, 2010). HAA can be present at up to 20 wt% and has been used for browning cheese, meat, sausages, poultry and fish; it could possibly be used as a precursor for glyoxal OHC-CHO, also produced by oxidation of ethylene glycol (Di Blasi *et al.*, 2010).

Glyoxal is used as a solubilizer and crosslinking agent in polymer chemistry, for example in the paper industry as a crosslinker for starch-based formulations and in textiles as a starting material with urea for wrinkle-resistant chemical treatments.

14.6.2 Phenols

A large number of the chemicals in bio-oil are phenolics, up to 50 wt%, derived from the degradation of lignin by pyrolysis (Venderbosch and Prins, 2010). Lignin is an amorphous polymer, composed of three aromatic constituents: *p*-hydroxyphenyl; guaiacyl and syringyl moieties – the ratios of these units differ, depending on the plant. In herbaceous plants hydroxycinnamic acids (*p*-coumaric and ferulic acids) are attached to lignin and hemicellulose by ester and ether bonds, forming lignin/phenolics-carbohydrate complexes. Pyrolysis bio-oil contains many of the same phenolics as the combustion of tobacco leaves, including phenol, dihydroxybenzenes, and their methyl-substituted derivatives, and hydroquinone, catechol and their methyl derivatives (McGrath *et al.*, 2009). At or below 350 °C, hydroquinone, catechol, guaiacol, and 3- and 4-methylcatechol are the predominant tobacco pyrolysis phenolic compounds formed. Above 350 °C, cresols are the predominate compound formed from tobacco components. Phenol and resorcinol are formed in both temperature regions. Water-extractable compounds, chlorogenic acid, rutin and scopoletin, are considered to be the precursors to hydroquinone, catechol and phenol formation. While the formation of cresols is similar, quinic acid and quinic acid derivatives are important precursors of hydroquinone, catechol and phenol.

Separated lignin can be used to produce vanillin, ferulic acid, vinyl guaiacol and optically active lignans, the dimers of monolignans (Buranov and Mazza, 2008). Phenolics are used in the production of resins and as aroma and flavouring agents in the food industry (McGrath *et al.*, 2009). The more valuable phenolic component is usually limited, in part due to only partial cracking of the original lignin, and contaminated with re-polymerized lignin and sugar fragments (Venderbosch and Prins, 2010).

14.7   Selected Phytochemicals from Pyrolysis Bio-Oils

The review by Mohan, Pittman, and Steele (2006) provided a chemical point of view of the fast pyrolysis of lignocellulosic feeds. The isolation and purification of specific single compounds (i.e., taxanes) usually requires the development of complex separation technologies for producing products from the whole bio-oil (Czernik and Bridgwater, 2004). Oasmaa *et al.* (2003a and b), Oasmaa *et al.* (2004) examined the effect of extractives on phase separation of pyrolysis of forestry residue and suggested improvements in storage stability and product quality of the liquid oil.

A recent successful development is the industrial processing of taxoids from *Taxus canadensis* harvested in eastern Canada as feedstocks for anticancer pharmaceuticals (Itokawa and Lee, 2003). Cass *et al.* (2001) describe a method of extracting Taxol® from the needles and whole clippings of *Taxus canadensis* by fast pyrolysis using a fluidized-bed reactor. The amount of Taxol® that was detected as being extracted was about 0.006%, based on the dry weight of the tissue mass subjected to pyrolysis. No more than 20% of the amount could be extracted using conventional organic extraction protocols and other taxanes could not be analyzed due to the phenolic compounds that were assumed to interfere in the detection. Abou-Zaid *et al.* (2007) developed a patent for a method of extracting and isolating a mixture of taxanes from a bio-oil composition of *Taxus canadensis,* overcoming the phenolics interference in collaboration with ENSYN Technologies Inc., Ottawa, Canada. The bio-oil composition is derived from the destructive distillation, or pyrolysis, of biomass obtained from *Taxus or Austrotaxus.* The method involves mixing a chromatographic resin with the bio-oil composition to form a bio-oil/resin mixture, and eluting the mixture of taxanes from the bio-oil/resin mixture using an eluting solvent. The method produces taxanes in a purified form from renewable parts of the yew tree in high yield without the requirement of large amounts of toxic and costly organic solvents.

14.8   Other Products

A number of bioproducts have been commercialized from pyrolysis of a variety of biomass. Wood flavour or liquid smoke is one of the most recognized from wood-derived bio-oil (Venderbosch and Prins, 2010). It is simply prepared by adding water to bio-oil, and is used for treating meat or food flavourings (Di Blasi *et al.*, 2010). Another common use is resin derived from lignans. The resins are currently used as a binding agent in foundry technologies (Di Blasi *et al.*, 2010) and lignin bio-oil may be used as a polymer to partially replace formaldehyde–phenol resins in particle-board construction (Venderbosch and Prins, 2010). A related application for this material could be adhesives (Di Blasi *et al.*, 2010) and as a wood preservative, since the phenolics have biocidal properties (Di Blasi *et al.*, 2010; Venderbosch and Prins, 2010). Along the same lines pesticides may be developed from the pyrolysis products. For example, tobacco bio-oil was active against microorganisms and insect pests (Booker *et al.*, 2010a, b) and pyroligneous acid had antibactericide and termite inhibitor properties (Lee *et al.*, 2010). When nicotine-free fractions of the tobacco bio-oil were tested in pesticide bioassays, these were found to still remain toxic to three common crop microorganisms and an insect pest, indicating a potential antibiotic and insecticide use. Acetic acid produced from pyroligneous acid (water, carboxylic acids, aldehydes, alcohols and pyrolytic lignin) have been used as preservatives to prevent fungal and mould growth and discourage termite feeding on wood (Lee *et al.*, 2010).

Other promising uses in agriculture include uses as a slow-release fertilizer. The reaction of bio-oil with ammonia, urea and other amino compounds leads to stable amides and amines, amongst other compounds, that can function as slow-release fertilizers, in soil conditioning, soil acidity control, control of excess Al and Fe, increasing phosphate availability and general plant growth promotion (Di Blasi *et al.*, 2010; Venderbosch and Prins, 2010).

Nicotine, still present in the tobacco bio-oil after fast pyrolysis temperatures as high as 450 °C, could be separated and used as a fumigant or in smoking cessation products (Booker *et al.*, 2010b). Other health care uses such as antioxidant, antimicrobial, anticancer, hypotensive, hypolipidemic and antileukemic activities, could be obtained from extracted polyphenols, tannins, proanthocyanidins (Hohtola, 2007; Matsuo *et al.*, 2010; Naczk *et al.*, 2007). Abundant sources for many of these compounds are from diverse plant materials, such as blueberry leaves and canola hulls, although the phenolic fractions in blueberry leaves had much greater antioxidant capacities (Naczk *et al.*, 2007). Northern species of *Vaccinium,* such as blueberry and cranberry, contain high concentrations of phenolic compounds in the leaves and berries, with high concentrations of flavanoids, such as proanthocyanidins and flavanols (Hohtola, 2007). Flavonoids have an important function to protect the plant against UV-radiation, the amounts of catechin, flavanol and hydroxycinnamic acids in the leaves are directly related to UV-B radiation and reduced levels of oxidative stress. Northern Hemisphere plants have levels of flavonoids that are linked positively to environmetal factors such as light irradiation and low temperature (Jaakola and Hohtola, 2010). Flavanols and flavones are conjugated to sugars (glucose, rhamnose, rutinose) and accumulated in plant vacuoles as glycosides; kaempferol and quercetin are found in many plant species, while isoflavones are found mainly in the Fabaceae family. Other body care, cosmetic and supplement ingredients could be obtained from the polyphenols found in grape, tomato and olive residues.

Other chemical products include acids and solvents. Carboxylic acids (calcium acetate and calcium formate) can be derived from the aqueous part of bio-oil up to 10 wt%. Potential uses could be for runway de-icing, sulfur dioxide removal during fuel combustion or as a catalyst during coal combustion (Venderbosch and Prins, 2010; Di Blasi *et al.*, 2010). An example of a Canadian product that combines bio-oil with lime (BioLime™), commercialized by Dyna-Motive Technologies Corporation, has been applied to clean flue-gas tunnels to remove sulfur and nitrogen oxides (Venderbosch and Prins, 2010). Other solvent applications could be during the production of lubricants from petroleum (Di Blasi *et al.*, 2010). Useful chemicals commonly produced are furfural and hydrogen. The former can act as an intermediate commodity chemical used for synthesizing more specialized chemical products (Di Blasi *et al.*, 2010).

14.9   Future Prospects

The application of both forestry and agricultural biomass in a biorefinery process will create value-added fuels and chemicals that will benefit both the economy and the environment. The next step is a thorough investigation and evaluation of the potential phytochemicals that can be extracted from the biomass. This includes developing new extraction/separation technologies, as was the case with Taxol®, and improving reactor technology to enable separation during pyrolysis, as is the promise of the MFR previously described in Section 14.3.3. We have identified several phytochemicals, such as phenolic acids, phenylpropanoids and flavonoids that can be isolated from biomass sources, but there are many others that show promise. These will be the compounds that have recognized biological activity and can survive the pyrolysis process. The challenge is to optimize the biorefinery to enable high-value chemical recovery, otherwise there will be slow progress for this industry.

References

Abou-Zaid, M.M., Helson, B.V., Beninger, C.W., and de Groot, P. (2000) Phenolics from deciduous leaves and coniferous needles as sources of novel control agents for lepidopteran forest pests, in *Phytochemicals and Phytopharmaceuticals Champaign, Ill* (eds F. Shahidi and C.-T. Ho), AOCS Press, pp. 398–417.

Abou-Zaid, M.M., Helson, B.V., Nozzolillo, C., and Arnason, J.T. (2001) Ethyl m-digallate from red maple, Acer rubrum L., as the major resistance factor to forest tent caterpillar, Malacosoma disstria Hbn. *Journal of Chemical Ecology.*, **27** (12), 2517–2527.

Abou-Zaid, M.M., Graham, R.G., Freel, B.A., and Boulard, D.C. (2007) inventors; Preparation Of Taxanes. Publication Number: WO/2007/045093. International Application No.: PCT/CA2006/001717.

Arnason, T.J., Hebda, R.J., and Johns, T. (1981) Use of plants for food and medicine by Native Peoples of eastern Canada. *Canadian Journal of Botany*, **59** (11), 2189–2325.

Beninger, C.W. and Abou-Zaid, M.M. (1997) Flavonol glycosides from four pine species that inhibit early instar gypsy moth (Lepidoptera: Lymantriidae) development. *Biochemical Systematics and Ecology.*, **25** (6), 505–512.

BIMAT (2011) Biomass Inventory Mapping and Analysis Tool (BIMAT). Agriculture and Agri-Food Canada. Available from http://www4.agr.gc.ca/AAFC-AAC/display-afficher.do?id=1226509218872&lang=eng.

Booker, C.J., Bedmutha, R., Vogel, T. *et al.* (2010) Experimental investigations into insecticidal, fungicidal, and bactericidal properties of pyrolysis bio-oil from tobacco leaves using a fluidized bed pilot plant. *Industrial and Engineering Chemistry Research*, **49**, 10074–10079.

Booker, C.J., Bedmutha, R., Scott, I.M. *et al.* (2010) Bioenergy II: characterization of the pesticide properties of tobacco bio-oil. *International Journal of Chemistry and Reactor Engineering*, **8**, Article 26.

Bridgwater, T. (2006) Biomass for energy. (In focus: The future of sustainable materials for industry.). *Journal of the Science of Food and Agriculture*, **86** (12), 1755–1768.

Briens, C., Piskorz, J., and Berruti, F. (2008) Biomass valorization for fuel and chemicals production - a review. *International Journal of Chemical Reactor Engineering*, **6**.

Broder, J.D. and Barrier, J.W. (1988) *Producing Fuels and Chemicals from Cellulosic Crops* (ed. J.E. Janick JaS), Indianapolis, Indiana.

Buranov, A.U. and Mazza, G. (2008) Lignin in straw of herbaceous crops. *Industrial Crops and Products. [Surveys.]*, **28** (3), 237–259.

Cass, B.J., Piskorz, J., Scott, D.S., and Legge, R.L. (2001) Challenges in the isolation of taxanes from Taxus canadensis by fast pyrolysis. *Journal of Analytical and Applied Pyrolysis*, **57** (2), 275–285.

Czernik, S. and Bridgwater, A.V. (2004) Overview of applications of biomass fast pyrolysis oil. *Energy and Fuels*, **18** (2), 590–598.

Demirbas, A. (2009) Biorefineries: current activities and future developments. *Energy Conversion and Management*, **50** (11), 2782–2801.

Di Blasi, C., Branca, C., and Galgano, A. (2010) Biomass screening for the production of furfural via thermal decomposition. *Industrial and Engineering Chemistry Research*, **49** (6), 2658–2671.

Fu, P., Hu, S., Xiang, J. *et al.* (2010) FTIR study of pyrolysis products evolving from typical agricultural residues. *Journal of Analytical and Applied Pyrolysis*, **88** (2), 117–123.

Harborne, J.B., Baxter, H., and Moss, G.P. (1999) *Phytochemical dictionary: a handbook of bioactive compounds from plants*, 2nd edn, Taylor & Francis, London.

Hohtola, A. (2007) Northern plants as a source of bioactive products, in *Physiology of Northern Plants Under Changing Environment* (eds. E. Taulavuori and K. Taulavuori), Research Signpost, Trivandrum, pp. 291–307.

Itokawa, H. and Lee, K.-H. (2003) *Taxus: the Genus Taxus*, Taylor & Francis, London; New York, NY.

Jaakola, L. and Hohtola, A. 2010. Effect of latitude on flavonoid biosynthesis in plants. *Plant, Cell and Environment*, **33** (8), 1239–1247.

Jones, N.P. (2000) *Quantitative ethnobotany of Eastern Canada's First Nations peoples and Relative Antifungal Activity of Selected Medicinal Plants*, University of Ottawa, Ottawa.

Jones, N.P., Arnason, J.T., Abou-Zaid, M. *et al.* (2000) Antifungal activity of extracts from medicinal plants used by first nations peoples of eastern Canada. *Journal of Ethnopharmacology*, **73** (1–2), 191–198.

Ladisch, M.R., Mosier, N.S., Youngmi, K.I.M. *et al.* (2010) Converting cellulose to biofuels. *Chemical Engineering Progress*, **106** (3), 56–63.

Lee, S.H., H'ng, P.S., Lee, A.N. *et al.* (2010) Production of pyroligneous acid from lignocellulosic biomass and their effectiveness against biological attacks. *Journal of Applied Sciences*, **10** (20), 2440–2446.

Matsuo, Y., Fujita, Y., Ohnishi, S. *et al.* (2010) Chemical constituents of the leaves of rabbiteye blueberry (Vaccinium ashei) and characterisation of polymeric proanthocyanidins containing phenylpropanoid units and A-type linkages. *Food Chemistry*, **121** (4), 1073–1079.

McCune, L.M. and Johns, T. (2002) Antioxidant activity in medicinal plants associated with the symptoms of diabetes mellitus used by the Indigenous Peoples of the North American boreal forest. *Journal of Ethnopharmacology*, **82** (2–3), 197–205.

McGrath, T.E., Brown, A.P., Meruva, N.K., and Chan, W.G. (2009) Phenolic compound formation from the low temperature pyrolysis of tobacco. *Journal of Analytical and Applied Pyrolysis*, **84** (2), 170–178.

Mohan, D., Pittman, C.U. Jr., and Steele, P.H. (2006) Pyrolysis of wood/biomass for bio-oil: A critical review. *Energy and Fuels*, **20** (3), 848–889.

Murphy, R., Woods, J., Black, M., and McManus, M. (2011) Global developments in the competition for land from biofuels. *Food Policy*, **36** (Suppl 1), S52–S61.

Naczk, M., Zadernowski, R., and Shahidi, F. (2007) Antioxidant capacity of phenolic extracts from selected food by-products. *Acta Horticulturae*, **956**, 184–194.

Nicol, R.W., Arnason, J.T., Helson, B., and Abou-Zaid, M.M. (1997) Effect of host and nonhost trees on the growth and development of the forest tent caterpillar, *Malacosoma disstria* (Lepidoptera: Lasiocampidae). *Canadian Entomologist*, **129** (6), 991–999.

Oasmaa, A., Kuoppala, E., Gust, S., and Solantausta, Y. (2003) Fast pyrolysis of forestry residue. 1. Effect of extractives on phase separation of pyrolysis liquids. *Energy and Fuels*, **17** (1), 1–12.

Oasmaa, A., Kuoppala, E., and Solantausta, Y. (2003) Fast pyrolysis of forestry residue. 2. Physicochemical composition of product liquid. *Energy and Fuels*, **17** (2), 433–443.

Oasmaa, A., Kuoppala, E., Selin, J.F. *et al.* (2004) Fast pyrolysis of forestry residue and pine. 4. Improvement of the product quality by solvent addition. *Energy and Fuels*, **18** (5), 1578–1583.

Omar, S., Lemonnier, B., Jones, N. *et al.* (2000) Antimicrobial activity of extracts of eastern North American hardwood trees and relation to traditional medicine. *Journal of Ethnopharmacology*, **73** (1–2), 161–170.

Rice-Evans, C. and Packer, L. (2003) *Flavonoids in health and disease*, 2nd, rev. and updat edn, Marcel Dekker, New York.

Spoor, D.C.A., Martineau, L.C., Leduc, C. *et al.* (2006) Selected plant species from the Cree pharmacopoeia of northern Quebec possess anti-diabetic potential. *Canadian Journal of Physiology and Pharmacology*, **84** (8–9), 847–858.

Stinson, J.M. (1981) TVA's Biomass Fuels Program. TVA Circular Z-120, Muscle Shoals, Alabama.

Ucar, S. and Ozkan, A.R. (2008) Characterization of products from the pyrolysis of rapeseed oil cake. *Bioresource Technology*, **99** (18), 8771–8776.

Venderbosch, R.H. and Prins, W. (2010) Fast pyrolysis technology development. *Biofuels, Bioproducts & Biorefining*, **4** (2), 178–208.

Wampler, T.P. (2007) *Applied Pyrolysis Handbook*, 2nd edn, CRC Press/Taylor & Francis, Boca Raton.

15

Char from Sugarcane Bagasse

K. Thomas Klasson

USDA-ARS Southern Regional Research Center,
New Orleans, Louisiana, USA

15.1 Introduction

Bagasse is the plant material remaining after sugar has been extracted from the sugarcane plant. It has been argued for many years that this and other plant byproducts represent an under-utilized by-product stream from agricultural operations. Many estimates exist of the amount of agricultural and forest resources that are available for alternative use. In a recent report, it was estimated that 129 and 176 million dry metric tons (tonnes) are currently available in the US for bioenergy and bioproducts from forestry and agriculture, respectively, but that a total of 1239 million tonnes could be available through high-yielding crops (Perlack *et al.*, 2005). Production of fuel ethanol from these vast quantities of lignocellulosic biomass has been an active research area for years. Ethanol has a tradition of being used as automobile fuel as far back as the Model T Ford and was in demand during WWI.

Solomon, Barnes, and Halvorsen (2007) wrote a recent review of existing and planned cellulosic ethanol facilities that shows recent pilot plants with capacities of less than $30,000 \, \text{m}^3 \, \text{yr}^{-1}$ constructed between 1985 and 2006, followed by a list of demonstration-scale plants starting in 2004. Eleven commercial plants were planned for start up in 2007–2008 for conversion of biomass, including bagasse but few have actually opened. Several companies are now targeting production dates 2012–2013 (Bevill, 2010; Bevill, 2011). The largest distillery producer in Brazil, Denini S/A Indústrias de Base, was the first company that built a bagasse-based pilot-scale ethanol facility (Solomon, Barnes, and

The contribution of K. Thomas Klasson has been written in the course of his official duties as a US government employee and is classified as a US government work, which is in the public domain in the United States of America.

Biorefinery Co-Products: Phytochemicals, Primary Metabolites and Value-Added Biomass Processing, First Edition.
Edited by Chantal Bergeron, Danielle Julie Carrier and Shri Ramaswamy.

Halvorsen, 2007), and while scale-up plans to a larger facility were later reported (Oliverio and Hilst, 2004; Olivério, 2006), commercialization has yet to take place. Hawaii, with a long history of sugarcane production, passed legislation that went into effect in 2006 (Reyes, 2006) to mandate ethanol/gasoline blends. However, the State still has no cellulosic fuel-ethanol facility (Solomon, Barnes, and Halvorsen, 2007) and appears to be focusing their efforts toward a sugar-based ethanol platform (Voegele, 2009).

While fuel ethanol is one of the main products proposed from excess biomass, there are a number of other products that could be considered, for example, chars of different qualities. Bagasse biochars could be used for applications such as:

- combustion fuel,
- soil conditioner,
- carbon sequestration, and
- in environmental management or industrial settings.

The pyrolysis of biomass to produce fuels and chemical feedstock has a long tradition. Direct combustion of biomass to recover its heating value may be useful for some operations, but its low density and variable moisture content makes it less suitable as a fuel than coal (Yaman, 2004). The pyrolysis (or charring) of the biomass to control the moisture and increase the energy density is therefore needed. This is the traditional process of making charcoal. Partial charring via torrefaction (pyrolysis at relative low temperatures; for example, 300 °C) is yet another way of removing moisture, increasing density of the material, and improving grinding properties.

Biochar is produced worldwide for solid fuel (charcoal) or industrial purposes, but it can cause a problem for the eco-system if green biomass (standing biomass) is converted to fuels without sustainable practices (FAO, 2006; Lehmann, Gaunt, and Rondon, 2006). The use of agricultural waste, animal manures in particular, offers a sustainable resource if the quality (e.g., carbon content) of the biochar can be improved. Pyrolysis of poultry litter, peanut hulls, or pine chips in steam or nitrogen atmospheres showed little difference in final carbon content when pyrolyzed at 400 or 500 °C (Gaskin *et al.*, 2008), but there were differences in the carbon content in chars generated from different raw materials. Pyrolysis at 350 °C resulted in a char with only 58.3% carbon content, while pyrolysis at 420 °C, or above, created chars with carbon contents of 85.7–96.5% (Mobarak, Fahmy, and Schweers, 1982). Increase in carbon content for the char when carbon dioxide is used as a pyrolysis gas has been noted when creating biochars from wood, olive stones, bagasse, straw, and wood. In some cases, the carbon content increased from 76 to 89% (Minkova *et al.*, 2000). Carbon content of char directly affects its heating value and several correlations have been made to calculate the heating value from either elemental (C, H, O, N, etc.) composition (Channiwala and Parikh, 2002) or from proximate (volatile matter, fixed carbon, ash) analysis (Cordero *et al.*, 2001).

Chars from biomass (or biochars) have long been used by humans in agriculture. It is well known that good land management is needed in order to maintain a sustainable agricultural practise. Poor management will result in waste lands and erosion, no matter where one looks as a casual observer (Lowdermilk, 1939). Balanced fertilization and other agricultural management practices are also advantageous for a well-developed subsurface microbial population (Ladd *et al.*, 1994; Goyal *et al.*, 1999). Charred biomass can help ameliorate plant nutrient availability (Lehmann *et al.*, 2003; Lehmann, 2007) and activated chars can help toxic soil and sediment recover (Chen *et al.*, 2006; Zimmermann *et al.*, 2008). For instance, activated char sprayed on rows of seeded soil, protected the plants from damage when later applying herbicides to control weeds on the field (Linscott and Hagin, 1967).

In a recent review, Lehmann (2007) also makes an argument that reversing climate change may be difficult without returning some of the atmosphere's carbon dioxide to the soil. One suggested way is to pyrolyze wood or grass (or other biomass) to produce gas or oil for energy and using the remaining chars as soil amendments. However, Lehmann (2007) recognizes that additional research is needed to evaluate these chars for use as carbon sinks or other environmental benefits. The benefits of using biochars as a way to mitigate climate change was even recognized and reported to the US Congress (Bracmort, 2009).

The majority of the research, both in the area of biochar as carbon sink and as soil conditioner, has been done on plant biomass (Lehmann, Gaunt, and Rondon, 2006; Lehmann, 2007; Chan *et al.*, 2007). Lehmann, Gaunt, and Rondon (2006), in their review, point out that biochars have been shown to suppress methane and nitrous oxide (important greenhouse gases) emissions from soil. The biochars also buffer the soil and retain ammonium, nitrate, and phosphate ions, but the environmental benefits of biochars as soil amendments, other than as a carbon sink, are poorly quantified (Lehmann, Gaunt, and Rondon, 2006). Evidence exists that char properties (including surface charge) may directly or indirectly influence soil microbial populations and nutrient transformations (DeLuca, MacKenzie, and Gundale, 2009). Glaser, Lehmann, and Zech (2002), in a review, noted that most studies conducted in the past showed improved biomass production in the presence of charcoal. However, Chan *et al.* (2007) showed that biochar (from grass, cotton trash, and plant prunings) did not improve growth unless nitrogen was added as a supplement. When nitrogen was added, growth improved substantially, compared with biochar-free controls.

While the above discussion focused on the advantages of biochar as fuel or as a soil conditioner for nutrient supply and retention, and as a carbon sink (to alleviate greenhouse gas emissions), chars can also be added to soil to make toxic contaminants less available. Lead-contaminated soil amended with bone chars showed less lead availability in the soil and less uptake by Chinese cabbage (Chen *et al.*, 2006). Also, lead and zinc were made less bio-available when phosphate-rich bone meal char was added to soil (Hodson, Valsami-Jones, and Cotter-Howells, 2000). In other studies, copper-contaminated activated-carbon-amended soil was able to sustain the same plant growth rate as clean soil (Bes and Mench, 2008). Likewise, sediments amended with activated carbon made organic contaminants less bio-available (Zimmermann *et al.*, 2008).

One of the best known uses of chars in the sugarcane industry is for the removal of color during sugar refining. Bone chars, as the name implies, are chars made from animal bones which are heated to 500–700 °C for several hours (Choy *et al.*, 2004). The use of this product has caused some concerns from vegetarians, who may be looking for alternative sugars (Yacoubou, 2007). Activated chars or activated carbons from a variety of biomass materials have been published in reviews by Heschel and Klose (1995) and Paraskeva, Kelderis, and Diamadopoulos (2008). These activated biomass char materials have the same potential use as activated carbons made from coal, such as contaminant removal from liquids and gases (Allen, Whitten, and McKay, 1998). Activated carbons in general have found their way into markets of water treatment, decolorization, solvent recovery, military uses, nuclear reactors, air treatment, domestic uses, precious metal recovery, catalysis, and so on (Allen, Whitten, and McKay, 1998; Baker *et al.*, 2007). However, the majority (82%) of the carbon is used for liquid applications. The market for activated carbon is not huge, but it is significant. The US demand for activated carbon in 2000 was 156 000 tonnes and increased about 1.3% per year during 2000–2005. The forecast for 2009 was 181 000 tonnes with a price history between 2000–2005 of $1650 to $2160 per tonne (Kirschner, 2006). It was recently predicted that if mercury control is fully implemented at coal-fired power plants, the market for activated carbons (or chars) might increase significantly (Klasson *et al.*, 2010).

15.2 Sugarcane Bagasse Availability

Sugarcane is grown in warm (tropical) areas around the world. The Food and Agriculture Organization of the United Nations estimates that the total world production is 1736 million tonnes and that it is grown on 24.2 million hectares (FAO, 2010). The main sugarcane-producing countries are Brazil, with an annual production of 645 million tonnes, India at 348 million tonnes, and China with 125 million tonnes. The United States produce 25 million tonnes grown on 351 000 hectares. Some of the other producers are listed in Table 15.1.

Sugarcane is harvested by hand or with mechanical means, and in developed countries, mechanical harvesters are the most common. The amount of residues from processing sugarcane (bagasse and tops) is not readily available and varies from country to country, from region to region, and from factory to factory. The bagasse is the fibrous residue that remains after the factory has extracted the sugars, while the tops (and leaves) are generally removed before processing. Koopmans and Koppejan (1997) collected information from other researchers and claim that in Asian countries, recommended values for gross residue generation is 0.29 kg of bagasse (at ~50% moisture) and 0.3 kg tops/leaves per kg sugarcane harvested. A recent article (Cerqueira, Filho, and da Silvia Meireles, 2007) suggests that, in Brazil, one metric ton of harvested sugarcane generates 280 kg bagasse (no moisture level was specified). This should be compared with earlier studies by Dasgupta *et al.* (1988), who claim that 270 kg dry bagasse is left after processing 1 tonne of crude sugarcane. When leaves and tops are included, Zandersons *et al.* (1999) quotes Bauer *et al.* (1998) who estimated that 316 kg of dry bagasse, leaves, and tops are generated from 1 tonne of sugarcane in Brazil.

Out of the generated bagasse, 75% is used as fuel for the evaporation of water and electricity generation in the sugarcane factory, according to Mobarak, Fahmy, and Schweers (1982). This stated fuel use is similar to the range, 70–90%, given by Ouensanga and Picard (1988). Marshall *et al.* (2000) and Ng *et al.* (2002) estimated that the bagasse use for fuel was 85% in the USA. In a workshop, a yet higher value (90%) was quoted by Aiman and

Table 15.1 Production of sugarcane and estimated sugarcane bagasse availability in selected countries.

Country	Sugarcane Production (million tonnes)	Bagasse Produced (million tonnes)	Excess Bagasse (million tonnes)
Brazil	645.3	174.2	34.8
India	348.2	94.0	18.8
China	124.9	33.7	6.7
Thailand	73.5	19.8	4.0
Pakistan	63.9	17.3	3.5
United States	25.0	6.8	1.4
World	1736	468.7	93.7

Stubington (1993). Some operations use essentially all the bagasse residues in the factory boilers (Junginger *et al.*, 2001) and return excess energy in the form of electricity to the grid (Ramjeawon, 2008). Assuming that 0.28 kg of dry bagasse is generated per kg raw sugarcane harvested, and that 80% of this is utilized for fuel, we can estimate that approximately 94 million tonnes of dry bagasse is available for other uses, such as char, in the world (see Table 15.1).

Ash from the combustion of bagasse at the sugarcane factories generates about 25 kg ash per tonne combusted bagasse and this ash is often used as a fertilizer in Brazil (Sales and Lima, 2010); it has also been proposed as a conditioner for calcareous soil (Khan and Qasim, 2008). In addition, it may be used in concrete production as a replacement for sand or cement (Payá *et al.*, 2002; Chusilp, Likhitsripaiboon, and Jaturapitakkul, 2009; Sales and Lima, 2010).

15.3   Thermal Processing in an Inert Atmosphere (Pyrolysis)

Pyrolysis is a process of thermally heating a material (often an organic material) in an inert atmosphere. By varying time, temperature and the inert gas, different decomposition products are obtained as different fractions are vaporized and reacted (Babu, 2008). For example, very fast pyrolysis processes may produce 75% of oil and only 12% char, while a slow pyrolysis process may produce 30% oil and 35% char (Bridgewater, 2007). A recent review of pyrolysis of biomass by Yaman (2004) gives a good overview of pyrolysis of various types of biomass and the report summarizes pyrolysis conditions and results from a large number of references.

Pyrolysis under some conditions is often referred to by other common names. In torrefaction, the material is heated to a relative low temperature (250–400 °C) for an extended period of time in an inert atmosphere. Water and low-boiling-temperature organics evaporate and the resulting char is less bulky and can serve as a fuel to replace (or augment) coal in burning or gasification (Deng *et al.*, 2009). The objective of torrefaction is to retain the energy value of the raw material in a solid form.

Flash or fast pyrolysis refers to fast heating (a few seconds) of finely ground organic materials in order to produce bio-oils. Here, the purpose is to maximize the amount of organic condensables, while minimizing decomposition so that the energy value in the raw material can be retained in a liquid form. These oils contain oxygenated species which are less desirable. Hydrogen gas was used to improve the composition of the oil during pyrolysis in a method called hydropyrolysis with little success in an high-pressure reactor (Rocha *et al.*, 1997). Further studies of hydropyrolysis of bagasse did not show significant improvements in the results (Pindoria *et al.*, 1999); however, other results on cellulose have shown some promise of reducing oxygen content in the bio-oil (Rocha, Luengo, and Snape, 1999).

In other pyrolysis methods, such as gasification, the organic material is heated to higher temperatures (greater than 700 °C) with the intention of breaking down the material into small molecules and obtaining a synthesis gas. The composition of the gas depends on the raw material and the treatment conditions, but the gas generally consists of H_2, CO, CO_2, CH_4, and minor amounts of other gases. The gases can be combusted for heat or allowed to react over a catalyst to produce a variety of products using Fisher–Tropsch reactions or microorganisms (Klasson *et al.*, 1992). The purpose of gasification is to maximize the decomposition of the organic raw material and retain the energy in a gaseous form.

While the above general methods exist, there is a variety of methods that operate in between the extreme conditions described above. These methods were developed when the desired product was to have an alternative use (other than fuel energy). It is also true that in each of the extreme cases, co-products are formed which may have value if they could be further processed. These co-products are the less desirable forms of streams; for example, while torrefaction seeks to retain the biomass in a solid form, both liquid and gas streams are produced as co-products. In the case of fast pyrolysis, the solid and gas streams are considered co-products. Similarly, gasification generates small amounts of char as a co-product. It is important to note that small alterations of the extreme methods may improve the value of those co-products, while not having a detrimental effect on the overall process.

The thermal degradation rate of sugarcane bagasse has been described in the literature (Aiman and Stubington, 1993; Garcìa-Pèrez *et al.*, 2001). Aiman and Stubington (1993) summarized previous research, did their own experiments and tried to define the simplest model that would accurately describe the thermal degradation. Garcìa-Pèrez *et al.* (2001) took a similar approach, but estimated individual degradation rates for hemicellulose, cellulose, and lignin present in the bagasse for a more complete model, which will be described later. Other considerations, beyond just the degradation of the carbohydrates, are important in selection of the pyrolysis conditions. Keown *et al.* (2005) showed that the rate of heating is also important and that extremely fast heating rates, of the order of 1000–10,000 °C s^{-1}, to temperatures above 600–700 °C, can cause a significant release of sodium, potassium, manganese, and calcium from bagasse into the gas stream.

15.4   Technology for Converting Char to Activated Char

An example of a pyrolysis method that operates at "intermediate" conditions is activated carbon manufacture. Here, a slow pyrolysis at a temperature between torrefaction and gasification conditions is followed by activation at a temperature slightly above the pyrolysis temperature in the presence of some activating gas. The purpose of activation is to increase the surface area and create a surface that is active. There are different activation methods, such as steam, acid, salt, carbon dioxide (Toles *et al.*, 2000a; Toles *et al.*, 2000b; Lima, Marshall, and Wartelle, 2004; Zeng, Jin, and Guo, 2004), just to mention a few. The properties of the activated carbons (or chars) are highly dependent upon the feed source and the thermochemical process by which they are pyrolyzed and activated.

Steam activation is by far the most common method of activation. In this method, the biomass is pyrolyzed in one stage and later (in the same or in a different furnace) is activated by steam. Typical conditions involve pyrolysis under nitrogen at 700 °C and steam activation at 850 °C, each for an hour (Ng *et al.*, 2002). The steam activation can also be followed by oxidation, either chemical or physical, to modify the surface area and its charge so that it can sorb cationic metals (Johns, Marshall, and Toles, 1998). Other methods to promote sorption of metals involve incorporating sulfur gases into the steam and thus the surface of the activate char (Krishnan and Anirudhan, 2002).

While steam is the most common activation agent, when it was desired to create granular chars (or activated char) from bagasse, binders such as sugarcane molasses, sugarbeet molasses, and corn syrup were mixed with bagasse and pressed into briquettes, which were then pyrolyzed (at 700 °C) and crushed. This material was the activated at 800 °C in a mixed N_2 and CO_2 atmosphere (Pendyal *et al.*, 1999a, 1999b).

Other methods involve different activation strategies. In order to create an activated char suitable for Cr (VI) (an anion) uptake, Mise and Shantha (1993) created an activated carbon from bagasse by pyrolysis (800–850 °C) using a pure CO_2 atmosphere. They also studied soaking the bagasse in CaI_2 or $MgCl_2$ before the pyrolysis. Pre-soaking the bagasse before pyrolysis was a method also investigated by Amin (2008), who soaked the bagasse in either 50% $ZnCl_2$ or 28% H_3PO_4 before pyrolysis in an oxygen-free atmosphere at 600 °C for 1 h. Higher levels of H_3PO_4 and other inorganic acids (H_2SO_4, HCl, and HNO_3) were used by Girgis, Khalil, and Tawfik (1994), who found that H_3PO_4 was the best acid to use for acid-activation of bagasse during pyrolysis carried out at 500 °C.

15.5   Char and Activated-Char Characterization and Implications for Use

The amount of char that can be obtained from pyrolysis depends on the pyrolysis conditions and the composition of the starting material. Bagasse has been characterized as soft biomass for the purpose of making activated chars and in the same class as rice straw and rice hulls (Ahmedna, Marshall, and Rao, 2000a; Pendyal *et al.*, 1999a). An example of hard biomass would be pecan shells (Ahmedna, Marshall, and Rao, 2000a).

As the pyrolysis takes place, the dry bagasse loses weight due to release of volatiles and later by the decomposition of the remainder. The characterization of the material throughout the pyrolysis process is usually thermogravimetric (monitoring sample weight over time). Using very small samples (7 mg), Drummond and Drummond (1996) determined that bagasse begins to release volatiles and decompose slightly below 300 °C and is fully charred at about 500 °C, with no further loss of weight up to 900 °C. The char yield was reported as approximately 10% (of starting material) and similar to results obtained with silver birch wood. It should be noted that the result was obtained by using a relatively fast heating rate (60 °C min^{-1}), heating the dry material to temperatures between 300 to 900 °C, and holding the sample at the final temperature for a short time (0–0.5 min) before recording the final weight.

Using another small-sample thermogravimetric method, Aiman and Stubington (1993) continuously monitored the weight of the bagasse sample while raising the temperature 5, 10, 20, or 50 °C min^{-1} from ambient to 800 °C. They found that after the initial moisture was lost, most mass was lost between T_i (temperate indicating initial loss of volatiles or bound moisture) and T_f ("final" temperature after rapid mass loss), with T_i being in the range of 180–220 °C and T_f in the range of 366–422 °C, depending on the overall heating rate; both T_i and T_f increased with faster heating rates. When the pyrolysis temperature reached T_f, the mass loss slowed down, but steadily continued until the experiment was ended at 800 °C. Using this method, they were able to determine the pyrolysis kinetics. Their mathematical expression predicted fairly well the rapid mass loss between T_i and T_f but was never extended to the slow mass loss, past T_f. No real definition of T_c was given other than a point on the curve where the slope increased. The weight-loss curve compares well with similar curves produced earlier by Ouensanga and Picard (1988), who performed thermogravimetric studies with a fixed heating rate of about 5 °C min^{-1}. Nassar, Ashour, and Wahid (1996) noted that the mass loss of bagasse under pyrolysis showed exothermic behaviour between about 450 to 700 °C, releasing energy as the biomass decomposed. A typical thermogravimetric curve for sugarcane bagasse is shown in Figure 15.1. In Figure 15.1, T_i, T_c, and T_f are also indicated.

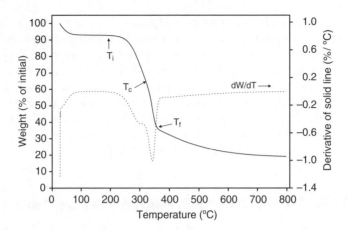

Figure 15.1   *Thermogravimetric curve for the pyrolysis of partially dried bagasse at a heating rate of 5 °C min⁻¹ (Klasson, previously unpublished results). T_i, T_c, and T_f as defined by Aiman and Stubington (1993). Also shown are the derivative values (dashed line) of the main curve.*

The derivative of the curve is also shown as a dashed line, the shape of this (dashed) curve between 200 and 400 °C has mainly been attributed to two overlapping curves of hemicellulose and cellulose degradation. The short plateau at about 300 °C (before T_c) corresponds to the maximum hemicellulose degradation rate and the beginning of cellulose degradation, and the sharp dip around 350 °C is related to the maximum cellulose degradation (Garcìa-Pèrez *et al.*, 2001). The exact temperatures of maximum hemicellulose and cellulose degradation are a function of the heating rate; for example, a higher heating rate shifts these temperatures up. Garcìa-Pèrez *et al.* (2001) used curves like these to determine the kinetics of the bagasse degradation during charring. The total weight loss rate (dVM/dt) was found to be composed of the individual components [hemicellulose (HC), cellulose (C), and lignin (L)] weight-loss rates. The kinetic expression used was defined as follows:

$$\frac{dVM}{dt} = X_{HC}\frac{dHC}{dt} + X_C\frac{dC}{dt} + (1 - X_{HC} - X_C)\frac{dL}{dt} \tag{15.1}$$

where VM is the total volatile matter and X_i is the fraction of volatile matter in component i. Then the individual degradation rates were defined by a simple expression,

$$\frac{dHC}{dt} = A_{HC}\exp\left(\frac{E_{HC}}{RT}\right)(1 - HC) \tag{15.2}$$

for hemicellulose and similar expressions for the cellulose and lignin components. Values for the activation energy (E) were determined to be 105, 225, and 26 kJ mol⁻¹ for hemicellulose, cellulose, and lignin, respectively. The values for log(A) were found to depend slightly on heating rate and varied between 9.2–9.4, 19.3–19.5, and 1–1.7 min⁻¹ for hemicellulose, cellulose, and lignin, respectively. It was also determined as part of the kinetic study that only 33.5–36, 44.5–48, and 18.5–20% of the hemicellulose, cellulose, and the lignin fractions were volatile (below 600 °C), depending on the heating rate.

The above discussion represents data collected with a controlled heating rate and very short hold times at the final temperature, which is very useful for thermal degradation studies. Data from another type of bagasse pyrolysis were published by Shinogi and Kanri (2003), who increased temperature slowly ($2\,°C\,min^{-1}$) and held it at preset final temperatures for 2 h. The data were presented as yield (percentage material remaining from dry bagasse) as a function of final pyrolysis temperature. The resulting trend is very similar to the thermogravimetric data discussed above. Because they used larger samples, the method used by Shinogi and Kanri (2003) allowed them to characterize the final pyrolytic products, with some very noteworthy results. The total carbon content increased substantially with pyrolysis temperature for the bagasse char, while the nitrogen content was mainly unaffected. The bulk density of the pyrolysis bagasse did not change substantially for the pyrolyzed products; it remained fairly constant at $0.1\,g\,mL^{-1}$, but a noticeable decrease was noted at about 400 °C pyrolysis temperature. This result was not unique for pyrolyzed bagasse, but was also found true for rice husks, cow biosolids, and activated sludge that, while having different final densities, experienced the same type of density reduction.

As previously mentioned, fast pyrolysis for the purpose of producing bio-oils also produces a significant yield of left-over char. In fast pyrolysis of bagasse, high heating rates (100–500 °C min^{-1}) were used to reach 400 to 800 °C final temperature and the final holding time was kept to 1–8 min (Tsai, Lee, and Chang, 2006). The result was similar to that previously reported, with char yields of 20–30% reported at final temperatures above 500 °C. Higher heating rates to the same temperature (500 °C) resulted in a slightly higher yield of bagasse char. A longer holding time at the final temperature (500 °C) showed, not surprisingly, a decrease in the char yield. The calorific value of the remaining solids increased as the yield decreased.

The characterization of char and activated char provides valuable information needed to predict suitability for use. The intended use also, to a large extent, dictates which of the characterization methods are most suitable. Some of these methods are standardized, while others are unique and application specific.

In the science of biomass char as a fuel, one of the basic methods of characterization has been borrowed from the characterization of coal; for example, proximate and ultimate analysis. Proximate analysisgives moisture content, volatile content (when heated to 950 °C), the fixed carbon at that point, and ash. Ultimate analysis gives the elemental composition (e.g., carbon, hydrogen, oxygen, nitrogen, and sulfur). Based on several references (see Table 15.2) bagasse (dry) contains 82–87% volatile matter (VM), 1–7% ash, and 10–16% fixed carbon (FC). Cane trash is similar in its composition (Keown *et al.*, 2005; Björkman and Strömberg, 1997). The elemental composition of (dry) bagasse is also listed in Table 15.2.

During pyrolysis, the moisture and part (or all) of the volatile matter escapes, leaving behind a char that is richer in its carbon content. The hydrogen and oxygen is substantially lower and the nitrogen content is similar to (or slightly higher than) the starting material. Proximate and ultimate characterization for sugarcane bagasse chars are listed in Table 15.3. The char yield and composition varies, depending on the pyrolysis conditions.

The suitability of bagasse char as a fuel, either by itself or for blending with fossil fuel, depends (amongst other things) on its heating value. Some of the experimentally determined heating values for bagasse char have been listed in Table 15.3. While heating value is a standard measurement, many literature references do not contain the data, but the same references often contain results from proximate analysis and elemental composition. Calculation of heating value from proximate analysis of biomass and its corresponding chars has

Table 15.2   *Characterization data for sugarcane bagasse (units in %).*

Reference	Proximate (moisture-free)			Ultimate/Elemental (moisture-free)					
	VM	Ash	FC	C	H	N	O	S	Cl
Aiman and Stubington (1993)	85.5	3.9	10.6	46.9	5.6	0.22	47.3	NA	NA
Asadullah *et al.* (2007)	68–70	1.3	28.7–30.7	48.6	6.0	0.2	38.9	0.05	0.05
Channiwala and Parikh (2002)	83.7	3.2	13.2	45.5	6.0	0.15	45.2		
Das *et al.* (2004)	84.4	1.9	13.3	56.3	7.8	0.89	27.5	NA	NA
Drummond and Drummond (1996)	86.5	1.6	11.9	46.3	6.3	0.2	47.2	NA	NA
Garcìa-Pèrez, Chaala, and Roy (2002)	82.1	1.6	16.3	49.6	6.0	0.5	43.8	<0.1	
Keown *et al.* (2005)		6.9		49.7	6.1	0.31	43.8	0.4	0.03
Nassar, Ashour, and Wahid (1996)	84.4	3.4	12.3	51.7	6.3	NA	42.0	0	0
Ouensanga and Picard (1988)				48.6	6.3	NA	45.1	NA	NA
Raveendran, Ganesh, and Khilar (1995)	84.2	2.9		43.8	5.8	0.4	47.1	NA	NA
Tsai, Lee, and Chang (2006)		5.2		58.1	6.0	0.69	34.6	0.19	0.36

been published by Cordero *et al.* (2001), who developed a correlation based on experimental data from three types of wood waste, olive stones, almond shells, and straw. Chars were created via pyrolysis at 300–700 °C. The proximate analyses of the experimental chars, together with data from commercial chars, were used to develop the following simple model to predict higher heating value (HHV):

$$HHV(\text{in MJkg}^{-1}) = 0.3543*FC + 0.1708*VM \tag{15.3}$$

where *FC* and *VM* are the fixed carbon content and the volatile matter in weight percent (on a dry basis). A later-developed correlation included data from a larger number of samples and literature information and suggested that an equation such as

$$HHV(\text{in MJkg}^{-1}) = 0.3536*FC + 0.1559*VM - 0.0078*ASH \tag{15.4}$$

would be better (Parikh, Channiwala, and Ghosal, 2005). In this equation, *ASH* refers to the percentage ash (on a dry basis) of the material. Proximate analysis is often more available than elemental composition; however if the ultimate composition is known, Channiwala and Parikh (2002) suggested that the following equation can be used to predict HHV of chars and other materials:

$$HHV(\text{in MJkg}^{-1}) = 0.3491*C + 1.1783*H + 0.1005*S - 0.1034*O \\ - 0.015*N - 0.0211*ASH \tag{15.5}$$

where *C*, *H*, *S*, *O*, and *N* refer to the carbon, hydrogen, sulfur, oxygen, and nitrogen weight percentages (dry basis). This equation was derived from a large data set and the authors claim it to be reliable over large ranges of compositions (even for high ash materials). A correlation to calculate basic elemental compositions (C, H, and O) in biomass from proximate analysis has been proposed by Parikh, Channiwala, and Ghosal (2007); it was suggested that this relationship could be used in combination with the above to calculate HHV, but there appears to be little advantage in using this over the relationship in Equation 15.4, above. Using the data

Table 15.3 *Characterization data for char made from sugarcane bagasse (units in %, unless otherwise noted).*

Reference	Char Yield	Proximate (moisture-free)			Ultimate/Elemental (moisture-free)					HHV (MJ kg^{-1})
		VM	Ash	FC	C	H	N	O	S	
Asadullah et al. (2007)	22.9–77[a]									
Erlich et al. (2006)	24.5–26.4[b]									
	18.1–18.5[c]									
Garcìa-Pèrez, Chaala, and Roy (2002)	19.4[d]	18.9	6.7	74.4	85.6	2.9	1.3	10.2[e]		36
	25.6[f]	15.4	5.5	79.1	81.5	3.1	0.8	14.6[e]		30[g]
Mobarak, Fahmy, and Schweers (1982)	47.2[h]		6.9		58.3	4.99	0.47	36.1[i]	0.13	22[g]
	26.3[j]		7.9		85.7	3.1	0.37	10.6[i]	0.2	32[g]
	22.9[k]		17.6		91.9	3.17	0.55	4.21[i]	0.17	35[g]
	19.6[l]		18.3		91.9	1.96	0.48	5.47[i]	0.19	33[g]
	22.1[m]		16.1		96.5	0.95	0.31	2.06[i]	0.18	34[g]
Nassar, Ashour, and Wahid (1996)	21[n]									
Raveendran, Ganesh, and Khilar (1995)	20.3[o]									
Tsai, Lee, and Chang (2006)	32[p]				71.4	3.32	1.77	15.5		27[g]
Tsui and Juang (2010)	16[q]		6.0		79.8	1.16	0.13	18.7	0.22	27[g]
Zandersons et al. (1999)	43.8[r]	45.5	4.8	49.7						25[s]
	28.1[t]	28.2	14.0	57.8						25[s]
	27.3[u]	21.3	21.5	57.2						23[s]
	23.0[v]	15.0	9.4	75.7						29[s]
	20.2[w]	13.3	22.1	64.6						25[s]
	20.7[x]	10.4	12.8	76.8						29[s]

[a] 50 °C min^{-1} to 300–600 °C with a 0-min hold of 200-g sample.
[b] 20 °C min^{-1} to 450 °C with a 0-min hold of ~3-g sample. Yield varied depending on initial pellet density.
[c] 20 °C min^{-1} to 800 °C with a 0-min hold of ~3-g sample. Yield varied depending on initial pellet density.
[d] 12 °C min^{-1} heating rate to 500 °C (under vacuum) with 60-min hold of 80-g sample.
[e] Includes ash.
[f] 2.4 °C min^{-1} to 530 °C with a 60-min hold of 20-kg sample.
[g] Calculated based on elemental analysis (Channiwala and Parikh et al., 2002).
[h] 350 °C. Unknown rate and hold.
[i] Oxygen content not given in reference but calculated by difference.
[j] 420 °C. Unknown rate and hold.
[k] 520 °C. Unknown rate and hold.
[l] 620 °C. Unknown rate and hold.
[m] 720 °C. Unknown rate and hold.
[n] 5 °C/min heating rate to 560 °C with a 0-min hold of 18-mg sample.
[o] Isothermal at 500 °C until no further gas generation (10- to 25-g samples).
[p] 200 °C min^{-1} heating rate to 500 °C and 8-min hold of 10-g sample.
[q] 20 °C min^{-1} to 400 °C with a 1-h hold of 50-g sample.
[r] Unknown rate to 280–320 °C with 45-min heating and hold time of 2.4–3.4 kg sample.
[s] Calulated based on proximate analysis (Parikh, Channiwala, and Ghosal, 2005).
[t] Unknown rate to 350–400 °C with 45-min heating and hold time of 2.4–3.4 kg sample.
[u] Unknown rate to 375–450 °C with 45-min heating and hold time of 2.4–3.4 kg sample.
[v] Unknown rate to 430–500 °C with 45-min heating and hold time of 2.4–3.4 kg sample.
[w] Unknown rate to 475–550 °C with 45-min heating and hold time of 2.4–3.4 kg sample.
[x] Unknown rate to 520–600 °C with 45-min heating and hold time of 2.4–3.4 kg sample.

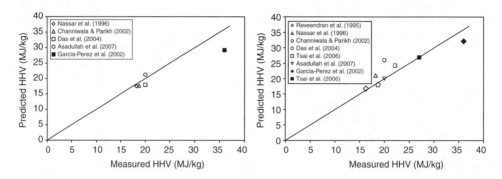

Figure 15.2   *Comparison of predicted HHV for dry bagasse (open symbols) and bagasse char (filled symbols) using proximate analysis values and Equation 15.4 (Parikh, Channiwala, and Ghosal, 2005) on the left and using ultimate analysis values and Equation 15.5 (Channiwala and Parikh, 2002) on the right. Data were taken from Tables 15.1 and 15.2.*

presented in Tables 15.1 and 15.2, we can generate the graphs shown in Figure 15.2. As noted, heating values for bagasse and bagasse chars can be estimated with reasonable accuracy using results from either proximate or ultimate analyses.

The suitability of using bagasse biochar as a soil amendment for the purpose of carbon sequestration is difficult to determine. No method exists which has proven to predict the longevity of carbon in char; it has been suggested that about 50% of the carbon in the raw biomass is relatively stable after charring, but it depends on pyrolysis conditions (Lehmann, Gaunt, and Rondon, 2006). A method of determining the portion of the mass of char which is mobile and immobile (or resident) was proposed by McLaughlin and Shields (2010), who named it the McShields biochar characterization procedure. In this method, which is similar to proximate analysis, the char sample is heated to 200 °C until bone dry. This material is then heated for 4 h at 450 °C in an inert atmosphere to drive off the mobile matter, and the remaining material is ashed at 550 °C in an air atmosphere to combust the resident CHNO and determine ash content. Together with ultimate analysis (CHNO) of the different fractions, mobile and resident CHNO can be calculated.

Other methods have also been proposed to determine the longevity of biochar. For example, Kawamoto, Ishimaru, and Imamura (2005) suggested that an accelerated oxidation in an ozone-rich atmosphere could be produced to predict stability. Based on a first-order reaction rate as a function of ozone concentration, the half-life of biochar at normal atmospheric ozone concentrations could be calculated. No studies on bagasse char were performed but, interestingly enough, it was suggested that wood char created at 400 °C was significantly more stable than char created at 1000 °C.

While no specific method was developed for determining biochar carbon resiliency, Keiluweit *et al.* (2010) performed extensive analysis of wood and grass biochars, created at different temperatures, to determine the carbon (mainly aromatic) structure. They noted that there are four types of charring stages:

1. Transitional char, where the crystalline structures of the precursor materials (cellulose, etc.) are preserved
2. Amorphous char, where the developing aromatics are randomly mixed

3. Composite char, where poorly ordered graphene stacks are embedded in amorphous phases
4. Turbostratic char, where graphitic crystallites dominate.

No claims were made regarding which carbon fractions were more stable and it has been pointed out that experimental results on biochar degradation are sometimes very different (Lehmann, Gaunt, and Rondon, 2006), with microbial and abiotic processes acting to degrade different parts of the biochar.

If the char or activated char is intended to be used to adsorb chemical compounds from waste streams or processing streams, surface area and surface charge (or chemistry) are important characterization parameters, together with other physical properties such as bulk density, hardness, and pH.

The change in surface area from a char to an activated char can be significant. For example, char from fast pyrolysis of corn stover, guayule bagasse, and soybean straw had surface areas of below $3 \, m^2 \, g^{-1}$, while their steam-activated counterparts had a significant Brunauer–Emmett–Teller (BET) surface area of 455–$793 \, m^2 \, g^{-1}$ (Lima, Boateng, and Klasson, 2010a). While the fast pyrolysis method is not intended to produce significant char, it did produce 17–22%, char, but when this char was steam-activated the final yield (based on initial feedstock weight) dropped to 5.4–7.4%. Table 15.4 lists surface areas for chars and activated chars made from sugarcane bagasse. The difference in surface areas between a char and an activated char can be noted again from the data by Darmstadt *et al.* (2001), who showed surface area for a bagasse char to be $529 \, m^2 \, g^{-1}$ and $1529 \, m^2 \, g^{-1}$ for its steam-activated counterpart. It should be noted that these values are significantly higher than the others listed in Table 15.4.

Surface area (as calculated via the BET relationship for nitrogen adsorption) was negligible for bagasse chars made at low temperature, but started to increase for those made between 500 and 600 °C (Shinogi and Kanri, 2003). The measured surface area was about $83 \, m^2 \, g^{-1}$ at 800 °C, which was the maximum pyrolysis temperature studied. The lack of surface area for low-temperature pyrolytic products is also seen for bamboo char (Shenxue, 2004) and wood char (McLaughlin and Shields, 2010; Lehmann, 2007), who noted that surface area as a function of pyrolysis temperature peaks at about 500–700 °C and drops off on either side of this. As bagasse is a softer biomass material, Pendyal *et al.* (1999a) used binders (e.g., sugar crop molasses, corn syrup, etc.) to create briquettes before pyrolysis and activation. Others have found that binder plays a significant role in the properties of the char, but it is the base material that determines the properties of the activated product (Morgan and Fink, 1946). In Pendyal *et al.*'s 1999a study, it was found that the binder had a significant influence on the final properties of the activated carbon. For example, surface areas were highest when cane or beet molasses were used as binders, but the bulk density and the hardness were improved when corn syrup (or coal tar) served as a binder.

While surface area is important, pore structure is also important. Micropores have pore widths of $<2 \, nm$, mesopores have pore widths between 2–50 nm (20–500 Å), and macropores have widths of $>50 \, nm$ (Baker *et al.*, 2007). As the majority of the adsorption of gaseous nitrogen (often used for the characterization of activated chars) takes place in the micropores, the micropore volume is often given in conjunction with the surface area. The extent of micropore development depends on pyrolysis and activation conditions (see Table 15.4). Often the iodine number is listed for commercial activated carbons, but it is rare to find it for experimental carbons. The iodine number is a measure of how much iodine an activated carbon (or char) can adsorb under specific conditions and it is an indication on how good a carbon may

Table 15.4 Surface area and pore volumes of bagasse char and activated chars under different thermochemical pyrolysis conditions.

Reference	Pyrolysis Condition/Activation	BET (m^2g^{-1})	Total Pore Volume (mL g^{-1})	Micropore Volume (mL g^{-1})
Ahmedna et al. (2000b)	700 °C, 1 h, N_2	314–337 (dependent on binder)		
	900 °C, 4 h, 13% CO_2/87% N_2			
	700 °C, 1 h, N_2	78–180 (dependent on binder)		
Darmstadt et al (2001)	900 °C, 20h, 13% CO_2/87% N_2			
	500 °C, 1 h, vacuum	529[a]		0.23
	500 °C, 1 h, vacuum	1579 (1947[a])		0.69
	850 °C, 4 h, 100% steam			
Das et al. (2004)	500 °C, vacuum, N_2 for cooling	98		
Girgis, Khalil, and Tawfik (1994)	50% H_3PO_4, over night	973	0.968	0.246
	500 °C, 3 h			
	50% H_2SO_4, over night	406	0.193	0.173
	500 °C, 3 h			
	50% HCl, over night	192	0.303	0.068
	500 °C, 3 h			
	50% HNO_3, over night	62	0.042	0
	500 °C, 3 h			
Johns, Marshall, and Toles (1998)	750 °C, 1 h, N_2	490		
	850 °C, 5 h, 13%CO_2/67%N_2			
	750 °C, 1 h, N_2	365		
	800 °C, 12 h, steam/N_2			
	750 °C, 1 h, N_2	396		
	850 °C, 5 h 13%CO_2/67%N_2			
	300 °C, 5 h, 38%O_2/62%N_2			
	750 °C, 1 h, N_2	245		
	850 °C, 5 h 13%CO_2/67%N_2			
	23 °C, 7 d, 0.5 M $(NH_4)_2S_2O_8$/1 M H_2SO_4			

Reference	Conditions	Specific surface[a]		
	750 °C, 1 h, N_2	419		
	800 °C, 12 h, steam/N_2			
	500 °C, 5 h, 38%O_2/62%N_2			
	750 °C, 1 h, N_2	326		
	800 °C, 12 h, steam/N_2			
	23 °C, 7 d, 0.5 M $(NH_4)_2S_2O_8$/1 M H_2SO_4			
Kalderis et al. (2008)	2% $ZnCl_2$, 80–100 °C, 2 h	864	0.43	0.44
Marshall et al. (2000)	700 °C, 0.5 h, CO_2			
	750 °C, 1 h, N_2	66.5–298 (dependent on binder)		
Pendyal et al. (1999a)	900 °C, 4–20h, 13% CO_2/87% N_2			
	700 °C, 1 h, N_2	135–440 (dependent on binder)		
Pendyal et al. (1999b)	800 °C, 6 h, 13% CO_2/87% N_2			
	700 °C, 1 h, N_2	195–524 (dependent on binder)	0.132–0.314 (dependent on binder)	0.084–0.232 (dependent on binder)
Shinogi and Kanri (2003)	800 °C, 6 h, 13% CO_2/87% N_2			
	600 °C, 2 h, no sweep gas	70		
	800 °C, 2 h, no sweep gas	83		
Tsui and Juang (2010)	400 °C, 1 h, N_2	3	0.0035	0.00037

[a] Specific surface are measured using the Dubinin and Radushkevich (Darmstadt et al., 2001) method.

be in the high-capacity range (Manes, 1998). The iodine number has the same order of magnitude as the BET surface area; for example, commercial products have a typical range of BET surface areas of 500–2500 $m^2 g^{-1}$ and iodine numbers of 500–1200 (Baker *et al.*, 2007). Iodine numbers in the range of 12–264 have been measured for bagasse chars (Das, Ganesh, and Wangikar, 2004; Tsui and Juang, 2010). For activated bagasse char an iodine number of 1140 (Darmstadt *et al.*, 2001) was measured.

Activation of bagasse chars causes changes to the adsorption surface. Ahmedna, Marshall, and Rao (2000b) determined via titration four classes of surface oxides: carbonyls, lactones, phenols, and carbonyls on CO_2-activated bagasse chars. The presence of oxides on the surface often caused the material to be negatively charged, making it suitable for adsorption of positively charged metal ions (Toles, Marshall, and Johns, 1999; Wartelle and Marshall, 2001). In these types of experiments, Boehm titration is often used as a simple chemical method to determine the surface active groups (Salame and Bandosz, 2001). Darmstadt *et al.* (2001) studied the surface of bagasse char extensively using ESCA (electron spectroscopy for chemical analysis) and SIMS (second ion mass spectroscopy). They determined that condensed aromatic rings were present on the surface. The size of the aromatic structures was larger than five rings and very few aliphatic chains were attached to the rings. Organic oxygen levels on the surface of the chars was 8.8%, and predominantly C=O or C–O–C, not C–OH. FTIR (Fourier transform infrared) spectroscopy has also shown the presence of C=O groups in bagasse char (Tsui and Juang, 2010).

The literature indicates that the acidity (or pH) of bagasse char can range from slightly acidic to slightly alkaline. Shinogi and Kanri (2003) noted that acidity/alkalinity of the pyrolysis bagasse was around pH 4 for pyrolysis products created below 400 °C, but increased to about pH 9 between 400 and 600 °C, after which it was stable. The pH of CO_2-activated bagasse char (with binder) was around pH 8, but if the bagasse char was activated for five times longer, the pH of the final activated char dropped to approximately 6.5 (Ahmedna, Marshall, and Rao, 2000b). Kalderis *et al.* (2008), using $ZnCl_2$ impregnation and pure CO_2 activation, also produced chars with a pH of 6.5. Marshall *et al.* (2000) found that the pH of activated bagasse char depended on the binder which was used for pelletilization (before charring). Cane and beet molasses as binders produced char with pH of about 10, but when corn syrup or coal tar was used the pH of the activated char dropped to slightly above pH 7. Using a slightly different activation process, bagasse with the same binders produced activated chars with an acidity of pH 6 (Pendyal *et al.*, 1999a). The acidity of the char or activated char is important in, for example, soil applications. Application of an alkaline char (not from bagasse) caused a significant rise in soil pH in some cases (Chan *et al.*, 2008; Novak *et al.*, 2009), but was less of an influence in other cases (Chan *et al.*, 2007).

The hardness of CO_2-activated bagasse (with binder) char was significant, with in excess of a 90% survival rate when abrasion testing was done (Ahmedna, Marshall, and Rao, 2000b). Shorter exposure to CO_2 produced a softer material, but this could be made stronger by chemical oxidation with $(NH_4)_2S_2O_8$, resulting in a char with an 87% survival rate (Johns, Marshall, and Toles, 1998). Air oxidation did not produce the same effect. Not surprisingly, the binder used to pelletize bagasse before charring affected the hardness of the char – coal tar and corn syrup (incorporated at 33%) were found to be the binders that produced the hardest (90% survival rate) activated chars (Pendyal *et al.*, 1999a). The hardness/abrasion numbers (very similar to the "survival rate" value) for commercial carbons are in the range of 50–100 (Baker *et al.*, 2007).

15.6 Uses of Bagasse Char and Activated Char

As previously mentioned, char or activated chars can be used in many applications, but far more uses of activated bagasse-based chars (or carbons) have been investigated.

15.6.1 Fuel

Production of fuel briquettes and granulated products from bagasse-based chars were proposed by Zandersons *et al.* (1999). The authors suggested that about 30 MJ kg^{-1} are available as energy in the volatile products given off by bagasse during pyrolysis, and that this amount of energy is sufficient to provide heat in the drying and pyrolysis process. However, they suggested that the drying and pyrolysis would have to be in two separate stages, a heat-up and pyrolysis stage at 350 °C, followed by a charcoal glowing stage at 450–500 °C. Higher temperatures or prolonged heating decreased the char yield to unsatisfactory levels.

15.6.2 Soil Conditioning and Carbon Sequestration

While the use of charcoal or biochar in agricultural applications have long been part of our culture, specific details about sugarcane-bagasse-based char soil amendments are lacking. Lima *et al.* (2010b) reports that sugarcane growth studies have begun on a small scale using chars from both sugarcane trash and bagasse, and blending it into soil at a rate of 4%. Slow pyrolysis was used to create the chars, either at 350 or 700 °C in a nitrogen atmosphere for 1 h. Based on the early results, the growth of the plants was influenced more by the addition of fertilizer than of char. While specific cost estimation was not done for the application, it was suggested the biochar could be available for as little as $550 per tonne ($0.55 per kg).

15.6.3 Environmental and Industrial Applications

The majority of published reports of bagasse biochar use have been for environmental and industrial applications. Decolorization is an intricate part of the refining of raw sugar to produce white sugar and typically, sugar solution is passed through columns with a sorbent to remove the color (Ahmedna, Marshall, and Rao, 2000c). The color compounds come both from the cane itself and from the refinery operations and can be classified as phenolics, flavinoids, caramels, melanoidins, and alkaline degradation products. Amino acids, hydroxy acids, aldehydes, iron and reduced sugars may not have color but may form colored compounds (Ahmedna, Marshall, and Rao, 2000c). Bone char, activated carbon, ion exchange materials, and precipitants are used to remove the color and ash. Of these, bone char is the preferred choice in most cane sugar refinery operations in the US (Ahmedna, Marshall, and Rao, 2000c; Yacoubou, 2007). It was pointed out by Ahmedna, Marshall, and Rao (2000c) that bone char has a bulk cost of $1.20–1.30 per kg, but that ion exchange materials (that also work well for color and ash removal) are significantly more expensive ($11–44 per kg). The same research group did cost estimations for producing steam-activated carbons from bagasse with molasses binder (Ng *et al.*, 2002; Ng *et al.*, 2003). Capital cost was estimated at $4.3 million for a plant producing about 2000 kg material per day at a cost of $3.12 per kg of carbon. It should be noted that activated carbons remove color better than bone char, but are less effective at removing minerals (Ahmedna, Marshall, and Rao, 2000c).

Activated chars made from bagasse and corn-syrup binder were used to determine the feasibility of removing color in two different types of molasses tests (Ahmedna, Marshall, and Rao, 2000b). The bagasse-based activated carbons performed quite well and removed color as efficiently as commercial carbons marketed to sugar industries even though their surface area and surface charge was about half of the commercial products. In other experiments (Pendyal *et al.*, 1999b), sugar cane (or beet) molasses also worked well as a binder when the activated chars were tested in color removal tests. Decolorization of molasses waste water was suggested by Bernardo, Egashira, and Kawasaki (1997) as an application for steam-activated bagasse-based carbons. They showed that melanoidin (a compound in the molasses waste water) was effectively adsorbed onto the char material and that, after use, the materials could be reactivated and used again. Interestingly, the regenerated material was actually better than the virgin activated char, but no explanation was given for this finding.

The potential for removing waste-water dyes from textile companies, dye manufacturing industries, tanneries, paper and pulp mills, and so on, using activated chars from sugarcane bagasse pith (short fibers from the core of the stalk) was studied by Amin (2008). In this research, chars from three pyrolysis/activation strategies were compared – pyrolysis only, soaking in $ZnCl_2$ before pyrolysis, or soaking in H_3PO_4 before pyrolysis. The results indicated that the H_2PO_4-activated char was superior to the others when it came to removing an anionic dye (reactive orange) from solution. These results held true over a large pH range (pH 1 to pH 9). The $ZnCl_2$-activated char performed slightly worse.

Using $ZnCl_2$ impregnation, Kalderis *et al.* (2008) obtained an activated char with a surface area of $864\,m^2\,g^{-1}$ and the material showed capacity for removing arsenic (as arsenate), humic acid, phenol, chemical oxygen-demand compounds, and color. The authors suggested that the material would be useful in landfill leachate treatment. Just as arsenic (as an anionic ion) was removed by bagasse-based activated (using $ZnCl_2$) carbons, chromate could be effectively removed by CO_2-, CaI_2-, or $MgCl_2$-activated carbons (Mise and Shantha, 1993). Pollutants such as phenol and naphthalene (one hydrophilic and one hydrophobic compound) were used as test materials for comparing adsorption capacities for an unactivated char by Tsui and Juang (2010). And while the surface area of the char was almost non-existent, the material was quite capable of adsorbing both phenol and naphthalene. Using a binder, sugarcane bagasse was made into activated chars for adsorbing metals and organics by Johns, Marshall, and Toles (1998). When steam was used as an activation strategy, the activated char adsorbed slightly more organics (methanol, acetonitrile, acetone, 1,4-dioxane, benzene, toluene), as compared to the CO_2-activated material.

By modifying the activation strategy, Johns, Marshall, and Toles (1998) and Marshall *et al.* (2000) showed that bagasse-based activated carbons could be tailor made to, in one case, be useful for sugar decolorization and in another case to adsorb organics (benzene toluene, 1,4-dioxane, acetonitrile, acetone, and methanol) and cationic metals (e.g., lead, copper, cadmium, zinc, and nickel). It should be noted that metal (cations) adsorption is not a very common trait to have for commercial coal-based activated carbons. A unique activated carbon of bagasse pith was created by Krishnan and Anirudhan (2002), who combined steam, SO_2, and H_2S in order to change the surface properties of a bagasse char. The surface area was slightly lower than when steam was used by itself, but the addition of the sulfur gases resulted in a charred material which had a sulfur content of almost 9%. They found it suitable for uptake of heavy metals, such as lead, mercury, cadmium, and cobalt. Hydrochloric acid could then be used to recover the metals. Krishnan and Anirudhan (2002) suggested that this specialized carbon could be made for $0.07 per kg but did not provide a detailed cost analysis.

15.7   Conclusions

Unused sugarcane bagasse represents an underutilized resource in sugarcane growing regions of the world. This is a renewable resource that can be used in a thermochemical process to create chars, which could be incorporated back into agricultural activities. The practice is likely to improve soil quality, reduce the need for fertilizer, and sequester some of the carbon in the soil, thereby reducing the overall carbon emissions from agricultural activities. The bagasse chars could also be used as fuel. Further processing of the chars to create higher-value products may be warranted and these products could be used for detoxification of soil, environmental applications, and by industry as replacements for coal-based products.

References

Ahmedna, M., Marshall, W.E., and Rao, R.M. (2000a) Production of granular activated carbons from select agricultural by-products and evaluation of their physical, chemical and adsorption properties. *Bioresource Technology*, **71**, 113–123.

Ahmedna, M., Marshall, W.E., and Rao, R.M. (2000b) Surface properties of granular activated carbons from agricultural by-products and their effects on raw sugar decolorization. *Bioresource Technology*, **71**, 103–112.

Ahmedna, M., Marshall, W.E., and Rao, R.M. (2000c) Granular activated carbons from agricultural by-products: Preparation, properties, and application in cane sugar refining. Bulletin No. 869, Louisiana State University AgCenter Research & Extension, Baton Rouge, LA.

Aiman, S. and Stubington, J.F. (1993) The pyrolysis kinetics of bagasse at low heating rates. *Biomass and Bioenergy*, **5** (2), 113–120.

Allen, S.J., Whitten, L., and McKay, G. (1998) The production and characterization of activated carbons: A review. *Developments in Chemical Engineering and Mineral Processing*, **6** (5), 231–261.

Amin, N.K. (2008) Removal of reactive dye from aqueous solutions by adsorption onto activated carbons prepared from sugarcane bagasse pith. *Desalination*, **223**, 152–161.

Asadullah, M., Rahman, M.A., Ali, M.M., *et al.* (2007) Production of bio-oil from fixed bed pyrolysis of bagasse. *Fuel*, **86**, 2514–2520.

Babu, B.V. (2008) Biomass pyrolysis: a state-of-the-art review. *Biofuels Bioprod Bioref*, **2** (5), 393–414.

Baker, F.S., Miller, C.E., Repik, A.J., and Tolles, E.D. (2007) Carbon, activated, in *Kirk-Othmer Encyclopedia of Chemical Technology*, vol. **4**, 5th edn, John Wiley & Sons, Hoboken, NJ, pp. 741–761.

Bauer, A., Rosillo-Calle, F., Cortez, L., and Bajay, S. (1998) Electricity from sugarcane in Brazil, in *Proceedings of the 10th European Conference and Technology Exhibition on Biomass for Energy and Industry* (eds H. Kopetz, T. Weber, W. Palz, P. Chartier, and G.L. Ferrero), C.A.R.M.E.N., Würzgurg, Germany, pp. 341–348.

Bernardo, E.C., Egashira, R., and Kawasaki, J. (1997) Decolorization of molasses' wastewater using activated carbon prepared from cane bagasse. *Carbon*, **35** (9), 1217–1221.

Bes, C. and Mench, M. (2008) Remediation of copper-contaminated topsoils from a wood treatment facility using *in situ* stabilisation. *Environmental Pollution (Barking, Essex: 1987)*, **156**, 1128–1138.

Bevill, K. (2010) Cellulosic producers advance projects. *Ethanol Producer Magazine*, **16** (11), 34–35.

Bevill, K. (2011) Biofuels are the new oil. *Ethanol Producer Magazine*, **17** (2), 40.

Björkman, E. and Strömberg, B. (1997) Release of chlorine from biomass at pyrolysis and gasification conditions. *Energy & Fuels*, **11**, 1026–1032.

Bracmort, K.S. (2009) Biochar: Examination of an emerging concept to mitigate climate change. CRC Report for Congress R40186, Congressional Research Service, Washington, DC.

Bridgewater, T. (2007) Biomass Pyrolysis. In: IEA Bioenergy Annual Report 2006. pp. 4–19.

Cerqueira, D.A., Filho, G.R., and da Silvia Meireles, C. (2007) Optimization of sugarcane bagasse cellulose acetylation. *Carbohydrate Polymers*, **69** (3), 579–582.

Chan, K.Y., Van Zwieten, L., Meszaros, I., *et al.* (2007) Agronomic values of greenwaste biochar as a soil amendment. *Australian Journal of Soil Research*, **45**, 629–634.

Chan, K.Y., Van Zwieten, L., Meszaros, I., *et al.* (2008) Using poultry litter biochars as soil amendments. *Australian Journal of Soil Research*, **46**, 437–444.

Channiwala, S.A. and Parikh, P.P. (2002) A unified correlation for estimating HHV of solid, liquid and gaseous fuels. *Fuel*, **81**, 1051–1063.

Chen, S.-B., Zhu, Y.-G., Ma, Y.-B., and McKay, G. (2006) Effect of bone char application of Pb bioavailability in a Pb-contaminated soil. *Environmental Pollution (Barking, Essex: 1987)*, **139** (3), 433–439.

Choy, K.K.H., Ko, D.C.K., Cheung, C.W., *et al.* (2004) Film and intraparticle mass transfer during the adsorption of metal ions onto bone char. *Journal of Colloid and Interface Science*, **271**, 284–295.

Chusilp, N., Likhitsripaiboon, N., and Jaturapitakkul, C. (2009). Development of bagasse ash as a pozzolanic material in concrete. *Asian Journal on Energy and Environment*, **20** (3), 149–159.

Cordero, T., Marquez, F., Rodriguez-Mirasol, J., and Rodriguez, J.J. (2001) Predicting heating values of lignocellulosics and carbonaceous materials from proximate analysis. *Fuel*, **80**, 1567–1571.

Darmstadt, H., Garcia-Perez, M., Chaala, A., *et al.* (2001) Co-pyrolysis under vacuum of sugar cane bagasse and petroleum residue. Properties of the char and activated char products. *Carbon*, **39**, 815–825.

Das, P., Ganesh, A., and Wangikar, P. (2004) Influence of pretreatment for deashing of sugarcane bagasse on pyrolysis products. *Biomass and Bioenergy*, **27**, 445–457.

Dasgupta, A., Weston, R.F., and Nemerow, N.L. (1988) Anaerobic digestion of lignocellulosic residues from sugar cane processing, in *Energy from Biomass Wastes*, vol. **XI** (ed. D.L. Klass), Institute of Gas Technology, Chicago, IL, pp. 613–635.

DeLuca, T.H., MacKenzie, M.D., and Gundale, M.J. (2009) Biochar effects on soil nutrient transformations, in *Biochar for Environmental Management: Science and Technology* (eds J. Lehmann and S. Joseph), Earthscan, London, UK, pp. 251–270.

Deng, J., Wang, G., Kuang, J., *et al.* (2009) Pretreatment of agricultural residues for co-gasification via torrefaction. *Journal of Analytical and Applied Pyrolysis*, **86**, 331–337.

Drummond, A.-R.F. and Drummond, I.W. (1996) Pyrolysis of sugar cane bagasse in a wire-mesh reactor. *Industrial & Engineering Chemistry Research*, **35**, 1263–1268.

Erlich, C., Björnbom, E., Bolado, D., *et al.* (2006) Pyrolysis and gasification of pellets from sugar cane bagasse and wood. *Fuel*, **85**, 1535–1540.

FAO (2006) Global forest resources assessment 2005: Progress towards sustainable forest management. FAO Forestry Paper 147. Food and Agriculture Organization of the United Nations, Rome, Italy.

FAO (2010) Food Agriculture Organization of the United Nations, FAOSTAT Database for production. Web accessed 10/05/2010.

Garcìa-Pèrez, M., Chaala, A., Yang, J., and Roy, C. (2001) Co-pyrolysis of sugarcane bagasse with petroleum residue. Part I: Thermogravimetric analysis. *Fuel*, **80**, 1245–1258.

Garcìa-Pèrez, M., Chaala, A., and Roy, C. (2002) Vacuum pyrolysis of sugarcane bagasse. *Journal of Analytical and Applied Pyrolysis*, **65**, 111–136.

Gaskin, J.W., Steiner, C., Harris, K., *et al.* (2008) Effect of low-temperature pyrolysis condition on biochar for agricultural usc. *Transactions of the ASABE*, **51** (6), 2061–2069.

Girgis, B.S., Khalil, L.B., and Tawfik, T.A.M. (1994) Activated carbons from sugar cane bagasse by carbonization in the presence of inorganic acids. *Journal of Chemical Technology and Biotechnology (Oxford, Oxfordshire: 1986)*, **61**, 87–92.

Glaser, B., Lehmann, J., and Zech, W. (2002) Ameliorating physical and chemical properties of highly weathered soils in the tropics with charcoal - a review. *Biology and Fertility of Soils*, **35**, 219–230.

Goyal, S., Chander, K., Mundra, M.C., and Kapoor, K.K. (1999) Influence of inorganic fertilizers and organic amendments on soil organic matter and soil microbial properties under tropical conditions. *Biology and Fertility of Soils*, **29** (2), 196–200.

Heschel, W. and Klose, E. (1995) On the suitability of agricultural by-products for the manufacture of granular activated carbon. *Fuel*, **74** (12), 1786–1791.

Hodson, M.E., Valsami-Jones, É., and Cotter-Howells, J.D. (2000) Bonemeal additions as a remediation treatment for metal contaminated soil. *Environmental Science & Technology*, **34**, 3501–3507.

Johns, M.M., Marshall, W.E., and Toles, C.A. (1998) Agricultural by-products as granular activated carbons for adsorbing dissolved metals and organics. *Journal of Chemical Technology and Biotechnology (Oxford, Oxfordshire: 1986)*, **71**, 131–140.

Junginger, M., Faaij, A., van den Broek, R., *et al.* (2001) Fuel supply strategies for large-scale bio-energy projects in developing countries. Electricity generation from agricultural and forest residues in Northeastern Thailand. *Biomass and Bioenergy*, **21** (4), 259–275.

Kalderis, D., Koutoulakis, D., Paraskeva, P., *et al.* (2008) Adsorption of polluting substances on activated carbons prepared from rice husks and sugarcane bagasse. *Chemical Engineering Journal*, **144**, 42–50.

Keiluweit, M., Nico, P.S., Johnson, M.G., and Kleber, M. (2010) Dynamic molecular structure of plant biomass-derived black carbon (biochar). *Environmental Science & Technology*, **44**, 1247–1253.

Keown, D.M., Favas, G., Hayashi, J., and Li, C.-Z. (2005) Volatilisation of alkali and alkaline earth metallic species during the pyrolysis of biomass: differences between sugar cane bagasse and cane trash. *Bioresource Technology*, **96**, 1570–1577.

Khan, M.J. and Qasim, M. (2008) Integrated use of boiler ash as organic fertilizer and soil conditioner with NPK in calcareous soil. *Songklanakarin Journal of Science and Technology*, **30** (3), 281–289.

Kirschner, M. (2006) Activated carbon. *Chemical Market Reporter*, **270** (2), 34.

Klasson, K.T., Ackerson, M.D., Clausen, E.C., and Gaddy, J.L. (1992) Bioconversion of synthesis gas into liquid or gaseous fuels. *Enzyme and Microbial Technology*, **14**, 602–608.

Klasson, K.T., Lima, I.M., Boihem, L.L. Jr., and Wartelle, L.H. (2010) Feasibility of mercury removal from simulated flue gas by activated chars made from poultry manures. *Journal of Environmental Management*, **91**, 2466–2470.

Koopmans, A. and Koppejan, J. (1997) Agricultural and forest residues - generation, utilization and availability. Presented at the Regional Consultation on Modern Application of Biomass Energy, Jan 6–10, Kuala Lampur, Malaysia.

Krishnan, K.A. and Anirudhan, T.S. (2002) Uptake of heavy metals in batch systems by sufurized steam activated carbon prepared from sugarcane bagasse pith. *Industrial & Engineering Chemistry Research*, **41**, 5085–5093.

Kawamoto, K., Ishimaru, K., and Imamura, Y. (2005) Reactivity of wood charcoal with ozone. *Journal of Wood Science*, **51**, 66–72.

Ladd, J.N., Amato, M., Li-Kai, Z., and Schultz, J.E. (1994) Differential effects of rotation, plant residue and nitrogen fertilizer on microbial biomass and organic matter in an Australian alifisol. *Soil Biology & Biochemistry*, **26** (7), 821–831.

Lehmann, J. (2007) Bio-energy in the black. *Frontiers in Ecology and the Environment*, **5** (7), 381–387.

Lehmann, J., Gaunt, J., and Rondon, M. (2006) Bio-char sequestration in terrestrial ecosystems - a review. *Mitigation and Adaptation Strategies for Global Change*, **11**, 403–427.

Lehmann, J., Pereire da Silva, J.P. Jr., Steiner, C., *et al.* (2003) Nutrient availability and leaching in an archaeological Anthrasol and Ferralsol of the Central America basin: fertilizer, manure and charcoal amendments. *Plant and Soil*, **249**, 343–357.

Linscott, D.L. and Hagin, R.D. (1967) Protecting Alphalfa seedlings from a triazine with activated charcoal. *Weeds*, **15** (4), 304–306.

Lima, I.M., Marshall, W.E., and Wartelle, L.H. (2004) Hardwood-based granular activated carbon for metals remediation. *Journal of the American Water Works Association*, **96** (7), 95–102.

Lima, I.M., Boateng, A.A., and Klasson, K.T. (2010a) Physicochemical and adsorptive properties of fast-pyrolysis bio-chars and their steam activated counterparts. *Journal of Chemical Technology and Biotechnology (Oxford, Oxfordshire: 1986)*, **85** (11), 1515–1521.

Lima, I., White, P., Klasson, T., and Uchimiya, M. (2010b) Biochars from sugarcane trash and sugarcane bagasse as soil amendment. Presented at SWRM and SERMACS 2010, 66th Southwest and 62nd Southeastern Regional Meeting of the American Chemical Society, New Orleans, LA, Nov. 30-Dec. 4.

Lowdermilk, W.C. (1939) Conquest of the land through 7,000 years. USDA Bulletin no. 99, US Department of Agriculture Natural Resources Conservation Service, Washington, DC.

Manes, M. (1998) Activated carbon adsorption fundamentals, in *Encyclopedia of Environmental Analysis and Remediation*, John Wiley & Sons, Hoboken, NJ, pp. 26–68.

Marshall, W.E., Ahmedna, M., Rao, R.M., and Johns, M.M. (2000) Granular activated carbons from sugarcane bagasse: production and uses. *International Sugar Journal*, **102** (1215), 147–151.

McLaughlin, H. and Shields, F.E. (June 27–30 2010) Schenkel and Shenxue revisited. Presented at the Biochar2010 U.S. Biochar Initiative Conference, Ames, IA.

Minkova, V., Marinov, S.P., Zani, R., *et al.* (2000) Thermochemical treatment of biomass in a flow of steam or in a mixture of steam and carbon dioxide. *Fuel Processing Technology*, **62**, 45–52.

Mise, S.R. and Shantha, G.M. (1993) Adsorption studies of chromium (VI) from synthetic aqueous solution by activated carbon derived from bagasse. *The Journal of Environmental Science and Health*, **A28** (10), 2263–2280.

Mobarak, F., Fahmy, Y., and Schweers, W. (1982) Production of phenols and charcoal from bagasse by a rapid continuous pyrolysis process. *Wood Science and Technology*, **16** (1), 59–66.

Morgan, J.J. and Fink, C.E. (1946) Binders and base materials for active carbon. *Industrial & Engineering Chemistry*, **38** (2), 219–228.

Nassar, M.M., Ashour, E.A., and Wahid, S.S. (1996) Thermal characteristics of bagasse. *Journal of Applied Polymer Science*, **61**, 885–890.

Ng, C., Bansode, R.R., Marshall, W.E., *et al.* (2002) Process description and product cost to manufacture sugarcane bagasse-based granular activated carbon. *International Sugar Journal*, **104** (1245), 401–408.

Ng, C., Marshall, W., Rao, R.M., *et al.* (2003) *Granular Activated Carbons from Agricultural by-Products: Process Description and Estimated Cost of Production*, Bulletin No. 881 Louisiana State University AgCenter Research & Extension, Baton Rouge, LA.

Novak, J.M., Lima, I., Xing, B., *et al.* (2009) Characterization of designer biochar produced at different temperatures and their effects on a loamy sand. *Annals Environmental Science*, **3**, 195–2006.

Olivério, J.L. (2006) Technological evolution of the Brazilian sugar and alcohol sector: Dedini's contribution. *International Sugar Journal*, **108** (1287), 120–129.

Oliverio, J.L. and Hilst, A.G.P. (2004) DHR-DEDINI Hidrólise Rápida (DEDINI Rapid Hydrolysis)-Revolutionary process for producing alcohol from sugar cane bagasse. *International Sugar Journal*, **106** (1263), 168–172.

Ouensanga, A. and Picard, C. (1988) Thermal degradation of sugar cane bagasse. *Themochim Acta*, **125**, 89–97.

Paraskeva, P., Kelderis, D., and Diamadopoulos, E. (2008) Production of activated carbons from agricultural by-products. *Journal of Chemical Technology and Biotechnology (Oxford, Oxfordshire: 1986)*, **83**, 581–592.

Parikh, J., Channiwala, SA., and Ghosal, G.K. (2005) A correlation for calculating HHV from proximate analysis of solid fuels. *Fuel*, **84**, 487–494.

Parikh, J., Channiwala, S.A., and Ghosal, G.K. (2007) A correlation for calculating elemental composition from proximate analysis of biomass materials. *Fuel*, **86**, 1710–1719.

Payá, J., Monzó, J., Borrachero, M.V., Díaz-Pinzón, L. and Ordóñez, L.M. (2002) Sugar-cane bagasse ash (SCBA): studies on its properties from reusing in concrete production. *Journal of Chemical Technology and Biotechnology*, **77**, 321–325.

Pendyal, B., Johns, M.M., Marshall, W.E., *et al.* (1999a) The effect of binders and agricultural by-products on physical and chemical properties of granular activated carbons. *Bioresource Technology*, **68**, 247–254.

Pendyal, B., Johns, M.M., Marshall, W.E., *et al.* (1999b) Removal of sugar colorants by granular activated carbons made from binders and agricultural by-products. *Bioresource Technology*, **69**, 45–51.

Perlack, R.D., Wright, L.L., Turhollow, A.F., *et al.* (2005) Biomass as Feedstock for a Bioenergy and Bioproducts Industry: The Technical Feasibility of a Billion-Ton Annual Supply, DOE/GO-102005-2135. US Department of Energy, Office of Scientific and Technical Information, Oak Ridge, TN, pp. 5–33.

Pindoria, R.V., Chatzakis, I.N., Lim, J.-Y., *et al.* (1999) Hydropyrolysis of sugar cane bagasse: effect of sample configuration on bio-oil yields and structures from two bench-scale reactors. *Fuel*, **78**, 55–63.

Ramjeawon, T. (2008) Life cycle assessment of electricity generation from bagasse in Mauritius. *Journal of Cleaner Production*, **16**, 1727–1734.

Raveendran, K., Ganesh, A., and Khilar, K.C. (1995) Influence of mineral matter on biomass pyrolysis characteristics. *Fuel*, **74** (2), 1812–1822.

Reyes, B.J. (2006) *Isle ethanol producers still year away*, Honolulu Star-Bulletin, Honolulu, HI, **11** (23), January 23.

Rocha, J.D., Brown, S.D., Love, G.D., and Snape, C.E. (1997) Hydropyrolysis: a versatile technique for solid fuel liquefaction, sulfur speciation and biomarker release. *Journal of Analytical and Applied Pyrolysis*, **40–41**, 91–103.

Rocha, J.D., Luengo, C.A., and Snape, C.E. (1999) The scope for generating bio-oils with relatively low oxygen contents via hydropyrolysis. *Organic Geochemistry*, **30**, 1527–1534.

Salame, I.I. and Bandosz, T.J. (2001) Surface chemistry of activated carbons: combining the results of temperature-programmed desorption, Boehm, and potentiometric titrations. *Journal of Colloid and Interface Science*, **240**, 252–258.

Sales, A. and Lima, S.A. (2010) Use of Brazilian sugarcane bagasse ash in concrete as sand replacement. *Waste Management*, **30**, 1114–1122.

Shenxue, J. (May 2004) *Training manual of bamboo charcoal for producers and consumers*, Bamboo Engineering Research Center, Nanjing Forestry University, Jiangsu Province, China.

Shinogi, Y. and Kanri, Y. (2003) Pyrolysis of plant, animal and human waste: physical and chemical characterization of the pyrolytic products. *Bioresource Technology*, **90**, 241–247.

Solomon, B.D., Barnes, J.R., and Halvorsen, K.E. (2007) Grain and cellulosic ethanol: history, economics, and energy policy. *Biomass and Bioenergy*, **31**, 416–425.

Toles, C.A., Marshall, W.E., and Johns, M.M. (1999) Surface functional groups on acid-activated nutshell carbons. *Carbon*, **37**, 1207–1214.

Toles, C.A., Marshall, W.E., Wartelle, L.H., and McAloon, A. (2000a) Steam- or carbon dioxide-activated carbons from almond shells: physical, chemical and adsorptive properties and estimated cost of production. *Bioresource Technology*, **75**, 197–203.

Toles, C.A., Marshall, W.E., Johns, M.M., *et al.* (2000b) Acid-activated carbons from almond shells: physical, chemical and adsorptive properties and estimated cost of production. *Bioresource Technology*, **71**, 87–92.

Tsai, W.T., Lee, M.K., and Chang, Y.M. (2006) Fast pyrolysis of rice straw, sugarcane bagasse and coconut shell in an induction-heating reactor. *Journal of Analytical and Applied Pyrolysis*, **76**, 230–237.

Tsui, L. and Juang, M.-A. (2010) Effects of composting on sorption capacity of bagasse-based chars. *Waste Management*, **30**, 995–999.

Voegele, E. (2009) Sugarcane economics. Ethanol Producer Magazine, March.

Wartelle, L.H. and Marshall, W.E. (2001) Nutshells as granular activated carbons: physical, chemical and adsorptive properties. *Journal of Chemical Technology and Biotechnology (Oxford, Oxfordshire: 1986)*, **76**, 451–455.

Yacoubou, J. (2007) Is your sugar vegan? *Vegetarian Journal*, **26** (4), 15–19.

Yaman, S. (2004) Pyrolysis of biomass to produce fuels and chemical feedstocks. *Energy Conversion and Management*, **45**, 651–671.

Zandersons, J., Gravitis, J., Kokorevics, A., *et al.* (1999) Studies of the Brazilian sugarcane bagasse carbonization process and product properties. *Biomass and Bioenergy*, **17**, 209–219.

Zeng, H., Jin, F., and Guo, J. (2004) Removal of elemental mercury from coal combustion flue gas by chloride-impregnated activated carbon. *Fuel*, **83** (1), 143–146.

Zimmermann, J.R., Bricker, J.D., Jones, C., *et al.* (2008) The stability of marine sediments at a tidal basin in San Francisco Bay amended with activated carbon for sequestration of organic contaminants. *Water Research*, **42**, 4133–4145.

Index

Note: Page numbers in *italics* refer to Figures; those in **bold** to Tables

---

Biorefinery Co-Products: Phytochemicals, Primary Metabolites and Value-Added Biomass Processing, First Edition.
Edited by Chantal Bergeron, Danielle Julie Carrier and Shri Ramaswamy.
© 2012 John Wiley & Sons, Ltd. Published 2012 by John Wiley & Sons, Ltd.